PHYSICS OF SOLIDS IN
INTENSE MAGNETIC FIELDS

PHYSICS OF SOLIDS IN INTENSE MAGNETIC FIELDS

*Lectures presented at the First Chania Conference
held at Chania, Crete, July 16-29, 1967*

Edited by

E. D. Haidemenakis

Ecole Normale Supérieure
Laboratoire de Physique
and
Faculté des Sciences
Université de Paris
Paris, France

⨁ Springer Science+Business Media, LLC **1969**

Library of Congress Catalog Card Number 68-20272

ISBN 978-1-4899-5510-4   ISBN 978-1-4899-5508-1 (eBook)
DOI 10.1007/978-1-4899-5508-1

© 1969 Springer Science+Business Media New York
Originally published by Plenum Press in 1969.
Softcover reprint of the hardcover 1st edition 1969

To the people of Chania

PREFACE

Over one hundred scientists from a number of Western European countries, Australia, Canada, Israel, Japan, and the United States gathered on the island of Crete to discuss the physics of solids in intense magnetic fields. It was six years ago that many of the same participants were discussing the production and technology of high fields at the MIT Conference. This volume contains, in large measure, the proceedings of the Crete gathering.

The topics discussed during the first week were concerned mainly with electrons in Landau levels, with a number of comprehensive reviews; during the last part of the meeting many reports on the up-to-date progress in the field were presented. The subject matter began at a low level and rapidly proceeded to an advanced stage.

The opening lecture was given by E. Burstein on the interaction of electromagnetic waves with single-particle and collective excitations of free carriers in a magnetic field and on the measurement of these phenomena using infrared techniques and the photon–phonon interaction in the Raman effect. The quantum theory of electrons in crossed electric and magnetic fields, with emphasis on relativistic effects and on interband processes, was presented by W. Zawadzki. B. Lax discussed quantum magnetooptics, interband magnetoreflection and magnetoabsorption, cyclotron resonance, nonlinear multiple-photon effects, and other physical processes encountered in lasers in magnetic fields. A. B. Pippard explained his theory on magnetic breakdown involving cyclotron orbits in metals.

The lectures continued with G. Landwehr's talk on Shubnikov–de Haas and quantum-limit effects in semiconductors such as indium antimonide. Oscillatory effects were also discussed by Y. Shapira, who spoke on giant oscillations in ultrasonic attenuation in metals, such as gallium, and their use in Fermi–Dirac distribution functions, and on ultrasonic attenuation in type-II superconductors. W. L. McLean dealt with quantum oscillations in helicon propagation in tin. The magnetoquantum (photon) electric effects in solid-state plasmas, including the amplification phenomenon, were discussed by Y. Sawada.

The last speaker of the first week was J. Zak, who lectured on the application of group theory to a Bloch electron in a magnetic field. In a panel discussion B. Lax, E. Burstein, A. Pippard, and E. Haidemenakis

reviewed the first week's subject matter and suggested areas of future research in the field.

The second week was devoted to more recent research in the field,
which included superconductors, magnetospectroscopy, modulation techniques, magnetism, various magnetooptical effects, and recent methods in
the production of very high magnetic fields. W. A. Runciman spoke on
group theory and on the Zeeman effect in rare-earth-doped crystals. Lead
salts and magnetooptical studies of band population effects were discussed
by E. D. Palik. Some interesting modulation techniques were presented by
J. Mavroides, who also illustrated his results in band structure studies.
S. Williamson and E. Saur described various aspects of type-II superconductors, and N. Kurti talked about high magnetic fields and low temperatures. The topic of magnetism in ferro, antiferro, and ferri magnets
was covered by S. Foncr.

Laser magnetospectroscopy and its application to the study of magnetooptical effects was discussed by S. Iwasa. P. R. Wallace developed a multi-
carrier model for the propagation of magnetoplasmas in semiconductors
such as PbTe. Finally, the Faraday effect discussed by G. Sacerdoti
and by A. Van Itterbeek for pulsed fields. The last lectures were those of
H. Knoepfel, who discussed megaoersted fields by implosion techniques,
and A. Rabinovitch, who gave a seminar on the tight-binding approximation applied to a Bloch electron in a magnetic field. The lectures were reviewed by Professor Lax, who also gave the closing address of the meeting.

The directors of the Chania meeting were Pierre Aigrain of the Ecole
Normale Supérieure, also Director of Higher Education in France, Eli
Burstein of the University of Pennsylvania and Chairman of its Physics
Department, and Benjamin Lax, Director of the National Magnet Labora-
tory at MIT. The chairman was Epimenides D. Haidemenakis of the Ecole
Normale Supérieure and the Faculté des Sciences de Paris.

The sponsors of this meeting include: NATO Scientific Affairs Division,
the Public Power Corporation of Greece, AGARD Greek Delegation,
Greek Bureau of International Conventions, IBM World Trade Corporation, International Telephone and Telegraph Corporation, Hellenic In-
dustrial Development Bank, Commercial Bank of Greece, Olympic Airways,
Ionian and Popular Bank of Greece, Hellenic Electric Railways, Chandris
Lines, National Organization of Greek Handicrafts, Telefunken, Efthi-
miades Lines, Agence Paris-Athenes, Phillips Greece, Metaxas Ouzo,
Papastratos Cigarettes, K. Lampathakis, and Viochym Fruitjuice.

We wish to take this opportunity to extend once more our sincere
gratitude to everyone involved in organizing this meeting. Above all we

wish to thank the Governor, the Mayor, the local authorities, Archbishop Irineos, the professional associations, the Ladies of Lykion, and all the people of Chania for the outstanding hospitality which they provided to the participants and their families.

E. D. Haidemenakis

CONTENTS

Contributors xix

Chapter 1

The Interaction of Acoustic and Electromagnetic Waves ("Son et Lumière") with Plasmas in a Magnetic Field

    by E. Burstein 1

Introduction 1
Single-Particle Excitations 3
  Spherical Energy Bands 3
    Landau ($\triangle L = 0$) Excitations 5
    $\triangle L \neq 0$ Cyclotron Excitations 10
    $\triangle \sigma \pm 1$ Spin Transitions 13
    $\triangle L = \pm 1$, $\triangle \sigma = \pm 1$ Combined Cyclotron and Spin Excitations 15
  Ellipsoidal Energy Bands 16
  ω versus $\mathbf{q}$ Dispersion Curves 17
  Center of Cyclotron Orbit Effects 22
  Crossed Electric and Magnetic Fields 25
Collective Excitation 27
  The $\mathbf{H}_0 = 0$ Case 27
  The $\mathbf{H}_0 \neq 0$ Case 28
    $\mathbf{Q} \mid\mid \mathbf{H}_0$ Collective Excitations 28
    $\mathbf{Q} \perp \mathbf{H}_0$ Collective Excitations 28
  Coupled Longitudinal Excitations 30
Interaction Matrix Elements 31
The Propagation of $\mathbf{E}$ and $\mathbf{A}$ Waves 34
The Dielectric Constant of a Plasma 35
  The $\mathbf{H}_0 = 0$ Case 36
  The $\mathbf{H}_0 \neq 0$. Faraday Configuration 41
  The $\mathbf{H}_0 \neq 0$. Voigt Configuration 47
Nonlocal Effects 49
Helicon–Phonon Coupling 53
Experimental Observation of Magnetooptical and Magnetoacoustical Phenomena .. 55
  Landau ($\triangle L = 0$) Resonance ($\mathbf{q} \mid\mid \mathbf{H}_0$) 56
  Doppler-Shifted Cyclotron Resonance ($\mathbf{q} \mid\mid \mathbf{H}_0$) 57
  Cyclotron Resonance ($\mathbf{q} \perp \mathbf{H}_0$) 58
  Displacement of the Center of Orbit 58
Acknowledgment 59
References 59

Chapter 2

Intraband Collective Effects and Magnetooptical Properties of Many-Valley Semiconductors

    by P. R. Wallace 61

Introduction 61
Normal Modes and the Dielectric Tensor 62

Helicons 64
Calculation of Dispersion Relations in a Magnetic Field 65
Calculation of the Dielectric Constant 68
Dielectric Tensor for Germanium and Lead Telluride 72
   [100] Direction 73
   [111] Direction 78
Interaction with Phonons 81
  Modes with Polarization Along B_0 82
   The Case $\omega_4{}^2 < \omega_c{}^2 < \omega_T{}^2$ 82
   The Case $\omega_4 < \omega_T < \omega_c < \omega_L, \omega_p$ 83
  Modes with Polarization Perpendicular to B_0 84
  Magnetic Field in the [111] Direction: Voigt Configuration 88
  Optical Effects 88
  References 89

Chapter 3

Effects of Band Population on the Magnetooptical Properties of the Lead Salts

by E. D. Palik and D. L. Mitchell

by E. D. Palik and D. L. Mitchell 90

Introduction 90
  General 90
  Optical Effects 91
Band Population Effects – Phenomenological 93
  Burstein–Moss Effect 93
  Interband Magnetoabsorption 96
  Faraday Rotation 98
  Derivation 101
  Dispersion Relations 105
  Comparison with Experiment: PbS 107
  p-PbS 110
  PbSe and PbTe 110
Quantum Limit – Conductivity Model 111
  Relations among Conductivity, Dielectric Constant, and Optical Constants 111
  Conductivity Formulation 114
  Oscillations in Interband Rotation 117
  Magnetoplasma 119
References 123

Chapter 4

The Coupling of Collective Cyclotron Excitations and Longitudinal Optic Phonons in Polar Semiconductors

by Sato Iwasa 126
References 135

Chapter 5

Application of Gas Lasers to High-Resolution Magnetospectroscopy

by Sato Iwasa 136

Introduction 136
Laser Spectrometer 138
Graphite 138
Bismuth 141
Arsenic 143
References 144

Chapter 6

Quantum Magnetooptics at High Fields

    by Benjamin Lax and Kenneth J. Button 145

Introduction 145
Cyclotron Resonance 148
  Classical Treatment 149
  Cyclotron Resonance in Indium Antimonide 151
  Quantum Effects in Germanium and Silicon 154
Interband Phenomena 156
  Direct Interband Magnetoabsorption 157
  Indirect Transition 163
  Semimetals 167
  Multiphoton Absorption 175
    For $E \parallel H$ 176
    For $E \perp H$ 177
Conclusion 180
References 182

Chapter 7

Magneto-Quantum-Electric Effect

    by Yasuji Sawada 184

Introduction 184
Theoretical 185
Experiment 193
Discussion 197
Appendix 204
References 205

Chapter 8

Electron Band Structure Studies Using Differential Optical Techniques
and High Magnetic Fields

    by J. G. Mavroides 206

Introduction 206
Methods of Differential Optical Modulation 207
  Electric Field Modulation 207
  Piezomodulation 210
  Thermal Modulation 212
Experimental Techniques 214
Theoretical Considerations 215
Magnetooptical Modulation Experiments 220
Conclusion 231
Acknowledgments 231
References 231

Chapter 9

High Field Magnetospectroscopy and Band Structure of Cd_3As_2

    by E. D. Haidemenakis 233

Properties and Preparation 233
Experimental 233
Zero-Field Reflection 236
Interband Magnetoabsorption and Magnetoreflection ... 241
References 244

Chapter 10

Effects of High Magnetic Fields on Electronic States in Semiconductors—The Rydberg Series and the Landau Levels

　　by H. Hasegawa 246

Introduction 246
Preliminaries 248
Spectra in the High Field Limit 251
Continuation Between the Rydberg and Landau Spectra 257
　Donor States in InSb 257
　Exciton Spectra in Layer-Type Semiconductors 259
Exciton Band in a Magnetic Field 265
Acknowledgments 269
References 269
Notes Added in Proof 270

Chapter 11

Faraday Rotation in Solids

　　by A. Van Itterbeek 271

Experimental Method and Results Obtained for Faraday Rotation 280
Acknowledgments 284
References 284

Chapter 12

A Technique for the Measurement of the Faraday Effect in Pulsed
Magnetic Fields at Low Temperatures

　　by G. Sacerdoti 285

Acknowledgments 300
Notes Added in Proof 300
References 300

Chapter 13

Electrons in a Magnetic Field

　　by W. Zawadzki 301

Classical Description 301
Quantum Mechanical Description 305
References 310

Chapter 14

Bloch Electrons in Crossed Electric and Magnetic Fields

　　by W. Zawadzki 311

Electron Motion in Crossed Fields 312
Interband Magnetooptical Effects in Crossed Fields 314
The Two-Band Model 319
Intraband Magnetooptical Effects in Crossed Fields 321
Magnetic- and Electric-Type Motion 325
References 327

Chapter 15

Group-Theoretical Approach to the Problem of a Bloch Electron in a
Magnetic Field

    by J. Zak 329

Introduction 329
A Free Electron in a Magnetic Field 329
A Bloch Electron in a Magnetic Field 333
References 336

Chapter 16

Tight-Binding Approximation for a Bloch Electron in a Magnetic
Field

    by A. Rabinovitch 337

References 343

Chapter 17

The Zeeman Effect in Crystals

    by W. A. Runciman 344

Introduction 344
Theoretical Background 345
Cubic Centers 350
    Cubic Centers with an Even Number of Electrons 350
    Cubic Centers with an Odd Number of Electrons 351
Noncubic Centers in Cubic Crystals 354
    Noncubic Centers with an Even Number of Electrons 354
    Noncubic Centers with an Odd Number of Electrons 354
Discussion 356
Acknowledgments 358
References 358

Chapter 18

Magnetic Breakdown

    by A. B. Pippard 359

Experimental Consequences of Magnetic Breakdown 370
    The de Haas–van Alphen Effect 370
    Transport Effects 372
References 377

Chapter 19

Ultrasonic Propagation in High-Field Superconductors

    by Y. Shapira 378

Some Properties of High-Field Superconductors 378
HFS Versus Superconductors 379
Magnetic-Field Effects on Ultrasonic Propagation in the Normal State ... 381
Magnetic-Field Effects on Ultrasonic Propagation in the Mixed State 382
References 386

Chapter 20

Giant Quantum Oscillations in Ultrasonic Absorption

by Y. Shapira 387

The Physical Origin of the Giant Quantum Oscillations 387
The Line Shape of the GQO 390
Information which can be Obtained from GQO 393
    The Period 393
    The g-Factor 393
    The Cyclotron Mass 393
    Matrix Elements of the Electron–Phonon Coupling 394
    The Electron Distribution Function 394
References 396

Chapter 21

Helicon Propagation in Metals: Quantum Oscillations in Tin

by W. L. McLean 397

The Basic Viewpoint for Metals 397
Experimental Methods 400
Effects of the Topology of the Fermi Surface on the Wave Propagation 402
Magnetic Breakdown Effects 407
References 413

Chapter 22

The Shubnikov-de Haas Effect and Quantum-Limit Phenomena in Semiconductors

by G. Landwehr 415

Introduction 415
Oscillatory Effects 416
    The Period 416
    Deviations in the Periodicity 419
Quantitative Theory 422
    The Theory of Adams and Holstein 422
    Oscillatory Magnetoresistance in Longitudinal Fields 426
Experimental Methods 427
Experimental Results 428
    Bismuth 428
    Graphite 428
    Indium Antimonide 429
    Indium Arsenide 431
    Gallium Antimonide 431
    Germanium and Silicon 431
    Tellurium 432
    Mercury Telluride 433
    Mercury Selenide 434
    Lead Telluride, Lead Selenide, and Lead Sulfide 435
    Bismuth Telluride 436
    Tin Telluride 437
    Cadmium Arsenide 437
Quantum-Limit Effects 438
Notes Added in Proof 441
References 441

Chapter 23

Hall Effect and Transverse Voltages in Type-II Superconductors in the Mixed State

    by S. J. Williamson and J. Baixeras 444

    Longitudinal Voltages 447
    Transverse Voltages 447
Acknowledgments 452
References 453

Chapter 24

Experiments on the Critical Behavior of Type-II Superconductors in High Magnetic Fields

    by E. Saur 454

Introduction 454
Experimental Results on the Critical Behavior of Type-II Superconductors 455
    Critical Field Curves 455
    Quenching Curves of Bulk Materials 456
        Nb_3Sn 456
        V_3Si 459
        V_3Ga 461
        NbN 461
    Quenching Curves of Materials with Artificial Structure 461
Comparison of Experimental Results with Recent Theories of Type-II
    Superconductors 461
Acknowledgments 465
References 465

Chapter 25

Megaoersted Fields and their Relation to the Physics of High Energy Density

    by H. Knoepfel 467

Introduction 467
The Diffusion of Magnetic Fields into a Metallic Conductor with Constant
    Conductivity 468
The Interaction of MOe Fields with a Metallic Conductor 471
The Exploding Metal Tube 475
The Acceleration of Metallic Conductors through MOe Fields 477
Conclusions 482
References 482

CONTRIBUTORS

J. Baixeras 444
L. C. I. E., Fontenay-aux-Roses, France

E. Burstein 1
University of Pennsylvania, Philadelphia, Pennsylvania

K. J. Button 145
Massachusetts Institute of Technology, Cambridge, Massachusetts

E. D. Haidemenakis 233
Ecole Normale Supérieure, Paris, France

H. Hasegawa 246
Kyoto University, Kyoto, Japan

S. Iwasa 126, 136
Massachusetts Institute of Technology, Cambridge, Massachusetts

H. Knoepfel 467
Comitato Nazionale per l'Energia Nucleare, Frascati (Roma), Italy

G. Landwehr 415
Universität Wuerzburg, Wuerzburg, Germany

B. Lax 145
Massachusetts Institute of Technology, Cambridge, Massachusetts

J. G. Mavroides 206
Massachusetts Institute of Technology, Cambridge, Massachusetts

W. L. McLean 397
Rutgers University, New Brunswick, New Jersey

D. L. Mitchell 90
Naval Research Laboratory, Washington, D. C.

E. D. Palik 90
Naval Research Laboratory, Washington, D. C.

A. B. Pippard 359
Cavendish Laboratory, Cambridge, England

A. Rabinovitch 337
Israel Institute of Technology, Haifa, Israel

W. A. Runciman 344
Atomic Energy Research Establishment, Harwell, Berkshire, England

G. Sacerdoti 285
Comitato Nazionale per l'Energia Nucleare, Frascati (Roma), Italy

E. Saur 454
Institut für Angewandte Physik, Universität Giessen, West Germany

Y. Sawada 184
Osaka University, Toyanaka, Japan

Y. Shapira 378, 387
University of Pennsylvania, Philadelphia, Pennsylvania

A. Van Itterbeek 271
Institute for Low Temperatures and Applied Physics, Leuven, Belgium

P. R. Wallace 61
McGill University, Montreal, Canada

S. J. Williamson 444
North American Aviation Science Center, Thousand Oaks, California

J. Zak 329
Massachusetts Institute of Technology, Cambridge, Massachusetts

W. Zawadski 301, 311
Polish Academy of Sciences, Warsaw, Poland

Chapter 1

THE INTERACTION OF ACOUSTIC AND ELECTROMAGNETIC WAVES ("SON ET LUMIÈRE") WITH PLASMAS IN A MAGNETIC FIELD*

E. Burstein

Physics Department
University of Pennsylvania
Philadelphia, Pennsylvania

INTRODUCTION

In this chapter I will be concerned with the nature of the single-particle and collective excitations of plasmas in solids and their interactions with acoustic (A) waves and electromagnetic (E) waves in a magnetic field under conditions $\hbar\omega > \Delta\mathscr{E}$ and $\hbar q > \Delta p$, where $\Delta\mathscr{E} = \hbar/\Delta t$ is the uncertainty in the energy of the excitation and $\Delta p = \hbar/\Delta x$ is the uncertainty in the momentum of the excitation. Under these conditions the interactions can be treated semiclassically in terms of elementary processes in which (1) a phonon or photon of frequency ω and wave vector $\mathbf{q}$ is absorbed (or emitted) and an electron undergoes a transition from an initial state of energy $\mathscr{E}$ and wave vector $\mathbf{k}$ to a final (unoccupied) state of energy $\mathscr{E}' = \mathscr{E} + \hbar\omega$ and wave vector $\mathbf{k}' = \mathbf{k} + \hbar\mathbf{q}$ in the same Landau subband (a Landau transition) or in another Landau subband (a cyclotron transition); or (2) a phonon or photon is absorbed (or emitted) and a quantum of collective excitation of frequency $\Omega = \omega$ and wave vector $\mathbf{Q} = \mathbf{q}$ is created or annihilated. The investigation of the interactions of A and E waves with plasmas provides useful information about the nature of the single-particle and

* Work supported in part by the U. S. Office of Naval Research.

collective excitations of the plasma as well as about the energy band structure of semiconductors and the Fermi surface of metals.

The plasma is considered as an assembly of electrons (or holes) immersed in a uniform background of positive (or negative) charge whose density is equal to the average density of the electrons. In compensated plasmas the densities of the electrons and holes are equal and the background is considered to be neutral. We will be concerned primarily with plasmas which are degenerate, and which therefore have a Fermi surface. Since plasmas in solids at low temperatures are degenerate at relatively low charge carrier densities, this does not constitute a major limitation.

The physics underlying the interaction of A waves (phonons) and E waves (photons) with free carriers in the presence of an externally applied magnetic field have much in common, and the various magnetic–acoustical effects have their conterparts in magnetooptical effects. The difference in the two types of phenomena stem from (1) the considerable difference in the magnitude of the phonon–electron and photon–electron coupling constants, (2) the marked difference in the velocities and in the range of frequencies, and (3) the fact that A waves are "longitudinal" as well as "transverse," whereas E waves are "transverse."

The interactions of A and E waves with the single-particle and collective excitations take place via the electric fields of the waves. In the case of A waves the electric fields arise from the deformation potential (semiconductors and semimetals), the piezoelectric effect (semiconductors) and from induction effects (metals) [1,2]. They are however much smaller than the electric fields of E waves, i.e., the energy of the A waves is predominantly mechanical. The magnitudes of the electric-dipole matrix elements are accordingly quite small. Consequently, the absorption of phonons by the free carrier excitations causes an attenuation of the waves but does not lead to any appreciable dispersion of the velocity of the waves. In the case of E waves the magnitude of the electric-dipole matrix elements are quite large, and the absorption of photons by the free carrier excitations produces strong anomalous dispersion effects. The frequency range of A waves is relatively small, and it is not possible to use A waves to investigate the collective excitations of a plasma except in the case of relatively low-density plasmas. On the other hand, there are excitations which interact with A waves but not with E waves—for example, Landau excitations in the Faraday configuration.

In the various discussions that follow the A and E waves will be assumed to be plane waves having a spatial and temporal dependence given by $\exp[i(\mathbf{q} \cdot \mathbf{r} - \omega t)]$. In addition, we make no distinction between RF, microwave, infrared, and visible frequencies since, apart from the differences in the experimental techniques that are required in the different frequency ranges, such distinctions do not, as such, enter into the underlying physics.

SINGLE-PARTICLE EXCITATIONS

Spherical Energy Bands

In the absence of a magnetic field the energy and wave function of an electron in a spherical energy band are given by

$$\mathscr{E}(\mathbf{k}) = \frac{\hbar^2(k_x{}^2 + k_y{}^2 + k_z{}^2)}{2m^*} \tag{1a}$$

and

$$\psi_n(\mathbf{r}) = F_n(\mathbf{r})U_n(\mathbf{r}) = \exp[i(\mathbf{k} \cdot \mathbf{r})]U_n(\mathbf{r}) \tag{1b}$$

respectively, where $\hbar k_x$, $\hbar k_y$, and $\hbar k_z$ are the x, y, and z, components of the momentum of the electron; m^* is the effective mass of the electron; $U_n(\mathbf{r})$ is the Bloch function for the energy band n at $\mathbf{k} = 0$; and $F_n(\mathbf{r}) = \exp[i(\mathbf{k} \cdot \mathbf{r})]$ is the wave function of the free electron.

The energy levels of electrons in a magnetic field are discussed by Zawadzki in Chapter 13 of this volume. For a magnetic field $\mathbf{H}_0$ directed along the z axis the energy and wave function of the electron, neglecting spin, are given by

$$\mathscr{E}(L, k_z) = \hbar\omega_c(L + \tfrac{1}{2}) + \hbar^2 k_z{}^2/2m^* \tag{2a}$$

and

$$\psi_n(L, \mathbf{r}) = F_n(L, \mathbf{r})U_n(\mathbf{r}) = \exp[i(k_y y + k_z z)]\phi_L(x - x_0)U_n(\mathbf{r}) \tag{2b}$$

respectively where $\omega_c = eH_0/m^*c$ is the cyclotron frequency; L is the Landau quantum number; $x_0 = -(c\hbar/eH_0)k_y$ is the center of the cyclotron orbit along the x axis; and $\phi_L(x - x_0)$ is the harmonic oscillator function of the variable $(eH_0/c\hbar)^{1/2}(x - x_0)$. (The expression for $\psi_n(L, \mathbf{r})$ is based on the Landau (or asymmetric) gauge, $\mathbf{A} = [0, H_0 x, 0]$.) We note that the dependence of the energy on momentum involves only $\hbar k_z$, the component of the momentum along the magnetic field, and that the center of orbit is determined by $\hbar k_y$.

When the spin of the electron is taken into account the expression for the energy contains an additional term, $g\beta H_0\sigma$, where g is the effective g-factor, $\beta = e\hbar/2mc$ is the Bohr magneton, and $\sigma = \pm\tfrac{1}{2}$ is the z component of the spin angular momentum.

$$\mathscr{E}(L, k_z, \sigma) = \hbar\omega_c(L + \tfrac{1}{2}) + \frac{\hbar^2 k_z^2}{2m^*} + g\beta H_0\sigma \qquad (3)$$

Similar expressions hold for the energy and the wave function of a hole in a spherical valence band.

The motion of the electron is quantized in the x–y plane and it is quasi-continuous in the z direction, the direction of the magnetic field. The energy levels form a series of "one-dimensional subbands" [3] (Fig. 1). Each energy level corresponds to a large number of degenerate states having different k_y values and therefore different centers of orbit. The degeneracy per unit area of the solid transverse to the magnetic field is equal to $(eH_0/c\hbar)$.

The electrons are also characterized by an uncertainty in energy, $\Delta\mathscr{E} = \hbar/\tau(L, \mathbf{k}, \sigma)$ where $\tau(L, \mathbf{k}, \sigma)$ is the lifetime of the electron in the $(L, \mathbf{k}, \sigma)$ state, and by an uncertainty in momentum $\Delta k = 1/l(L, \mathbf{k}, \sigma)$ where $l(L, \mathbf{k}, \sigma)$ is the mean free path of the electron. For the purposes of the present discussion we will assume that $l(L, \mathbf{k}, \sigma) = v(L, \mathbf{k}, \sigma)\tau(L, \mathbf{k}, \sigma)$ where $v(L, \mathbf{k}, \sigma)$ is the velocity of the electron; namely, we assume that both $\Delta\mathscr{E}$ and $\Delta(\hbar k)$ are determined by the lifetime, $\tau(L, \mathbf{k}, \sigma)$. The wave func-

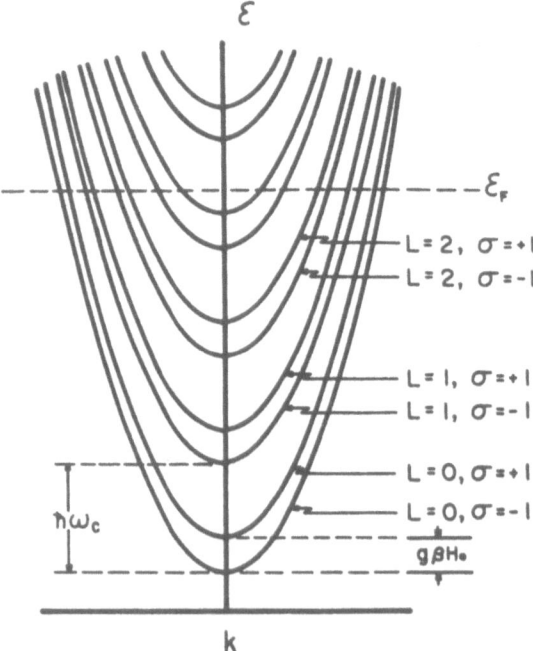

Fig. 1. One-dimensional "spherical" energy bands (Landau subbands) for an electron in a magnetic field.

tion of the electron determines the selection rules and the transition probabilities for the various single-particle excitations that can be induced by A and E waves. The conditions for observing well-defined single-particle transitions are that $\hbar\omega > \Delta\mathscr{E}$ and $q > \Delta k$. They correspond to $\omega\tau(L, \mathbf{k}, \sigma) > 1$ and $ql(L, \mathbf{k}, \sigma) > 1$.

The single-particle excitations which can occur include (1) $\Delta L = 0$ Landau (intra-sub-band) transitions, (2) $\Delta L \neq 0$ cyclotron (inter-subband) transitions, (3) $\Delta\sigma = \pm 1$ electron spin transitions, and (4) $\Delta L \neq \pm 1$ and $\Delta\sigma = \pm 1$ combined cyclotron and spin transitions.

Landau (ΔL = 0) Excitations

In a Landau excitation, an electron makes an "intra-sub-band" transition from an initial state $(L, \mathbf{k}, \sigma)$ to a final unoccupied state $(L, \mathbf{k}', \sigma)$ in the same Landau subband, and a quantum (photon or phonon) having an energy $\hbar\omega(\mathbf{q})$ and a momentum $\hbar\mathbf{q}$ is absorbed (or emitted) in the process (Fig. 2). The changes in energy and momentum of the electron are given by

$$\Delta\mathscr{E} = \mathscr{E}(L, k_z', \sigma) - \mathscr{E}(L, k_z, \sigma) = \frac{\hbar^2(k_z'^2 - k_z^2)}{2m^*} = \hbar\omega(\mathbf{q}) \qquad (4a)$$

and

$$\Delta k_x = q_x, \qquad \Delta k_y = q_y, \qquad \text{and} \qquad \Delta k_z = q_z = q\cos\theta \qquad (4b)$$

respectively where θ is the angle between the direction of propagation and the direction of the magnetic field, i.e., the angle between $\mathbf{q}$ and the z axis. The change in energy, $\Delta\mathscr{E}$, is positive for transitions in which a quantum is absorbed and it is negative for transitions in which a quantum is emitted. On introducing $k_z' = k_z + q_z = k_z + q\cos\theta$ into the expression for $\Delta\mathscr{E}$, we obtain

$$\Delta\mathscr{E} = \frac{\hbar^2(\mathbf{k}\cdot\mathbf{q})_z}{m^*} + \frac{\hbar^2 q_z^2}{2m^*} = \hbar q v(L, k_z, \sigma)\cos\theta + \frac{\hbar^2 q^2}{2m^*}\cos^2\theta = \hbar\omega(\mathbf{q}) \qquad (5)$$

where $v(L, k_z, \sigma) = \partial\mathscr{E}/\partial\hbar k = \hbar k_z/m^*$ is the "group" velocity of the electron in the *initial* state. Excitations for which $q_z/2k_z \ll 1$ correspond to forward scattering and those for which $q_z/2k_z \gg 1$ correspond to backward scattering. The two types of transitions are illustrated in Fig. 2. In the case of forward scattering excitations the expression for $\Delta\mathscr{E}$ takes the simple approximate form

$$\Delta\mathscr{E} = \hbar\omega(\mathbf{q}) \approx \hbar q v(L, k_z, \sigma)\cos\theta = \hbar\omega_D \qquad (6)$$

where $qv(L, k_z, \sigma)\cos\theta = \omega_D$ represents a Doppler shift in frequency due

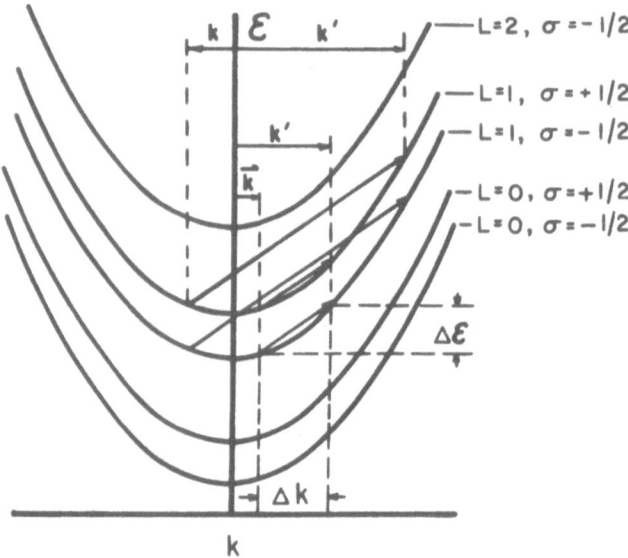

Fig. 2. Forward and back scattering $\Delta L = 0$ Landau transitions of a free electron in a magnetic field.

to the motion of the electron along the magnetic field. On dividing by $\hbar q$, one obtains

$$\omega(q)/q = v_W(\omega) = v(L, k_z, \sigma) \cos \theta \tag{7}$$

where $v_W(\omega)$ is the phase velocity of the phonon (or photon). In the Faraday ($\mathbf{q} \parallel \mathbf{H}_0$) configuration the $\Delta L = 0$ transitions which can be excited are those for which the velocity of the electron is equal to the phase velocity of the phonon (or photon). The $\Delta L = 0$ transitions do not take place in the Voigt ($\mathbf{q} \perp \mathbf{H}_0$) configuration (Fig. 3), since the absorption of a quantum does not change k_z, the component of $\mathbf{k}$ along $\mathbf{H}_0$.

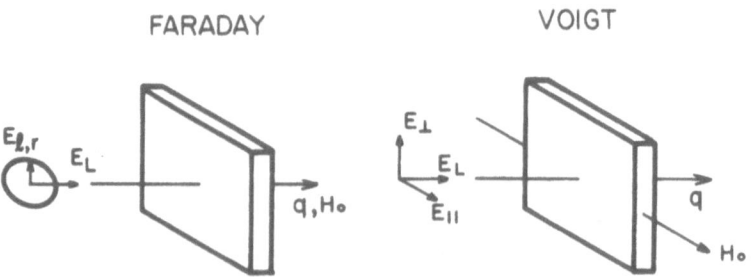

Fig. 3. The Faraday ($\mathbf{q} \parallel \mathbf{H}_0$) and the Voigt ($\mathbf{q} \perp \mathbf{H}_0$) configurations.

The Fermi surface of the plasma places a further restriction on the allowed Landau excitations, namely, only those transitions can be excited in which the initial (or final) state lies below $\mathscr{E}_F$, the Fermi energy, and the final (or initial) state lies above $\mathscr{E}_F$. When the energy of the quantum absorbed (or emitted) is small compared to the Fermi energy, i.e., $\hbar\omega \ll \mathscr{E}_F$, the $\Delta L = 0$ excitation which can be excited involve electrons "at the Fermi surface" and the expression for the change in energy of the electron takes the form

$$\Delta\mathscr{E} = \frac{\hbar^2 q k_F(L, \sigma)}{m^*}\cos\theta + \frac{\hbar^2 q^2 \cos^2\theta}{2m^*}$$

$$= \hbar q v_F(L, \sigma)\cos\theta + \frac{\hbar^2 q^2 \cos^2\theta}{2m^*} \tag{8}$$

where $k_F(L, \sigma)$ is the Fermi momentum of an electron in the (L, σ) sub-band and $v_F(L, \sigma)$ is the corresponding Fermi velocity, given by

$$v_F(L, \sigma) = [(2/m^*)(\mathscr{E}_F - L\hbar\omega_c + g\beta H\sigma)]^{1/2} \tag{9}$$

$\mathscr{E}_F$ is measured relative to the band edge of the lowest ($L = 0$, $\sigma = -\frac{1}{2}$) subband. We note that $v_F(L, \sigma)$ has its largest value in the $L = 0$, $\sigma = -\frac{1}{2}$ subband and that its value in the other subbands decreases with increasing magnetic field, going to zero when $L\hbar\omega_c + g\beta H\sigma = \mathscr{E}_F$.

In the case of forward scattering excitations $\Delta\mathscr{E}$ is given by

$$\Delta\mathscr{E} \approx \hbar q v_F(L, \sigma)\cos\theta = \hbar\omega(\mathbf{q}) \tag{10}$$

and the condition for the occurance of the excitation becomes

$$\omega(q)/q = v_W = v_F(L, \sigma)\cos\theta \tag{11}$$

Thus in the Faraday ($\mathbf{q} \parallel \mathbf{H_0}$) configurations, Landau excitations can be excited when the velocity of the phonon (or photon) is equal to velocity of the electron in one of the Landau subbands. Since the Fermi velocity depends on L, σ, and H_0, this condition can be satisfied, for a given v_W and a given subband, only for one particular value of H_0, provided $v_F(L = 0$, $\sigma = -\frac{1}{2}) > v_W$. As the velocity of E waves is almost always larger than the Fermi velocity of the electrons, $\Delta L = 0$ excitations generally cannot, even apart from selection rules, take place via the absorption of photons. The $\Delta L = 0$ excitations can take place via the absorption of phonons, as the requirement $v_F > v_W$ is satisfied for A waves.

The $\Delta L = 0$ single-particle excitations in a magnetic field are actually quite similar to the single-particle excitations of electrons in the absence of a magnetic field. For $H_0 = 0$ the expression for $\Delta\mathscr{E}$ is

$$\Delta\mathscr{E} = \hbar\mathbf{q} \cdot \mathbf{v}_F(\mathbf{k}) + \frac{\hbar^2 q^2}{2m^*} \tag{12}$$

where $v_F(\mathbf{k}) = \hbar k_F/m^* = (2\mathscr{E}F/m^*)^{1/2}$. For forward scattered excitations $\Delta\mathscr{E}$ is given by

$$\Delta\mathscr{E} \approx \hbar\mathbf{q} \cdot \mathbf{v}_F(\mathbf{k}) \tag{13}$$

We see that the expressions for $\Delta\mathscr{E}$ are similar to those for $\mathbf{H}_0 \neq 0$. The main difference between the $\mathbf{H}_0 = 0$ and the $\mathbf{H}_0 \neq 0$ Landau excitations is that in the $\mathbf{H}_0 = 0$ case the electrons at the Fermi surface have a quasi-continuous range of velocities, from zero to v_F, whereas in the $\mathbf{H}_0 \neq 0$ case they have a discrete set of values $v_F(L, \sigma)$.

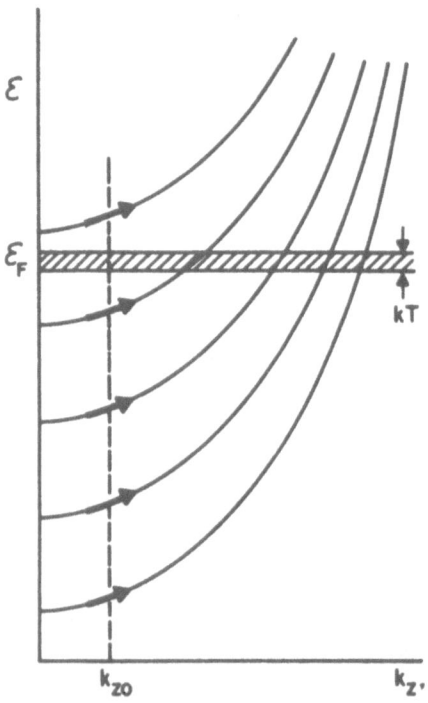

Fig. 4. The forward-scattering $\Delta L = 0$ transitions which would be possible in the absence of a Fermi surface but which are not allowed in the presence of a Fermi surface.

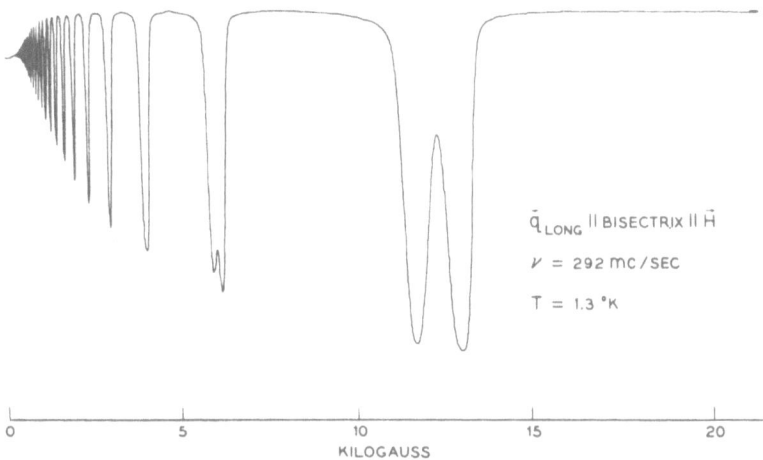

Fig. 5. Giant quantum attenuation of acoustic waves in bismuth showing the spin splitting of the attenuation peaks. From Sawada *et al.* ([5]).

The absorption of phonons by $\Delta L = 0$ excitations is illustrated in Fig. 4. The momenta of the electrons in each (L, σ) subband, whose velocities are equal to the velocity of the A waves, is independent of L and σ and of the magnitude of the magnetic field. The excitations which would occur in each subband in the absence of the restrictions of the Fermi surface are indicated by arrows. However, transitions can only take place for values of H_0 at which the energy states involved lie at the Fermi surface. As the magnetic field is increased the vertical separation between the Landau subbands, $\hbar\omega_c$, becomes larger. The absorption of phonons by the $\Delta L = 0$ excitations, which takes place as the appropriate energy states in each (L, σ) subband reaches the Fermi surface, yields a series of sharp attenuation peaks which are periodic in $(1/H_0)$. This form of A-wave attenuation, called $\Delta L = 0$ giant quantum attenuation (GQA), was first predicted theoretically by Gurevich *et al.* ([4]). It has been observed experimentally in various metals and semimetals.

From Eq. (9) one obtains the following expression for the $\Delta(1/H_0)$ period of the $\Delta L = 0$ GQA peaks:

$$\Delta\left(\frac{1}{H_0}\right) = \frac{e\hbar/m^*c}{\mathscr{E}_F - m^*v} = \frac{e\hbar/m^*c}{m^*/2(v_F{}^2 - v_W{}^2)} = \frac{4m\beta}{m^{*2}(v_F{}^2 - v_W{}^2)} \tag{14}$$

For high-density plasmas where $v_F \gg v_W$, the $\Delta(1/H_0)$ period of the peaks is the same as that for the de Hass–van Alphen oscillations in magnetic susceptibility, namely,

$$\Delta\left(\frac{1}{H_0}\right) = \frac{e\hbar}{m^*c\mathscr{E}_F} = \Delta\left(\frac{1}{H_0}\right)_{\text{dHvA}} \tag{15}$$

The period is independent of σ. However, the position of the $\Delta L = 0$ peaks does depend on σ, as well as on L, being given by

$$\left(\frac{1}{H_0}\right)_{L,\sigma} = \frac{2(e\hbar/m^*c)L + g\beta\sigma}{m^*[v_F^2(L,\sigma) - v_W^2]} \tag{16}$$

Consequently there is a spin splitting of the attenuation peaks which is readily observed at the higher fields (Fig. 5) (^{5}).

$\Delta L \neq 0$ Cyclotron Excitations

In these excitations the electron makes a transition from an initial state (L, k, σ) in a given subband to a final unoccupied state (L', k', σ) in another subband with the same σ (Fig. 6). The $\Delta L = \pm 1$ excitations are the ones which occur in Faraday and Voigt cyclotron resonance, whereas $\Delta L = \pm 2, \pm 3, \ldots$, as well as $\Delta L = \pm 1$ excitations occur in Azbel–Kaner cyclotron resonance (AKCR). The changes in energy and momentum of the electron resulting from the absorption (or emission) of a quantum are given, respectively, by

$$\Delta\mathscr{E} = \Delta L\hbar\omega_c + \frac{\hbar^2[k_z'(L')^2 - k_z(L)^2]}{2m^*} = \hbar\omega(q) \tag{17a}$$

and

$$\Delta\mathbf{k} = \Delta\mathbf{q} \tag{17b}$$

On introducing $k_z' = k_z + q_z$ into $\Delta\mathscr{E}$, we obtain

$$\Delta\mathscr{E} = \hbar\omega = \Delta L\hbar\omega_c + \hbar q_z v(L, k_z, \sigma) + \frac{\hbar^2 q_z^2}{2m^*}$$

$$= \Delta L\hbar\omega_c + \hbar q v(L, k_z, \sigma)\cos\theta + \frac{\hbar^2 q^2}{2m^*}\cos^2\theta \tag{18}$$

For excitations where $q_z/2k_z \ll 1$, the expression for $\Delta\mathscr{E}$ reduces to the simple approximate form

$$\Delta\mathscr{E} \approx \Delta L\hbar\omega_c + \hbar q_z v(L, k_z, \sigma) = \Delta L\hbar\omega_c + \hbar q v(L, k_z, \sigma)\cos\theta \tag{19}$$

in which $qv(L, k_z, \sigma)\cos\theta$ represents the Doppler shift in frequency due to the motion of the electron along the z axis. The first term can be positive or negative depending on the sign of ΔL. The Doppler shift term can be

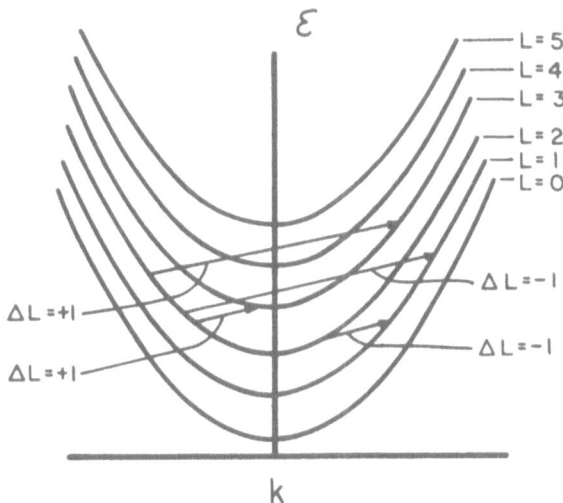

Fig. 6. Forward- and back-scattering $\Delta L = \pm 1$ cyclotron transitions of a free electron in a magnetic field.

positive or negative depending on the direction of the electron motion relative to q_z. Cyclotron resonance in which there is an appreciable Doppler shift in frequency is called Doppler-shifted cyclotron resonance (DSCR). The Doppler shift in frequency is equal to zero in the Voigt ($\mathbf{q} \perp \mathbf{H_0}$) configuration and equal to $qv(L, k_z, \sigma)$ in the Faraday ($\mathbf{q} \parallel \mathbf{H_0}$) configuration.

In cyclotron resonance both $\Delta L = -1$ and $\Delta L = +1$ excitations can occur when a quantum is either absorbed ($\Delta\mathscr{E}$ positive) or emitted ($\Delta\mathscr{E}$ negative). This is illustrated in Fig. 6 for forward scattering ($q_z/2k_z \ll 1$) and for backward-scattering ($q_z/2k_z \gg 1$) transitions.

In degenerate plasmas, only those $\Delta L = \pm 1, \pm 2, \pm 3, \ldots$ excitations can occur in which the initial (or final) state lies below the Fermi surface and the final (or initial) state lies above the Fermi surface. When the energy of the quantum is small compared to the Fermi energy, i.e., $\hbar\omega \ll \mathscr{E}_F$, the cyclotron excitations which can occur involve electron at the Fermi surface, and the expression for $\Delta\mathscr{E}$ takes the form

$$\Delta\mathscr{E} = \hbar\omega(\mathbf{q}) = \Delta L\hbar\omega_c + \hbar q v_F(L, \sigma) \cos\theta + \frac{\hbar^2 q^2}{2m^*} \cos^2\theta \qquad (20)$$

and for $q_z/2k_z \ll 1$, it reduces to the simple approximate form

$$\Delta\mathscr{E} = \hbar\omega(\mathbf{q}) \approx \Delta L\hbar\omega_c + \hbar q v_F(L, \sigma) \cos\theta \qquad (21)$$

which can also be written as

$$\omega(q)/q = v_W \approx (\Delta L \omega_c/q) + v_F(L, \sigma) \cos \theta \qquad (22)$$

Since the velocity of E waves is generally much larger than the Fermi velocity, the Doppler shift in frequency, $qv_F(L, \sigma) \cos \theta$, is usually quite small in optically excited transitions. An exception to this occurs in the case of helicon and Alfvén waves in semimetals and metals at microwave (and lower) frequencies, where the Doppler shift in frequency can be quite appreciable. The velocity of A waves, on the other hand, is generally smaller than the Fermi velocity even for moderate-density plasmas and the Doppler shift in frequency is relatively large. When v_W is very much larger than v_F, which is generally the case for E waves, the excitations involve essentially "*vertical*" transitions, and $\hbar\omega \approx |\Delta L| \hbar\omega_c$. When v_W is very much smaller than v_F, as is the case for A waves, the excitations involve essentially "*horizontal*" transitions, and (for small H_0) $\hbar\omega \approx \hbar qv_F(L, \sigma) \cos \theta$.

The DCSR absorption of phonons involving "*horizontal*" $\Delta L = \pm 1$ transitions should in principle lead to a series of attenuation peaks with increasing magnetic field. This is illustrated in Fig. 7. The values of k_z

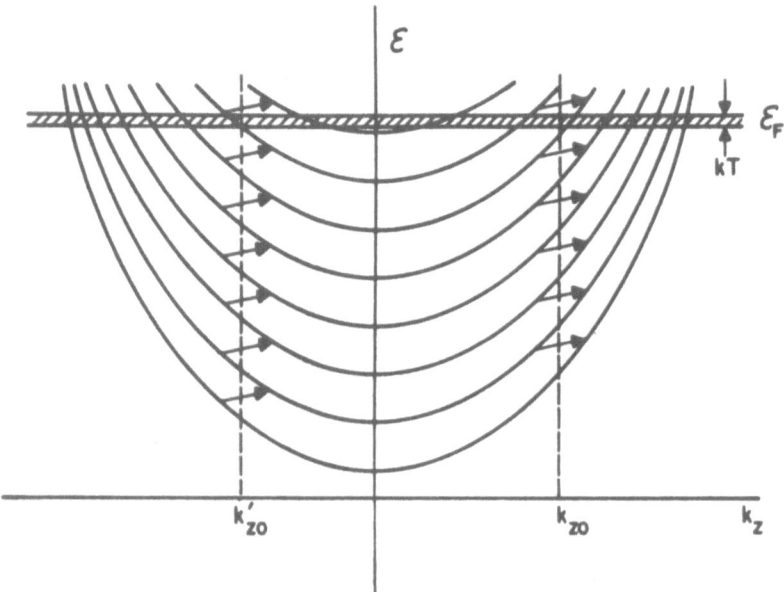

Fig. 7. The forward-scattering $\Delta L = \pm 1$ transitions which would be possible in the absence of a magnetic field but which are not allowed in the presence of a Fermi surface.

for the electrons in each subband which satisfy the requirement for ΔL $= \pm 1$ transitions are independent of L and σ, but do depend on the relative directions of q and $v(L, k_z, \sigma)$, and on the magnitude of the magnetic field. The forward-scattering excitations, which would occur in each subband in the absence of a Fermi surface, are indicated by arrows. For electrons with a negative k_z the excitations involve $\Delta L = +1$ transitions, whereas for electrons with positive k_z the excitations involve $\Delta L = -1$ transitions. The k_z values at which the excitation can occur are different for the ΔL $= +1$ and $\Delta L = -1$ transitions and, furthermore, they increase as the magnetic field is increased. The magnetic fields at which the appropriate energy states involved lie at the Fermi surface are different for the two types of transition. The absorption of phonons should thus lead to two series of attenuation peaks periodic in $(1/H_0)$. The possible existence of this type of A-wave attenuation, called $\Delta L = \pm 1$ GQA, was proposed by Langenberg et al. ([6]) and by Gantsevich et al. ([7]). It has not yet been observed experimentally. The corresponding $\Delta L = \pm 1$ GQA of E waves has been proposed ([8,9]) but the situation for E waves is complicated even further by anomalous dispersion effects.

$\Delta\sigma \pm 1$ Spin Transitions

In $\Delta\sigma = \pm 1$ excitations the electron makes a spin-flipping transition from an initial state $(L, \mathbf{k}, \sigma)$ to a final state $(L, \mathbf{k}', \sigma \pm 1)$ (Fig. 8). The expression for the change in energy of the electron is similar to that for the $\Delta L = \pm 1$ cyclotron excitations, namely,

$$\Delta\mathscr{E} = \Delta\sigma g\beta H_0 + \hbar q_z v(L, k_z, \sigma) + \hbar^2 q_z^2/2m^* = \hbar\omega(\mathbf{q}) \qquad (23)$$

in which the term $\Delta\sigma g\beta H_0$ appears in the place of $\Delta L\hbar\omega_c$. For $q_z/2k_z \ll 1$, the expression for $\Delta\mathscr{E}$ reduces to the approximate form

$$\Delta\mathscr{E} \approx \Delta\sigma g\beta H + \hbar q v(L, k_z, \sigma) \cos\theta = \frac{\Delta\sigma m^* g}{2m} \hbar\omega_c + \hbar q v(L, k_z, \sigma) \cos\theta \qquad (24)$$

in which, as in the case of cyclotron transitions, $q v(L, k_z, \sigma) \cos\theta$ represents the Doppler shift in frequency due to the motion of the electron along the z axis. The Doppler shift can be positive or negative depending on the direction of $\mathbf{v}$ relative to q_z. The Doppler shift in frequency is equal to zero for the Voigt configuration and has its largest value for the Faraday configuration. Both $\Delta\sigma = +1$ and $\Delta\sigma = -1$ excitations can occur when a quantum is absorbed or emitted, as shown in Fig. 8.

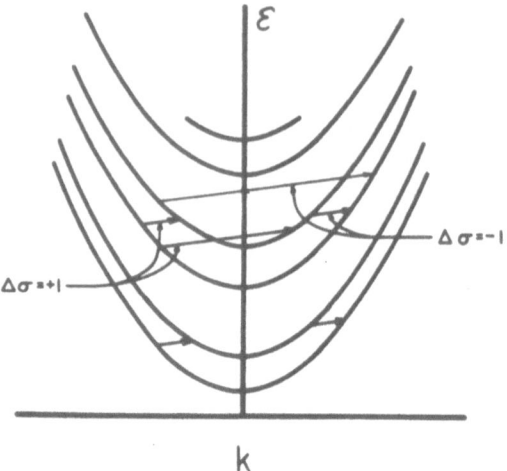

Fig. 8. Forward- and back-scattering $\Delta\sigma = \pm 1$ spin-flipping transitions of a free electron in a magnetic field.

In degenerate plasmas the expression for $\Delta\mathscr{E}$ for small $\hbar\omega$ and for $q_z/2kz \ll 1$, takes the form

$$\Delta\mathscr{E} = \hbar\omega(\mathbf{q}) \approx \Delta\sigma gm^* \hbar\omega_c/2m + \hbar q v_L(L, \sigma) \cos\theta \qquad (25)$$

which can also be written as

$$\frac{\omega(q)}{q} = v_W \approx \frac{\Delta\sigma gm^*}{2m}\left(\frac{\omega_c}{q}\right) + v_F(L, \sigma) \cos\theta \qquad (26)$$

These expressions for Doppler shifted electron spin resonance are similar to the corresponding ones for Doppler-shifted cyclotron resonance, with the first term containing the factor $(\Delta\sigma gm^*/2m)$ in place of ΔL. As pointed out in the case of DSCR, the Doppler shift in frequency is usually quite small for optically induced spin transitions. An appreciable Doppler shift may occur, however, for helicon and Alfvén waves at microwave and lower frequencies in metals and semimetals. The Doppler shift in frequency may be expected to be large for phonon-induced spin transitions. In general, for $v_W \gg v_F$ the transitions are "vertical" and $h\omega \approx g\beta H_0$, whereas for $v_W \ll v_F$ the transitions are horizontal and (for small H_0) $h\omega \approx hqv_F(L, \sigma)$ $\cos\theta$. The Doppler-shifted spin resonance absorption of phonons, involving horizontal $\Delta\sigma = \pm 1$ transitions, may be expected to yield a series of attenuation peaks similar in their general features to the $\Delta L = \pm 1$ GQA peaks.

$\Delta L = \pm 1$, $\Delta\sigma = \pm 1$ Combined Cyclotron and Spin Excitations

In these excitations the electron makes transition from an initial state $(L, \mathbf{k}, \sigma)$ to a final state $(L', \mathbf{k}', \sigma')$ (Fig. 9). The $\Delta\mathscr{E}$ for the excitations is given by

$$\Delta\mathscr{E} = \Delta L\hbar\omega_c + \Delta\sigma g\beta H_0 + \hbar q_z v(L, k_z, \sigma) + \hbar^2 q_z^2/2m = \hbar\omega(\mathbf{q}) \quad (27)$$

and when $q_z/2k_z \ll 1$ the expression for $\Delta\mathscr{E}$ reduces to the approximate form

$$\Delta\mathscr{E} \approx \Delta L\hbar\omega_c + \Delta\sigma g\beta H_0 + \hbar q v(L, k_z, \sigma)\cos\theta$$

$$= (\Delta L + \Delta\sigma m^* g/2m)\hbar\omega_c + \hbar q v(L, k_z, \sigma)\cos\theta \quad (28)$$

As in the case of cyclotron excitations and spin excitations, $\Delta\sigma = +1$ and -1 and $\Delta L = +1$ and -1 excitations can occur when a quantum is absorbed or emitted.

In degenerate plasmas the expression for $\Delta\mathscr{E}$ (for small $\hbar\omega$ and

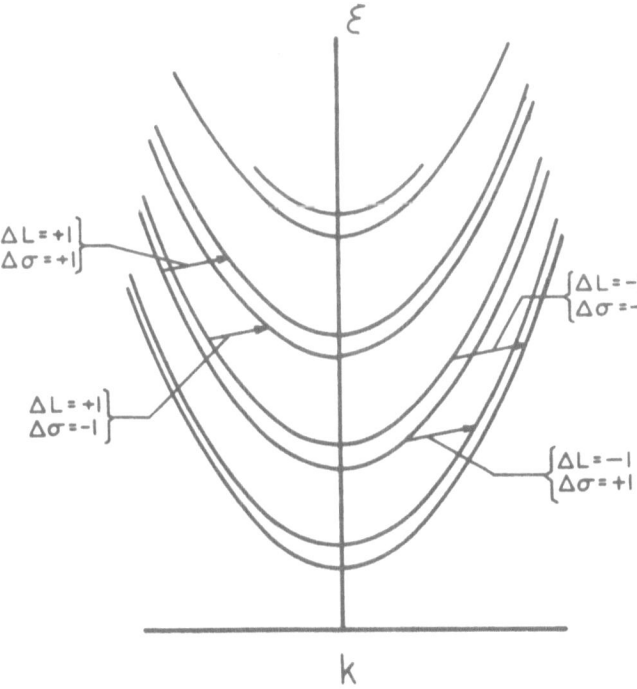

Fig. 9. Combined (cyclotron plus spin-flipping) transitions of a free electron in a magnetic field.

$q_z/2k_z \ll 1$) takes the form

$$\Delta\mathscr{E} = \hbar\omega(\mathbf{q}) \approx (\Delta L + \Delta\sigma m^*g/2m)\hbar\omega_c = \hbar q v_F(1, \sigma)\cos\theta \qquad (29)$$

Apart from selection rules and the magnitudes of $\Delta\mathscr{E}$ the same considerations that apply for cyclotron and spin excitations also apply to the combined cyclotron and spin excitations.

Ellipsoidal Energy Bands

In the absence of a magnetic field the energy and wave function of an electron in an ellipsoidal energy band whose major axis is oriented at an angle to the y and z axes (Fig. 10), are given by

$$\mathscr{E}(\mathbf{k}) = \frac{\hbar^2}{2m_0}[\alpha_{11}k_x{}^2 + \alpha_{22}ky^2 + \alpha_{33}k_z{}^2 + 2\alpha_{23}k_yk_z] \qquad (30a)$$

and

$$\psi(\mathbf{r}) = [\exp(i\mathbf{k}\cdot\mathbf{r})]U_n(\mathbf{r}) \qquad (30b)$$

respectively, where the α_{ij} are the components of the inverse mass tensor, $\alpha = \mathbf{m}^{*-1}$, which characterize the shape and orientation of the ellipsoid.

In the presence of a magnetic field directed along the z axis, the energy and wave function of the electron, neglecting spin, are given by

$$\mathscr{E}(L, k_z) = \hbar\omega_c(L + \tfrac{1}{2}) + \left(\alpha_{33} - \frac{\alpha_{23}^2}{\alpha_{22}}\right)\frac{\hbar^2k_z{}^2}{2m_0} \qquad (31a)$$

and

$$\psi(L, \mathbf{r}) = \exp[i(k_yy + k_zz)]\varphi_L(x - x_0)U_n(\mathbf{r}) \qquad (31b)$$

respectively, where $\omega_c = eH_0\alpha_{11}\alpha_{22}/m_0c$ and x_0, the center of the cyclotron orbit along the x axis, is given by

$$x_0 = \frac{-e\hbar}{cH_0}\left(k_y + \frac{\alpha_{23}}{\alpha_{22}}k_z\right) \qquad (32)$$

The velocity of the electron is given by

$$v(L, k_z) = \frac{\hbar k_z}{m_0}\left(\alpha_{33} - \frac{\alpha_{23}^2}{\alpha_{22}}\right) \qquad (33)$$

The quantities ω_c, x_0 and $v(L, k_z)$ are dependent upon the direction of the magnetic field relative to the axes of the ellipsoid. Apart from this directional dependence on the magnetic field the expressions for the energies of the

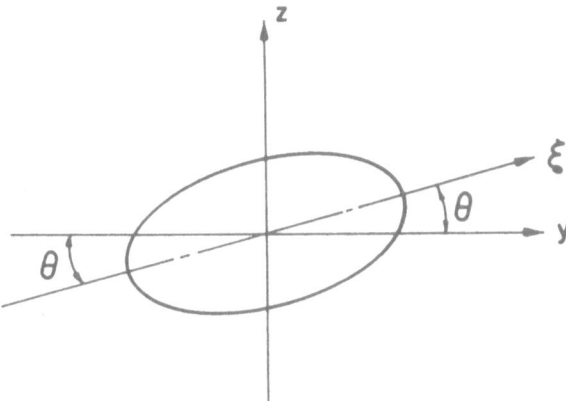

Fig. 10. Schematic diagram of an ellipsoidal energy band whose principal axis ξ is tilted in the yz plane.

single-particle excitations in terms of ω_c and $v(L, k_z)$ are the same as those for the electron in a spherical energy band. However, the center of the cyclotron orbit of the electron in the ellipsoidal band depends on k_z as well as on k_y, whereas the center of the cyclotron orbit of an electron in a spherical band depends only on k_y.

ω versus q Dispersion Curves

The various types of single-particle excitations have characteristic ω versus $\mathbf{q}$ ($\Delta\mathscr{E}$ versus $\Delta\hbar\mathbf{k}$) curves. These are illustrated in Figs. 11, 12, and 13 for $H_0 = 0$, $\Delta L = 0$ and $\Delta L = \pm 1$ excitations, respectively. The curves for $\Delta\sigma = \pm 1$ excitations (not shown) are quite similar to those for the $\Delta L = \pm 1$ excitations.

The ω versus $\mathbf{q}$ curves for $H_0 = 0$ excitations (Fig. 11) form a quasi-continuous area bounded by two curves, one corresponding to forward scattering of an electron at the Fermi surface, which goes through the origin, and the other corresponding to backward scattering of the electron, which intersects the abscissa ($\omega = 0$) at $q = 2k_F$. The slopes of the two curves at $\omega = 0$ are equal to $+v_F$ and increase with increasing ω. The two curves represent the excitation of electrons at the Fermi surface having velocities $\pm v_F$. The quasicontinuous distribution of points in the area between the two curves represent the excitations of electrons having components of motion perpendicular to the direction of $\mathbf{q}$.

The ω versus $\mathbf{q}$ curves for $\Delta L = 0$ Landau excitations at a given magnetic field (Fig. 12) consists of two discrete curves for each Landau subband

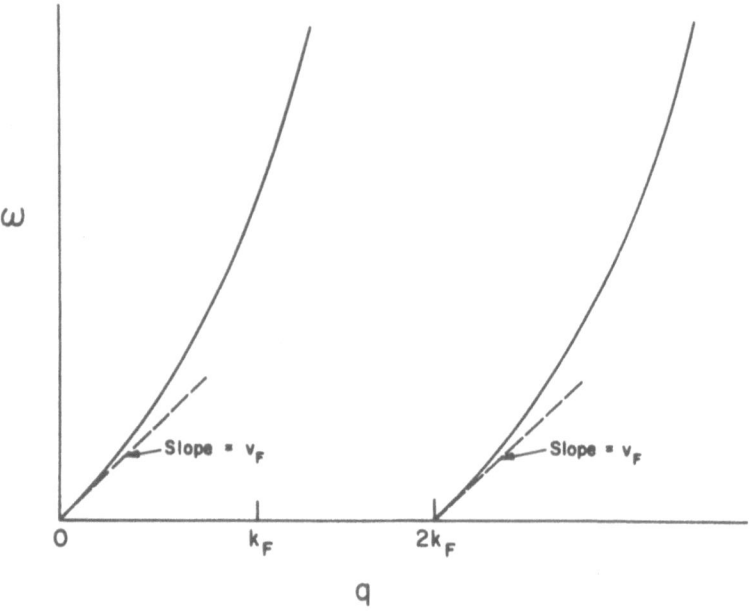

Fig. 11. Plot of $\omega = \Delta\mathscr{E}/\hbar$ versus $\mathbf{q} = \Delta\mathbf{k}$ for the $\mathbf{H}_0 = 0$ intraband excitations of an electron at the Fermi surface.

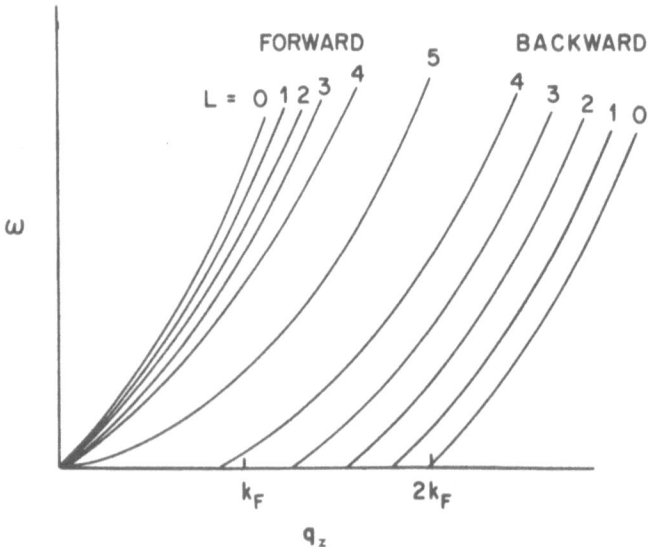

Fig. 12. Plot of $\omega = \Delta\mathscr{E}/\hbar$ versus $q_z = \Delta k_z$ for the $\Delta L = 0$ Landau excitations of an electron at the Fermi surface.

L, σ having states at the Fermi surface, one curve corresponding to forward scattering which goes through the origin, and a second curve, corresponding to backward scattering which intercepts the abscissa at $2k_F(L, \sigma)$. As in the case of the $H_0 = 0$ excitations, the slopes of the two curves at $\omega = 0$ are, for a given subband, equal to $v_F(L, \sigma)$. The slopes are largest for the lowest energy subband ($L = 0$, $\sigma = -\frac{1}{2}$) and smallest for the highest energy subband with states at the Fermi surface. The two curves coalesce and have zero slopes at $\omega = 0$ and $\mathbf{q} = 0$ for a subband when its band edge coincides with the Fermi level at the particular value of the magnetic field. With increasing magnetic field the Fermi velocity in each subband (with the exception of the $L = 0$, $\sigma = -1$ subband) decreases, and pairs of curves successively disappear as the subbands rise above the Fermi surface. At fields such that $\hbar\omega_c > \mathscr{E}_F$, the only curves that appear are those for the lowest ($L = 0$, $\sigma = -1$) subband.

The corresponding ω versus $\mathbf{q}$ curves for $\Delta L = \pm 1$ excitations (Fig. 13) are somewhat more complicated than those for the $\Delta L = 0$ excitations. Forward-scattering Landau excitations only occur for positive $\mathbf{q} \cdot \mathbf{v}_F(L, \sigma)$ whereas forward-scattering cyclotron excitations can occur for negative as

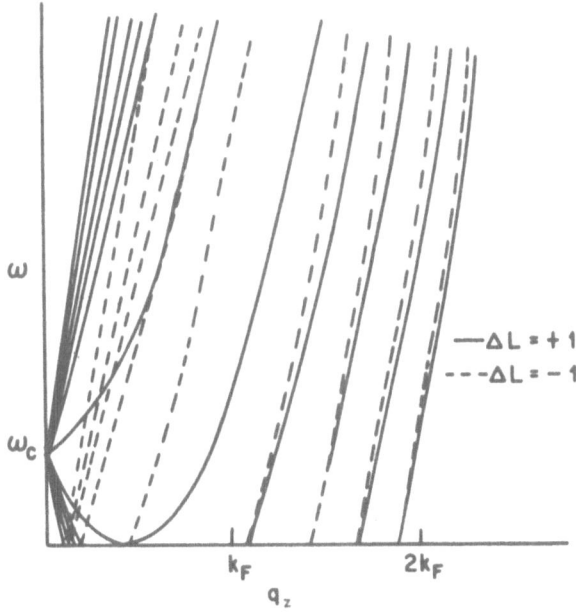

Fig. 13. Plot of $\omega = \Delta\mathscr{E}/\hbar$ versus $q_z = \Delta k_z$ for the $\Delta L = \pm 1$ cyclotron excitations of an electron at the Fermi surface.

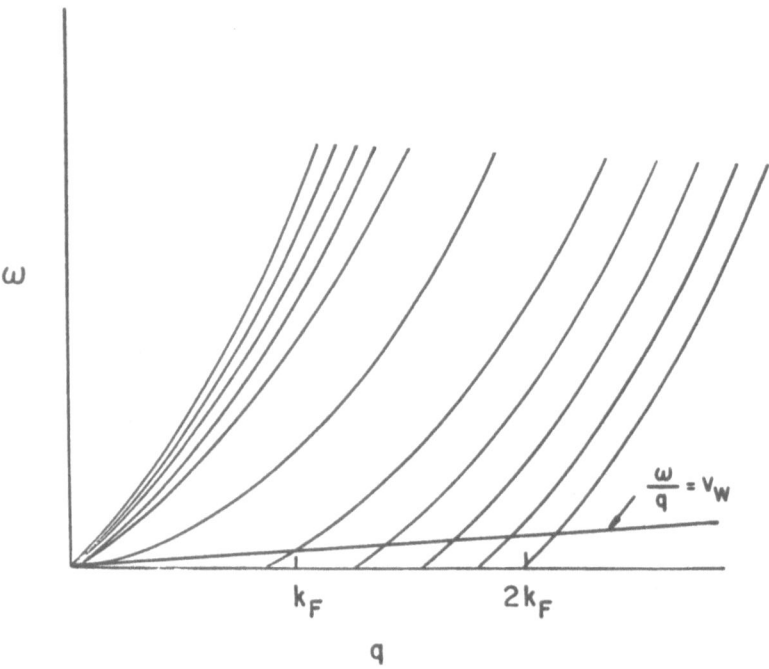

Fig. 14. The ω versus $\mathbf{q}$ curves for A waves and for the $\Delta L = 0$ Landau excitations of electrons at the Fermi surface. The intersections of the curves represent the ω, $\mathbf{q}$ values at which phonons will be absorbed by the electrons.

well as positive $\mathbf{q} \cdot \mathbf{v}_F(L, \sigma)$. For $\mathbf{q} \approx 0$ the cyclotron excitations correspond to vertical $\Delta L = \pm 1$ transitions and $\omega = \Delta \mathscr{E}/\hbar = \omega_c$. As $\mathbf{q}$ increases $\Delta \mathscr{E}/\hbar$ increases for electrons at the Fermi surface which have positive $\mathbf{q} \cdot \mathbf{v}_F(L, \sigma)$ but decreases for electrons which have negative $\mathbf{q} \cdot \mathbf{v}_F(L, \sigma)$, going to zero when $\mathbf{q} \cdot \mathbf{v}_F(L, \sigma) = -\omega_c$.

Forward-scattering $\Delta L = +1$ excitations with $\mathbf{q} \cdot \mathbf{v}_F(L, \sigma)$ negative do not occur for $q > \omega_c/v_F(L, \sigma) \cos \theta$. Forward scattering $\Delta L = -1$ excitations (dashed curves) occur for $\mathbf{q} \cdot \mathbf{v}_F(L, \sigma)$ positive and $q < \omega_c/v_F(L, \sigma) \cos \theta$, but do not occur at smaller values of $\mathbf{q}$. Both $\Delta L = +1$ and -1 excitations occur in backscattering. At $\mathbf{q} = 0$ the slopes of the curves for the different subbands are equal to $\pm v_F(L, \sigma)$. With increasing magnetic field ω_c increases, the Fermi velocity in each subband (other than the $L = 0$, $\sigma = -1$ subband) decreases, and, as in the case of $\Delta L = 0$ excitations, the curves for the subbands successively disappear as the subbands rise above the Fermi surface. For strong magnetic fields such that $\hbar \omega_c > \mathscr{E}_F$

the only curves that appear are those for the lowest ($L = 0$, $\sigma = -1$) subband.

For a given magnetic field, the ω and q_z values of the A and E waves that can induce single-particle excitations are determined by the intersections of the ω versus q_z curves of the A and E waves with the ω versus q_z curves of the single-particle excitations. [The situation in the case of E waves is actually more complicated, since the interactions of E waves with the excitations leads to a strong anomalous dispersion.] This is illustrated in Figs. 14 and 15 for the interaction of an A wave with $\Delta L = 0$ and $\Delta L = \pm 1$ single-particle excitations of a moderately low density plasma. The A waves, unlike the E waves, exhibit no appreciable dispersion in velocity, and their ω versus q_z curves are essentially straight lines having slopes $\omega/q = v_W$. In the situation shown in the figures, $v_W \ll v_F$ and the ω versus q_z curve for the A wave is relatively horizontal. On the same scale the corresponding ω versus q_z curve for E waves would be essentially a vertical line.

At a given magnetic field the absorption of phonons by the single-particle excitations are seen to occur at a number of discrete (ω, q_z) values.

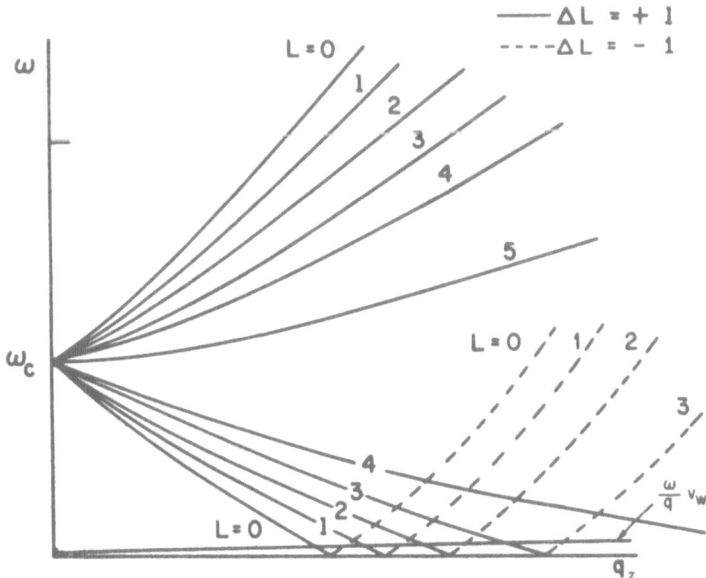

Fig. 15. The ω versus q_z curves for A waves and for the $\Delta L = \pm 1$ cyclotron excitations of electrons at the Fermi surface. The intersections of the curves represent the ω, q_z values at which phonons will be absorbed by the electrons.

One also notes the small but finite difference in the (ω, q_z) values at which phonons are absorbed by the $\Delta L = +1$ and the $\Delta L = -1$ cyclotron excitations. It is not generally possible at the present state of the art to generate A waves of sufficiently high frequency (and wave vector) to interact with the backward-scattering excitations. Moreover, it is difficult to vary the frequency of the waves in a continuous fashion over any appreciable range of frequencies. It is much more convenient and practical to fix ω and to vary the magnitude of the applied magnetic field. When the magnetic field is varied the intercept of the (ω, q_z) value of the A wave with the discrete ω versus q curves of the various subbands occurs only at one specific value of the magnetic field for each subband. These correspond to the values of the magnetic field at which peaks occur in the $\Delta L = 0$ and $\Delta L = \pm 1$ GQA of A waves.

Center of Cyclotron Orbit Effects

The center of the cyclotron orbit of an electron depends on its momentum in the plane of the orbit. For an electron in a spherical energy band with $\mathbf{H}_0$ along the z axis, the x coordinate of the center of orbit, based on the gauge $\mathbf{A} = [0, H_0 x, 0]$, is given by

$$x_0 = -\frac{c\hbar}{eH_0} k_y \tag{34}$$

When the electron absorbs a phonon or a photon and undergoes a single-particle excitation in which k_y changes by $\Delta k_y = q_y$ the center of orbit changes by an amount Δx_0 given by

$$\Delta x_0 = (-c\hbar/eH_0)\Delta k_y = -(c\hbar/eH_0)q_y \tag{35}$$

The sign of Δx_0 will depend on the sign of $\Delta k_y = k_y' - k_y$ and also on the sign of H_0 (Fig. 16).

The shift of the center of orbit does not depend on the choice of gauge. In general, one chooses the gauge most appropriate to the configuration of $\mathbf{q}$ and $\mathbf{H}_0$ being used. Thus for situations where the electron undergoes a transition in which k_x is changed, one would most conveniently use the gauge $\mathbf{A} = [-H_0 y, 0, 0]$. In this gauge k_x is a good quantum number and the y coordinate of the center of orbit is given by

$$y_0 = (c\hbar/eH_0)k_x \tag{36}$$

Thus the shift of the center of orbit when a quantum is absorbed having a component of momentum along x is given by

$$\Delta y_0 = (c\hbar/eH_0)\Delta k_x = (c\hbar/eH_0)q_x \tag{37}$$

For ellipsoidal energy bands the x coordinate of the center of orbit is

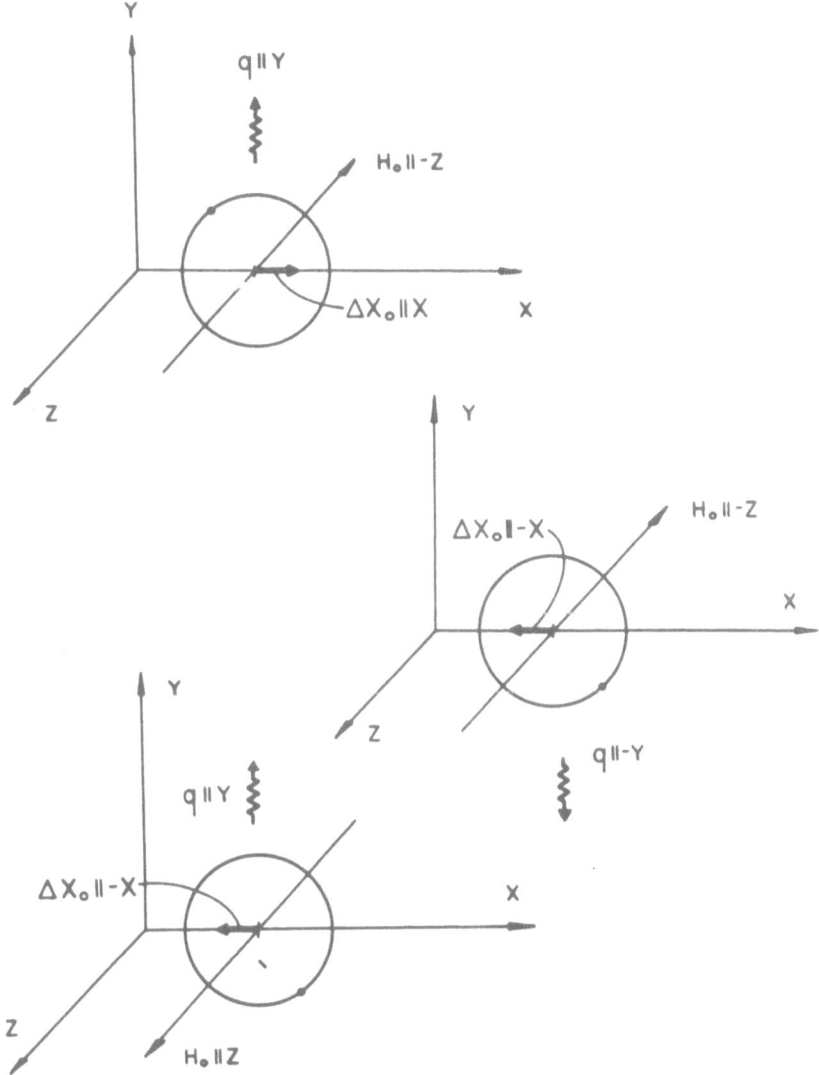

Fig. 16. The displacements of the center of cyclotron orbits 1, which accompany the absorption of a quantum by an electron for (a) H $\parallel$ $-$Z and q $\parallel$ Y, (b) H_0 $\parallel$ $-$Z and q $\parallel$ $-$Y, and (c) H_0 $\parallel$ Z and q $\parallel$ Y.

given by

$$x_0 = \frac{-c\hbar}{eH_0}\left(k_y + \frac{\alpha_{23}}{\alpha_{22}}k_z\right) \tag{38}$$

A shift of the center of orbit occurs when the electron absorbs a phonon or a photon and makes a transition in which either k_y or k_z, or both k_y and k_z, change (Fig. 17):

$$\Delta x_0 = \frac{-c\hbar}{eH_0}\left(\Delta k_y + \frac{\alpha_{23}}{\alpha_{22}}\Delta k_z\right) \tag{39}$$

One notes that in the case of spherical energy bands a shift in the center of orbit cannot occur in the Faraday configuration, i.e., when $\mathbf{q}$ is parallel to $\mathbf{H}_0$. For an arbitrary direction of $\mathbf{q}$ relative to $\mathbf{H}_0$ the shift of the center of orbit takes place in a direction perpendicular to the plane of $\mathbf{q}$ and $\mathbf{H}_0$. In the case of ellipsoidal bands a shift in the center of orbit can occur in the Faraday configuration provided the axes of the ellipsoid are tilted from the direction of $\mathbf{H}_0$. For $\mathbf{q} \parallel \mathbf{H}_0$ the displacement of the center of orbit takes place in a direction perpendicular to $\mathbf{q}$ and to the plane of tilt.

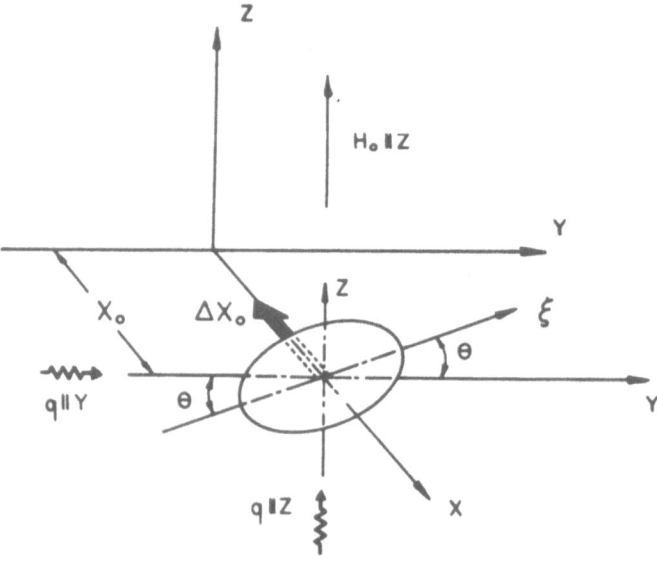

Fig. 17. The displacement of the center of cyclotron orbit which accompanies the absorption of a quantum by an electron in an ellipsoidal energy band whose principal axis is tilted in the yz plane for $\mathbf{H}_0 \parallel \mathbf{Z}$ and $\mathbf{q} \parallel \mathbf{Y}$ and for $\mathbf{H}_0 \parallel \mathbf{Z}$ and $\mathbf{q} \parallel \mathbf{Z}$.

Crossed Electric and Magnetic Fields

The energy levels of an electron (or hole) in crossed electric and magnetic fields have been discussed by Zawadzki in Chapter 13 of this volume. For H_0 along the z direction and a static electric field E_0 in the x direction the energy and wave function of an electron in a spherical band, neglecting spin, are given by ([10])

$$\mathscr{E}(L, k_z, E_0) = \hbar\omega_c\left(L + \frac{1}{2}\right) + \frac{\hbar^2 k_z^2}{2m^*} - \frac{cE_0}{H_0}\hbar k_y - \frac{m^*c^2E_0^2}{2H_0^2} \quad (40a)$$

and

$$\psi_m(L, E_0\mathbf{r}) = \exp[i(k_y y + k_z z)]\phi_{LE}(x - x_0)U_n(\mathbf{r}) \quad (40b)$$

respectively, where x_0 is the center of the cyclotron orbit; x_0 is given by

$$x_0 = \frac{-c\hbar}{eH_0}k_y + \frac{cE_0}{\omega_c H_0} = \frac{-c\hbar}{eH_0}k_y + \frac{v_D}{\omega_c} \quad (41)$$

and $\mathbf{v}_D = c\mathbf{E}_0 \times \mathbf{H}_0/H_0^2$ is the carrier drift velocity. Thus the application of a static electric field perpendicular to H_0 causes a decrease in the energy of the electron and an average shift in the center of orbit. The term $m^*c^2E_0^2/2H_0^2 = m^*v_D^2/2$ is the average kinetic and potential energy associated with the drift motion.

In crossed electric and magnetic fields the single-particle Landau and cyclotron excitation energies for electrons in a degenerate plasma are given by

$$\Delta\mathscr{E} = \hbar q_z v_F(L, \sigma) + \frac{\hbar^2 q_z^2}{2m^*} + eE\,\Delta x_0 = \hbar\omega(\mathbf{q}) \quad (\Delta L = 0)$$

$$\Delta\mathscr{E} = \Delta L\hbar\omega_c + \hbar q_z v_F(L, \sigma) + \frac{\hbar^2 q_2^2}{2m^*} + eE\,\Delta x_0 = \hbar\omega(\mathbf{q}) \quad (\Delta L \neq 0)$$

$$(42)$$

The change in the center of orbit which accompanies the change in momentum of the electron,

$$\Delta x_0 = \frac{-c\hbar}{eH_0}\Delta k_y = \frac{-c\hbar}{eH_0}q_y \quad (43)$$

is the same as in the absence of the static electric field. The term $eE_0\,\Delta x_0$ which appears in the expression for $\Delta\mathscr{E}$ represents the change in the potential energy of the electron resulting from the shift in the center of orbit. (An analogous expression for $\Delta\mathscr{E}$ holds for the case where the electrons are in an ellipsoidal band). Thus as a result of the applied electric field perpendicular to H_0 the excitation energies become dependent on $\Delta k_y = q_y$.

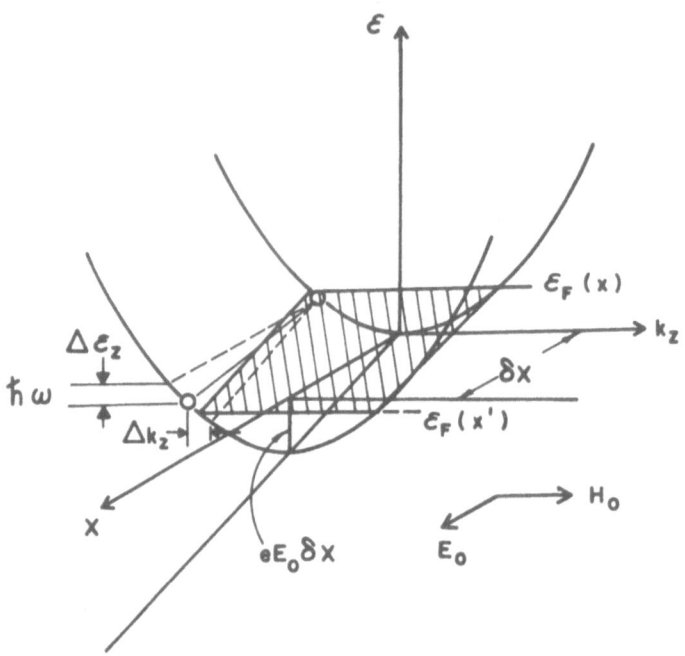

Fig. 18. The $\Delta L = 0$ transitions of an electron at the Fermi surface
in crossed electric and magnetic fields. The gain in energy arising
from the displacement of the center of orbit of the electron in the
direction of $\mathbf{E}_0$ permits the emission of phonons.

The effect of the crossed static electric field can also be visualized in
terms of its effect on the Fermi level. In the presence of $\mathbf{E}_0$ the Fermi level
is tilted and the energy of an electron at a position x is larger by an amount
eE_0x relative to that of an electron at the origin. When the electron undergoes a transition in which there is a change Δx_0 in its center of orbit, its
energy will be increased (or decreased depending on the sign of Δx_0) by an
amount $eE \Delta x_0$. At $\mathbf{E}_0 = 0$ the electrons at the Fermi surface can only
undergo transition in which a quantum is absorbed. With increasing $\mathbf{E}_0$,
a point is reached at which the electrons, because of the shift in the Fermi
level linear in x, can undergo transitions in which a quantum may either be
emitted or absorbed. At still higher $\mathbf{E}_0$ the electron can only undergo transitions in which a quantum is emitted (Fig. 18). The condition for an emission
of a quantum by a single particle $\Delta L = 0$ excitation in the Voigt configuration is given by

$$\Delta \mathscr{E} = \hbar\omega(q) = eE_0\Delta x_0 = c\hbar q E_0/H_0 = \hbar q v_D, \quad \omega/q = cE_0/H_0 = v_D \quad (44)$$

The emission of a quantum can occur at an applied electric field for which the drift velocity of the electron is equal to the velocity of the quantum. A further discussion of these aspects is given by Y. Sawada in Chapter 7 of this volume.

COLLECTIVE EXCITATION

The collective excitations of the electrons in a plasma are longitudinal harmonic oscillator modes which arise from long-range Coulomb interactions between the electrons. The fluctuations in the charge density sets up a macroscopic polarization field, E_P. The resulting force on the electrons $F = eE_P$ serves as a restoring force which determines the resonance frequency of the harmonic-oscillator-type collective excitations. The collective excitations are characterized by a wave vector Q and a frequency $\Omega(Q)$, and by macroscopic polarization fields $E_P = E_P \exp[i(Q \cdot r - \Omega t)]$ directed along Q. They can be treated as quanta having energies $\mathscr{E} = \hbar \Omega$ and momenta $p = \hbar Q$.

The $H_0 = 0$ Case

Let us consider collective excitations with Q along the z direction in which the displacements of the electrons are given by $z = z_0 \exp[i(Q \cdot r - \Omega t)]$. The displacements of the electrons sets up a longitudinal polarization (or dipole moment per unit volume), $P_L = P_z$, and a macroscopic polarization field, $E_P = -4\pi P_z = E_z$:

$$P_z = nez + \chi_0 E_z = nez + \chi_0(-4\pi P_z) = nez/\varepsilon_0 \tag{45}$$

where ez is the dipole moment set up by the displacement of one electron, n is the number of electrons per unit volume, and χ_0 and $\varepsilon_0 = 1 + 4\pi\chi_0$ are, respectively, the electric susceptibility and the dielectric constant of the medium in the absence of the plasma. The term $\chi_0 E_z$ is the polarization of the medium induced by the macroscopic field arising from the displacement of the electrons.

In the limit of long wavelength ($Q \approx 0$) the equation of motion of the electrons, which are moving in phase with one another under the influence of the macroscopic polarization field, is given by

$$m^* \ddot{z} = eE_z = -4\pi e P_z \tag{46}$$

On introducing the expression for P_z we obtain

$$m^* \ddot{z} + (4\pi ne^2/\varepsilon_0)z = 0 \tag{47}$$

where $4\pi ne^2/\varepsilon_0$ represents the restoring force constant, $m\omega_0^2$, of the collective excitation. Thus the resonance frequency of the collective excitations, called plasmons, in the absence of a magnetic field, is given by $\Omega_p = (4\pi ne^2/m^*\varepsilon_0)^{1/2} = \omega_P$. We will retain the more familar designation ω_P for the plasma frequency. The collective excitations are called plasmons.

The plasma frequency actually depends on $\mathbf{Q}$ and in the case of a degenerate plasma the expression for $\omega_P(q)$ has the form [11]

$$\omega_P(\mathbf{Q}) = \omega_P[1 + \tfrac{3}{10}(Qv_F/\omega_P)^2] \tag{48}$$

The $H_0 \neq 0$ Case

The character of the collective excitations in the presence of a magnetic field is determined by the relative directions of $\mathbf{Q}$ and $\mathbf{H}_0$. There are two limiting types of collective excitations corresponding to (1) $\mathbf{Q} \parallel \mathbf{H}_0$ and (2) $\mathbf{Q} \perp \mathbf{H}_0$. The excitations with $\mathbf{Q}$ at an arbitrary angle relative to $\mathbf{H}_0$ have an intermediate character between these two types.

$Q \parallel H_0$ Collective Excitations

For $\mathbf{Q}$ and $\mathbf{H}_0$ along the z direction, the equations of motion of the electrons are

$$\begin{aligned} m^*\ddot{x} &= e(\mathbf{v} \times \mathbf{H}_0)_x/c = e\dot{y}H_0/c \\ m^*\ddot{y} &= e(\mathbf{v} \times \mathbf{H}_0)_y/c = -e\dot{x}H_0/c \\ m^*\ddot{z} &= eE_p = -4\pi eP_z = -4\pi ne^2/\varepsilon_0 \end{aligned} \tag{49}$$

We note that the only force acting on the electron in the z direction is that of the macroscopic polarization field $eE_z = -4\pi eP_z$. The Lorentz force of the magnetic field acts only in the xy plane. Thus the resonance frequency of the collective excitations for $\mathbf{Q} \parallel \mathbf{H}_0$ is the same as that for $\mathbf{H}_0 = 0$, namely, $\Omega_p = (4\pi ne^2/m^*\varepsilon_0)^{1/2} = \omega_P$. The equations of motion in the xy plane yield ω_c for the resonance frequency of the single-particle excitations.

$Q \perp H_0$ Collective Excitations

For $\mathbf{H}_0$ along z and $\mathbf{Q}$ along x the equations of motion of the electrons are

$$\begin{aligned} m^*\ddot{x} &= \frac{e(\mathbf{v} \times \mathbf{H}_0)_x}{c} + eE_x = \frac{e\dot{y}H_0}{c} - 4\pi eP_x = \frac{e\dot{y}H_0}{c} - \frac{4\pi ne^2 x}{\varepsilon_0} \\ m^*\ddot{y} &= \frac{e(\mathbf{v} \times \mathbf{H}_0)_y}{c} = -\frac{e\dot{x}H_0}{c} \\ m^*\ddot{z} &= 0 \end{aligned} \tag{50}$$

The x and y motion of the electrons are acted upon by the Lorentz force of the magnetic field. There is an additional restoring force acting on the electrons in the x direction due to the macroscopic polarization field. On solving the equations of motion in the xy plane, one finds that the resonance frequency of the collective excitation is given by

$$\Omega_c{}^2 = \omega_c{}^2 + \frac{4\pi n e^2}{m^* \varepsilon_0} = \omega_c{}^2 + \omega_P{}^2 \tag{51}$$

The collective excitations for $\mathbf{Q} \perp \mathbf{H_0}$ have been called coupled cyclotron–plasmon modes. We prefer to call them collective cyclotron excitations. It is the macroscopic field of the cyclotron motion of the electrons which produces the collective motion, rather than a coupling of plasmons with cyclotron excitation modes.

Another perhaps more familiar situation, in which there is a similar increase in resonance frequency resulting from the additional restoring force of the macroscopic polarization field, is that of the longitudinal optical vibration modes in a polar crystal ([12]). Consider the case of a diatomic cubic crystal. The equations of motion for the long-wave ($\mathbf{Q} \approx 0$) transverse optical (TO) and longitudinal optical (LO) vibrations are

$$\bar{m}\ddot{u}_T + \bar{m}\omega_0{}^2 u_T = 0$$
$$\bar{m}\ddot{u}_L + \bar{m}\omega_0{}^2 u_L = e^* E_L = 4\pi P_L \tag{52}$$

where u_T and u_L are the transverse and longitudinal relative displacements of the $+$ and $-$ ions and have a spatial and temporal dependence of the form $\exp[i(\mathbf{Q} \cdot \mathbf{r} - \Omega t)]$; $\bar{m}$ is the reduced mass of two ions; $\bar{m}\omega_0{}^2$ is the restoring force constant; e^* is the effective charge of the ions; and P_L is the longitudinal polarization given by

$$P_L = 4\pi N e^* u_L + \chi_0 E_L = 4\pi N e^* u_L / \varepsilon_0 \tag{53}$$

where $e^* u_L$ is the dipole moment per unit cell set up by the relative longitudinal displacements of the ions and N is the number of unit cells per unit volume. On introducing the expression for P_L the equation of motion for the longitudinal optical vibrations becomes

$$\bar{m}\ddot{u}_L + \bar{m}\omega_0{}^2 u_L + \frac{4\pi N e^{*2}}{\varepsilon_0} u_L = 0 \tag{54}$$

Thus as a result of the macroscopic polarization field there is an additional restoring force $-4\pi N e^{*2} u_L / \varepsilon_0$ acting on the ions. The resonance frequency of the LO vibration modes (LO phonons) is given by

$$\Omega_L{}^2 = \omega_0{}^2 + \frac{4\pi Ne^{*2}}{\bar{m}\varepsilon_0} = \omega_T{}^2 + \Omega^2 = \omega_L{}^2 \qquad (55)$$

where $\Omega = (4\pi Ne^{*2}/\bar{m}\varepsilon_0)^{1/2}$ is the effective plasma frequency of the atoms and $\omega_T = \omega_0$ is the resonance frequency of the TO vibration modes (TO phonons). We will retain the designation ω_L for the frequency of the $\mathbf{Q} \approx 0$ LO phonons.

Coupled Longitudinal Excitations

When more than one longitudinal mode is present the macroscopic polarization field acts to couple the modes. Thus in the case of a plasma in a polar crystal the polarization field couples the plasmons and the LO phonons. In the presence of a magnetic field with $\mathbf{Q} \perp \mathbf{H_0}$ the polarization field couples the collective cyclotron excitations with the LO phonons.

In the absence of a magnetic field the equation of motion for the plasmons and the LO phonon are (for $\mathbf{Q} \parallel z$)

$$m^*\ddot{z} = eE_L = -4\pi eP_L, \qquad \bar{m}\ddot{u}_L + \bar{m}\omega_T{}^2 u_L = e^*E_L = -4\pi e^*P_L \quad (56)$$

where P_L is given by

$$P_L = (nez/\varepsilon_0) + (Ne^*u_L/\varepsilon_0) \qquad (57)$$

On solving the equations of motion one finds that the resonance frequencies of the two coupled plasmon–LO phonon modes are given by the roots of the equation [13]

$$2\Omega_\pm{}^2 = \omega_P{}^2 + \omega_L{}^2 \pm [(\omega_P{}^2 + \omega_L{}^2)^2 - 4\omega_P{}^2\omega_T{}^2]^{1/2} \qquad (58)$$

In the high-frequency (Ω_+) mode the contributions to P_L from the displacements of the electrons and the atoms are in phase, whereas in the low frequency (Ω_-) mode the contributions to P_L are out of phase.

In the presence of a magnetic field with $\mathbf{Q} \perp \mathbf{H_0}$ the coupling of the collective cyclotron excitations with the LO phonons via the macroscopic field yields two coupled modes whose resonance frequencies are given by the two roots of the equation [14]

$$2\Omega_\pm{}^2 = \Omega_c{}^2 + \omega_L{}^2 \pm [(\Omega_c{}^2 + \omega_L{}^2)^2 - 4(\omega_P{}^2\omega_T{}^2 + \omega_c{}^2\omega_L{}^2)]^{1/2} \qquad (59)$$

The nature of the coupled plasmon–LO phonon modes and the coupled collective cyclotron–LO phonon modes and their interaction with EM radiation are discussed by S. Iwasa in Chapter 5 of this volume.

The $\mathbf{Q} \perp \mathbf{H}_0$ collective cyclotron excitations of a compensated plasma also involve a coupling of longitudinal excitations via the macroscopic polarization field, namely, a coupling of the collective cyclotron excitations of the electrons with the collective cyclotron excitations of the holes. The resonance frequencies of the two coupled electron–hole modes are given by the roots of the equation ([15])

$$2\Omega_{c\pm}^2 = \Omega_{ce}^2 + \Omega_{ch}^2 \pm [(\Omega_{ce}^2 + \Omega_{ch}^2) - 4\omega_{ce}\omega_{ch}(\omega_P{}^2 + \omega_{ce}\omega_{ch})]^{1/2} \qquad (40)$$

where

$$\Omega_{ce}^2 = \omega_{Pe}^2 + \omega_{ce}^2, \qquad \Omega_{ch}^2 = \omega_{Ph}^2 + \omega_{ch}^2, \qquad \omega_P{}^2 = \omega_{Pe}^2 + \omega_{Ph}$$

In the limit $\omega^2 > \omega_{ce}, \omega_{ch}$ the resonance frequencies are given approximately by

$$\Omega_{c+}^2 \approx \Omega_{ce}^2 + \Omega_{ch}^2 \approx \omega_P{}^2, \qquad \Omega_{c-}^2 \approx \omega_{ce}\omega_{ch} \qquad (61)$$

In the high-frequency (Ω_+) mode the contributions of the electrons and the holes to $E_L = -4\pi P_L$ are in phase, and in the low-frequency (Ω_-) mode, more familiarly known as the hybrid resonance frequency, the contributions are out of phase. When the effective masses of the electrons and holes are equal to one another the contributions to the polarization field in the low frequency mode cancel and $\Omega_{c-}^2 = \omega_{ce}^2 = \omega_{ch}^2$.

INTERACTION MATRIX ELEMENTS

The probability that a phonon or a photon will be absorbed by the single particle and collective excitations is proportional to $|\mathbf{M} \cdot \mathbf{E}|^2$ where $\mathbf{M} = e\mathbf{r}$ is the electric moment of the excitation, and $\mathbf{E} = \mathbf{E} \exp[i(\mathbf{q} \cdot \mathbf{r} - \omega t)]$ is the electric field of the A and E waves. The interaction matrix elements for the absorption of phonons or photons have the form

$$\mathcal{H}_{fi} = \langle \psi_i | \mathcal{H}_{\text{int}} | \psi_f \rangle \qquad (62)$$

where ψ_i and ψ_f are the wavefunctions of the initial and final states, and $\mathcal{H}_{\text{int}} = \mathbf{M} \cdot \mathbf{E}$ is the interaction Hamiltonian which connects the initial and final states.

The matrix elements for the single-particle excitations of an electron in a magnetic field are given by

$$\begin{aligned} \mathcal{H}_{Fi} &= \langle U_n | U_n \rangle \langle F_n(L, k_y, k_z) | \mathbf{M} \cdot \mathbf{E} | F_n(L', k_y', k_z') \rangle \\ &= \langle U_n | U_n \rangle \langle \exp[i(k_y y + k_z z)]\phi_L(x - x_0) | \mathbf{M} \cdot \mathbf{E} | \exp[i(k_y' y \\ &\quad + k_z' z)]\phi_{L'}(x' - x_0') \rangle \end{aligned} \qquad (63)$$

where $\langle U_n | U_n \rangle = 1$ for intraband transitions. From the character of the free-electron wave functions $F_n(L, k_y, k_z)$ one finds that the matrix elements for $\Delta L = 0$ excitations are nonzero only when $\mathbf{E}$ has a component parallel to $\mathbf{H}_0$ [3], and that the matrix elements for $\Delta L = \pm 1$ excitations are non-zero only when $\mathbf{E}$ has a component perpendicular to $\mathbf{H}_0$. The conservation of momentum relations $\Delta k_y = q_y$ and $\Delta k_z = q_z$ also follow naturally from the form of the matrix elements. In the electric dipole approximation, $(\exp i[\mathbf{q} \cdot \mathbf{r}] \approx 1)$ which is applicable when there is no appreciable variation of the electric field over the spatial extent of the wave function of the electron, the selection rule for $\mathbf{E} \perp \mathbf{H}_0$ is that $\Delta L = \pm 1$. This selection rule is always applicable in the Faraday ($\mathbf{q} \parallel \mathbf{H}_0$) configuration for electrons in spherical energy bands since the spatial extent of the wave function along the direction of the magnetic field is relatively small. However, in the Voigt ($\mathbf{q} \perp \mathbf{H}_0$) configuration the excitation is accompanied by a shift in the center of the orbit of the electron, and the harmonic oscillator waves in the initial and final states, $\phi_L(x - x_0)$ and $\phi_{L'}(x' - x_0')$, do not have the same centers of orbit. Consequently, for finite $\mathbf{q}$ the selection rules for cyclotron excitations becomes $\Delta L = \pm 1, \pm 2, \pm 3, \ldots$. There is of course another major difference between the cyclotron excitations that take place in the Faraday configuration and those that take place in the Voigt configuration. In the Faraday configuration the excitation depends on $\mathbf{q}$, i.e., they exhibit a Doppler shift in energy, whereas in the Voigt configuration the excitation energies are independent of $\mathbf{q}$ since the energy of the electron is independent of its momentum in the plane perpendicular to $\mathbf{H}_0$.

Because of their transverse character, $\mathbf{E}$ waves propagating in the direction of $\mathbf{H}_0$ cannot excite $\Delta L = 0$ transitions as this requires a component of $\mathbf{E}$ parallel to $\mathbf{H}_0$. However, $\Delta L = 0$ transitions can be excited in the Faraday configuration by longitudinal $\mathbf{A}$ waves. The $\Delta L = 0$ transitions can also be excited by $\mathbf{E}$ waves propagating at an angle ($\theta(\mathbf{q}, \mathbf{H}_0) \neq 0°$ or $90°$) to the magnetic field direction. It should be noted that in the case of an ellipsoidal energy band whose major axis is tilted away from the direction of $\mathbf{H}_0$ the plane of the orbit in real space is not perpendicular to $\mathbf{H}_0$ (Fig. 19). In this situation, the $\Delta L = \pm 1, \pm 2, \pm 3, \ldots$ selection rules are applicable in the Faraday configuration as well as in the Voigt configuration [16].

The collective excitations of a plasma are harmonic-oscillator-type modes and therefore have harmonic-oscillator-type electric dipole matrix elements. The probability that a phonon or a photon will be absorbed by a collective excitation is proportional to $|\mathbf{M} \cdot \mathbf{E}|^2$, where $\mathbf{M}$ is the electric moment of the excitation. The equations for the conservation of energy

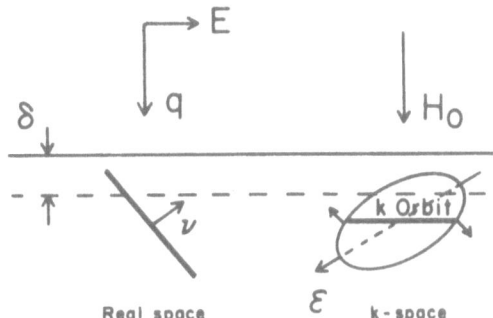

Fig. 19. Schematic diagram showing an electron in an ellipsoidal energy band whose major axis is tilted away from the direction of the magnetic field. Although the normal to the plane of the orbit in **k** space is parallel to H_0, the normal to the plane of the orbit in real space is tilted away from the direction of the magnetic field. Consequently, an electron near the surface moves into and out of the "skin depth," δ, of the metal.

and momentum take the form $h\omega = \hbar\Omega$ and $hq = h\mathbf{Q}$. Since $\mathbf{Q}$ is always in the direction of $\mathbf{q}$, this means that E waves and transverse A waves can only excite collective excitations which have a transverse component of the electric moment and that longitudinal A waves can only excite collective excitations having a longitudinal component of the electric moment.

The interaction of A and E waves with the single-particle excitations are classified as local or nonlocal on the basis of their dependence on wave vector. Local interactions do not depend on $\mathbf{q}$, whereas nonlocal interactions do. A nonlocal effect can arise either from the dependence of the excitation energy on $\mathbf{q}$, as in the case of Landau excitations and Doppler-shifted cyclotron excitations, or from the dependence of the matrix elements on wave vector, as in the case of cyclotron excitations in the Voigt configuration where the electric dipole approximations break down for finite $\mathbf{q}$.

The electric fields associated with A waves are quite small, and consequently the strength of the "acoustical" matrix elements are relatively weak. The macroscopic response of the plasma does not lead to any appreciable variation in velocity of the A waves. The absorption of phonons by the single-particle and collective excitations of the plasma manifests itself primarily as a weak attenuation of the A waves. The wave vector of the A waves, $\mathbf{q}(\omega) = \mathbf{q}_1(\omega) + i\mathbf{q}_2(\omega)$, is predominantly real, and its magnitude is determined by the nature of the host medium rather than by the plasma.

In the case of E waves the electric field is large and the "optical" matrix elements are correspondingly strong. As a consequence, the dielectric constant of the plasma exhibits a strong anomalous dispersion, and there are frequency regions in which the wave vector of the E waves is predominantly imaginary and the waves are nonpropagating.

THE PROPAGATION OF E AND A WAVES

The wave equation which determines the propagation of EM radiation in a nonmagnetic medium is given by

$$\mathbf{q} \times \mathbf{q} \times \mathbf{E} + \frac{\omega^2}{c^2} \tilde{\varepsilon}(\omega, \mathbf{q}) \cdot \mathbf{E} = 0 \tag{64}$$

where $\mathbf{E}$ is the electric field vector, $\mathbf{E}(\omega, q) = \mathbf{E} \exp[i(\mathbf{q} \cdot \mathbf{r} - \omega t)]$; $\omega/c = q_0$ is the wave vector of the radiation in free space; and $\tilde{\varepsilon}(\omega, \mathbf{q}) = \varepsilon_1(\omega, \mathbf{q}) + i \varepsilon_2(\omega, \mathbf{q})$ is the complex dielectric constant, a second rank tensor, defined by the relation $\mathbf{D}(\omega, \mathbf{q}) = \tilde{\varepsilon}(\omega, \mathbf{q})\mathbf{E}(\omega, \mathbf{q})$. The wave vector dependence (or spatial dispersion) of $\tilde{\varepsilon}(\omega, \mathbf{q})$ arises from the wave vector dependence of the oscillator strengths $f_i(\mathbf{q})$ of the single-particle and collective excitations, whose magnitudes are determined by the matrix elements $|\langle \psi_i | H_{\text{int}} | \psi_j \rangle|$, and from the wave vector dependence of the excitation energies $\Delta \mathscr{E}(\mathbf{q})$.

For isotropic media ε_1 and ε_2 have only one independent component, i.e., they behave as scalars, and the wave equation reduces to the form

$$q^2\mathbf{E} - (\omega^2/c^2)\tilde{\varepsilon}(\omega, \mathbf{q})\mathbf{E} = 0 \tag{65}$$

The ω versus $\mathbf{q}$ dispersion relation is given by

$$q^2 c^2 = \omega^2 \varepsilon(\omega, \mathbf{q}) = \omega^2 \eta^2 \tag{66}$$

where $\tilde{\eta} = \eta + i\varkappa$ is the complex refractive index ($\varepsilon_1 = \eta^2 - \varkappa^2$ and $\varepsilon_2 = 2\eta\varkappa$). The velocity of the propagating modes is independent of polarization and direction, and the modes are pure transverse.

In the case of anisotropic media, where in general ε_1 and ε_2 have six independent components, the wave equation has the form

$$q^2\mathbf{E} - (\mathbf{q} \cdot \mathbf{E})\mathbf{E} - (\omega^2/c^2)\tilde{\varepsilon}(\omega, \mathbf{q})\mathbf{E} = 0 \tag{67}$$

For propagation in directions along the principal axes of the dielectric constant tensor the propagating modes are pure transverse and the term

$(\mathbf{q} \cdot \mathbf{E})\mathbf{E}$ is equal to zero. On the other hand, for propagation in a general direction the electric field of the propagating modes will have a longitudinal component. One obtains a dispersion equation involving the components of $\tilde{\varepsilon}(\omega, \mathbf{q})$ which is quadratic in η^2. For each direction of propagation there are two values of η^2 corresponding to two propagating modes whose polarizations are orthogonal.

The wave equation which determines the propagation of A waves has the form

$$\varrho(\omega^2/q^2)\mathbf{u} - \tilde{\mathsf{D}}(\omega, \mathbf{q})\mathbf{u} = 0 \tag{68}$$

where $\mathbf{u}$ is the displacement of a point in the medium, $\mathbf{u}(\omega, \mathbf{q}) = \mathbf{u} \times \exp[i(\mathbf{q} \cdot \mathbf{r} - \omega t)]$; ϱ is the density of the medium; and $\tilde{\mathsf{D}}(\omega, \mathbf{q}) = \mathsf{D}_1(\omega, \mathbf{q}) + i\,\mathsf{D}_2(\omega, \mathbf{q})$ is the complex "dynamical matrix". The matrix $\tilde{\mathsf{D}}(\omega, \mathbf{q})$ is related to the stiffness elastic constants c_{imnj} by the Christoffel equation

$$\sum_j \mathsf{D}_{ij}u_j = \sum_j \sum_{mn} c_{imnj}l_m l_n u_j \tag{69}$$

where l_m and l_n are the direction cosines of $\mathbf{q}$.

In general, there are three propagating modes, two transverse and one longitudinal, whose displacement vectors, $\mathbf{u}$, are mutually orthogonal. For propagation along high symmetry directions, for example, along $\langle 100 \rangle$, $\langle 110 \rangle$, and $\langle 111 \rangle$ directions in a cubic crystal, the modes are pure transverse and pure longitudinal, and the associated electric fields are pure transverse and pure longitudinal. For general directions of propagation the modes are quasitransverse and quasilongitudinal, and the associated electric fields have both transverse and longitudinal components.

In the next section we consider the contributions to $\tilde{\varepsilon}(\omega, \mathbf{q})$ arising from the single-particle and collective excitations. The corresponding contributions to $\tilde{\mathsf{D}}(\omega, \mathbf{q})$ are quite small and will not be discussed further other than to note that the small changes in the velocity of the A waves as well as the attenuation of the A waves which result from the absorption of phonons by the single-particle excitations are observable.

THE DIELECTRIC CONSTANT OF A PLASMA

The electromagnetic properties of a medium are investigated experimentally by measuring the reflection, absorption and transmission of E waves through the medium which is generally in the form of a slab. The general features of the relation, power absorption, and transmission spectra, apart from the frequency region in which they occur, are independent of the specific nature of the medium. They are determined mainly by the strength of the interaction of the E waves with the "electric dipole oscillators" of the medium, i.e., with the optical phonons and excitons and with the single-particle and collective excitations of the free carriers. The strength of the interaction is determined by the effective plasma frequency of the oscillators, $\Omega_{0j} = [4\pi n_j e_j^2 f_j(\mathbf{q})/m_j]^{1/2}$ [where n_j, e_j, $f_j(\mathbf{q})$, and m_j are the density, charge, oscillator strength, and reduced (or effective) mass of the oscillators]. For small Ω_{0j}, which occurs when n_j or f_j are very small, the medium will exhibit a resonance absorption maximum (or transmission minimum) at the excitation frequency of the oscillators. On the other hand, for large Ω_{0j} the medium will exhibit a strong anomalous dispersion and the spectra will not have the appearance of a resonance spectrum.

The $H_0 = 0$ Case

In the absence of a magnetic field the dielectric constant of a polar isotropic crystal containing a degenerate plasma has the general form ([11])

$$\tilde{\varepsilon}(\omega, \mathbf{q}) = \varepsilon_0(\omega, \mathbf{q}) - \frac{4\pi e^2}{m^*} \sum_{k < k_F} \frac{1}{[(\omega + \hbar \mathbf{k} \cdot \mathbf{q})/m^*]^2 - (\hbar q^2/2m^*)^2 + (i\omega/\tau)}$$

$$+ \sum_j \frac{\Omega_{0j}^2}{\omega_{Tj}^2(\mathbf{q}) - \omega^2 - (i\omega/\tau_j)} \tag{70}$$

where $\varepsilon_0(\omega, \mathbf{q})$ is the dielectric constant in the absence of the electron plasma and of infrared-active optical phonons; ω_{Tj} is the transverse resonance frequency of the optical vibration mode of type j; and τ_j is the lifetime of the mode. The second term represents the contribution of the electron plasma and the third term represents the contribution from the infrared-active optical phonons. In the discussion that follows, we will write e_{Tj}^{*2} in place of $e_j^2 f_j$ (where e_{Tj}^* is the effective change of the TO phonon) and $m_j = \bar{m}_j$. The contribution of the plasma to $\tilde{\varepsilon}(\omega, \mathbf{q})$ comes from Drude–Zener type interactions of the E waves with single-particle interactions.

In the limit of long waves ($\mathbf{q} \approx 0$) which holds for many of the optical phenomena, Eq. (70) becomes

$$\tilde{\varepsilon}(\omega) = \varepsilon_0(\omega) - \frac{\omega_{p0}^2}{\omega^2 + (i\omega/\tau)} + \sum_j \frac{\Omega_{0j}^2}{\omega_{Tj}^2 - \omega^2 - (i\omega/\tau_j)} \tag{71}$$

where $\omega_{p0}^2 = \varepsilon_0 \omega_p^2 = 4\pi n e^2/m^*$ and $\Omega_{0j}^2 = 4\pi n_j^{*2}/\bar{m}_j$. The corresponding dispersion relation for the propagating E modes is given by

$$q^2 c^2 = \omega^2 \tilde{\varepsilon}(\omega) = \omega^2 \left[\varepsilon_0(\omega) - \frac{\omega_{p0}^2}{\omega^2 + (i\omega/\tau)} + \sum \frac{\Omega_{0j}^2}{\omega_{Tj}^2 - \omega^2 - (i\omega/\tau_j)} \right] \tag{72}$$

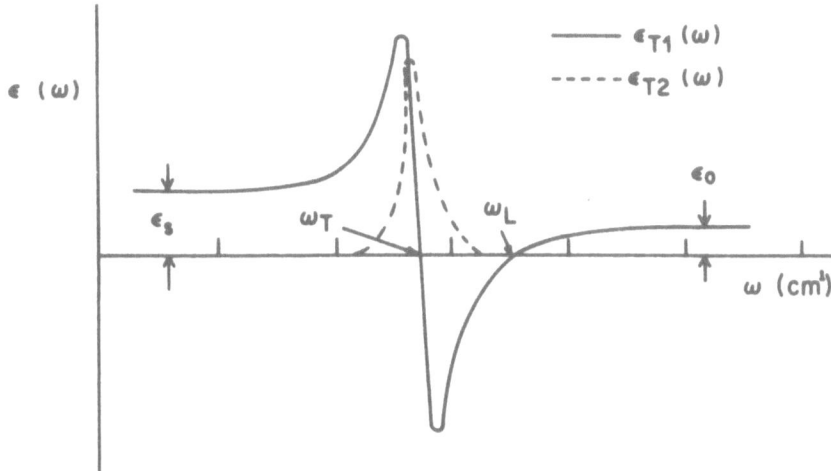

Fig. 20. Curves of $\varepsilon_1(\omega)$ and $\varepsilon_2(\omega)$ versus ω for a crystal having an infrared-active optical vibration made.

The curves of $\varepsilon_1(\omega)$ and $\varepsilon_2(\omega)$ versus ω for a cubic diatomic crystal, in the absence of a plasma, are illustrated in Fig. 20. The frequency at which $\varepsilon_2(\omega)$ is a maximum corresponds to ω_T, the frequency of the $\mathbf{q} \approx 0$ TO phonons. The dielectric anomaly $\varepsilon_1(\omega) = 0$ occurs at ω_L the frequency of the $\mathbf{q} \approx 0$ LO phonons. We note that $\varepsilon_1(\omega)$ is negative at frequencies in the range $\omega_T < \omega < \omega_L$. This range of frequencies is determined by the effective plasma frequency of the optical phonons,

$$\omega_L{}^2 - \omega_T{}^2 = \frac{4\pi n e^{*2}}{\bar{m}\varepsilon_0} = \frac{\Omega_0{}^2}{\varepsilon_0} = \Omega^2 \tag{73}$$

In this range of frequencies the wave vector $\tilde{\mathbf{q}}(\omega)$ is predominantly imaginary and the E waves do not propagate through the crystal. The magnitude of $\tilde{\mathbf{q}}(\omega)$ is given by

$$\tilde{\mathbf{q}}(\omega) = (2\pi/\lambda_0)(\eta(\omega) + i\varkappa(\omega)) = (2\pi/\lambda_0)\tilde{\eta}(\omega) \tag{74}$$

When $\varepsilon_1(\omega) = \eta(\omega)^2$ is negative, $\varkappa(\omega)$ is greater than $\eta(\omega)$, and in regions where $\varepsilon_2(\omega) = 2\eta(\omega)\varkappa(\omega) \approx 0$, this means that $\eta(\omega) = 0$.

The normal incidence reflectance, $R_n(\omega)$ is given by

$$R_n(\omega) = \frac{(\eta - 1)^2 + \varkappa^2}{(\eta + 1)^2 + \varkappa^2} \tag{75}$$

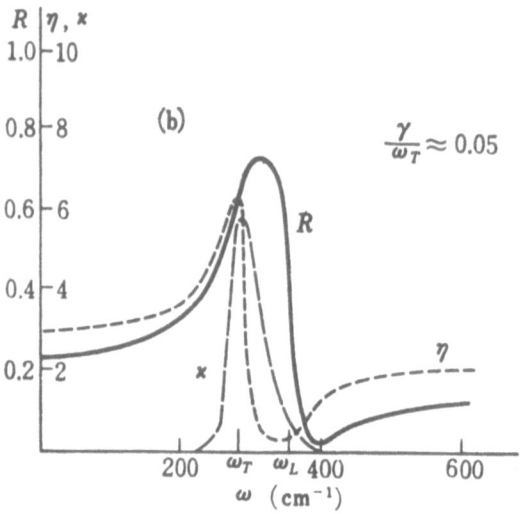

Fig. 21. Curves of R, η, and $\varkappa$ versus ω for a crystal having an infrared-active optical mode of vibration.

When $\varkappa(\omega)$ is much greater than $\eta(\omega)$ the reflectance is approximately equal to unity. The region of negative $\varepsilon_1(\omega)$, therefore, corresponds to a region of "metallic" reflectivity. The curve of $R_n(\omega)$ versus ω is shown in Fig. 21.

The dispersion equation for the propagating waves in the medium is given in the limit $\tau \to \infty$ by

$$q^2 c^2 = \omega^2 \left[\varepsilon_0 + \frac{\Omega_0{}^2}{\omega_T{}^2 - \omega^2} \right]. \tag{76}$$

The ω versus $\mathbf{q}$ curve for the interaction of E waves with the $\mathbf{q} \approx 0$ TO phonon modes is shown in Fig. 22. The propagating modes are coupled photon–TO phonon modes. They are one form of the coupled photon–electric dipole excitation modes called polaritons. For $\omega < \omega_T$, the polariton modes are largely photon in character, i.e., the electric field is large and the atomic displacements are small. On the other hand, at large $\mathbf{q}$ and $\omega \approx \omega_T$ the polariton modes are predominantly phonon in character, i.e., large atomic displacements and *very* small electric field. No propagating modes occur in the range $\omega_T < \omega < \omega_L$ where $\varepsilon_1(\omega)$ is negative and $\mathbf{q}$ is imaginary. At $\omega = \omega_L$ the polariton is largely phonon in character, and at $\omega > \omega_L$ it is predominantly photon in character. There are two values of ω for each $\mathbf{q}$ value, corresponding to two normal modes having different proportions of photon and phonon in character. We note also that for $\omega \gg \omega_L$ the slope of the ω versus $\mathbf{q}$ curve corresponds to $\omega^2/q^2 = c^2/\varepsilon_0$,

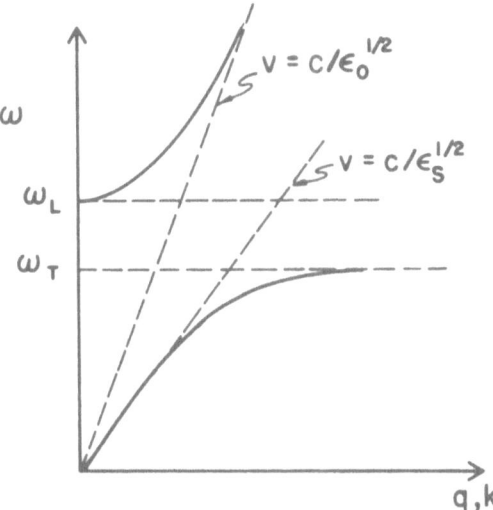

Fig. 22. The dispersion (ω versus $\mathbf{q}$) curve for the propagating *EM* radiation modes (polaritons) in a crystal having an infrared-active optical mode of vibration.

and for $\omega \ll \omega_T$ the slope corresponds to $\omega^2/q^2 = c^2/\varepsilon_S$, where ε_S is the so-called static ($\omega = 0$) dielectric constant of the medium,

$$\varepsilon_S = \varepsilon_0 + (\Omega_0{}^2/\omega_T{}^2) \tag{77}$$

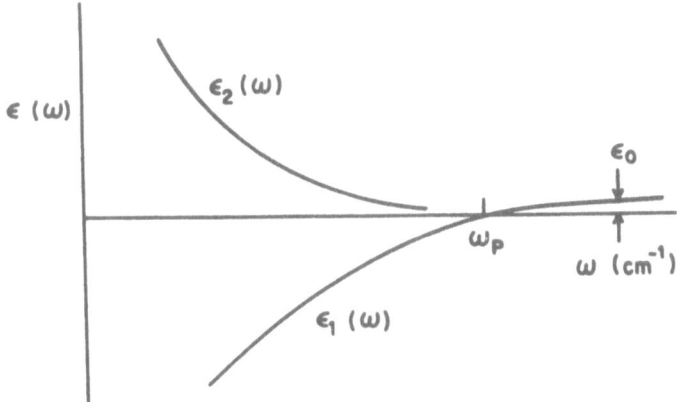

Fig. 23. Curves of $\varepsilon_1(\omega)$ and $\varepsilon_2(\omega)$ versus ω for an electron plasma.

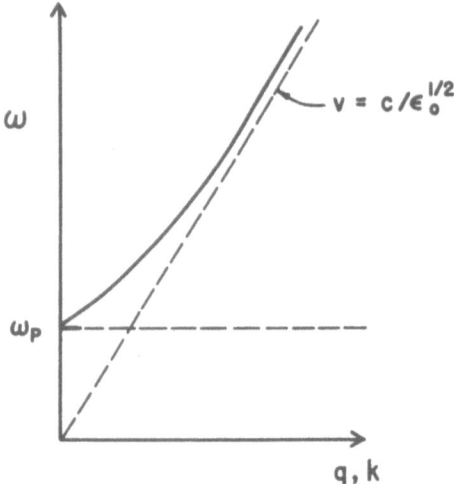

Fig. 24. The dispersion curve for the propagating *EM* radiation mode of an electron plasma.

The curves of $\varepsilon_1(\omega)$ and $\varepsilon_2(\omega)$ versus ω and the corresponding ω versus **q** curve in the absence of optical phonons are shown in Figs. 23 and 24,
respectively. We note that the transverse resonance frequency is equal to
zero and that the dielectric anomaly, $\varepsilon_1(\omega) = 0$, occurs at $\omega = \omega_p$. $\varepsilon_1(\omega)$
is negative at all frequencies $\omega < \omega_p$. Thus at frequencies below ω_p, **q**
is predominantly imaginary and the medium exhibits a metallic reflectivity
(Fig. 25).

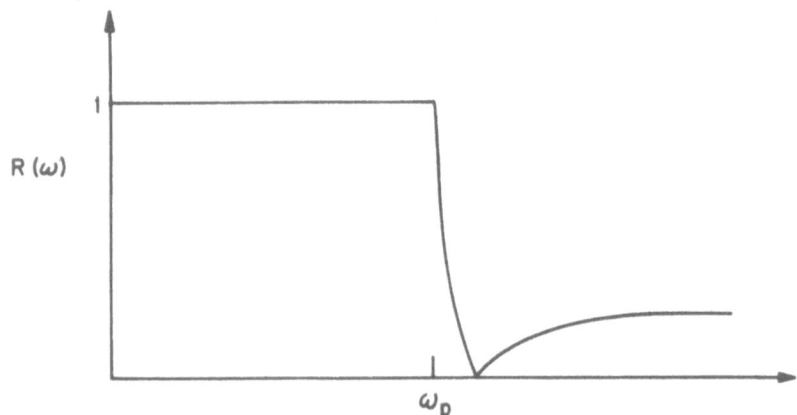

Fig. 25. Curve of $R(\omega)$ versus ω for an electron plasma.

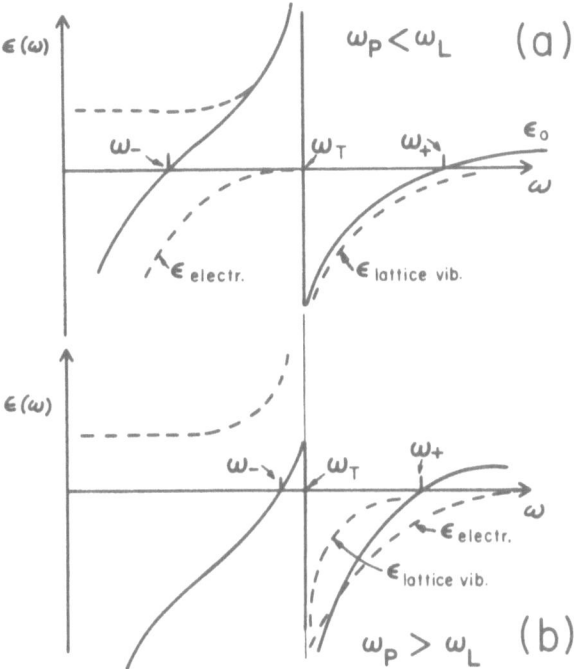

Fig. 26. Curve of $\varepsilon_1(\omega)$ versus ω for a crystal having both an electron plasma and an infrared-active optical mode of vibration (a) for $\omega_P < \omega_L$ and (b) for $\omega_P > \omega_L$.

The $\varepsilon_1(\omega)$ and $\varepsilon_2(\omega)$ curves and the corresponding ω versus $\mathbf{q}$ dispersion curves for a polar diatomic crystal containing a plasma $(\omega_P > \omega_L)$ are shown in Figs. 26 and 27, respectively. One notes that there are two dielectric anomalies. These occur at the frequencies of the coupled plasmon–LO phonon modes discussed on p. 30.

$H_0 \neq 0$. Faraday Configuration

The $\mathbf{q} \approx 0$ dielectric constant of an electron plasma in a magnetic field for E waves propagating in the Faraday $(\mathbf{q} \parallel \mathbf{H_0})$ configuration can be derived in a straightforward manner from the equations of motion of the electron in a magnetic field and the relation $\mathbf{P}(\omega) = \chi(\omega)\,\mathbf{E}(\omega) = (\varepsilon(\omega) - 1)\,\mathbf{E}(\omega)/4\pi$. One obtains the following expression ([17]):

$$\tilde{\varepsilon}(\omega)_{l,r} = \varepsilon_0(\omega) - \frac{\omega_{P0}^2}{\omega[\omega \mp \omega_c + (i\omega/\tau)]} = \eta(\omega)_{l,r}^2 \tag{78}$$

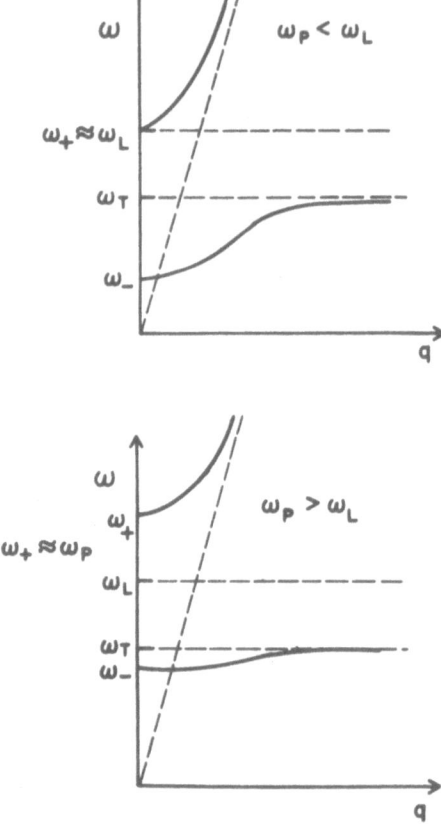

Fig. 27. The dispersion curves of the propagating *EM* modes in a crystal having an electron plasma and an infrared-active optical mode of vibration (*a*) for $\omega_P < \omega_L$ and (*b*) for $\omega_P > \omega_L$.

where l and r designate, respectively, left and right circularly polarized modes ($\mathbf{E}_l = E_x + iE_y$ and $\mathbf{E}_r = E_x - iE_y$). The l and r subscripts are interchanged for holes.

The curves of $\tilde{\varepsilon}(\omega)_{l,r}$ and $R_n(\omega)_{l,r}$ are shown in Figs. 28 and 29, respectively, for $\omega_c < \omega_P$. We note that $\varepsilon_2(\omega)_l$ exhibits a resonance peak at $\omega = \omega_c$ and that $\varepsilon_2(\omega)_r$ does not exhibit any resonance. Accordingly the l mode is called the cyclotron-active mode. For this case, $\varepsilon_1(\omega)_l$ exhibits a dielectric anomaly at $\omega = \omega_P + \omega_c/2$ and $\varepsilon_1(\omega)_r$ exhibits a dielectric anomaly at $\omega = \omega_P - \omega_c/2$. Furthermore, $\varepsilon_1(\omega)_l$ is positive at frequencies $\omega < \omega_c$

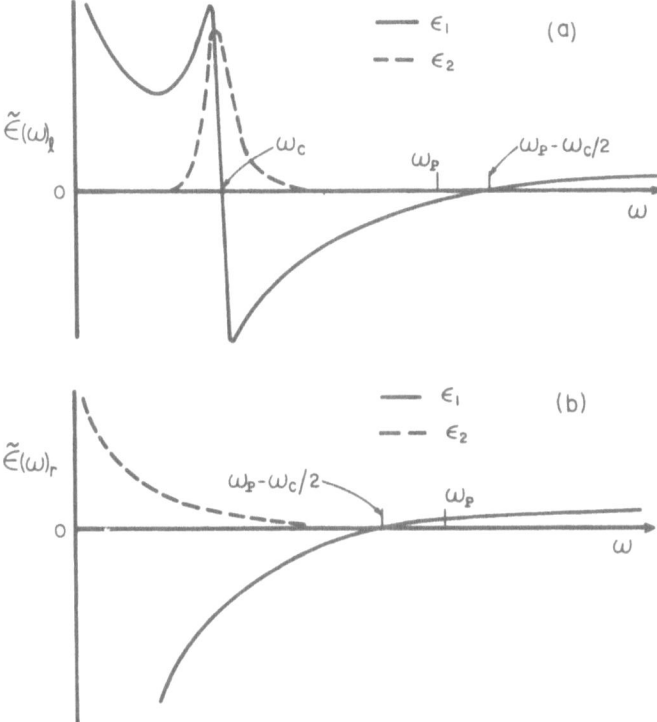

Fig. 28. Curves of ε_1 and ε_2 versus ω for an electron plasma in a magnetic field in the Faraday configuration (a) for the cyclotron-active *EM* mode and (b) for the cyclotron-inactive *EM* mode.

as well as frequencies $\omega > \omega_P + \frac{1}{2}\omega_c$, whereas $\varepsilon_1(\omega)_r$ is positive for frequencies $\omega > \omega_P + \frac{1}{2}\omega_c/2 + \frac{1}{2}\omega_c$.

The dispersion relation is given by

$$q_{l,r}^2 c^2 = \omega^2 \tilde{\varepsilon}(\omega)_{l,r} = \omega^2 \left\{ \varepsilon_0 - \frac{\omega_{P0}^2}{\omega[\omega \mp \omega_c + (i\omega/\tau)]} \right\} \qquad (79)$$

At frequencies $\omega \ll \omega_c$ the expression for $\tilde{\varepsilon}(\omega)_{l,r}$ is given to a good approximation by

$$\tilde{\varepsilon}(\omega)_{l,r} = \pm \frac{\omega_{P0}^2}{\omega\omega_c} \pm \frac{i\omega_{P0}^2}{\omega\omega_c} \left(\frac{1}{\omega_{c\tau}} \right) \qquad (80)$$

The propagating *l* modes, i.e., $\varepsilon_1(\omega)_l$ positive, are known as helicon waves [18,19]. Their attenuation is proportional to $(1/\omega_c\tau)$. The dispersion relation

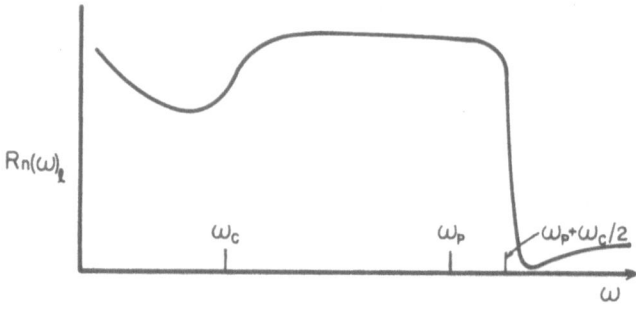

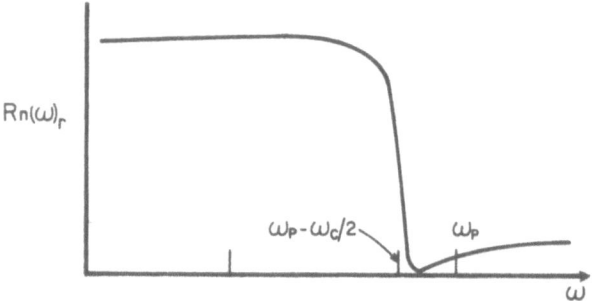

Fig. 29. Curve of $R(\omega)$ versus ω for an electron plasma in a magnetic field in the Faraday configuration (a) for the cyclotron-active EM mode and (b) for the cyclotron-inactive EM mode.

for the helicon waves, in the limit $\tau \to \infty$, has the form

$$\omega = (\omega_c/\omega_{P0}^2)q^2c^2 \tag{81}$$

Thus for helicon waves the frequency varies quadratically with the wave vector.

The ω versus $\mathbf{q}$ dispersion curves for l and r modes (in the limit $\tau \to \infty$) are shown in Fig. 30. We note that for the cyclotron-inactive r modes the form of the ω versus $\mathbf{q}$ curve is similar to that for the plasma in the absence of a magnetic field (Fig. 24).

For plasmas having several types of carriers the expression for $\tilde{\varepsilon}(\omega)_{l,r}$ takes the form

$$\tilde{\varepsilon}(\omega)_{l,r} = \varepsilon_0(\omega) - \sum \frac{\omega_{0j}^2}{\omega[\omega \pm \omega_c, + (i\omega/\tau_j)]} \tag{82}$$

where $\omega_{cj} = e_jH_0/m_j^*c$, $\omega_{P0_j}^2 = 4\pi n_je^2/m_j^*$, and e_j is $-e$ for electrons and

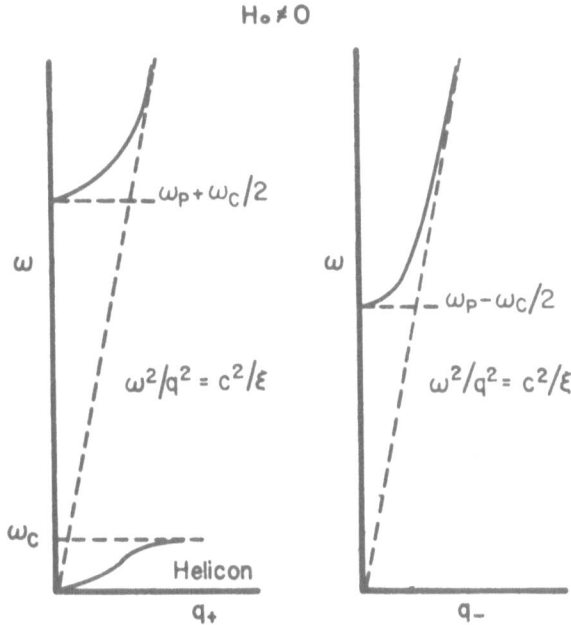

Fig. 30. The dispersion curves for the cyclotron-active *EM*
mode and the cyclotron-inactive *EM* modes in an electron
plasma in a magnetic field in the Faraday configuration.

$+e$ for holes. In the particular case of a compensated plasma the expression
for $\tilde{\varepsilon}(\omega)_{l,r}$ becomes

$$\tilde{\varepsilon}(\omega)_{l,r} = \varepsilon_0(\omega) \frac{\omega_{P0e}^2}{\omega[\omega \mp \mid \omega_{ce} \mid + (i\omega/\tau_e)]} - \frac{\omega_{P0h}^2}{\omega[\omega \mp \mid \omega_{ch} \mid + (i\omega/\tau_h)]} \quad (83)$$

and at low frequencies such that $\omega \ll \omega_{ce}$, ω_{ch} it reduces in the limit
$\tau \to \infty$ to

$$\varepsilon(\omega)_{l,r} = \varepsilon_0 \pm \frac{\omega_{P0}^2}{\omega \mid \omega_{ce} \mid} \mp \frac{\omega_{P0h}^2}{\omega \mid \omega_{ch} \mid} + \frac{\omega_{P0c}^2}{\omega_{ce}^2} + \frac{\omega_{P0h}^2}{\omega_{ch}^2}$$

$$\pm \frac{4\pi e}{H_0\omega} (n_e - n_h) + \frac{\omega_{P0e}^2}{\omega_{ce}^2} + \frac{\omega_{P0h}^2}{\omega_{ch}^2} \quad (84)$$

Since $n_e = n_h$ in compensated plasma, the first two terms cancel and $\varepsilon(\omega)_{l,r}$
is given by

$$\varepsilon(\omega)_{l,r} = \varepsilon_0 + \frac{\omega_{P0e}^2}{\omega_{ce}^2} + \frac{\omega_{P0h}^2}{\omega_{ch}^2} \quad (85)$$

$\varepsilon(\omega)_{l,r}$ is positive and independent of ω and of the polarization of the modes. The dispersion relation for $\omega < \omega_{ce}, \omega_{ch}$ is given by

$$q^2c^2 = \omega^2\left(\frac{\omega_{Poe}^2}{\omega_{ce}^2} + \frac{\omega_{Poh}^2}{\omega_{ch}^2}\right) \tag{86}$$

The frequency of the propagating modes, which are called Alfvén waves [20], varies linearly with the wave vector, and their velocity is independent of their state of polarization. The attenuation of Alfvén waves is proportional to $(1/\omega\tau)$. Their attenuation increases as ω decreases, unlike that of helicon waves, which is independent of ω.

The ω versus $\mathbf{q}$ curves for the Alfvén waves are shown in Fig. 31. We note that the degeneracy of the modes is split at frequencies approaching

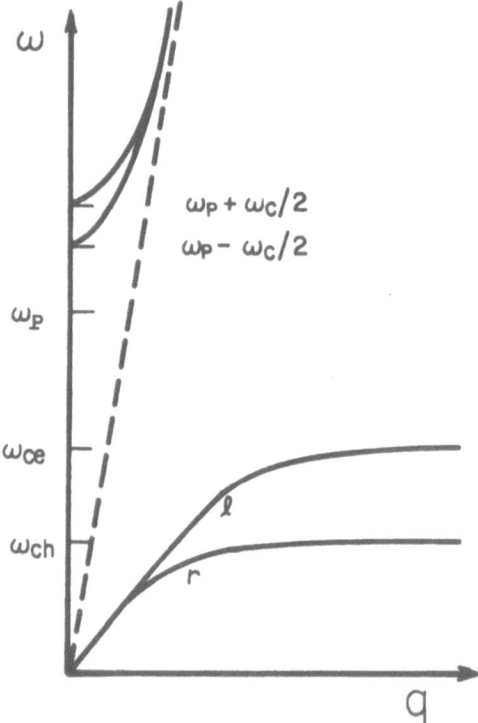

Fig. 31. The dispersion curves for the electron-cyclotron-active and hole-cyclotron-active *EM* modes in a compensated plasma in a magnetic field in the Faraday configuration.

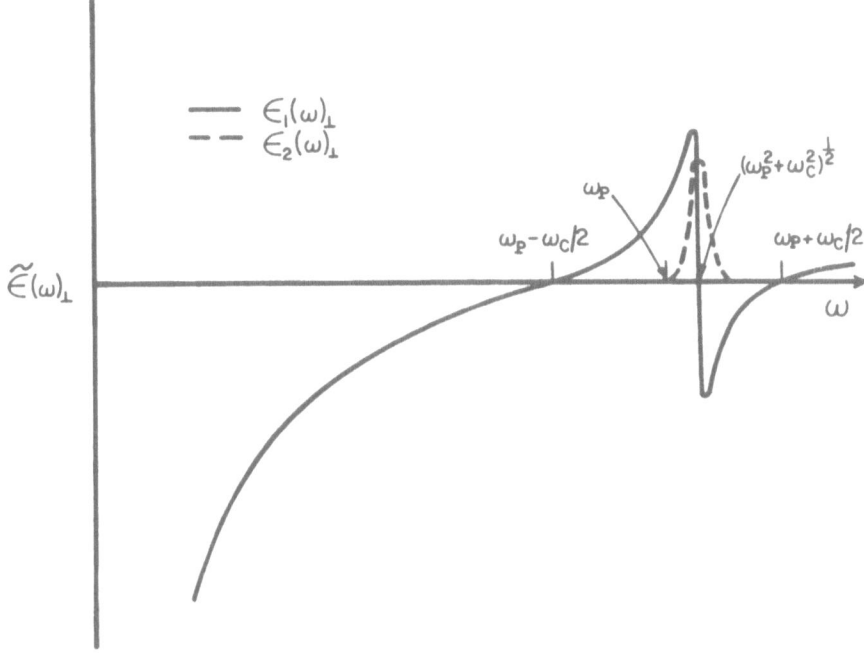

Fig. 32. Curves of $\varepsilon_1(\omega)$ and $\varepsilon_2(\omega)$ versus ω of an electron plasma in a magnetic field in the Voigt configuration for $\mathbf{E} \perp \mathbf{H_0}$.

the lower of the two cyclotron frequencies. Since l modes are cyclotron-inactive for holes and the r modes are cyclotron-inactive for electrons, the l and r curves do not exhibit any anomalous dispersion effects at ω_{ch} and ω_{ce}, respectively. In their recent investigation of the propagation of Alfvén waves in bismuth Kawamura and co-workers ([21]) have observed an attenuation of l modes at ω_{ch}. They attribute this effect to the admixture of the l and r modes arising from nonlocal effects.

It should be noted that the helicon and Alfvén waves are polariton modes in which photons are coupled with cyclotron excitations. The modes have both electromagnetic and cyclotron excitation character, the proportions of each varying with wave vector.

$H_0 \neq 0$. Voigt Configuration

The dielectric constant of an electron plasma in a magnetic field for E waves propagating in the Voigt ($\mathbf{q} \perp \mathbf{H}$) configuration is given in the limit

$\tau \to \infty$ [15]

$$\varepsilon_1(\omega)_\perp = \eta(\omega)_\perp{}^2 = \varepsilon_0 + \frac{\omega_{P0}^2(\omega_{P0}^2 - \omega^2)}{\omega^2(\omega_{P0}^2 + \omega_c{}^2 - \omega^2)} \qquad (87a)$$

$$\varepsilon_1(\omega)_\parallel = \eta(\omega)_\parallel{}^2 = \varepsilon_0 - (\omega_{P0}^2/\omega^2) \qquad (87b)$$

where the $\perp$ mode is linearly polarized, having $\mathbf{E} \perp \mathbf{H_0}$ and the $\parallel$ mode is linearly polarized with $\mathbf{E} \parallel \mathbf{H_0}$. We note that $\varepsilon(\omega)_\perp$ exhibits a resonance at the frequency of the collective cyclotron excitation $\Omega_c = (\omega_P^2 + \omega_c{}^2)^{1/2}$ and that the expression for $\varepsilon_1(\omega)_\parallel$ is the same as that for a plasma in the absence of a magnetic field.

The $\varepsilon_\perp(\omega)_\perp$ versus ω curve and the corresponding ω versus $\mathbf{q}$ curve are shown in Figs. 32 and 33, respectively. There are two dielectric a-nomalies, one at $\omega = \omega_P + \frac{1}{2}\omega_c$ and one at $\omega = \omega_P - \frac{1}{2}\omega_c$, the frequencies at which dielectric anomalies occur for the l and r modes, respectively,

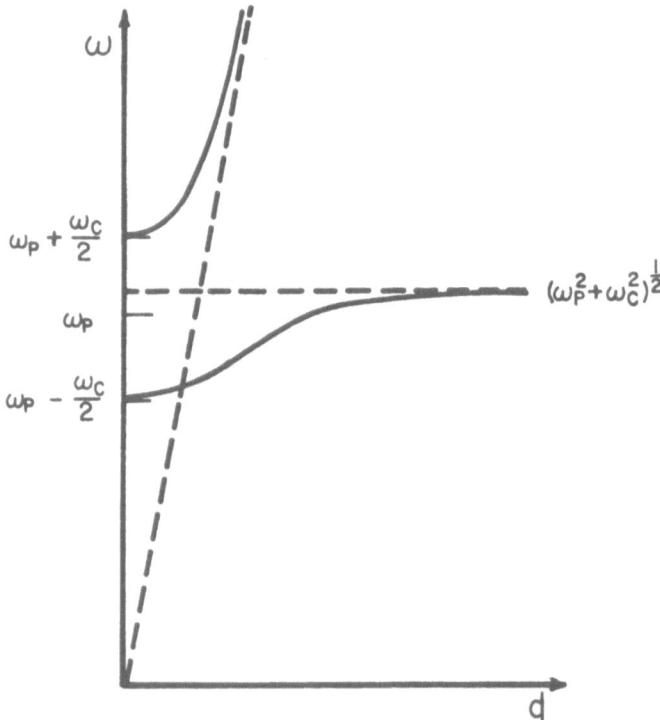

Fig. 33. Dispersion curves of the propagating E modes in an electron plasma in a magnetic field in the Voigt configuration for $\mathbf{E} \perp \mathbf{H_0}$.

in the Faraday configuration. One notes also that $\varepsilon_1(\omega)_\perp$ is negative at all frequencies below $\omega_P - \frac{1}{2}\omega_c$.

In the case of a compensated plasma, the expressions for $\varepsilon_1(\omega)_\perp$ and $\varepsilon_1(\omega)_\parallel$ have the form $(^{15})$

$$\varepsilon_1(\omega)_\perp = \varepsilon_0 - \frac{\omega_{P0}^2(\omega^2 - \omega_{P0}^2 - \omega_{ce}\omega_{ch})}{(\omega^2 - \omega_{ce}^2)(\omega^2 - \omega_{ch}^2) - \omega_{P0}^2(\omega^2 - \omega_{ce}\omega_{ch})} \qquad (88a)$$

$$\varepsilon_1(\omega)_\parallel = \varepsilon_0 - (\omega_{P0}^2/\omega^2) \qquad (88b)$$

where $\omega_{P0}^2 = \omega_{P0}^2\varepsilon_0 + \omega_{P0}^2\varepsilon_0 = \omega_P^2\varepsilon_0$. For $\omega_P > \omega_c$, $\tilde{\varepsilon}_1(\omega)_\perp$ has two resonances, one at $\Omega_{c+} = (\omega_P^2 + \omega_{ce}^2 + \omega_{ch}^2)^{1/2}$ and one at $\Omega_{c-} = (\omega_{ce}\omega_{ch})^{1/2}$, corresponding to the two collective cyclotron excitation frequencies.

At low frequencies, i.e., $\omega < \omega_{ce}$, ω_{ch} the expression for $\varepsilon_1(\omega)_\perp$ becomes

$$\varepsilon_1(\omega)_\perp = \frac{\omega_{P0}^2}{\omega_{ce}\omega_{ch}} \qquad (89)$$

Thus at low frequencies $\varepsilon_\perp(\omega)_\perp$ for the compensated plasma is independent of frequency, and the dispersion relation is linear, as in the case of the low-frequency propagating modes of the compensated plasma in the Faraday configuration. However, there is only one propagating Alfvén mode at low frequencies in the Voigt configuration.

NONLOCAL EFFECTS

When the electric field, $E(\omega, q) = E \exp i(q \cdot r - \omega t)$, of the E waves does not vary appreciably over the spatial extent, R, of the electric dipole oscillator, its spatial variation can be neglected. This corresponds to the "electric dipole approximation" $\exp(iq \cdot R) = 1$. Under these circumstances the matrix elements are independent of q. On the other hand, when the electric field varies over the spatial extent of the oscillator, the matrix elements and therefore the oscillator strengths will exhibit a dependence on q. In some situations the spatial dispersion may actually lead to transitions which would otherwise not be allowed in the electric dipole approximation. Such a situation occurs in Azbel'–Kaner-type cyclotron resonance (AKCR) which is observed in metals and in semiconductors having high carrier densities. The AKCR experiments are carried out in the Voigt configuration at frequencies $\omega < \omega_P$ where $\varepsilon_1(\omega)$ is negative and therefore q is predominantly imaginary. As can be seen from the expressions for $\varepsilon_1(\omega)_\perp$ in the Voigt configurations [Eqs. (87a) and (88b)] the single-particle $\Delta L \pm 1$ cyclotron resonance at $\omega = \omega_c$ does not appear in the electric-dipole (long-wave)

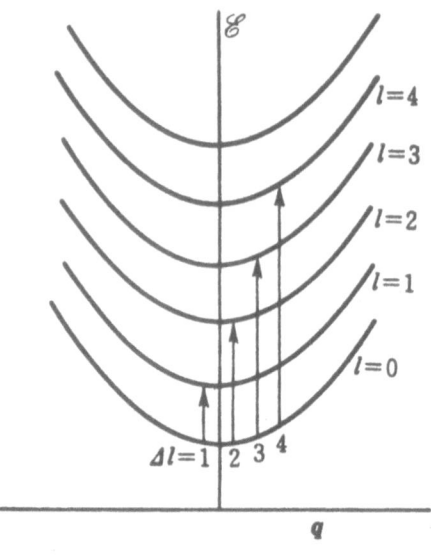

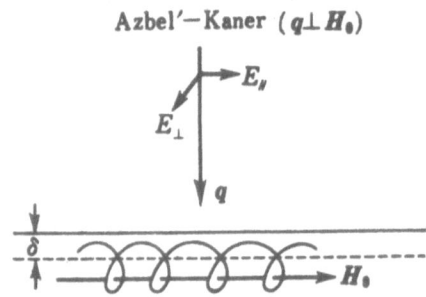

Fig. 34. The configuration of q, H_0, and E for Azbel–Kaner cyclotron resonance and the $\Delta L = 1, 2, 3, 4 \ldots$ transitions which occur in Azbel'–Kaner cyclotron resonance.

approximation. When the skin depth, δ, is comparable to or smaller than the cyclotron orbit (Fig. 34), single particle cyclotron resonances involving $\Delta L = \pm 1, \pm 2, \ldots$ are allowed for $E \parallel H_0$ and for $E \perp H_0$ [22].

The wave vector dependence of the resonance frequency $\omega_{Tj}(q)$ arises from the dependence of $\Delta\mathscr{E}$ on q. Thus in the Faraday configuration the cyclotron resonance frequency is given by

$$\omega(q) = \pm\omega_c + q \cdot v_F(L, \sigma) + \frac{\hbar q^2}{2m^*} \tag{90}$$

Since q is generally small, and therefore $q/2k \ll 1$, at optical frequencies, the cyclotron resonance frequency is given to a good approximation by

$$\omega(q) \approx \pm \omega_c \pm \mathbf{q} \cdot \mathbf{v}_F(L, \sigma) \tag{91}$$

The relative importance of the Doppler shift is determined by the ratio $qv_F/\omega_c(q) = v_F/v_W$. At infrared frequencies v_W is generally much larger than v_F, and the Doppler shift in frequency is relatively small, i.e., the optical transitions are essentially vertical. At microwave and lower frequencies, particularly in the case of high-density plasmas, the velocities of helicon waves (in uncompensated plasmas) and Alfvén waves (in compensated plasma) may be smaller than v_F and there may be an appreciable Doppler shift in frequency. In the case of helicon waves the ratio is given by

$$\frac{qv_F}{\omega_c(q)} \approx \frac{v_F \omega_{P0}}{c\omega_c} \left(\frac{\omega}{\omega_c}\right)^{1/2} \tag{92}$$

and for Alfvén waves the ratio is given in the case of the hole resonance by

$$\frac{qv_F}{\omega_{ch}} \approx \frac{v_F}{c} \frac{\omega}{\omega_{ch}} \left(\frac{\omega_{P0e}^2}{\omega_{ce}^2} + \frac{\omega_{P0h}^2}{\omega_{ch}^2}\right)^{1/2}$$

Thus for both helicon and Alfvén waves the relative importance of the Doppler shift increases with increase in ω_P (i.e., with increase in carrier density) and with decrease in ω_c (i.e., with decrease in magnetic field), which in turn implies with decrease in ω.

When the dependence of the cyclotron excitation frequency on wave vector is taken into account, the expression for $\bar{\varepsilon}(\omega, \mathbf{q})$ of an electron plasma for $\mathbf{q} \parallel \mathbf{H}_0$ has the form [23]

$$\bar{\varepsilon}(\omega, \mathbf{q})_{l,r} = \varepsilon_0 - \frac{4\pi e^2}{m^*\omega} \sum_{\substack{L,\sigma \\ v(L,\sigma)<v_F}} \frac{1}{\omega[(\omega \mp \omega_c - qv(L,\sigma) + (i/\tau)]} \tag{93}$$

Since the Doppler shifts are quite small and a large number of Landau subbands participate simultaneously in the cyclotron resonance absorption of photons, one can to a good approximation neglect the existence of the subbands. Under these circumstances one replaces the summation by an integral over the velocity distribution of the electrons. The resulting expression for $\bar{\varepsilon}(\omega, \mathbf{q})_{l,r}$ has the form [23,24]

$$\bar{\varepsilon}(\omega, \mathbf{q})_{l,r} = \varepsilon_0 - \frac{\omega_{P0}^2}{\omega} \int \frac{f(v)d^3v}{\omega \mp \omega_c - \mathbf{q} \cdot \mathbf{v} + (i/\tau)} \tag{94}$$

where $f(v)$ is the velocity distribution function. For a degenerate plasma and $\omega < \omega_c$ one obtains the following expression for $\varepsilon_1(\omega, \mathbf{q})_l$ [25]:

$$\varepsilon_1(\omega, \mathbf{q})_l = \frac{\omega_{P0}^2}{\omega\omega_c}\left(1 + \frac{1}{5}\frac{q^2 v_F^2}{\omega_c^2}\right) \tag{95}$$

The corresponding dispersion relation for helicon wave propagation is given by

$$q^2 c^2 = \omega \frac{\omega_P^2}{\omega_e}\left(1 + \frac{1}{5}\frac{q^2 v_F^2}{\omega_c^2}\right) \tag{96}$$

The ω versus $\mathbf{q}$ curves for helicon waves in an uncompensated plasma and for Alfvén waves in a compensated plasma under DSCR conditions are shown schematically in Figs. 35 and 36.

It should be pointed out that the propagation of helicon (and Alfvén waves) and the phenomenon of Doppler-shifted cyclotron resonance manifest themselves differently when the measurements are carried out as a function of frequency than when they are carried out as a function of magnetic field. When the experiments are carried out as a function of frequency, propagation of the cyclotron-active helicon mode occurs in the region $0 < \omega < \omega_c - qv_F \cos\theta$ and in the region $\omega > \omega_P + \frac{1}{2}\omega_c$. When ω_c is smaller than $qv_F \cos\theta$ the helicon waves are strongly attenuated over the

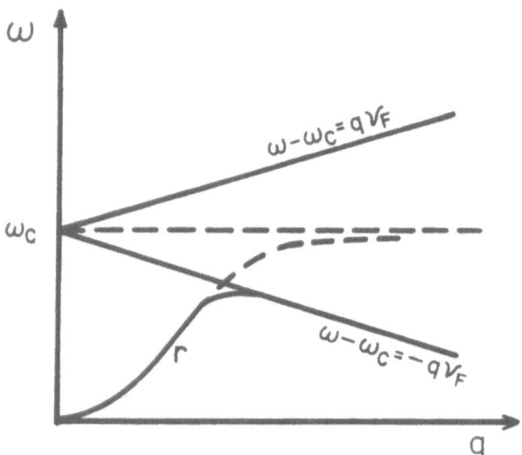

Fig. 35. The $\omega = \Delta\varepsilon/\hbar$ versus $\mathbf{q} = \Delta k$ curves for an electron at the Fermi surface of an electron plasma in the presence of a magnetic field in the Faraday configuration and the corresponding dispersion curve of the propagating coupled EM–electron-cyclotron excitation (polariton) modes.

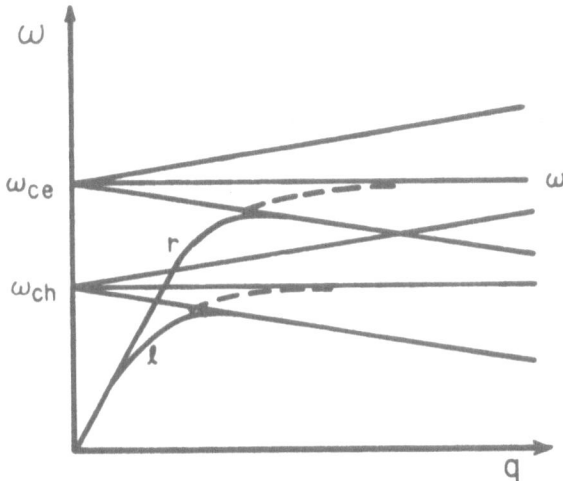

Fig. 36. The $\omega = \Delta\varepsilon/\hbar$ versus $\mathbf{q} = \Delta\mathbf{k}$ for electrons and holes at the Fermi surface of a compensated plasma in the presence of a magnetic field in the Faraday configuration and the corresponding dispersion curve of the propagating coupled *EM*–electron-cyclotron excitation and coupled *EM*–hole-cyclotron excitation modes.

frequency range $0 < \omega < qv_F \cos\theta$, and since $\varepsilon_1(\omega, \mathbf{q})_l$ is negative in the approximate frequency region $\omega_c < \omega < \omega_p + \tfrac{1}{2}\omega_c$ propagation only occurs at frequencies $\omega > \omega_p + \tfrac{1}{2}\omega_c$. When the experiments are carried out as a function of magnetic field the helicon waves are nonpropagating at magnetic fields in the range $0 < \omega_c < \omega + qv_F \cos\theta$. There is an onset of propagation at $\omega_c = \omega + qv_F \cos\theta$, the so-called Kjeldaas edge ([26]). When $qv_F \cos\theta > \omega$ which occurs in high-density plasmas, the onset of propagation corresponds to $\omega_c \approx qv_F \cos\theta$, and the threshold magnetic field is given by

$$\omega_c = \left(\frac{\omega\omega_p^2 v_F^2}{c^2}\right) \tag{97}$$

Thus measurements of the threshold field can, in principle, provide information about the curvature of the Fermi surface ([27]).

HELICON-PHONON COUPLING

Under appropriate conditions the velocity of helicon waves may be equal to the velocity of the A waves and there should be a strong coupling between them. The frequency of helicon waves varies quadratically with

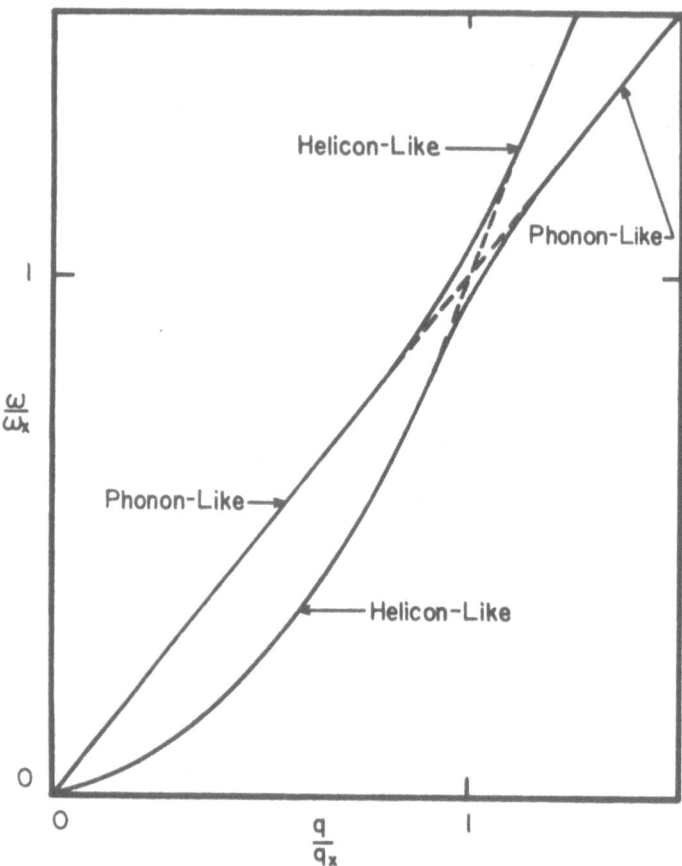

Fig. 37. Dispersion curves for the propapating helicon and phonon
modes for crystal with an electron plasma in the presence of a magnetic field. The ω_X and q_X are the frequency and wave vector at
which the dispersion curves of phonons and helicon waves would
cross in the absence of any interactions between the waves. From
Langenberg and Bok ([28]).

wave vector [Eq. (81)]. Its quadratic dispersion curve crosses the linear
dispersion curve at a point where the phase velocities of the two waves are
equal, i.e., where $v_{HW} = v_{AW} = \omega/q$. The crossover frequency and the wave
vector are given respectively, by

$$\omega_X = \frac{\omega_P{}^2}{\omega_c} \left(\frac{v_{AW}}{c} \right)^2 \tag{98a}$$

$$q = \frac{\omega_P}{c} \left(\frac{\omega}{\omega_c} \right)^{1/2} \tag{98b}$$

The dispersion curves of the coupled helicon–phonon modes is shown in Fig. 37. For $v_{AW} = 10^5$ cm/sec, $n = 10^{18}/cm^3$ and $H_0 = 10^3$ G, the crossover frequency is about 1 MHz.

As noted earlier, the helicon waves are coupled photon–cyclotron excitation (polariton) modes which have both electromagnetic and cyclotron excitation (electron) character. An interaction between the helicon waves and phonons can therefore take place either via their electric fields or via electron–phonon interactions. The existence of a coupling between helicon and transverse acoustic waves was proposed simultaneously by Langenberg and Bok [28] and by Quinn and Rodriguez [29] and demonstrated by Grimes and Buchsbaum [30]. A coupling of helicon waves to longitudinal waves at oblique magnetic fields has been predicted by Skobov and Kaner [31]. More recently, Schilz [32] has demonstrated the bulk generation of transverse and longitudinal A waves by helicon waves in PbTe. He finds appreciable difference in the dependence of the amplitudes of the TA and LA waves generated on the angle between H_0 and q.

A coupling may also be expected between Alfvén waves and A waves when their velocities are equal. The dispersion relation for Alfvén waves is, like that of the A waves, linear. Although their velocity is independent of frequency, it does depend on the magnetic field. The magnetic field at which the velocity of the Alfvén wave is equal to that of the A waves and their linear dispersion curves coincide is given by the condition

$$\omega_c = \omega_p v_{AW}/c \tag{99}$$

The mechanism of the interaction between Alfvén waves and phonons should be similar to that for the helicon–phonon interaction.

EXPERIMENTAL OBSERVATION OF MAGNETOOPTICAL AND MAGNETOACOUSTICAL PHENOMENA

The various types of measurements which one can carry out are as follows:

1. Attenuation in bulk samples. This is applicable to A waves, and also to E waves in frequency regions where propagating modes exist.

2. Reflection. This is particularly applicable to E waves in frequency regions where $\varepsilon(\omega)$ is negative and the medium exhibits metallic reflectivity.

3. Normal and oblique incidence transmission in thin films. In fre-

quency regions where $\varepsilon(\omega)$ is negative (or the absorption constant is high) such measurements yield resonance-type transmission spectra.

4. Birefringence [circular or elliptical birefringence $(n_l - n_r)$ in the Faraday configuration and linear birefringence $(n_\perp - n_\parallel)$ in the Voigt configuration]. This is applicable to both E and A waves. The effects are much smaller in magnetoacoustic phenomena, but are nevertheless observable in high-density plasmas.

5. Velocity change. This is applicable to E and A waves. The changes in velocity of A waves are quite small but are nevertheless observable in high-density plasmas. The measurements are carried out using either a Fabry–Perot configuration or a Rayleigh interferometer configuration, in which the waves are split into two paths.

The various measurements can in principle be carried out either as a function of frequency $[F(\omega)]$ at fixed magnetic field or as a function of magnetic field $[F(\mathbf{H_0})]$ at fixed frequency. The $F(\omega)$ measurements are readily carried out for E waves at infrared and higher frequencies, but are rather difficult to carry out at microwave and lower frequencies, where the use of cavities and resonant circuits severely limits the range of frequencies in a given experimental set-up. The range of frequency variation is also rather limited in the case of A waves, where piezoelectric transducers or microwave cavities are used to generate the waves.

We now consider the various types of phenomena that occur and the conditions under which they are observed experimentally.

Landau ($\Delta L = 0$) Resonance ($q \parallel H_0$)

The Landau resonances have been observed for A waves in the form of $\Delta L = 0$ GQA peaks which are periodic in $1/H_0$. The condition for observing well separated peaks has been derived by Gurevich et al. ([2]). It is given by

$$ql_F\left(\frac{\hbar\omega_c}{\mathscr{E}_F}\right)^{1/2} > 1$$

where $l_F = v_F\tau$ is the mean free path of the electrons at the Fermi surface in the lowest ($L = 0$, $\sigma = -\frac{1}{2}$ Landau subband. This condition is more stringent than $ql > 1$ but less stringent than $\omega\tau > 1$, since the latter corresponds to $ql_F(L, \sigma)\tau > 1$ where $l_F(L, \sigma) = (v_W/v_F)l_F$. The lowest magnetic field at which distinct GQA peaks appear is given approximately by $\hbar\omega_c \approx \mathscr{E}_F(1/ql_F)^2$. The maximum magnetic field at which the GQA peaks occur corresponds to $gBH_0 \approx \mathscr{E}_F$. It is the field at which the spin splitting of the Landau subbands is equal to the Fermi energy.

When the condition for observing distinct peaks is not met one generally observes oscillations in the attenuation which have the appearance of Shubnikov–de Haas-type oscillations. Both types of oscillation have the same period in $1/H_0$. The SdH type oscillations are generally attributed to the oscillations in τ which accompany the variation in the density of states as the Landau subbands rise periodically (in $1/H_0$) above the Fermi level with increasing magnetic field.

Helicon and Alfvén waves do not interact with the $\Delta L = 0$ single-particle excitations in the Faraday configuration since the electric field of the waves is transverse and therefore $\mathbf{M} \cdot \mathbf{E} = 0$. However, they can in principle interact with the $\Delta L = 0$ excitations when $\mathbf{q}$ is tilted away from the direction of $\mathbf{H}_0$ by an angle $\theta < 90°$. In this configuration the electric field of the waves has a longitudinal component and $\mathbf{M} \cdot \mathbf{E} \neq 0$. However, it is still necessary to satisfy the requirement that $v_W < v_F$. This may be possible for helicon and Alfvén waves at low frequencies in high-density plasmas.

Doppler-Shifted Cyclotron Resonance (q ‖ H₀)

It should be possible to observe $\Delta L = \pm 1$ resonances in the form of GQA peaks for both A and E waves. The condition for observing well-separated peaks [7] is $ql_F(\hbar\omega c/\mathscr{E}_F) > 1$. This condition is more stringent than the one for $\Delta L = 0$ GQA. The lowest magnetic field at which distinct peaks may be observed is given approximately by $\hbar\omega_c = \mathscr{E}_F/ql_F$. Higher magnetic fields are needed to observe the onset of $\Delta L = \pm 1$ peaks than for $\Delta L = 0$ peaks. The maximum field at which $\Delta L = \pm 1$ resonances occur is given by $\omega_c \approx \omega + qv_F$. It is much lower than the corresponding field for $\Delta L = 0$ resonances. Thus far no experimental observation of $\Delta L = \pm 1$ GQA peaks has been reported. Although $\Delta L = \pm 1$ GQA peaks are not observed, the Doppler-shifted $\Delta L \pm 1$ excitations do cause an attenuation of A waves and a variation in the power absorption (surface impedance) of E waves when the experiments are carried out at microwave or lower frequencies. In the case of A waves where $\omega < qv_F$ the attenuation is found to exhibit a sharp decrease at a field corresponding to $\omega_c = \omega + qv_F \approx qv_F$, the maximum field at which the $\Delta L \pm 1$ transitions can be excited, i.e., the Kjeldaas edge. In the case of E waves at microwave frequencies, the theory predicts a Doppler-shifted cyclotron resonance peak in the imaginary part of the surface impedance at a magnetic field $\omega_c \approx \omega + qv_F$. Such a peak was observed experimentally in bismuth by Kirsch [24].

By tilting $\mathbf{q}$ away from $\mathbf{H}_0$, it is possible to decrease the Doppler shift in frequency to the point where $\omega > qv_F \cos \theta$ even for A waves. Under these conditions it is possible to observe an onset of attenuation of A waves at $\omega_c = \omega - qv_F \cos \theta$ and a subsequent cessation of attenuation at $\omega = \omega + qv_F \cos \theta$.

The condition for observing a well-defined Doppler-shifted cyclotron resonance is $\omega_c \tau > 1$, which corresponds to $(\omega + qv_F)\tau > 1$. In the case of A waves where qv_F is generally larger than ω the condition for a well-defined resonance becomes $qv_F \tau = ql > 1$. In the case of E waves where qv_F is generally smaller than ω the condition for a well-defined resonance becomes simply $\omega\tau > 1$.

Cyclotron Resonance ($q \perp H_0$)

Since there is no dependence of $\Delta\mathscr{E}$ on $\mathbf{q}$ in the Voigt configuration the cyclotron resonances do not exhibit a Doppler shift in frequency. Furthermore, at $\omega_c < \omega_P$ no resonance should appear at $\omega = \omega_c$ in the electric-dipole approximation. As shown by Smith et al. ([15]), the longitudinal electric field which is present at $\omega = \omega_c$ has the same magnitude as the transverse electric field, but is out of phase by $90°$. The resultant of the two fields is circularly polarized but rotating in the cyclotron-inactive sense. A resonance does occur at $\omega = (\omega_P{}^2 + \omega_c{}^2)^{1/2} = \Omega_c$ corresponding to the frequency of the collective cyclotron excitations.

The single-particle $\Delta L = \pm 1, \pm 2, \ldots$ excitations at ω equal to ω_c, $2\omega_c, \ldots$ are observed in AKCR for both E and A waves. In the case of the E waves the small skin depth, δ, corresponds to an effective wave vector, $q \approx 2\pi/\delta$. The condition for observing well-defined AKCR are that $\omega_c \tau > 1$ and $qR > 1$, where R is the radius of the cyclotron orbit.

Displacement of the Center of Orbit

The lateral displacements of the centers of orbit of the electrons that occur when quanta having components of momentum in the plane of the cyclotron orbit in real space are continuously absorbed constitutes a DC current. The accumulation of a surface charge density, Σ, sets up a DC electric field which is observable. This phenomenon which is called the magneto-quantum-electric (MQE) effect has been observed in $\Delta L = 0$ GQA of A waves in bismuth ([33,34]). It is called a quantum effect because the magnitude of the displacements of the centers of orbit, Δx_0 is determined by the momentum, $\hbar\mathbf{q}$, of the phonons or photons and not by the intensity of the A or E waves. A detailed discussion of the MQE effect is presented by Y. Sawada in Chapter 7 of this volume.

In crossed electric and magnetic fields the displacement of the center of orbit in a direction opposite to that of the applied electric field increases the kinetic energy of the electron by an amount $eE\Delta x_0$. This energy can be emitted as a phonon (or a photon) having $\hbar\omega = eE\,\Delta x_0$ and $q = eH_0 \times \Delta x_0/c\hbar$. The critical field E_{0c} at which this can occur corresponds to the field at which $v_D = cE_0/H_0 = v_W$. At lower fields the probability of phonon absorption is larger than the probability of phonon emission. At higher fields the probability of emission exceeds that for absorption and a spontaneous emission of phonons occurs. This has been observed in different materials. The amplification (stimulated emission) of A (or E) waves also becomes possible (this is discussed further in Chapter 7).

ACKNOWLEDGMENT

The treatment presented above is the outgrowth of many valuable discussions with Dr. Y. Sawada during his stay at the University of Pennsylvania as a graduate student and postdoctoral fellow.

REFERENCES

1. M. H. Cohen, M. J. Harrison, and W. A. Harrison, *Phys. Rev.* **117**: 939 (1964).
2. V. L. Gurevich, *Soviet Physics* JETP (*English Transl.*) **10**: 5 (1961) and **10**: 1090 (1961).
3. E. Burstein, G. S. Picus, R. F. Wallis, and F. Blatt, *Phys. Rev.* **113**: 15 (1959).
4. V. L. Gurevich, V. G. Skobov, and Yu. A. Firsov, *Soviet Physics JETP* (*English Transl.*) **13**: 552 (1961).
5. Y. Sawada, E. Burstein, and L. Testardi, "Proceeding of the International Conference on the Physics of Semiconductors, Kyoto, 1966," *Suppl. J. Phys. Soc. Jap.* **21**: 760 (1966).
6. D. H. Langenberg, J. J. Quinn, and S. Rodriguez, *Phys. Rev. Letters* **12**: 104 (1964).
7. S. V. Gantsevich and V. L. Gurevich, *Soviet Phys. JETP* (*English Transl.*) **18**: 403 (1964).
8. P. B. Miller, *Phys. Rev. Letters* **11**: 537 (1963).
9. J. J. Quinn, *Phys. Letters* **7**: 235 (1963).
10. H. N. Spector, *Phys. Letters* **10**: 163 (1964).
11. J. Bok in: *Dynamical Processes in Solid State Optics*, R. Kubo and H. Kamimura eds., Syokabo and W. A. Benjamin, Inc., 1967, p. 59.
12. E. Burstein in: *Dynamical Processes in Solid State Optics*, R. Kubo and H. Kamimura eds., Syokabo and W. A. Benjamin, Inc., 1967, p. 1.
13. B. Varga, *Phys. Rev.* **137A**: 1896 (1965).

14. S. Iwasa, Y. Sawada, E. Burstein, and E. Palik, "Proceeding of the International Conference on the Physics of Semiconductors Kyoto, 1966," Suppl. *J. Phys. Soc. Jap.* **21**: 742 (1966).
15. G. Smith, L. C. Hebel, and S. Buchsbaum, *Phys. Rev.* **129**: 154 (1963).
16. J. F. Koch and A. Kip, *Phys. Rev. Letters* **8**: 473 (1962).
17. S. J. Buchsbaum, *Proceedings of the International Conference on the Physics of Semiconductors Symposium on Plasma Effects in Solids*, (Dunod, Paris,1964,) p. 3.
18. P. Aigrain, *Proceedings of the International Conference on Semiconductor Physics*, Czechoslovak Academy of Sciences, Prague, 1961, page 224; O. V. Konstantinov and V. L. Perel, *Soviet Physics JETP (English Transl.)* **11**; 117 (1961).
19. R. Bowers, *Proceedings of the International Conference on the Physics of Semiconductors Symposium on Plasma Effects in Solids*, Dunod, Paris,1964, p. 19.
20. S. J. Buchsbaum and J. K. Galt, *Phys. Fluids* **4**: 1514 (1961).
21. H. H. Kawamura, S. Nagata, S. Takano, and J. Nakahara, *Phys. Letters* **23**: 642 (1966).
22. E. Burstein, P. J. Stiles, D. N. Langenberg, and R. F. Wallis, *Phys. Rev. Letters* **9**: 260 (1962), and *Proceedings International Conference on the Physics of Semiconductors, Exeter*, 1962, Institute of Physics and Physical Society, 1962, p. 345.
23. J. J. Quinn, *Arkiv Fysik* **26**: 93 (1964).
24. J. Kirsch, *Phys. Rev.* **133A**: 1390 (1964).
25. P. M. Platzmann and S. Buchsbaum, *Phys. Rev.* **132**: 2 (1963).
26. T. Kieldaas, *Phys. Rev.* **113**: 1473 (1959).
27. E. A. Stern, *Phys. Rev. Letters* **10**: 91 (1963).
28. D. N. Langenberg and J. Bok, *Phys. Rev. Letters* **11**: 549 (1963).
29. J. J. Quinn and S. Rodriguez, *Phys. Rev.* **133A**: 1589 (1964).
30. C. C. Grimes and S. J. Buchsbaum, *Phys. Rev. Letters* **12**: 356 (1964).
31. V. G. Skobov and E. A. Kaner, *Soviet Phys. JETP (English Transl.)* **19**: 189 (1964).
32. W. Schilz, *Phys. Rev. Letters* **20**: 104 (1968).
33. Y. Sawada, E. Burstein, W. Salaneck, and L. Testardi, *Phys. Rev. Letters* **18**: 776 (1967).
34. W. Salaneck, Y. Sawada, and E. Burstein, *Phys. Rev. Letters* **18**: 779 (1967).

Chapter 2

INTRABAND COLLECTIVE EFFECTS AND MAGNETOOPTICAL PROPERTIES OF MANY-VALLEY SEMICONDUCTORS

P. R. Wallace

Department of Physics
McGill University
Montreal, Canada

INTRODUCTION

The propagation of helicon waves in a semiconductor was first observed by Libchaber and Veilex in 1963 ([1]). They used n-type InSb with 10^{14}–10^{16} carriers/cm³. Because the energy surfaces of InSb are almost spherical, the magnetoplasma behavior of its electrons is similar to that of a free-electron gas, though modified by the small effective mass of the carriers and the sizeable dielectric constant of the material.

The situation is more complicated, and interesting new effects appear, in many-valleyed semiconductors with highly anisotropic ellipsoidal energy surfaces. The spectrum of normal modes of the electrons is more complex, as are, in consequence, the magnetooptical properties at microwave frequencies.

These differences are due to:

1. The existence of carriers with different effective masses, and thus of several cyclotron frequencies.

2. The ellipticity of the cyclotron orbits, leading to the presence of both circular polarizations.

3. The coupling of longitudinal and transverse motions when the external magnetic field and direction of propagation are oblique to the symmetry axes of the energy ellipsoids.

We shall pay particular attention to Ge and PbTe, for both of which the bottom of the conduction band lies at the Brillouin zone surface in the

[111] and similar directions, while energy surfaces in the neighborhood of these points are highly elongated prolate spheroids. The most interesting difference between them is that in PbTe the polar character of the crystal makes possible interaction between the magnetoplasma excitations and optical phonons.

Before proceeding to discuss in detail the spectra of collective normal modes of the electrons let us note that with carrier concentrations in the range 10^{14}–10^{18} these semiconductors are better plasmas than metals with much higher carrier densities. The condition that a substance be a plasma is that the mean interparticle spacing should be less than the Thomas–Fermi screening length.

If this is not the case, particles interact weakly except for close collisions of pairs. In a plasma each particle may be considered to move in the *average* field of all the other particles, and the residual interactions or fluctuating fields due to individual particle–particle collisions have a minor effect. In low-density systems, on the other hand, the latter tend to dominate. The screening length in a degenerate Fermi gas of particles of mass m^* in a medium of static dielectric constant ε_0 is

$$\lambda_s = \tfrac{1}{2}(\pi/3)^{1/6}a_0^{1/2}(m\varepsilon_0/m^*)^{1/2}N^{-1/6} \tag{1}$$

where a_0 is the Bohr radius $\hbar^2/me^2$. The volume encompassed within this screening length is $\tfrac{4}{3}\pi\lambda_s^3$ and the number of particles within this volume is $\tfrac{4}{3}\pi\lambda_s^3N$. The condition that it be greater than 1 is

$$\frac{1}{2}\left(\frac{\pi}{3}\right)^{3/2}\left(\frac{\varepsilon_0 m}{m^*}\right)^{3/2}(Na_0^3)^{1/2} \geq 1 \tag{2}$$

It is easily seen that in metals, in which $m/m^* \approx 1$ and $\varepsilon_0 \approx 1$, this condition is never really satisfied. Germanium, for which $\varepsilon_0 = 16$ and $m/m^* \approx 10$, is in this respect like a metal having $\sim 4 \times 10^6$ as many carriers. The case of PbTe is even more striking. In this material $m/m^* \approx 30$ and $\varepsilon_0 \approx 400$. Thus the effective enhancement of the number of carriers is $\sim 2 \times 10^{12}$. It is therefore probably the best solid-state plasma known.

NORMAL MODES AND THE DIELECTRIC TENSOR

For disturbances propagated along a symmetry direction of the crystal and with the external field $\mathbf{B}_0$ also along such a direction the dielectric tensor at long wavelength is found to be of the form

$$\varepsilon_{ij} = \varepsilon_1\delta_{ij} + (\varepsilon_3 - \varepsilon_1)n_in_j + i\varepsilon_2\eta_{ijk}n_k \tag{3}$$

where η_{ijk} is the familiar permutation symbol and $\mathbf{n}$ is the direction of the magnetic field. [If the crystal has such a symmetry that there is no other preferred direction, the tensor *must* be constructed from the vector $\mathbf{n}$, and (3) represents the most general such tensor.] From (3) it follows that

$$\mathbf{D} = \boldsymbol{\epsilon}\mathbf{E} = \varepsilon_1\mathbf{E} + (\varepsilon_3 - \varepsilon_1)\mathbf{n}(\mathbf{n} \cdot \mathbf{E}) - i\varepsilon_2\mathbf{n} \times \mathbf{E} \tag{4}$$

The dielectric tensor is not diagonal in a Cartesian representation, but it is in the "polarization representation" defined as follows: consider the "polarization operator" $i\mathbf{n}\times$; the eigenvalue equation is

$$i\mathbf{n} \times \mathbf{E} = \lambda\mathbf{E} \tag{5}$$

Taking the scalar product with $\mathbf{n}$, we see that for the component of $\mathbf{E}$ in the direction of $\mathbf{n}$ to be nonzero we must have $\lambda = 0$. On the other hand, in the transverse case ($\mathbf{E} \cdot \mathbf{n} = 0$) it follows on operating again with $i\mathbf{n} \times$ that

$$\lambda^2 = 1 \quad \text{or} \quad \lambda = \pm 1 \tag{6}$$

Taking $\mathbf{n}$ in the 3 direction and taking the 1 and 2 directions perpendicular to $\mathbf{n}$, $i\mathbf{n} \times$ has the matrix form

$$\begin{pmatrix} 0 & -i \\ i & 0 \end{pmatrix}$$

Thus the eigenvector corresponding to $\lambda = 1$ is proportional to $\binom{1}{i}$, while that corresponding to $\lambda = -1$ is $\binom{1}{-i}$. We shall call the first "right circularly polarized" (r.c.p.) about $\mathbf{n}$ and the second "left circularly polarized" (l.c.p.). These conventions are the opposite of those prevalent in classical optics, where the direction of rotation of the electric field *at a fixed time* is considered. We consider instead rotation *at a fixed point*. Since $i\mathbf{n} \times$ is proportional to the operator for angular momentum of a vector field about $\mathbf{n}$, our r.c.p. corresponds to *positive* angular momentum and l.c.p. to *negative* angular momentum.

Designating the r.c.p. and l.c.p. components of a vector respectively by E_+ and E_- and the component in the direction of $\mathbf{n}$ by E_3, we see that (4) takes the form

$$D_\pm = (\varepsilon_1 \mp \varepsilon_2)E_\pm = \varepsilon_\pm E_\pm \tag{7}$$

$$D_3 = \varepsilon_3 E_3 \tag{8}$$

Thus in the "polarization representation" the dielectric tensor is diagonalized.

HELICONS

Helicon waves are normal modes of the coupled system of electrons and photons (electromagnetic fields). Before presenting the systematic mathematical theory of such modes it may be useful to focus attention on the physical mechanism of helicon wave propagation. Let us outline how an electromagnetic wave of low velocity may be propagated through, say, a metal in a magnetic field, when it obviously cannot in the absence of the field.

The clue to the phenomenon is the fact discussed by Zawadski in Chapter 13 of this volume that in a magnetic field and in the absence of scattering electrons tend to move at right angles to an electric field. (This is as true for oscillating as for DC fields). Thus the electric field does no work, and may be propagated without loss. Consider a right circularly polarized electric field transverse to the external magnetic field, which we shall take to be in the z direction (Fig. 1). Now, if all fields and currents have the space and time dependence $\exp(i\mathbf{q} \cdot \mathbf{r} - \omega t)$, the Faraday–Maxwell equation $\nabla \times \mathbf{E} = -(1/c) \, \partial \mathbf{H}/\partial t$ becomes $\mathbf{q} \times \mathbf{E} = (\omega/c)\mathbf{H}$, so that if the electric field is in the x direction it is accompanied by an oscillating magnetic field in the y direction of magnitude $(cq/\omega)E$. There is also an electron current in the y direction which is determined by the Hall coefficient: for strong fields and weak scattering the Hall angle is nearly 90° and the current is $J = -E/HR$, with R the Hall coefficient. But this is exactly

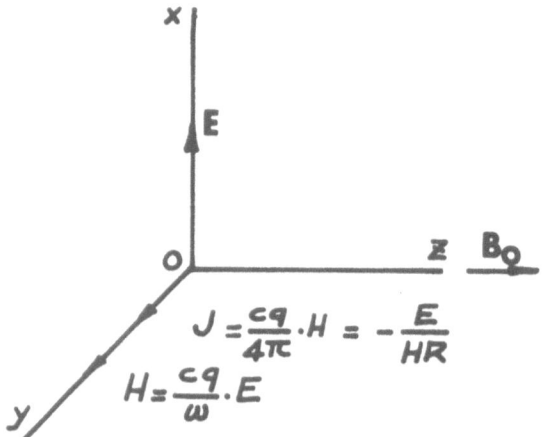

Fig. 1. Relation of fields and current in a helicon wave.

the current associated with the magnetic field through

$$\nabla \times \mathbf{H} = i\mathbf{q} \times \mathbf{H} = iq\mathbf{n} \times \mathbf{H} = q\mathbf{H} = (4\pi/c)\mathbf{J}$$

provided

$$-E/HR = (cq/4\pi)H = (c^2q^2/4\pi\omega)E$$

Thus we have a wave which propagates itself through the mechanism of the Hall current provided

$$\omega = -c^2q^2RH/4\pi$$

This is the dispersion relation for helicon propagation. It should be noted that it does not contain the mass, ·nor in fact anything characterizing the band parameters of the lattice, as is the case for the Hall coefficient. It depends only on the *number* of carriers. This turns out always to be true, even when there are several carriers and the effective-mass tensor is aniso-tropic. However, as we shall see, *other* normal modes which appear in the case of ellipsoidal bands *do* give information about band parameters.

We note that if the wave originally assumed had *left* circular polariza-tion, the magnetic field would be in the $-y$ direction, and there would no longer be self-consistency.

CALCULATION OF DISPERSION RELATIONS IN A MAGNETIC FIELD

To calculate the normal modes of the coupled system of electrons and electromagnetic field we proceed as follows: We can calculate the response of the electrons to an arbitrary *external* field of given frequency ω and wave-number vector $\mathbf{q}$ by the methods of conventional linear response theory. A convenient quantity to calculate is the current, and hence the conductivity tensor σ_{ij}. We then make an approximation of the Hartree type, in which we assume that the field acting on a given electron at a point is the *average* field of all the other electrons. This leads to a self-consistent system in which the collectivity of electrons generates in its motion the fields which in turn *determine* the electron motions. What is ignored in this approach is the effect of *fluctuations* in field felt by an electron (due, e.g., to strong correlation with another electron with which it makes a close collision).

Let us now see how this will work in practice. Writing Maxwell's equations in Fourier-transformed form, all quantities having the space and time variation $\exp i(\mathbf{q} \cdot \mathbf{r} - \omega t)$, they become (in c.g.s. units)

$$iq \cdot \varepsilon_0 \mathbf{E_q} = 4\pi \varrho_q \tag{9a}$$

$$\mathbf{q} \cdot \mathbf{H_q} = 0 \tag{9b}$$

$$i\mathbf{q} \times \mathbf{E_q} = (i\omega/c)\mathbf{H_q} \tag{9c}$$

$$i\mathbf{q} \times \mathbf{H_q} = (4\pi/c)\mathbf{J_q} - (i\omega\varepsilon_0/c)\mathbf{E_q} \tag{9d}$$

ε_0 is the dielectric constant of the lattice. The equation of continuity requires that

$$\varrho_q = (1/\omega)\mathbf{q} \cdot \mathbf{J_q} \tag{10}$$

If now ϱ and $\mathbf{J}$ represent a response to the fields which they produce,

$$\mathbf{J} = \boldsymbol{\sigma}\mathbf{E}$$

and (9a) and (9b) may be written

$$i\mathbf{q} \cdot \mathbf{D_q} = 0 \tag{11}$$

$$\mathbf{q} \times \mathbf{H_q} = -(\omega/c)\mathbf{D_q} \tag{12}$$

where

$$\mathbf{D_q} = \boldsymbol{\epsilon}\mathbf{E_q} = [\varepsilon_0 + (4\pi i/\omega)\boldsymbol{\sigma}]\mathbf{E_q} \tag{13}$$

If we now combine Eqs. (9c) and (11), we find that

$$\mathbf{q} \times (\mathbf{q} \times \mathbf{E}) = \frac{\omega}{c}\mathbf{q} \times \mathbf{H_q} = -\frac{\omega^2}{c^2}(\boldsymbol{\epsilon}\mathbf{E_q}) \tag{14}$$

Eliminating the components of $\mathbf{E}$, it is now possible to obtain the dispersion relation for the wave in terms of the dielectric tensor ε.

If the propagation vector $\mathbf{q}$ is in the direction of the external magnetic field $\mathbf{B_0}$, Eq. (14) takes a simple form in the polarization representation, namely,

$$q^2 E_\pm = (\omega^2/c^2)\varepsilon_\pm E_\pm$$

and

$$0 = \varepsilon_3 E_3$$

[these equations may be grouped as $\lambda_\beta^2 q^2 E_\beta = (\omega^2/c^2)\varepsilon_\beta E_\beta$, with $\beta = +, -,$ 3 specifying the polarization]. The normal modes are therefore given by

$$c^2 q^2/\omega^2 = \varepsilon_\pm \tag{15}$$

for the transverse circularly polarized modes and

$$\varepsilon_3 = 0 \tag{16}$$

for longitudinal ones.

If $\mathbf{q}$ is *normal* to $\mathbf{B}_0$ and is, for instance, in the 1 direction, Eq. (14) becomes

$$q E_x \mathbf{q} - q^2 \mathbf{E} = -(\omega^2/c^2)\mathbf{D} \tag{17}$$

Taking 3-components we obtain for polarization in the 3 direction

$$c^2 q^2 / \omega^2 = \varepsilon_3 \tag{18}$$

On the other hand, writing 1- and 2-components of (17) and eliminating E_x and E_y, we obtain for polarization transverse to the field

$$\frac{c^2 q^2}{\omega^2} = \frac{\varepsilon_+ \varepsilon_-}{\frac{1}{2}(\varepsilon_+ + \varepsilon_-)} = \frac{\varepsilon_+ \varepsilon_-}{\varepsilon_1} = \varepsilon' \tag{19}$$

Astrøm, in 1950 [2] worked out the dispersion relation for the case in which $\mathbf{q}$ makes an angle θ with $\mathbf{n}$:

$$\tan^2 \theta = -\frac{[1 - (\xi^2/\varepsilon_+)][1 - (\xi^2/\varepsilon_-)]}{[1 - (\xi^2/\varepsilon_3)][1 - (\xi^2/\varepsilon')]} \tag{20}$$

where

$$\xi^2 = c^2 q^2 / \omega^2 \tag{21}$$

is the square of the refractive index and

$$\varepsilon_1 = \tfrac{1}{2}(\varepsilon_+ + \varepsilon_-) \tag{22}$$

Equations (15) and (16) appear as special cases when $\theta = 0$, while (18) and (19) correspond to $\theta = \pi/2$. We shall concern ourselves principally with these two special cases.*

---

* A simple derivation of (20) is the following:

    Let $\mathbf{n}$ be a unit vector in the direction of propagation. Then writing Eq. (14) in terms of polarization components, we have

$$\xi^2 n_\beta (\mathbf{n} \cdot \mathbf{E}) = (\xi^2 - \varepsilon_\beta) E_\beta$$

where β is an index designating the polarization: $+$ and $-$ for right and left circular polarization, 3 for longitudinal. Thus

$$E_\beta = \frac{\xi^2 (\mathbf{n} \cdot \mathbf{E}_\mu)}{\xi^2 - \varepsilon_\beta} n_\beta$$

CALCULATION OF THE DIELECTRIC CONSTANT

The various dielectric constants for a free-electron gas have been calculated by Celli and Mermin [3] and are implicit in the earlier work of Quinn and Rodriguez [4].

To calculate them for ellipsoidal energy surfaces we first make use of the fact that when the energy surfaces are of the form

$$\mathcal{H} = E(\mathbf{k}) \tag{23}$$

in the absence of a magnetic field the effective Hamiltonian in the presence of the field is

$$\mathcal{H} = E[-i\nabla + (e\mathbf{A}/\hbar c] = E(\boldsymbol{\pi}) \tag{24}$$

If

$$E(\mathbf{k}) = (\hbar^2/2m)(\alpha_1 k_x^2 + \alpha_2 k_y^2 + \alpha_3 k_z^2) \tag{25}$$

it appears that we should be able to reduce the problem to that of free electrons by a change of scale in each of the principal directions. The various quantities appearing in the calculation scale in the following manner:

$$
\begin{aligned}
k_x' &= \sqrt{\alpha_1}\, k_x, & B_x' &= (\alpha_2\alpha_3)^{1/2} B_x \\
x' &= x/\sqrt{\alpha_1}, & E_x' &= \sqrt{\alpha_1}\, E_x \\
A_x' &= \sqrt{\alpha_1}\, A_x, & J_x' &= (1/\sqrt{\alpha_1})J_x \\
\sigma_{ij}' &= (\alpha_i\alpha_j)^{-1/2}\sigma_{ij} & \text{etc.}
\end{aligned}
\tag{26}
$$

If we consider a single spheroidal energy surface whose axis of symmetry is oblique to $\mathbf{B}_0$ we can first use axes corresponding to principal axes

But

$$\mathbf{n} \cdot \mathbf{E} = 2(n_+ E_- + n_- E_+) + n_3 E_3$$

where $E_\pm = \frac{1}{2}(E_x \mp iE_y)$, etc. Therefore, using the relation between E_β and n_β above we may write

$$\mathbf{n} \cdot \mathbf{E} = \left\{ 2n_+ n_- \xi^2 \left(\frac{1}{\xi^2 - \varepsilon_+} + \frac{1}{\xi^2 - \varepsilon_-} \right) + n_3^2 \frac{\xi^2}{\xi^2 - \varepsilon_3} \right\} \mathbf{n} \cdot \mathbf{E}$$

Therefore, since $\mathbf{n}$ makes an angle θ with $\mathbf{B}_0$,

$$1 = \frac{1}{2} \sin^2\theta \left(\frac{\xi^2}{\xi^2 - \varepsilon_+} + \frac{\xi^2}{\varepsilon^2 - \varepsilon_-} \right) + \cos^2\theta \frac{\xi^2}{\xi^2 - \varepsilon_3}$$

If we write the left-hand side as $\cos^2\theta + \sin^2\theta$ and use the resulting equation to solve for $\tan^2\theta$, (20) follows immediately.

of the spheroid and rescale to determine the conductivity tensor relative to these axes. We must then transform back to *fixed* axes relative to which (for convenience) the external magnetic field is along the 3 direction.

When this is done a quite complicated formula is obtained for the conductivity tensor. It is *not* of the simple form (3), and in particular has elements which couple circular polarization about $\mathbf{B}_0$ to linear polarization along $\mathbf{B}_0$. In particular, the x–z (or 1–3) component of the conductivity due to one such surface for the case in which the axis of the energy surface is along the [111] direction and $\mathbf{q}$ and $\mathbf{B}_0$ are along [100] is nonzero but contains the factor $(\alpha_T - \alpha_L)$, and so goes to zero when the surface becomes spherical. Thus, normal modes corresponding to propagation along $\mathbf{B}_0$ involve *longitudinal* as well as transverse fields, and are consequently accompanied by electron density fluctuations.

Let us now comment on the dielectric tensor for free electrons, which constitutes the starting point of the above calculations. By linear response theory it may be shown that the conductivity tensor, neglecting interband contributions as well as relaxation processes, is

$$\sigma_{ij} = -\frac{\varepsilon_\infty \omega_p{}^2}{4\pi i \omega}(\delta_{ij} + I_{ij}) \tag{27}$$

where

$$I_{ij} = \frac{m}{N}\sum_{\nu,\nu'}[f_0(E_{\nu'}) - f_0(E_\nu)]\frac{1}{E_{\nu'} - E_\nu - \hbar\omega}$$
$$\times \langle \nu' \mid V_i(\mathbf{q}) \mid \nu \rangle \langle \nu' \mid V_j(\mathbf{q}) \mid \nu \rangle^* \tag{28}$$

where ν stands for the quantum numbers of Landau states (n, k_y, k_z); f_0 is the Fermi distribution function, and $\mathbf{V}(\mathbf{q})$ is the operator

$$\mathbf{V}(\mathbf{q}) = \tfrac{1}{2}[\mathbf{V}\exp(i\mathbf{q}\cdot\mathbf{r}) + \exp(i\mathbf{q}\cdot\mathbf{r})\,\mathbf{V}] \tag{29}$$

In (27) ε_∞ is the high-frequency dielectric constant and $\omega_p{}^2$ is the square of the plasma frequency:

$$\omega_p{}^2 = 4\pi N e^2/m\varepsilon_\infty \tag{30}$$

Let the external magnetic field be in the z direction. Then the matrix elements are as follows:

$$\langle n'k_y'k_z' \mid V_x(\mathbf{q}) \pm iV_y(\mathbf{q}) \mid nk_yk_z \rangle$$
$$= i\left(\frac{2\hbar\omega_c}{m}\right)^{1/2}\delta(k_y', k_y + q_y)\delta(k_z', k_z + q_z)$$
$$\times \begin{pmatrix} (n+1)^{1/2}f_{n',n+1}(\sqrt{q_x{}^2 + q_y{}^2}) \\ n^{1/2}f_{n',n-1}(\sqrt{q_x{}^2 + q_y{}^2}) \end{pmatrix}\begin{matrix}(+)\\(-)\end{matrix} \tag{31}$$

The cyclotron frequency ω_c is eB_0/mc and

$$f_{n'n} = \left(\frac{n!}{n'!}\right)^{1/2} u^{(1/2)(n'-n)} e^{-(1/2)u} L_n^{(n'-n)}(u),\tag{32}$$

where

$$u = \frac{\hbar q_\perp^2}{2m\omega_c} = \frac{\hbar}{2m\omega_c} (q_x^2 + q_y^2)^{1/2}\tag{33}$$

and $L_n^{(n'-n)}$ is an associated Laguerre function. The matrix element of the velocity component along the field is

$$\langle n'k_y'k_z' \mid V_3(\mathbf{q}) \mid nk_yk_z\rangle$$
$$= \delta(k_y', k_y + q_y)\delta(k_z', k_z + q_z)\frac{\hbar}{2m} (k_z + k_z')f_{n'n}(q_\perp)\tag{34}$$

The function $f_{n'n}$ is an "overlap integral" of two oscillator functions:

$$f_{n'n}(q_\perp) = \int_{-\infty}^{\infty} u_{n'}\left(x + \frac{\hbar q_\perp}{m\omega_c}\right)u_n(x)\, dx\tag{35}$$

where the u_n are the harmonic oscillator functions. Consequently, in the $q_\perp = 0$ case it reduces to

$$f_{n'n} = \delta_{n'n}\tag{36}$$

Let us note the following features of the above results:

1. If the wave is propagated along the direction of the magnetic field, $q_\perp = 0$. Thus in (28) we have the selection rule

$$\Delta n = \pm 1\tag{37}$$

Since the energy levels are

$$E(n, k_z) = (n + \tfrac{1}{2})\hbar\omega_c + (\hbar^2 k_z^2/2m)\tag{38}$$

the denominators are as follows:

For $n' - n = 1$

$$\hbar(\omega_c - \omega) + (\hbar^2/2m)[(k_z + q)^2 - k_z^2]\tag{39}$$

This becomes zero, not for the cyclotron frequency ω_c, but for the Doppler-shifted frequency

$$\omega \approx \omega_c + (\hbar k_z/m)q$$

provided $q \ll k_z$. When we sum over the whole range of k we get a logarithmic singularity at

$$\omega = \omega_c - (\hbar k_{\max}/m)q$$

For $n' - n = -1$

$$-\hbar(\omega_c + \omega) + (\hbar^2/2m)[(k_z + q)^2 - k_z^2]$$

Neglecting the Doppler-shifting term and observing the ω in the denominator of (27), we see that this contribution simply corresponds to the transformation $\omega \to -\omega$ in the previous one. But a change of sign of ω merely corresponds to a reversal of polarization. Thus σ_{+-} and σ_{-+} correspond to opposite polarizations, a fact which may be verified directly.

This "Doppler-shifting" effect is in fact quite complex, due to Landau quantization. We observe that the conductivity depends on a *sum* over n of terms of the form

$$\int \frac{|\langle n | V_{\pm}(\mathbf{q}) | n + 1 \rangle|^2 [f_0(n + 1, k_z + q_z) - f_0(n, k_z)]}{\omega_c - \omega + (\hbar/m)k_z q_z}$$

for small enough q. At zero temperature the *limits* of k_z for a given n are (E_F is the Fermi energy)

$$k_z = \pm \{(2m/\hbar^2)[E_F - (n + \tfrac{1}{2})\hbar\omega_c]\}^{1/2} = \pm k^{(n)}$$

Therefore it is necessary to evaluate an integral of the form

$$\int_{-k^{(n)}}^{k^{(n)}} \frac{dk_z}{\omega_c - \omega + (\hbar/m)qk_z} = \frac{m}{\hbar q} \ln \frac{\omega_c - \omega + (\hbar q/m)k^{(n)}}{\omega_c - \omega - (\hbar q/m)k^{(n)}}$$

Consequently, the dielectric constant will have *logarithmic* singularities at frequencies

$$\omega_c - (\hbar q/m)\{(2m/\hbar^2)[E_F - (n + \tfrac{1}{2})\hbar\omega_c]\}^{1/2}$$

There will therefore be a sequence of resonances and intermediate pass bands equal in number to the number of Landau levels below the Fermi energy. They may, however, be difficult to detect.

The above discussion must be considered to be illustrative rather than exact, since the spin-energy should be included in the expressions for the limits of k_z. The situation will not, however, be changed qualitatively.

For small enough q (long enough wavelength) the Doppler-shifting effect will be negligible.

It is fairly easy to estimate the frequency at which the Doppler-shifting comes into play. Combining the condition $\omega_c - \omega = v_F q$ with the dispersion relation $c^2q^2 = \omega^2\varepsilon_+$, we obtain

$$\omega_c - \omega = (\omega/c)\sqrt{\varepsilon_+}\, v_F$$

But for $\omega_p \gg \omega, \omega_c$ we have $\varepsilon_+ \approx \omega_p{}^2/\omega(\omega_c - \omega)$, so we obtain

$$\omega_c - \omega = \frac{v_F}{c} \frac{\omega^{1/2}\omega_p}{(\omega_c - \omega)^{1/2}}$$

or $(\omega_c - \omega)^{3/2} = (v_F/c)\omega^{1/2}\omega_p$, so that

$$\omega_c - \omega \approx \left(\frac{v_F}{c}\,\omega_p\right)^{2/3}\omega^{1/3}$$

Alternatively, we may write

$$(\omega_c - \omega)/\omega_c \lesssim (v_F\omega_p/c\omega_c)^{2/3}$$

which in semiconductors will usually be quite small.

2. If the wave is propagated *perpendicular* to the external magnetic field, $q_z = 0$ and the Doppler-shifting term will not be present. However, in this case terms will appear corresponding to $\Delta n = \pm 2, \pm 3, \ldots$; i.e., there will be singularities at $\omega = n\omega_c$. The residues of these singularities will, however, decrease successively in the ratio $\hbar q_\perp{}^2/2m\omega_c$. Using the dispersion relation, we see that these terms will be negligible so long as

$$\frac{\hbar q_\perp{}^2}{2m\omega_c} = \frac{\omega}{\omega_c}\,\varepsilon\,\frac{\hbar\omega}{2mc^2} \ll 1$$

Thus, the $\Delta n = 2$ term will be negligible so long as

$$\hbar\omega_c\varepsilon/2mc^2 \ll 1$$

with ε the appropriate dielectric constant. It will then be important only in areas of very high dielectric constant, i.e., very close to resonances.

Each residue will also involve an expansion in powers of $\hbar q^2/2m\omega_c$.

Again, in the long-wavelength limit these terms may be neglected. So long as these terms *are* neglected, the dielectric tensor may be assumed to be the same for arbitrary directions of propagation.

DIELECTRIC TENSOR FOR GERMANIUM AND LEAD TELLURIDE

We have already noted that for propagation along the direction of the magnetic field when there is a single spheroidal energy surface whose axis is oblique to the field the normal modes are neither completely longitudinal nor completely transverse, but a mixture of the two. But when we have several such energy surfaces and the field is along a symmetry direction the dielectric tensor once again takes on the simple form (3), the coupling terms disappearing when the *sums* of the contributions to the conductivity from the various surfaces are added. This is true in particular for $\mathbf{B}_0$ along the [100] and [111] directions in germanium and lead telluride.

In addition to the contribution to σ from the intraband electron transitions one should add contributions from the lattice and from interband electron transitions. If we are dealing with low frequencies (i.e., frequencies such that $\hbar\omega$ is less than the energy difference for interband transitions and ω is less than the frequencies of the optical phonons) these contributions will be approximately constant. We shall designate their contribution as ε_0, the zero-frequency dielectric constant.

We shall ignore for the present the contribution of acoustic phonons.

Finally, we shall neglect losses. These be may taken into account by replacing ω *in the conductivity* by $\omega + (i/\tau)$, where τ is a relaxation time (it will not of course in reality be constant, but will be frequency- and field-dependent; however, most qualitative physical features can be represented by a constant τ). The condition for the neglect of losses is that $\omega_c\tau \gg 1$. We may then write the long-wavelength dielectric constants for propagation in the [100] and [111] directions.

[100] Direction

Let the parameters of the spheroidal energy surfaces be α_T and α_L, with α_T the coefficient of the square of the component of $\mathbf{k}$ perpendicular to the prolate axis and α_L the coefficient of the square of the component along it. In terms of these coefficients

$$\varepsilon_\pm = \varepsilon_0 - \frac{4\pi Ne^2}{3m}(2\alpha_T + \alpha_L)\frac{\omega \pm \omega_3}{\omega(\omega^2 - \omega_c^2)} \tag{40}$$

where

$$\omega_3 = \frac{\alpha_T(2\alpha_L + \alpha_T)}{2\alpha_T + \alpha_L}\frac{eB_0}{mc} = \frac{\alpha_T(2\alpha_L + \alpha_T)}{2\alpha_T + \alpha_L}\omega_0 \tag{41}$$

with ω_0 the cyclotron frequency of free electrons. The cyclotron frequency ω_c is

$$\omega_c = [\tfrac{1}{3}\alpha_T(2\alpha_L + \alpha_T)]^{1/2}\omega_0 \tag{42}$$

The plasma frequency of the conduction electrons is given by

$$\omega_p{}^2 = 4\pi Ne^2/m^*\varepsilon_\infty \tag{43}$$

where ε_∞ is the high-frequency dielectric constant and $1/m^* = (2\alpha_T + \alpha_L)/m$. Here ε_∞ corresponds to a frequency lower than that for interband transitions but sufficiently high that lattice contributions are negligible. Thus we may write

$$\varepsilon_\pm = \varepsilon_0 \left[1 - \frac{\varepsilon_\infty}{\varepsilon_0} \frac{\omega_p^2(\omega \pm \omega_3)}{\omega(\omega^2 - \omega_c^2)} \right] \tag{44}$$

The parallel component is

$$\varepsilon_3 = \varepsilon_0 \left[1 - \frac{\varepsilon_\infty}{\varepsilon_0} \frac{\omega_p^2}{\omega^2} \frac{\omega^2 - \omega_4^2}{\omega^2 - \omega_c^2} \right] \tag{45}$$

where

$$\omega_4^2 = \frac{3\alpha_T^2 \alpha_L}{2\alpha_T + \alpha_L} \omega_0^2 \tag{46}$$

Consider first the case of propagation in the direction of the field. The transverse modes are given by

$$c^2 q^2 / \omega^2 = \varepsilon_\pm$$

for the two polarizations, while the longitudinal ones have frequencies determined by $\varepsilon_3 = 0$. So far as the transverse modes are concerned, they may be represented diagrammatically. If we plot $\omega^2 \varepsilon_+$ for both positive and negative values of ω, the values on the *negative* side represent $\omega^2 \varepsilon_-$. The situation is shown schematically in Fig. 2 for the case $\omega_p^2 \gg \omega_c^2$. Normal modes of wave number $\mathbf{q}$ occur where a line $c^2 q^2$ above the ω axis cuts the curve shown. On the right-hand side we have, below ω_c, normal helicons, right circularly polarized.

For very long wavelength their dispersion relation is approximately

$$\omega = c^2 q^2 \omega_c^2 / \omega_p^2 \omega_3 \varepsilon_\infty \tag{47}$$

It may be verified that this does not depend on the mass parameters, but only on the number of carriers.

A new and interesting feature is the existence of helicon-like modes with *left* circular polarization. These have a *minimum* frequency which, for large ω_p, is approximately ω_3. If ω_p is not large compared to ω_c, this minimum is somewhat below ω_3, at the lowest positive root of the cubic equation

$$\omega^3 - \omega(\omega_c^2 + \omega_p^2) + \omega_3 \omega_p^2 = 0 \tag{48}$$

The reason for the appearance of this l.c.p. mode is the following: the electron currents in the plane perpendicular to the magnetic field are *elliptical*, and rotate in the positive direction. Thus the disturbance has a

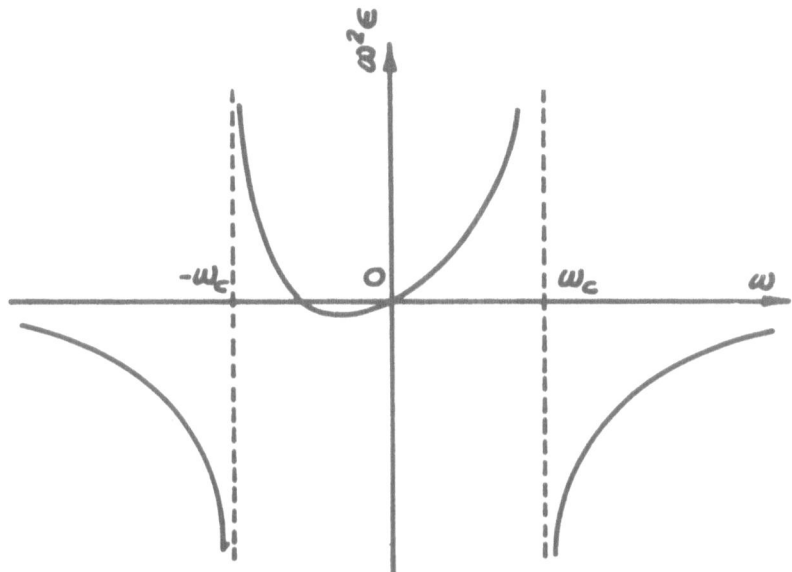

Fig. 2. Plot of $\omega^2\varepsilon_+$ for both positive and negative values of ω.

right elliptical polarization. But the ellipses intersected on the different energy surfaces differ in phase by $\pi/2$. Adding right *elliptical* polarizations in such a case, one may obtain either right or left *circular* polarizations. In fact, if the semimajor axes of the ellipse are a and b, a vector describing right elliptical polarization is $\binom{a}{ib}$. An elliptical polarization with axes interchanged is $\binom{b}{ia}$. These may be combined to give

$$\begin{pmatrix} a \\ ib \end{pmatrix} + \begin{pmatrix} b \\ ia \end{pmatrix} = (a + b)\begin{pmatrix} 1 \\ i \end{pmatrix},$$

which represents a right circular polarization. On the other hand, if they are combined in the form

$$\begin{pmatrix} a \\ ib \end{pmatrix} - \begin{pmatrix} b \\ ia \end{pmatrix} = (a - b)\begin{pmatrix} 1 \\ -i \end{pmatrix}$$

the resulting vector corresponds to left circular polarization. Put differently, the right elliptically polarized vector may be written as a combination of right and left circular polarizations:

$$\begin{pmatrix} a \\ ib \end{pmatrix} = \tfrac{1}{2}(a + b)\begin{pmatrix} 1 \\ i \end{pmatrix} + \tfrac{1}{2}(a - b)\begin{pmatrix} 1 \\ -i \end{pmatrix} \tag{49}$$

The longitudinal modes have frequencies given by

$$\omega^2(\omega^2 - \omega_c^2) - \bar{\omega}_p^2(\omega^2 - \omega_4^2) = 0 \tag{50}$$

where

$$\bar{\omega}_p^2 = (\varepsilon_\infty/\varepsilon_0)\omega_p^2 \tag{51}$$

The solutions of (50) are

$$\omega^2 = \tfrac{1}{2}\{\omega_c^2 + \bar{\omega}_p^2 \pm [(\omega_c^2 + \bar{\omega}_p^2)^2 - 4\bar{\omega}_p^2\omega_4^2]^{1/2}\} \tag{52}$$

If $\bar{\omega}_p\omega_4 \ll \omega_c^2 + \omega_p^2$, the smaller root of this equation is approximately

$$[\omega_p^2/(\omega_p^2 + \omega_c^2)]\omega_4^2 \tag{53}$$

We thus have a new longitudinal mode whose frequency is lower than the cyclotron frequency (since $\omega_4 < \omega_c$), and depends on the magnetic field. The origin of this mode is as follows: since the electrons associated with a given ellipsoid describe orbits around the magnetic field, their cyclotron motion is coupled to a longitudinal oscillation due to the ellipticity and tilting of the energy surface. If all of the spheroidal surfaces are taken into account, the fields produced by the transverse motions can destructively interfere, leaving only a longitudinal density fluctuation.

It is interesting to observe the correlation of these modes with those appearing when propagation is *perpendicular* to the field $\mathbf{B}_0$. In this case the normal modes are given by

$$c^2q^2/\omega^2 = \varepsilon_+\varepsilon_-/\varepsilon_1 = \varepsilon' \tag{54}$$

for polarization transverse to $\mathbf{B}_0$ (but *not* to the direction of propagation), and by

$$c^2q^2/\omega^2 = \varepsilon_3 \tag{55}$$

for a transverse mode linearly polarized in the direction of the magnetic field.

The latter modes are related to the longitudinal ones in the previous case, but since they are now transverse they do not involve a density fluctuation. There is a spectrum of frequencies running from the frequency of the "magnetic plasma" mode to the cyclotron frequency.

As for the modes transverse to the magnetic field, the lowest frequency corresponds to that for which ε_- changes sign from negative to positive. This will then be left circular about the direction of the magnetic field. On the other hand, it will have a longitudinal component, and thus will be a density fluctuation mode. However, as the frequency approaches the cyclotron frequency,

$$\varepsilon_-/(\omega_c - \omega_3) = (\varepsilon_+/\omega_c + \omega_3) \tag{56}$$

so that*

$$\mathbf{E} = \text{const}[(\omega_c - \omega_3)\mathbf{E}_+ - (\omega_c + \omega_3)\mathbf{E}_-] \tag{57}$$

$\mathbf{E}_+$ and $\mathbf{E}_-$ representing eigenvectors for r.c.p. and l.c.p. respectively. This is represented in matrix form by

$$(\omega_c - \omega_3)\begin{pmatrix} 1 \\ i \end{pmatrix} - (\omega_c + \omega_3)\begin{pmatrix} 1 \\ -i \end{pmatrix} = -2\begin{pmatrix} \omega_3 \\ -i\omega_c \end{pmatrix}$$

and so corresponds to an elliptical polarization.

If we look at the absorption resonances, they appear at the singularities of ε_+ and ε_- (that is, at ω_c), but also at the zeros of $\varepsilon_1 = \frac{1}{2}(\varepsilon_+ + \varepsilon_-)$. The latter is easily calculated to be

$$\varepsilon_1 = \varepsilon_0 - [\varepsilon_\infty \omega_p^2/(\omega^2 - \omega_c^2)]$$

which is zero at

$$\omega^2 = \omega_c^2 + \bar{\omega}_p^2$$

This is simply the "collective cyclotron resonance" discussed by Burstein, Sawada, Iwasa, and others in other chapters of this volume. The frequency ω_4 does not appear here.

This mode will be split and its frequency altered by interaction with longitudinal optical phonons, as will be seen later.

The nature of this resonance, in the absence of phonons, is easily deduced from the condition $\varepsilon_+ + \varepsilon_- = 0$. Then the field is determined by calculating components in the direction of propagation of the equation

$$\mathbf{q} \times (\mathbf{q} \times \mathbf{E}) = \boldsymbol{\epsilon}\mathbf{E} = 0$$

This leads to the equation

$$\varepsilon_+\mathbf{E}_+ + \varepsilon_-\mathbf{E}_- = \varepsilon_+(\mathbf{E}_+ - \mathbf{E}_-) = 0$$

which gives $E_y = 0$. The field is therefore polarized in the direction of propagation E_x.

This "collective cyclotron resonance" can also be discussed on the basis of a classical model in which one solves the coupled equations of motion of the precessing electrons. The difference between the dielectric constant approach and the approach which involves solving coupled equa-

---

* Taking components of $\mathbf{D} = 0$ in the direction of propagation gives $\varepsilon_1 E_1 + i\varepsilon_2 E_2 = 0$. In polarization representation this says that $\varepsilon_+ E_+ + \varepsilon_- E_- = 0$, from which (57) follows.

tions is the following: in the dielectric response approach each component (e.g., electrons, phonons) is considered to respond to the same average field. We therefore first calculate independently the response of each component to an arbitrary field. The field is then taken, for self-consistency, to be the total field produced by the motion of *all* the components.

We now consider the case in which the magnetic field is in the [111] direction.

[111] Direction

There are now two cyclotron frequencies. That at

$$\omega_{c1} = [\tfrac{1}{9}\alpha_T(8\alpha_L + \alpha_T)]^{1/2}\omega_0 \tag{58}$$

is associated with the energy surfaces whose axis is tilted with respect to the [111] direction. We then expect an l.c.p. mode below this frequency. The higher cyclotron frequency at

$$\omega_{c2} = \alpha_T\omega_0 \tag{59}$$

will appear only for r.c.p., i.e., on the $\omega > 0$ range for ε.

The explicit forms for the components of the dielectric tensor are:

$$\varepsilon_\pm = \varepsilon_0\left[1 - \frac{\bar{\omega}_p{}^2}{\omega}\left\{\frac{1}{4}\frac{1}{\omega \mp \omega_{c2}} + \frac{1}{12}\frac{(4\alpha_L + 5\alpha_T)\omega \pm (8\alpha_L + \alpha_T)\alpha_T\omega_0}{\omega^2 - \omega_{c1}^2}\right\}\right] \tag{60}$$

and

$$\varepsilon_3 = \varepsilon_0\left[1 - \frac{\bar{\omega}_p{}^2}{\omega^2}\left\{\frac{\alpha_L}{4} + \frac{1}{12}\frac{(\alpha_L + 8\alpha_T)\omega^2 - 9\alpha_L\alpha_T\omega_0{}^2}{\omega^2 - \omega_{c1}^2}\right\}\right] \tag{61}$$

First considering propagation along the direction of $\mathbf{B}_0$, we proceed as before to plot ε_+ for positive and negative ω, the negative side representing ε_- (Fig. 3). We have indicated, this time by intermittent lines, the form taken by ε when relaxation effects are taken into account; what is shown is then the real part of ε. The imaginary part, which describes the absorption of the wave, will have peaks near the cyclotron frequencies.

The Kramers–Kronig relations connecting the real and imaginary parts of a complex function,

$$f_r = -\frac{1}{\pi}P\int_{-\infty}^{\infty}\frac{f_i(\omega')}{\omega - \omega'}\,d\omega' \tag{62}$$

$$f_i = \frac{1}{\pi}P\int_{-\infty}^{\infty}\frac{f_r(\omega')}{\omega - \omega'}\,d\omega' \tag{63}$$

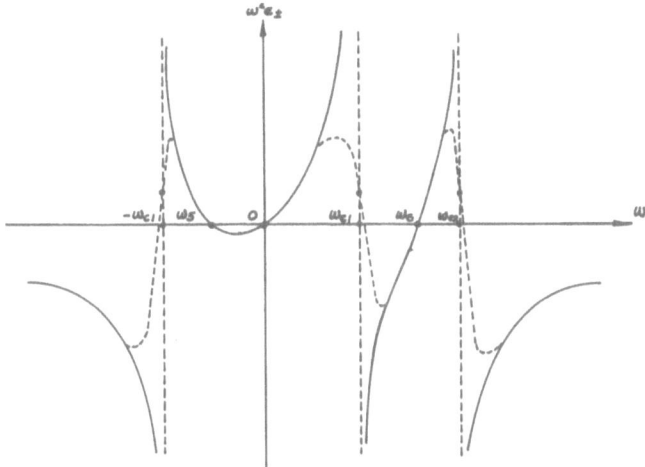

Fig. 3. Plot of ε_+ for positive and negative ω for propagation along the direction of B_0.

may be used to show that between any two cyclotron resonances there must be a transmission band ("P" stands for "principal part"). Thus there is a r.c.p. mode with frequencies running from ω_6 to somewhere below ω_{c2}. If the imaginary part of $\varepsilon_+ - \varepsilon_0$ has sharp and narrow peaks at ω_{c1} and ω_{c2}, i.e. if it contains two terms of Lorentzian form $A/[(\omega - \omega_{c1})^2 + \Gamma^2]$ and $B/[(\omega - \omega_{c2})^2 + \Gamma'^2]$, we may directly calculate their contribution to the real part to be

$$\frac{A\pi}{\Gamma} \frac{\omega_{c1} - \omega}{(\omega - \omega_{c1})^2 + \Gamma^2} + \frac{B\pi}{\Gamma'} \frac{\omega_{c2} - \omega}{(\omega - \omega_{c2})^2 + \Gamma'^2} \tag{64}$$

If Γ and Γ' are small compared to ω_{c1}, and ω_{c2} the first term predominates and is negative for $\omega \gtrsim \omega_{c1}$, while the second predominates and is positive for $\omega \lesssim \omega_{c2}$. Thus there must be a zero of the real part between the absorption peaks.

In general it will be necessary to determine ω_5 and ω_6 numerically. If $\bar{\omega}_p^2 \gg \omega_{c1}^2, \omega_{c2}^2$, they may be calculated explicitly:

$$\omega_5 = \frac{[\alpha_T^2(\alpha_T - \alpha_L)^2 + 12\alpha_L\alpha_T^2(2\alpha_T + \alpha_L)]^{1/2} - \alpha_T(\alpha_T - \alpha_L)}{2(2\alpha_T + \alpha_L)}\omega_0 \tag{65}$$

and

$$\omega_6 = \frac{[\alpha_T^2(\alpha_T - \alpha_L)^2 + 12\alpha_L\alpha_T^2(2\alpha_T + \alpha_L)]^{1/2} + \alpha_T(\alpha_T - \alpha_L)}{2(2\alpha_T + \alpha_L)}\omega_0 \tag{66}$$

We have, in this case, not only a l.c.p. mode, but also a higher branch r.c.p. one.

As for longitudinal modes, we again have a coupling between the plasma and cyclotron motions, leading to two longitudinal modes which will be, respectively, below the cyclotron frequency and above the plasma frequency in the absence of the field. Assuming still that $\omega_p^2 \gg \omega_c^2$, the frequency of the "magnetic" plasma mode is

$$\omega_4 \approx \left[\frac{\alpha_L \alpha_T (2\alpha_L + 7\alpha_T)}{3(2\alpha_T + \alpha_L)} \right]^{1/2} \omega_0 \tag{67}$$

If we now consider propagation *perpendicular* to $\mathbf{B}_0$, we see that the frequency of the "magnetic" plasma mode corresponds to the bottom of a band of modes linearly polarized along the magnetic field. The situation with respect to the modes circularly polarized about the magnetic field, however, contains new features, which appear when we plot

$$\varepsilon' = \varepsilon_+ \varepsilon_- / \varepsilon_1$$

This function starts off with *negative* values due to the fact that $\varepsilon_- < 0$ at low frequencies; thus there are no modes of this nature at low frequencies, corresponding to the well-known fact that there are no "helicons" for propagation perpendicular to $\mathbf{B}_0$. A band then starts at ω_5, where ε_- is zero, and at which there is therefore an l.c.p. mode. This band terminates near ω_{c1}, where ε_+, ε_-, ε_1, and ε' all become infinite. The ratio $\varepsilon_+ / \varepsilon_-$ may be calculated:

$$\frac{\varepsilon_+}{\varepsilon_-} = \frac{(4\alpha_L + 5\alpha_T)\omega_{c1} + (8\alpha_L + \alpha_T)\alpha_T \omega_0}{(4\alpha_L + 5\alpha_T)\omega_{c1} - (8\alpha_L + \alpha_T)\alpha_T \omega_0} \tag{68}$$

It follows that the wave at the top of the band has elliptical polarization given by

$$\begin{pmatrix} \varepsilon_+ + \varepsilon_- \\ i(\varepsilon_+ - \varepsilon_-) \end{pmatrix} = \begin{pmatrix} (4\alpha_L + 5\alpha_T)\omega_{c1} \\ i(8\alpha_L + \alpha_T)\alpha_T \omega_0 \end{pmatrix} \tag{69}$$

A second band starts where $\varepsilon_+ = 0$ between ω_{c1} and ω_{c2}, i.e., at $\omega = \omega_6$. Its polarization is right circular. This band terminates, however, not at ω_{c2}, but where $\varepsilon_1 = 0$ or $\varepsilon_+ = \varepsilon_-$. This corresponds to linear polarization along the direction of propagation. The disturbance is thus a pure density oscillation.

The absorption resonance at this point is a true hybrid resonance. The cyclotron motion of each type of carrier creates a density oscillation; the resulting field acts on the other carrier, thus creating a coupled motion. The hybrid frequency is the frequency of their coupled motion.

INTERACTION WITH PHONONS

It was shown by Yokota ([5]) that the contribution from optical phonons to the dielectric constant is, to a reasonable approximation, provided $qa \ll 1$, given by

$$- \frac{\varepsilon_\infty \Omega_p{}^2}{\omega^2 - \omega_T{}^2} \tag{70}$$

where $\Omega_p{}^2$ is the square of the phonon plasma frequency and is equal to $4\pi N_0 e^{*2}/\mu\varepsilon_\infty$, with N_0 the number of atoms per unit volume, e^* the effective ionic charge, and μ the reduced mass of the two atoms in the cell. In (70) ω_T is the frequency associated with the restoring force of the atoms in the lattice. The result is obtained by a simple classical mechanical treatment of the coupled motion of the two coupled ion lattices.

Actually, (70) represents only the contribution to the lattice dielectric constant from the optical modes. There are further contributions from the acoustic modes; the whole lattice contribution to the dielectric constant is of the form

$$\varepsilon_\infty \Omega_p{}^2 (\omega^2 - f(qa)] \, \frac{1}{\omega_2{}^2 - \omega_1{}^2} \left[\frac{1}{\omega_2{}^2 - \omega^2} - \frac{1}{\omega_1{}^2 - \omega^2} \right] \tag{71}$$

Here $f(qa)$ is a function which becomes very small as qa (a being the lattice parameter) approaches zero and ω_1 represents the acoustic phonon frequency, which, at long wavelength, is given approximately by $\omega_1 \approx v_s q$, where v_s is the velocity of sound. In the same approximation

$$\omega_2 = (\lambda/\mu)^{1/2}$$

where λ is the force constant for displacement of an ion of the lattice and μ is the reduced mass defined above.

For $\omega^2 \gg \omega_1{}^2$ and $f(qa)$ Eq. (71) reduces to Eq. (70). For $\omega^2 \ll \omega_2{}^2$ it is approximately

$$\varepsilon_\infty \frac{\Omega_p{}^2}{\omega_2{}^2} \, \frac{\omega^2 - f(qa)}{\omega^2 - \omega_1{}^2} \tag{72}$$

Equation (72) together with the contribution from the electrons enable us to calculate the frequencies of the coupled system of "helicons" and transverse phonons, which was studied both experimentally and theoretically for potassium by Grimes and Buchsbaum ([6]). We shall not, however, concern ourselves further with this problem. We shall consider, rather, the frequency region in which $\omega \approx \omega_c \approx \omega_T$. We may then add the phonon contribution (70) to that arising from electron conductivity.

Modes with Polarization Along B₀

Let us first look at the situation in which there is only one contributing cyclotron mass and consider the interaction with phonons. The relevant dielectric constant for longitudinal modes is

$$\varepsilon_3 = \varepsilon_\infty \left\{ 1 - \frac{\Omega_p{}^2}{\omega^2 - \omega_T{}^2} - \frac{\omega_p{}^2}{\omega^2} \frac{\omega^2 - \omega_4{}^2}{\omega^2 - \omega_c{}^2} \right\} \tag{73}$$

Since the longitudinal optical phonon frequency is given by

$$\omega_L{}^2 = \omega_T{}^2 + \Omega_p{}^2 \tag{74}$$

this may be written

$$\varepsilon_3 = \varepsilon_\infty \left\{ \frac{\omega^2 - \omega_L{}^2}{\omega^2 - \omega_T{}^2} - \frac{\omega_p{}^2}{\omega^2} \frac{\omega^2 - \omega_4{}^2}{\omega^2 - \omega_c{}^2} \right\} \tag{75}$$

Now consider normal modes for propagation along the direction of the field. These are given by $\varepsilon_3 = 0$; thus we have

$$\omega^2(\omega^2 - \omega_c{}^2)(\omega^2 - \omega_L{}^2) - \omega_p{}^2(\omega^2 - \omega_T{}^2)(\omega^2 - \omega_4{}^2) = 0 \tag{76}$$

This is a cubic equation, and the solution is therefore not transparent. A plot of $\omega^2\varepsilon_3$, however, makes clear the general character of the modes.

The Case $\omega_4{}^2 < \omega_c{}^2 < \omega_T{}^2$

For this case the plot is as shown in Fig. 4. The diagram is schematic. In PbTe ω_L is very considerably larger than ω_T, a fact which is associated with the exceptionally large *static* dielectric constant. The highest mode is then given roughly by

$$\frac{\omega^2 - \omega_L{}^2}{\omega^2} - \frac{\omega_p{}^2}{\omega^2} = 0$$

or

$$\omega^2 = \omega_L{}^2 + \omega_p{}^2$$

This mode represents a coupled oscillation of electrons and ions.

The one of lowest frequency, which would be at ω_4 for ω_p infinitely large, is lowered somewhat by the lattice and plasma interactions.

The intermediate mode represents a coupled phonon–cyclotron motion.

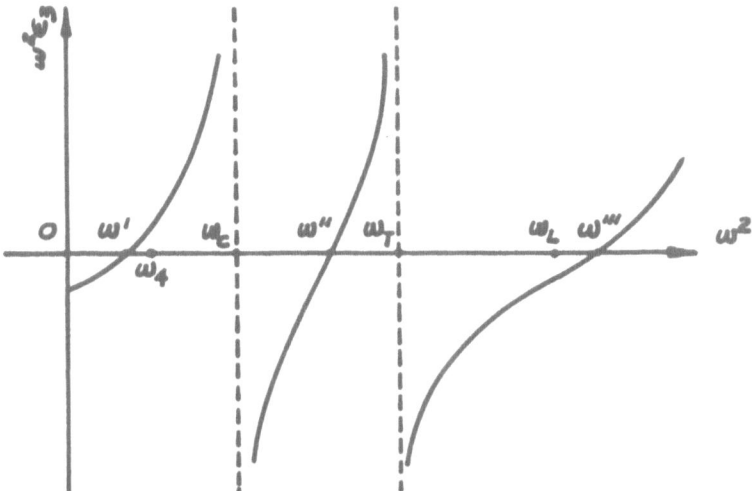

Fig. 4. Plot of $\omega^2 \varepsilon_3$ for normal modes for propagation along the direction of the field.

There would seem to be no simple way to detect these modes in the Faraday configuration. Suppose, however, we go to the Voigt configuration. In this case the dispersion relation becomes

$$c^2 q^2 = \omega^2 \varepsilon_3 = \varepsilon_\infty \left\{ \frac{\omega^2 - \omega_L{}^2}{\omega^2 - \omega_T{}^2} - \frac{\omega_p{}^2}{\omega^2} \frac{\omega^2 - \omega_4{}^2}{\omega^2 - \omega_c{}^2} \right\} \omega^2 \tag{77}$$

The same plot may now be used, and leads to the observation that transmission is possible for frequencies above ω' and terminated by the cyclotron resonance (a conclusion which would have to be modified, however, close to ω_c by virtue of the higher order terms in q^2); again in the frequency range ω'' to ω_T; and finally for frequencies above ω''. In these modes the electric field is polarized along the direction of the magnetic field, i.e. linearly polarized relative to the surface. They are therefore *transverse* waves, and are *not* accompanied by density oscillations. Related modes, appearing in the case of oblique incidence, would, however, be accompanied by density fluctuation.

There will, of course, be absorption resonances near ω_c and ω_T.

The Case $\omega_4 < \omega_T < \omega_c < \omega_L, \omega_p$

The situation is qualitatively the same as in the Fig. 4 except that the roles of ω_c and ω_T are now changed.

The evolution with increasing magnetic field of the normal modes for the Faraday configuration, or the bottom of the transmission bands for the Voigt configuration, is as follows*:

1. Until $\omega_4 = \omega_T$ (perhaps 16 kGs) the lowest band lies below ω_4. The band is terminated at the absorption resonance ω_c below about 10 kG; above that it is terminated at ω_T. For higher fields the transmission band remains below ω_T. For a range of frequencies below ω_T the reflectivity will be < 1.

2. The second transmission band in the Voigt configuration will begin *above* ω_4 for $\omega_4 > \omega_T$ and will terminate at the cyclotron resonance. After $\omega_4 = \omega_L$ ($\sim$64 kG) it will again start *below* ω_4, but will never be far from it.

3. The third resonance will be somewhat beyond $\omega_p[\sim(\omega_L{}^2 + \omega_p{}^2)^{1/2}]$, increasing slowly until $\omega_c \to \omega_p$, equality being reached at $\sim$120 kG for PbTe. Thereafter the transmission band appears beyond the cyclotron resonance, its lower limit being approximately $(\omega_c{}^2 + \omega_p{}^2)^{1/2}$.

Modes with Polarization Perpendicular to B_0

Though there is no important difference in the principles involved in the cases $\mathbf{E} \parallel \mathbf{B_0}$ and $\mathbf{E} \perp \mathbf{B_0}$, we shall now represent the situation in a slightly different way.

If we again consider the case of a single cyclotron resonance (i.e., propagation along the [100] direction) we may sketch diagrammatically ω^2 times that part of the dielectric constant associated with the free carriers, and which has the form

$$\varepsilon^{(1)} = -\frac{\varepsilon_\infty \omega_p{}^2(\omega + \omega_3)}{\omega(\omega^2 - \omega_c{}^2)} \tag{78}$$

[ω_3 being given by (41)]. This has the form shown in Fig. 2.

The contribution from the lattice (combined with that from electron interband transitions), namely,

$$\omega^2 \varepsilon^{(2)} = \varepsilon_\infty \omega^2 \frac{\omega_L{}^2 - \omega^2}{\omega_T{}^2 - \omega^2}$$

is shown in Fig. 5. This has now been plotted for $\omega < 0$ as well as for $\omega > 0$. Adding $\varepsilon^{(2)}$ to $\varepsilon^{(1)}$, we then have a representation of the whole dielectric constant for both r.c.p. and l.c.p. modes.

---

* We choose the values of the constants as follows: number of carriers: 10^{13}; $\varepsilon_\infty = 30$; $\omega_p = 5.5 \times 10^{13}$, $\varepsilon_0 = 480$; $\omega_L = 1.8 \times 10^{13}$; and $\omega_T = 4.5 \times 10^{12}$.

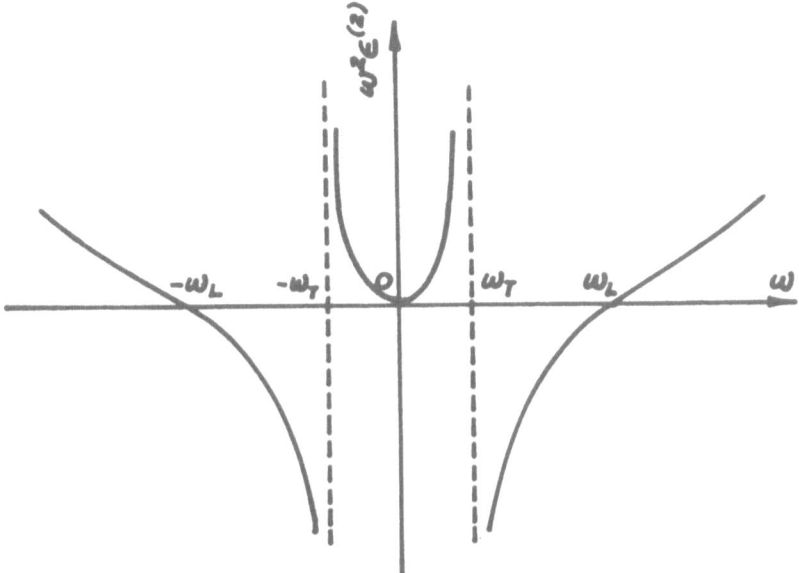

Fig. 5. Contribution of the lattice to the dielectric constant.

We note that as the magnetic field is changed there is simply an expansion of the horizontal scale of $\varepsilon^{(1)}$, while $\varepsilon^{(2)}$ stays fixed. This then makes it easy to see, schematically, how the critical frequencies evolve in the case of the Faraday configuration. For purposes of illustration, we sketch in Fig. 6 the situation for a low field (ω_c, $\omega_3 < \omega_T$) and in Fig. 7 for a higher

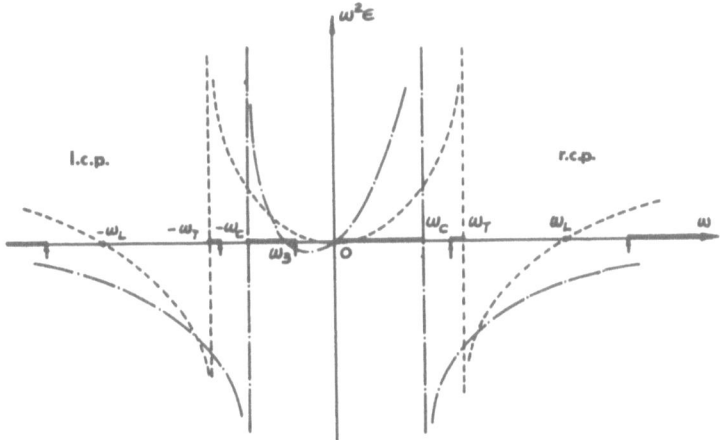

Fig. 6. Case of low magnetic field. The dotted line ----- gives the lattice contribution, the broken line —·—·—· that of the electrons.

field ($\omega_c > \omega_L > \omega_3 > \omega_T$). We superimpose the two components, and indicate by arrows the locations of the zeros of $\omega^2\varepsilon$. The absorption resonances are, of course, at the poles of the function. The scale is somewhat distorted for clarity of representation.

The curves in dotted lines represent the LO phonon contribution, while those in broken lines represent the electron contribution, r.c.p. for $\omega > 0$ and l.c.p. for $\omega < 0$. Those parts of the ω axis corresponding to frequencies for which propagating modes exist are darkened. We see first that for r.c.p. there is transmission in the helicon region, terminating below ω_c, and also below the reststrahl frequency ω_T. For l.c.p. the frequency at which transmission sets in is lowered somewhat due to the fact that the lattice has a "depolarizing" effect, i.e. does not distinguish the two polarizations. The other regions of transmission are not very different than in the r.c.p. case.

In the case represented in Fig. 7 the lowest r.c.p. mode is terminated (or rather, interrupted), not at the cyclotron frequency but at the reststrahl frequency ω_T. Transmission sets in again somewhat above ω_T, this time terminated by the cyclotron frequency. The final region of transmission is above ω_p.

In the l.c.p. case there is again a region of transmission just below ω_T. A second transmission band appears a bit beyond ω_3 and terminates at ω_c. A third, as in the r.c.p. case, lies near ω_p. The "mobility" of the electron contributions with magnetic field, coupled with the stationary character of the lattice contributions, makes possible the determination of the lattice

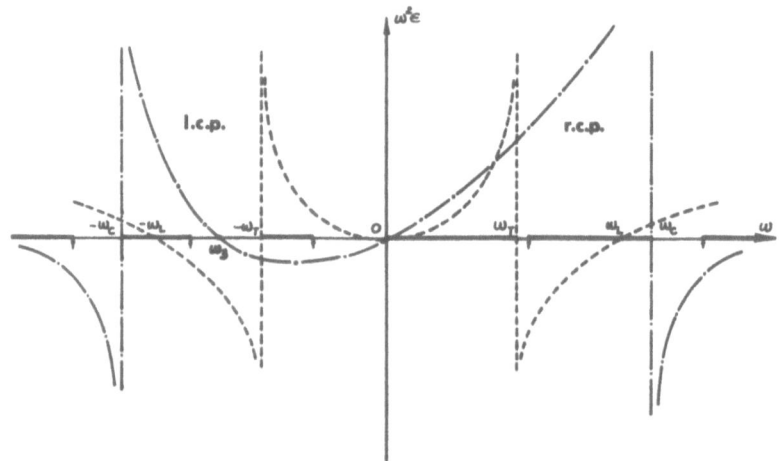

Fig. 7. Case of high magnetic field.

frequencies ω_T and ω_L. Since these are related by the Lyddane–Sachs–Teller relation

$$\omega_L{}^2/\omega_T{}^2 = \varepsilon_0/\varepsilon_\infty$$

it is then a straightforward task to deduce the ratio of the static and "high-frequency" dielectric constants.

Finally, let us consider the Voigt configuration. The normal mode condition is still given, for polarization along the magnetic field by

$$c^2q^2 = \omega^2\varepsilon_3$$

and for polarization *around* the magnetic field by

$$c^2q^2 = \omega^2\varepsilon' = \omega^2(\varepsilon_+\varepsilon_-/\varepsilon_1)$$

The zeros of ε_+ and ε_- have already been discussed, and these again represent the lower ends of transmission bands. One thing is changed, however, and that is the location of the resonances—in particular, those corresponding to zeros of ε_1 (collective cyclotron resonance). It is easily seen that

$$\varepsilon_1 = \varepsilon_\infty\left[\frac{\omega_L{}^2 - \omega^2}{\omega_T{}^2 - \omega^2} + \frac{\omega_p{}^2}{\omega_c{}^2 - \omega^2}\right] \tag{79}$$

The situation is sketched in Fig. 8 for $\omega_T < \omega_c < \omega_L$. The lattice contribution is given by the dotted line, the electron one by the broken line.

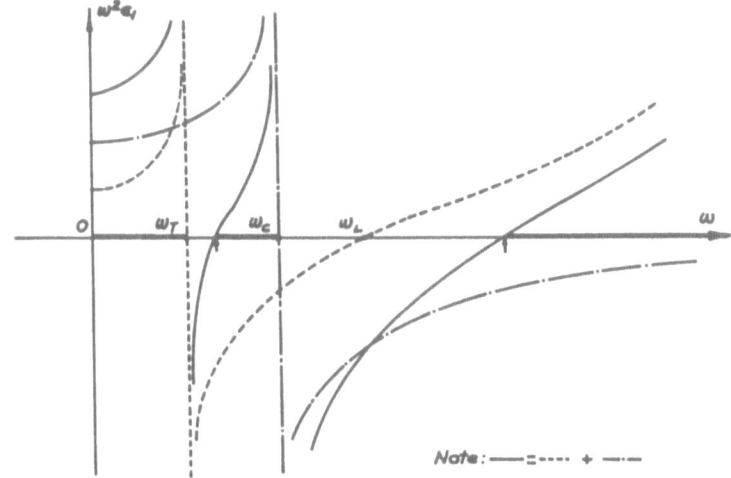

Fig. 8. The Voigt configuration.

Their sum is the full curve. The lower resonance is now displaced *below* ω_c, while the upper one may be located roughly as follows: the equation for the zeros is

$$(\omega^2 - \omega_L{}^2)(\omega^2 - \omega_c{}^2) - \omega_p{}^2(\omega^2 - \omega_T{}^2) = 0 \tag{80}$$

(this would of course be further modified by relaxation effects). Thus the sum of the roots is $\omega_L{}^2 + \omega_p{}^2 + \omega_c{}^2$. But since one root of $\omega^2 < \omega_c{}^2$, the other must be $> \omega_L{}^2 + \omega_p{}^2$. The larger the value of ω_p, the closer is the lower root to $\omega_c{}^2$ and the upper to

$$\omega^2 = \omega_L{}^2 + \omega_p{}^2 \tag{81}$$

MAGNETIC FIELD IN THE [111] DIRECTION: VOIGT CONFIGURATION

Although it would be exhausting to delineate all of the corresponding effects in the two-carrier case (i.e., magnetic field along the [111] direction), two particular phenomena are of interest. One is the influence of phonon interactions on the hybrid resonance; the other is the extension of the concept of collective cyclotron resonance (CCR) to multiple collective cyclotron resonance (MCCR). Since both relate to the zeros of ε_1, we need only look at this quantity. The hybrid resonance referred to earlier in the case of field along [111] is merely the lower of the two modes referred to above. For the two-carrier case it may be shown on extracting the electron contribution from (60) that

$$\varepsilon_1 = \varepsilon_\infty\left[\frac{\omega_L{}^2 - \omega^2}{\omega_T{}^2 - \omega^2} - \omega_p{}^2\left(\frac{1/4}{\omega^2 - \omega_{c2}} + \frac{1}{12}\frac{4\alpha_L + 5\alpha_T}{\omega^2 - \omega_{c1}^2}\right)\right] \tag{82}$$

For the case $\omega_{c1}, \omega_{c2} > \omega_T$, perusal of (82) indicates that one zero (resonance) lies between ω_T and ω_{c1}, a second between ω_{c1} and ω_{c2} (the electron hybrid resonance), and a third in the plasma region. For large ω_p the lowest lies close to ω_T, and therefore is primarily a phonon resonance. Because of larger coefficient of the $1/(\omega^2 - \omega_{c1}^2)$ term, the hybrid will be closer to ω_{c1} than to ω_{c2}. The highest one will lie, as before, beyond $(\omega_p{}^2 + \omega_L{}^2)^{1/2}$. All are, however, in a sense hybrid resonances, since they involve coupling of the two cyclotron motions, the lattice and the longitudinal electron motion.

OPTICAL EFFECTS

All of the critical frequencies mentioned may be observed by reflection or transmission at normal incidence in the Faraday and Voigt configurations. The reflection coefficient from a thick specimen is known from classical optics to be

$$R = \left| \frac{1 - (cq/\omega)}{1 + (cq/\omega)} \right|^2 = \left| \frac{1 - \sqrt{\varepsilon}}{1 + \sqrt{\varepsilon}} \right|^2 \tag{83}$$

where ε is the appropriate dielectric constant. This applies for complex as well as real values. Consequently, in regions of nonpropagation $R = 1$. Reflection minima come from $\varepsilon = 1$ or, in the complex case, $|1 - \sqrt{\varepsilon}|$ a minimum. Alternatively, the condition is $cq/\omega \approx 1$. Resonances are characterized by a sudden rise of R to 1. The minimum of the reflectivity will in general come near the lower end of a transmission band for materials of high dielectric constant.

Studies of the Faraday and Voigt effects should be particularly instructive, since they should show rapid variation near the onset of transmission bands. Thus for propagation in the [100] direction, for instance, the effect should be more or less normal up to ω_3, but behave quite differently when both polarizations can be propagated (beyond ω_3). For thick samples the material will act as a polarization filter in regions where only one polarization will propagate, but not in those in which both polarizations can be transmitted. The phonon coupling will tend to diminish the Faraday effect, since it tends to reduce the discrimination of the material between the two polarizations.

It is also worth noting that since penetration or transmission is largest near the bottom of a transition band, in the frequency range just above the bottom of an l.c.p. transmission band the left polarized mode will dominate the r.c.p. one. As a consequence, there will be a region of reversal in the Faraday effect. The transmitted wave will pass from right elliptical polarization to left, passing through linear polarization above but close to the frequency ω_3.

REFERENCES

1. A. Libchaber and R. Veilex, *Phys. Rev.* **127**: 774 (1963).
2. E. Astrøm, *Arkiv Fysik* **2**: 443 (1950).
3. V. Celli and M. D. Mermin, *Ann. Phys.* **30**: 249 (1964).
4. J. J. Quinn and S. Rodriguez, *Phys. Rev.* **128**: 2487 (1962).
5. I. Yokota, in: "Proceedings of the International Conference on Semiconductor Physics, Kyoto, 1966", *J. Phys. Soc. Japan Suppl.* **21**: 738 (1966).
6. CC. Grimes and S. J. Buchsbaum, Phys. Rev. Letters **12**: 357 (1964).

Chapter 3

EFFECTS OF BAND POPULATION ON THE MAGNETOOPTICAL PROPERTIERS OF THE LEAD SALTS

E. D. Palik and D. L. Mitchell

Naval Research Laboratory
Washington, D. C.

INTRODUCTION

General

Recent experimental and theoretical studies of the lead salts have developed the major features of the one-electron band structure. The valence and conduction band extrema are located at the L point of the Brillouin zone and have only the twofold Kramers degeneracy required by time-reversal symmetry. There are four equivalent valence bands and four equivalent conduction bands. For each compound the effective masses in the conduction and valence bands are nearly equal. The bands are somewhat anisotropic and nonparabolic. (The energy ellipsoids, oriented along the [111] directions, are illustrated in Fig. 1.) These features of the band structure make the lead salts very attractive for studying the dynamical effects due to free carriers and particularly the interband effects, since the complications arising from a complex valence or conduction band are avoided.

Several kinds of experiment have been used to study the lead salts—the transport and the optical experiments.* Since ordinary cyclotron resonance has not been observed in these materials, other techniques have had to be used. Among these are piezoresistance ([2,3]), the de Haas–van Alphen effect ([4]), magnetoresistance ([5,6]) and the Shubikov–de Haas effect ([7-10]), Azbel'–Kaner cyclotron resonance ([11-14]), dielectric anomaly phenomena associated with magnetoplasma effects ([15]), interband magnetoabsorption ([16-19]), interband and free-carrier Faraday rotation ([18,20-25]), and several

* For a review of work prior to 1959 see ([1]).

90

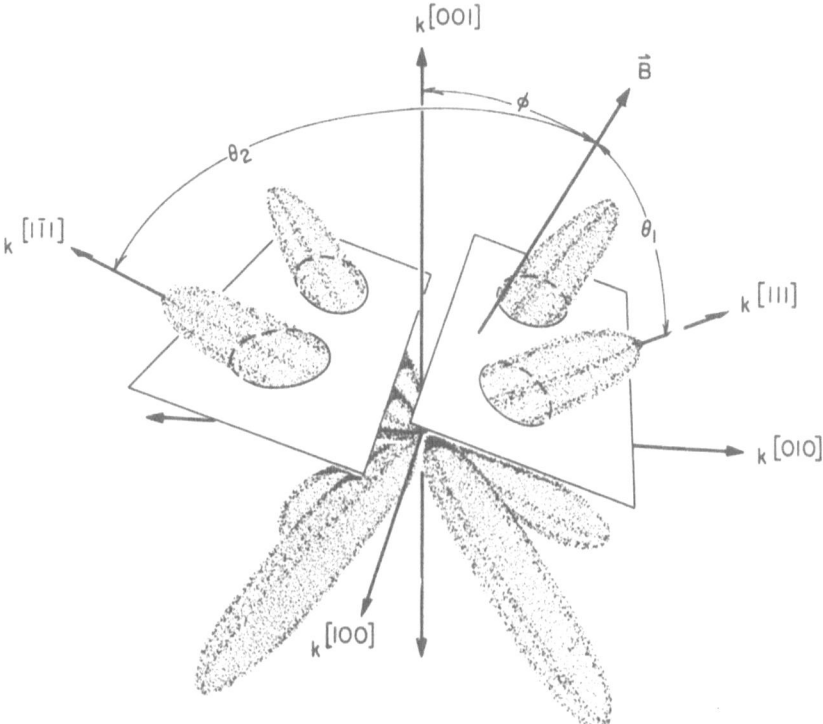

Fig. 1. Energy ellipsoids of revolution located along [111] directions in **k** space. The cyclotron resonance frequency for a given band depends on the average effective mass for motion perpendicular to the magnetic field. The effective mass m_i^θ for a given cross section is given by $(1/m_i^\theta)^2 = (1/m_i^T)^2 [\cos^2 \theta + (1/K_i) \sin^2 \theta]$, where θ is the angle between the ellipsoid axis and magnetic field, m_i^T is the transverse effective mass, and K_i is the ratio m_i^L/m_i^T for the ith band.

zero-field optical measurements at the band edge ([1,26-31]) and in the region of plasma reflection ([20,32-35]). The transport experiments have determined the number of equivalent bands and their symmetry and information about the effective masses, while the optical-type experiments have yielded some detail about the band gap, effective masses, and g-factors. Several theoretical calculations ([36-41]) have accounted for the major features of the band structure.

Optical Effects

Focusing attention on the optical and magnetooptical aspects, we show in Fig. 2 a composite picture of the reflection and transmission of two samples of PbS, one with no free carriers and the other with $N = 6 \times 10^{18}$

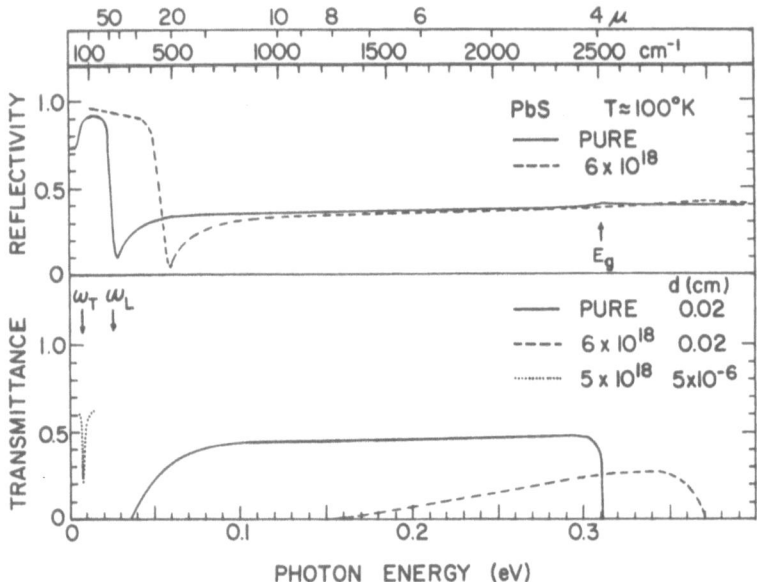

Fig. 2. Composite reflection and transmission spectra of two PbS samples illustrating dominant optical properties.

carriers/cm³. Several interesting features are evident. For pure PbS in reflection the prominent feature is the reststrahlen reflection in the far infrared. Analysis of reflectivity on the basis of a single classical oscillator [42] has yielded the transverse and longitudinal optical phonon frequencies, ω_T and ω_L, and the high- and low-frequency dielectric constants ε_∞ and ε_0. The transverse lattice vibration frequency ω_T has also been observed as an absorption line by transmission measurements in thin films [43,44]. Another feature of the reflection spectrum is the small peak near the optical band gap E_g [27]. The transmission spectrum shows the strong band edge absorption above E_g in the near infrared, transmission at intermediate wavelengths, and then absorption in the far infrared due to the tail of the strong lattice absorption centered at ω_T.

The addition of free carriers alters the optical behavior, as shown for $N = 6 \times 10^{18}$ carriers/cm³. A plasma reflection edge develops at $\omega_p{}^2 \approx 4\pi Ne^2/m_o{}^*\varepsilon_\infty$. This corresponds to the frequency at which the real part of the total dielectric constant goes to zero. A similar reflection edge appears at low energy below ω_T at another zero of the dielectric constant. In another description of the effect of the added free carriers longitudinal plasmon–phonon coupled modes occur near the two zeros in the dielectric constant ([45]). In Fig. 2 the carrier concentration N has been chosen sufficiently high so that the above approximation for ω_p is valid. In the case for carrier concentrations in the range 10^{17}–10^{18} carriers/cm³ the plasma frequency falls in a region near ω_T or ω_L, and the full frequency-dependent equations must be solved. For the case $N \ll 10^{17}$ cm⁻³ ω_p may be calculated by replacing ε_∞ by ε_0. In the transmission spectrum strong free-carrier absorption occurs at intermediate frequencies. The band edge absorption also shifts to higher frequency due to the Burstein–Moss effect ([46,47]), the filling of states in the conduction band up to the Fermi level, and consequent blockage of interband transitions. Interestingly, a sample with carriers may now be somewhat transparent just inside the band edge, whereas the pure sample is strongly absorbing in this region. Similar optical effects occur in PbSe and PbTe.

BAND POPULATION EFFECTS—PHENOMENOLOGICAL

Burstein–Moss Effect

The Burstein–Moss effect in the absorption edge has been studied in PbS at low temperatures ([18]). In pure material for simple, spherical, and parabolic bands the frequency-dependent interband absorption constant $\alpha_0(\omega)$ for direct, allowed transitions is given by

$$n_0(\omega)\alpha_0(\omega) = C(\hbar\omega - E_g)^{1/2}/\hbar\omega \tag{1}$$

where $\hbar\omega$ is the photon energy, $n_0(\omega)$ is the frequency-dependent index of refraction, and

$$C = \frac{2e^2 M}{m_0{}^2 c}\left(\frac{2m_{cv}^*}{\hbar^2}\right)^{3/2} |P_{cv}|^2 \tag{2}$$

is a constant factor characteristic of the particular band structure and is directly proportional to the transition probability through the momentum matrix element $|P_{cv}|^2$ and the valence and conduction band density of states through the reduced effective mass $(m_{cv}^*)^{-1} = (m_c{}^*)^{-1} + (m_v{}^*)^{-1}$. In Eq. (2) M is the number of equivalent bands, m_0 is the free-electron mass, and $m_c{}^*$ and $m_v{}^*$ are the conduction and valence band effective masses, respectively. The interband transitions are indicated in Fig. 3a.

When free carriers are added to the conduction band some interband transitions will be blocked, and the interband absorption constant $\alpha_N(\omega)$ may now be written ([18])

$$n_N(\omega)\alpha_N(\omega) = n_0(\omega)\alpha_0(\omega)\{1 + \exp[(E_F - E)/kT]\}^{-1} \tag{3}$$

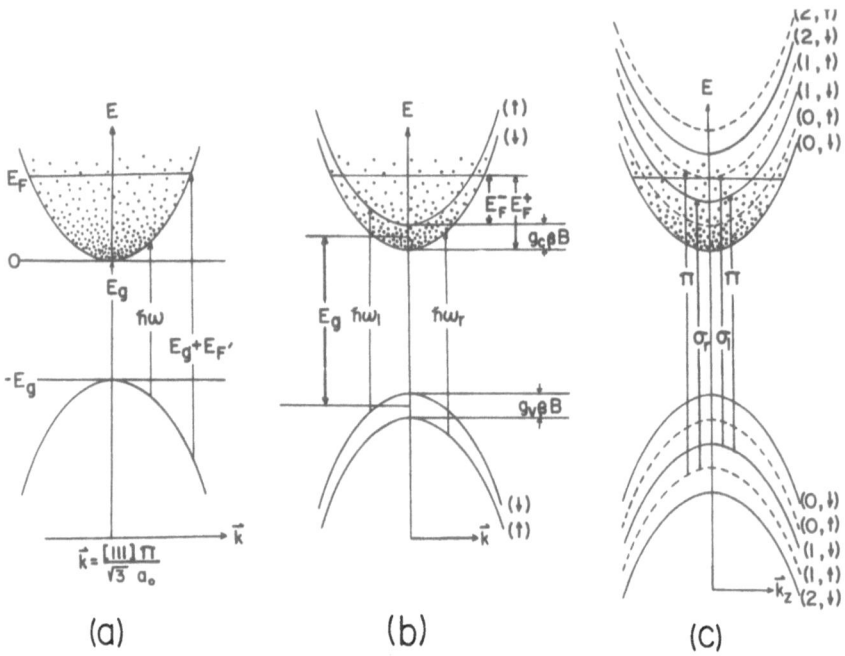

Fig. 3. (a) Simple parabolic valence and conduction band model with free carriers in the conduction bands. The stippling represents the occupation of conduction band states by free carriers with Fermi level E_F. (b) Rigid band model showing gyromagnetic splitting of the valence and conduction bands. Typical r.c.p. and l.c.p. interband transitions are shown. (c) Landau energy level diagram for quantum band model. The level ordering, determined by the gyromagnetic splitting, is specifically for PbS.

where at $0\,^\circ K$ $E_F = (h^2/2m_c^*)(3N/8\pi M)^{2/3}$ is the Fermi energy for a total carrier concentration N distributed in M equivalent bands, and E is the energy in the conduction band measured from the band edge. Since $\hbar\omega - E_g = (1 + m_c^*/m_v^*)E$, Eq. (3) may be rewritten as

$$n_N \alpha_N = n_0 \alpha_0 \left\{ 1 + \exp\left[\frac{E_F' - (\hbar\omega - E_g)}{\{1 + (m_c^*/m_v^*)\}kT} \right] \right\}^{-1} \qquad (4)$$

where $E_F' = [1 + (m_c^*/m_v^*)]E_F$ is the photon energy, relative to the band gap, at which the absorption is just half the value for the pure crystal.

In practice, the index of refraction in the interband region does not vary much with frequency or with the number of added free carriers. At low temperature and high carrier concentration Eq. (4) has been used to determine that in PbS there are four equivalent bands. Figure 4 shows the interband absorption constant of PbS at $77\,^\circ K$ for $N = 1.1 \times 10^{18}$

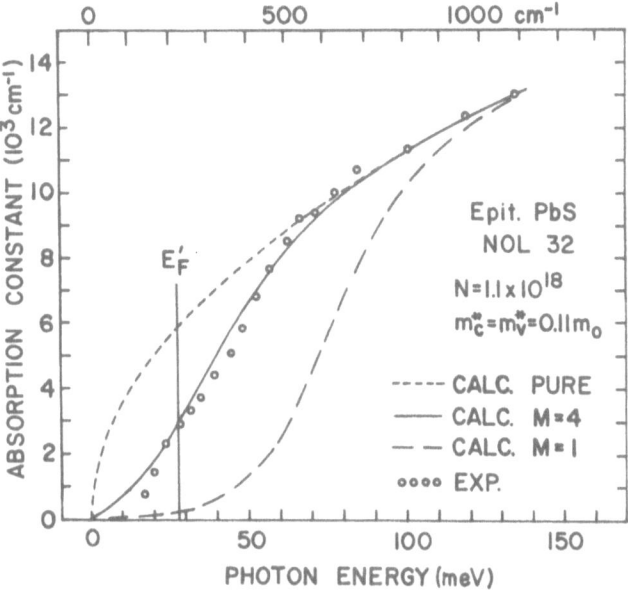

Fig. 4. The measured optical absorption edge in an epitaxial PbS sample at 77°K showing the Burstein–Moss shift. Photon energy is measured with respect to the strained band gap at 0.28 eV. Calculated curves for the absorption constant are shown with the assumption that the free carriers are in one parabolic and spherical band or in four equivalent bands located at the L points in the Brillouin zone.

cm^{-3}. Photon energies are measured relative to the energy gap, which at 77 °K for unstrained samples is found to be 0.31 eV. The pure crystal curve was calculated by fitting Eq. (1) to the known band gap E_g and the absorption constant at 0.12 eV inside the gap, where no Burstein–Moss effect was evident. The shape of the calculated absorption edge is in reasonable agreement with the shape measured at room temperature [1,26,28]. The Burstein–Moss edge was fitted with Eq. (4) with the assumption that all the carriers were in four bands. Also shown is a calculation placing all the carriers into one band. Similar measurements have been made in PbS for carrier concentrations ranging to above 10^{19} cm^{-3} [31]. Equation (4) provided a good fit to the band edge data up to 10^{19} carriers/cm^3. At higher concentrations, where there may be Coulomb effects with band tailing, Eq. (4) did not fit the data. The Burstein–Moss edge is not nearly so clear in room-temperature data, since the Fermi distribution is smeared out and the Fermi level is shifted downward, the temperature-dependent Fermi level being given by the usual analysis [48].

Interband Magnetoabsorption

Interband magnetoabsorption has also been studied in PbS ([16–18]). The conduction and valence Landau levels are given by

$$E_c = (l' + \tfrac{1}{2})\hbar\omega_c + m'g_c\beta B + (\hbar^2 k_z^2/2m_c^*)$$
$$E_v = -E_g - (l + \tfrac{1}{2})\hbar\omega_v + mg_v\beta B - (\hbar^2 k_z^2/2m_v^*)$$

$$(5)$$

where l' and l and the Landau quantum numbers for the subbands, $\omega_{c,v} = eB/m_{c,v}^* c$ are the cyclotron frequencies in the conduction and valence bands, m_c^* and m_v^* are defined as positive effective masses, $g_{c,v}$ are the algebraic effective g-factors for the conduction or valence electron, and β is the Bohr magneton $e\hbar/2m_0 c$ times Planck's constant. The quantum numbers m' and m are the "effective spin" quantum numbers with values $(\pm\tfrac{1}{2})$. The spin itself is not a good quantum number, since the large spin-orbit interaction introduces mixed-spin states for the valence and conduction bands. However, the two bands have only the twofold Kramers degeneracy and transform around symmetry axes like states with total angular momentum $J_{1/2}$, so that they may be treated the same as simple spin-degenerate bands.

The energy levels are shown in Fig. 3c for a fixed magnetic field and in Fig. 5 as a function of magnetic field at $k_z = 0$.

The energy separation for an interband transition at $k_z = 0$ is given by

$$E_{c,v} = E_g + (l' + \tfrac{1}{2})\hbar\omega_c + (l + \tfrac{1}{2})\hbar\omega_v + m'g_c\beta B - mg_v\beta B \qquad (6)$$

In the Faraday configuration (direction of propagation of radiation parallel to applied magnetic field) the selection rules for σ transitions require that $\Delta l = 0$, $\Delta m = \pm 1$ for left (l.c.p.) and right (r.c.p.) circularly polarized radiation, respectively. In the Voigt configuration (direction of propagation perpendicular to magnetic field) the selection rules for the π transitions require $\Delta l = 0$, $\Delta m = 0$, while for σ transitions they remain the same as above. These selection rules apply for a spherical band or a band with rotational symmetry around the direction of the magnetic field. Nonparabolicity of the bands will allow other weak transitions to occur for $\Delta l \neq 0$, and anisotropy will mix the left and right circular selection rules for a given transition giving rise to elliptically polarized transitions. In PbS and PbSe the bands are nearly spherical, so that for a given l there are two σ transitions in both the Faraday and Voigt configurations. The energy separation of these two lines is $|g_c + g_v|\beta B$. In the π spectrum the two π lines are separated by $|g_c - g_v|\beta B$.

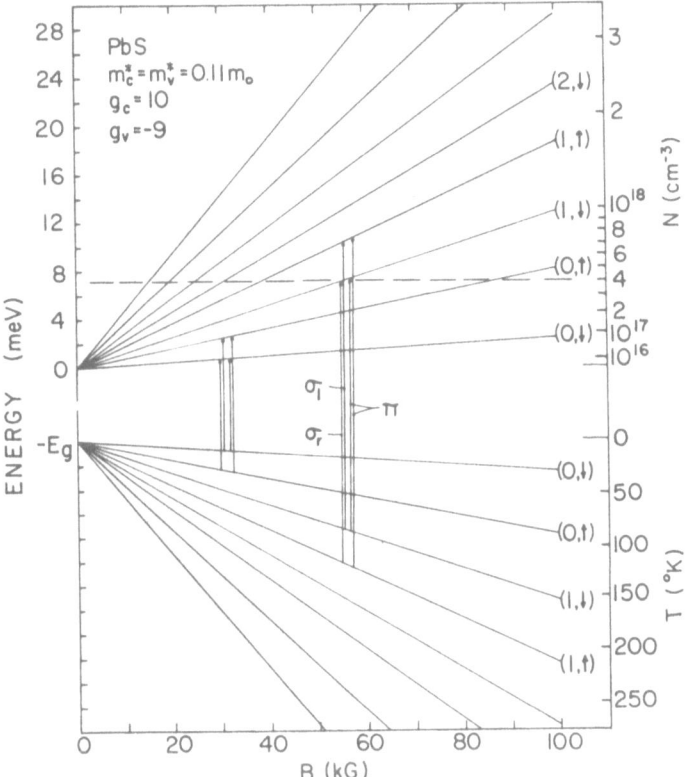

Fig. 5. Landau level dependence on magnetic field at $k_z = 0$. Typical σ and π transitions are shown. The position of the Fermi level is shown on the upper right-hand scale by indicating the carrier concentration. The dashed line is the Fermi level for 4×10^{17} carriers/cm³. Thermal energy kT may be obtained from the lower right-hand temperature scale.

The experimental σ and π spectra for the Voigt orientation are shown in Fig. 6. The σ spectrum consisted of lines evenly spaced in photon energy at fixed magnetic field, while the π spectrum consisted of double lines. The data are plotted in a fan chart in Fig. 7. The origin of the lines is shown in Fig. 3c. Since the σ spectra show no splitting experimentally, the g-factors are assumed to be nearly equal in magnitude and opposite in sign. This is subsequently verified by the sign and magnitude of the interband rotation in pure PbS. The splitting of the π lines then yields the magnitude of the g-factors. Closer examination of the σ spectrum in the Faraday configuration, in which σ(l.c.p.) and σ(r.c.p.) can be measured individually,

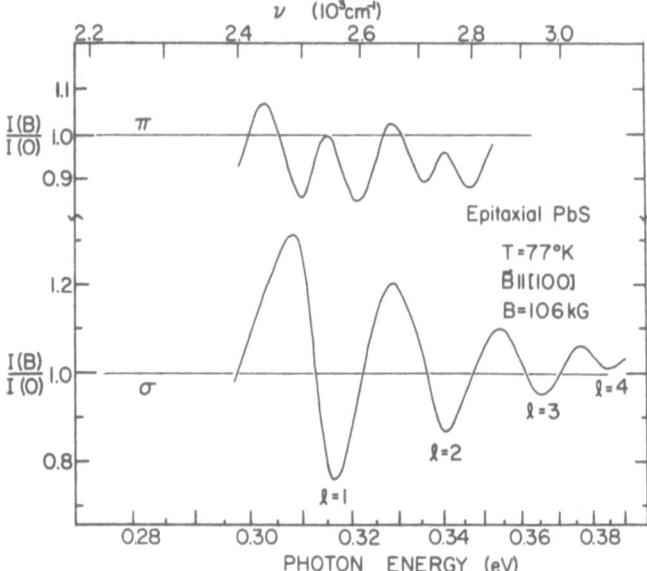

Fig. 6. Interband magnetooptical absorption lines for both σ and π transitions in the Voigt sample configuration. The PbS sample contained 3×10^{17} carriers/cm³.

gave a slight separation of the lines from which the g-factors were determined to be $g_v = -8.5$ and $g_c = 10.0$ at 77 °K. The reduced effective mass obtained from the line spacings in Fig. 7 was $m_{cv}^* = 0.0508 \, m_0$. The strained energy gap was found to be 0.28 eV, while the unstrained gap is 0.31 eV, at 77 °K. The two $l = 0$ lines were not clearly seen in these experiments, as these transitions were blocked by the free carriers, and the lines were weakened considerably. Interestingly, these lowest transitions are the strongest ones observed in recombination radiation experiments [49] and in lead salts laser experiments [50] done in a magnetic field, so that these experiments complement the interband magnetoabsorption. In Fig. 5 the position of the Fermi level is also noted as a function of carrier concentration, giving an idea of what transitions will be blocked.

Faraday Rotation

The next magnetooptical experiment to be discussed is Faraday rotation. A collection of room-temperature rotation data for PbS is shown in Fig. 8 [20,21]. At longer wavelengths the rotation θ_{FC} is linear in λ^2, the slope of the line giving the free-carrier effective mass. A Drude model

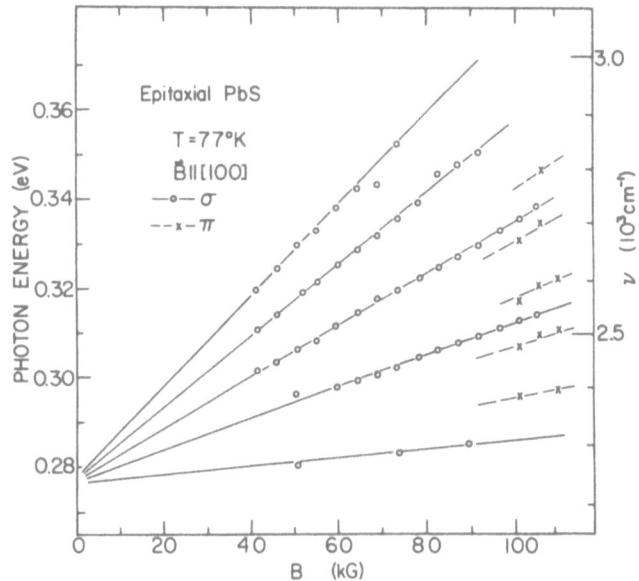

Fig. 7. The photon energy positions of the Landau transition are plotted versus magnetic field. The value for the energy gap obtained from extrapolation of the lines to zero field, and the value of the reduced effective mass obtained from the line spacing, are characteristic of the strained crystal.

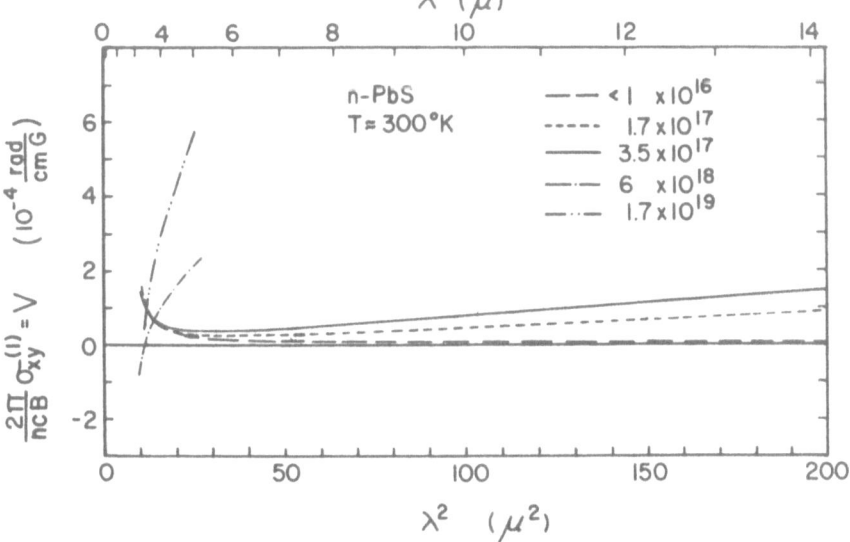

Fig. 8. Room-temperature interband and free-carrier Faraday rotation in PbS is shown for several carrier concentrations. Note the reversal of interband rotation at short wavelength.

based on a spherical effective mass ([51]) gives the rotation θ_{FC} as

$$n(\lambda)\theta_{FC} = -\frac{e^3 dN\lambda^2 B}{2\pi c^4 m_c^{*2}} \tag{7}$$

where d is the sample thickness, λ is the wavelength, and $n(\lambda)$ is the zero-field index of refraction at the wavelength λ. The effective mass is found to be about $0.17\, m_0$ in $10^{18}\, cm^{-3}$ n-type PbS. At shorter wavelengths the interband rotation reverses sign as the carrier concentration is increased. In Fig. 9 we show similar results as a function of temperature for a single

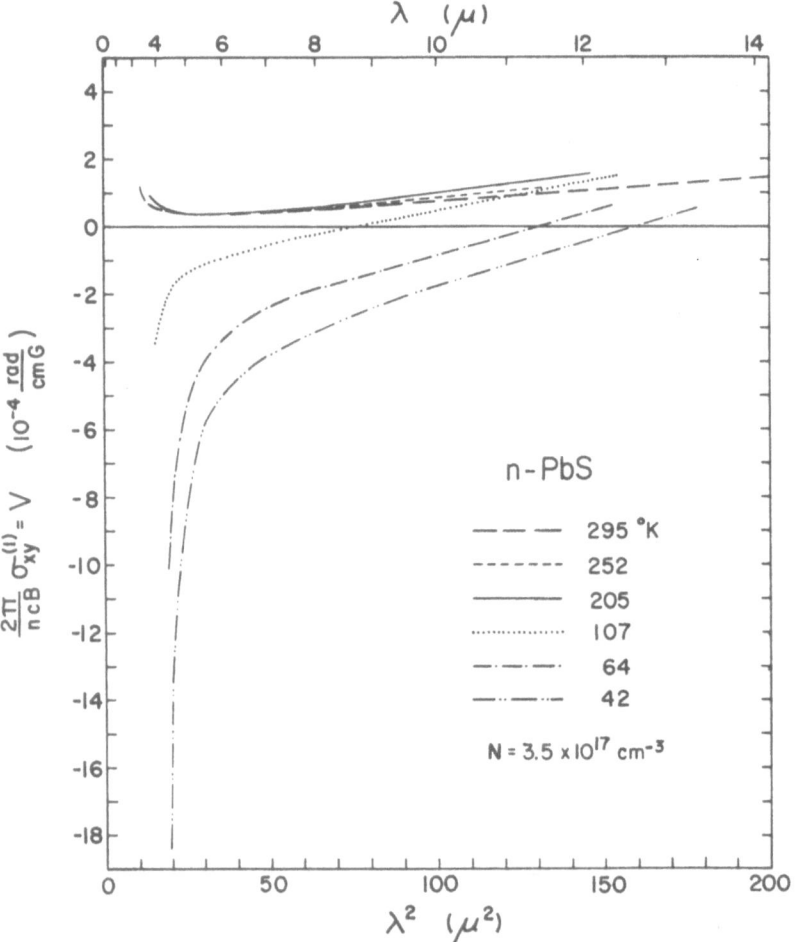

Fig. 9. Faraday rotation in a PbS sample at several temperatures. Note the reversal of interband rotation at short wavelengths.

sample of PbS with $N = 3.5 \times 10^{17}$ cm^{-3}. Again the rotation reverses sign as the temperature is lowered. The slope of the straight-line portion of the curve yields an effective mass of 0.11 m_0. However, the straight-line free-carrier contribution to the rotation does not extrapolate through the origin when $\lambda^2 \to 0$, as would be indicated by Eq. (7), which implies that there is a persistent residue or very slowly decreasing tail to the interband rotation as $\lambda \to \infty$, i.e., $\omega = 2\pi c/\lambda \to 0$. We will dwell on this aspect below.

Studies of the optical plasma reflection in PbS ([20,33]) indicate similar values for the effective masses and their temperature variation, but also imply some nonparabolicity of the conduction band, since the effective mass was noted to increase slightly with increasing concentration.

Derivation

We now concentrate on the interband Faraday rotation below the energy gap to explain the reversal in the sign of rotation with carriers and temperature. We first calculate the Burstein–Moss effect for magnetic fields small enough to satisfy the condition $\omega_c \tau < 1$. In this low-field regime Landau levels are not distinctly formed, and oscillatory effects due to Landau quantization are absent. In this case we consider two rigid conduction bands separated by the gyromagnetic ratio splitting $g_c \beta B$ and a similar separation of two valence bands by $g_v \beta B$ as shown in Fig. 3b. In this model the averaged magnetoabsorption above the band edge is the same as the averaged magnetoabsorption for a model with well-defined Landau levels ([52]). Assuming that the Fermi level stays constant with respect to the bottom of the zero-field conduction band, we can determine $\alpha_{r,l}$ absorption constants for the magneto-Burstein–Moss edge as follows.

Referring to Fig. 3b for a pure material, in analogy to Eq. (1) we may write

$$n_{0r,l}\alpha_{0r,l} = C'[\hbar\omega - E_g \mp \tfrac{1}{2}\hbar\gamma_{cv}\beta B]^{1/2}/\hbar\omega \qquad (8)$$

where $\gamma_{cv} = g_c - g_v$ and C' now contains a term $|P_{cv}^{r,l}|^2$. For this derivation the magnitudes of g_c and g_v are used, since the model in Fig. 3b is specifically for PbS with the signs already accounted for by the order of the bands. When free carriers are added we may write

$$n_{Nr,l}\alpha_{Nr,l} = n_{0r,l}\alpha_{0r,l}F_{\pm} = n_{0r,l}\alpha_{0r,l}[1 + \exp(E_F - E_{\pm})/kT]^{-1} \qquad (9)$$

where $F_{\pm}$ is the Fermi occupation function and $E_{\pm}$ is the energy in the conduction band measured from the bottom of the zero-field band as shown in Fig. 3b. Here $+$ goes with r and $-$ with l. Since

$$\hbar\omega_{r,l} = E_g \pm \frac{\hbar\gamma_{cv}\beta B}{2} + \frac{\hbar^2 k^2}{2}\left(\frac{1}{m_v^*} + \frac{1}{m_c^*}\right) \tag{10}$$

and

$$E_\pm = \frac{\hbar^2 k^2}{2m_c^*} \mp \frac{\hbar g_c \beta B}{2} \tag{11}$$

it follows that

$$E_\pm = \frac{\hbar\omega_{r,l} - E_g \pm \Gamma}{[1 + (m_c^*/m_v^*)]} \tag{12}$$

where

$$\Gamma = [g_v + (m_c^*/m_v^*)g_c]\beta B/2$$

Equation (9) represents the band edge absorption in a magnetic field when free carriers are present. Some typical calculated results are shown in Fig. 10a and 10b. Figure 10a shows the magneto-Burstein–Moss absorption edges with $N = 3.5 \times 10^{17}\,\text{cm}^{-3}$ at various temperatures. At room temperature the α_r and α_l curves are only slightly separated, while at 8 °K the l.c.p. and r.c.p. absorption edges are sharper and more widely separated

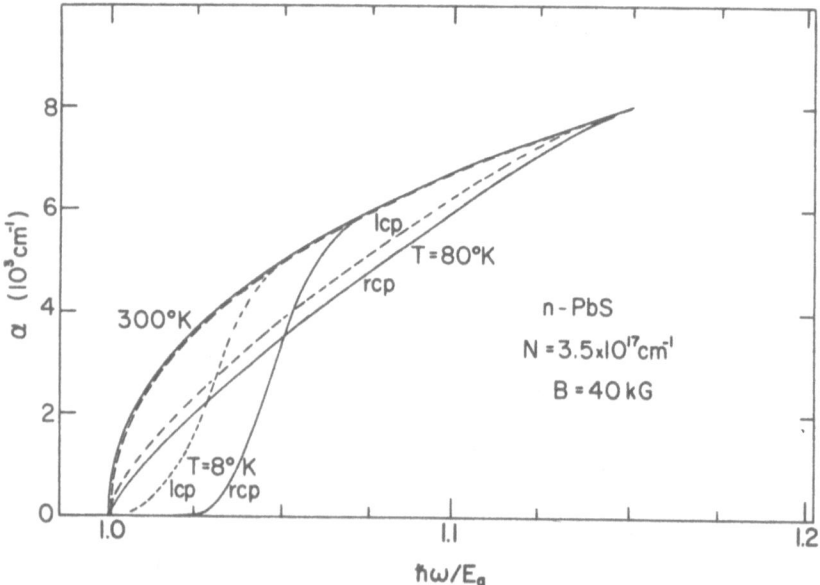

Fig. 10a. Band gap absorption in a magnetic field, characterized by $\alpha_{r,l}$, calculated for a single sample of PbS with $N = 3.5 \times 10^{17}\,\text{cm}^{-3}$ at three temperatures.

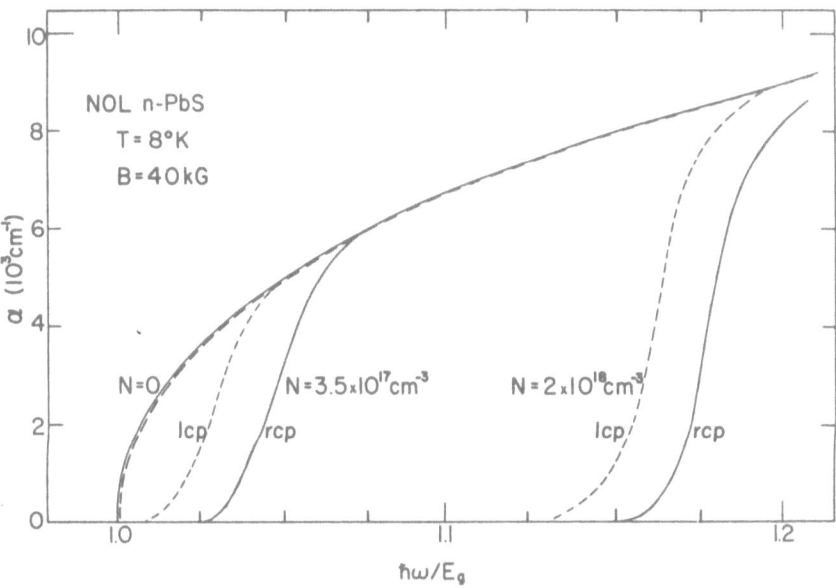

Fig. 10*b*. Band gap absorption $\alpha_{r,l}$ calculated at 8°K for three carrier concentrations.

and reversed in order. Note that the abscissa is normalized to E_g, and that although the calculation is for $B = 40$ kG to emphasize the separation of the r.c.p. and l.c.p. absorption edges, the absorption is plotted to start at E_g. In Fig. 10*b* the magnetoabsorption edges are shown at 8°K as a function of carrier concentration.

It follows at $\omega_r = \omega_l = \omega$ that

$$n_r\alpha_r - n_l\alpha_l = n_{0r}\alpha_{0r}F_+ - n_{0l}\alpha_{0l}F_-$$

$$= \frac{C}{\hbar\omega}\left[\frac{x + \sqrt{\varepsilon}}{1 + \exp(y + \delta)} - \frac{x - \sqrt{\varepsilon}}{1 + \exp(y - \delta)}\right] \qquad (13)$$

where $x = \hbar\omega - E_g$, $\varepsilon = \hbar\gamma_{cv}\beta B/2$,

$$y = \frac{E_F' - \left\{\hbar\omega - \left[E_g - \left(1 + \frac{m_c^*}{m_v^*}\right)\frac{\hbar\omega_c}{2}\right]\right\}}{[1 + (m_c^*/m_v^*)]kT}$$

$\delta = \Gamma\{[1 + (m_c^*/m_v^*)]kT\}^{-1}$, and the assumption $P_{cv}^r/\omega_r = P_{cv}^l/\omega_l$ is made ([53]). Equation (13) may be simplified, with certain assumptions, to reveal the origin of reversal in rotation more clearly. With the assumptions that $x \gg \varepsilon$ and $\delta \ll 1$ it can be shown, after much algebraic manipulation, that

$$n_r\alpha_r - n_l\alpha_l = 4\sqrt{2}\,M\,\frac{m_{cv}^{3/2}}{a_0}\,\frac{|P_{cv}|^2}{m_0 E_g}\left(\frac{E_g}{m_0 c^2}\right)^{1/2}\frac{\beta B}{2\hbar\omega}$$

$$\times \left\{ \frac{(g_c - g_v)[(X-1)^{-1/2} - 4(X-1)^{1/2}/X]}{1 + \exp\{[E_F{}' - (\hbar\omega - E_g)]/kT'\}} \right.$$

$$\left. - \frac{E_g[g_v + (m_c{}^*/m_v{}^*)g_c](X-1)^{1/2}}{2kT'\cosh^2\{[E_F{}' - (\hbar\omega - E_g)]/2kT'\}} \right\} \tag{14}$$

where $X = \hbar\omega/E_g$, a_0 is the Bohr radius, and $E_F{}' = E_F[1 + m_c{}^*/m_v{}^*)]$. The first term represents the contribution to $(n_r\alpha_r - n_l\alpha_l)$ from the left and right circular edges in a pure crystal multiplied by the Fermi occupation function. Additional carriers tend to drive this term to zero. The second term arises from the difference in the left and right circular Burstein–Moss absorption edges. This term is responsible for the reversal of the rotation and the finite contribution in the low-frequency limit. It should be noted

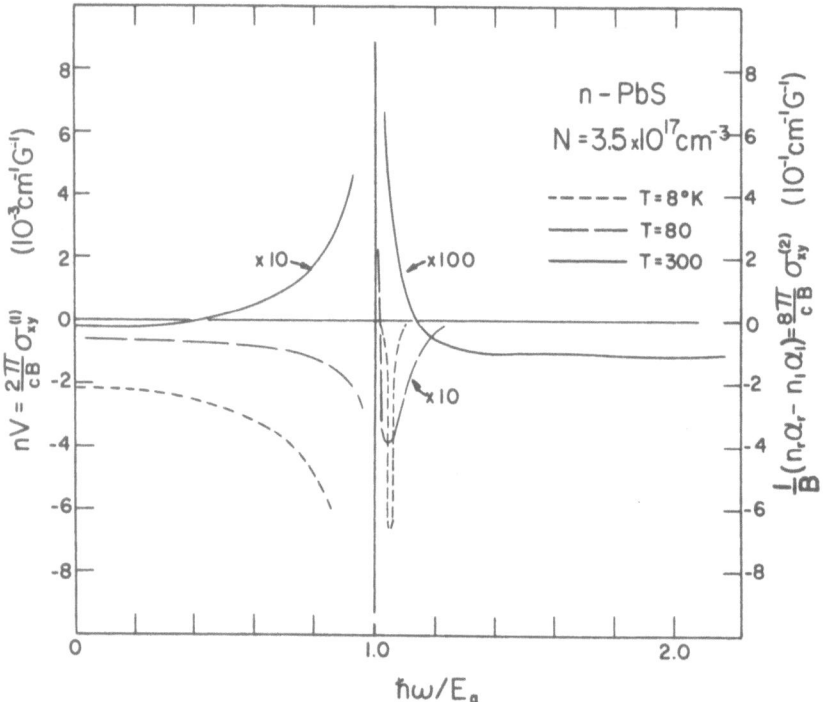

Fig. 11. Calculated difference in interband r.c.p. and l.c.p. absorption above the band gap E_g for a single sample of PbS at three temperatures and calculated interband Faraday rotation below E_g.

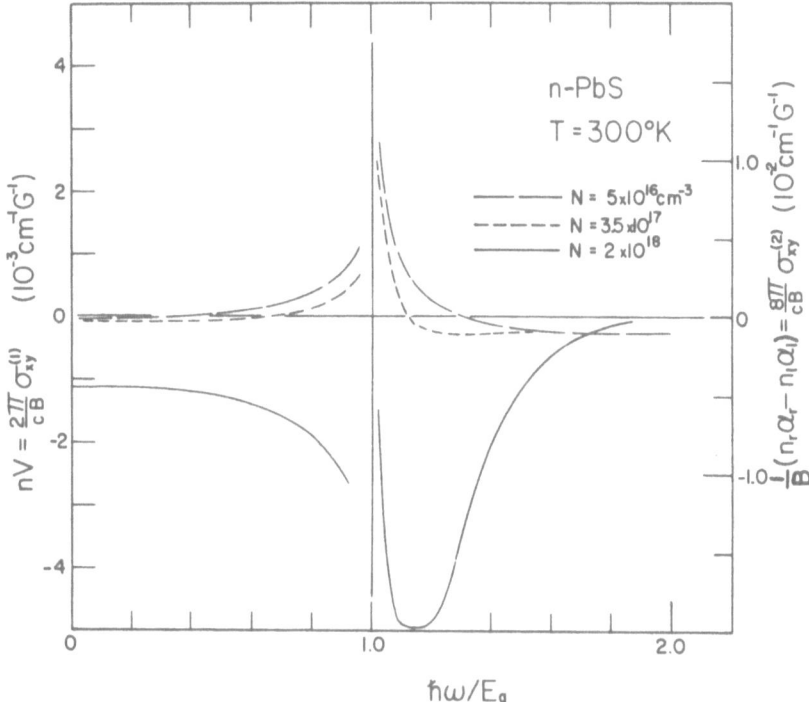

Fig. 12. Calculated difference is interband r.c.p. and l.c.p. absorption above the band gap E_g for three carrier concentrations at room temperature and calculated interband Faraday rotation below E_g.

that different combinations of the g-factors occur in the two terms. Measurements of the circularly polarized absorption for a pure sample and one with added carriers will yield the g-factors separately.

The significant features of Eq. (14) are shown in Figs. 11 and 12. Here $(n_r\alpha_r - n_l\alpha_l)/B$ is plotted as a function of temperature (Fig. 11) and carrier concentration (Fig. 12). The absorption difference, which in each case is peaked sharply positive for pure PbS, acquires a broader, negative peak as the temperature is lowered in Fig. 11 or as the carrier concentration is raised in Fig. 12. It is of interest to note in Figs. 11 and 12 that the abscissas are normalized to E_g and that the magnetoabsorption effects start at E_g in this low-field approximation.

Dispersion Relations

The Faraday rotation in a solid arises from the difference between the indices of refraction for r.c.p. and l.c.p. radiation. In absorbing regions of the spectrum there will also be an ellipticity produced by a difference in the optical absorption constants for r.c.p. and l.c.p. This difference is called circular dichroism. The circular dichroism and Faraday rotation are not independent, but are related by dispersion relations which are similar in form but different in detail from the usual Kramers–Kronig dispersion relations for the zero-field optical constants. The appropriate relations for the magnetic field case were developed in unpublished theses ([54,55]). The dispersion relations for the conductivity and for the dielectric response were subsequently published ([56,57]). The dispersion relations may be expressed in terms of any of the equivalent sets of optical constants, as will be noted in the section beginning on p. 112. These include the set of optical constants n and $\varkappa$, or the set of real and imaginary parts of the dielectric constant, or the set of real and imaginary parts of the conductivity. The dispersion equation relating the dichroism to the Faraday rotation is given by

$$[n_r^2(\omega) - n_l^2(\omega)] - [\varkappa_r^2(\omega) - \varkappa_l^2(\omega)]$$

$$= \frac{2c}{\pi\omega} \int_0^\infty \frac{\omega'[n_r(\omega')\alpha_r(\omega') - n_l(\omega')\alpha_l(\omega')]d\omega'}{\omega'^2 - \omega^2} \tag{15}$$

where $\alpha = 4\pi\varkappa/\lambda = 2\omega\varkappa/c$ is the absorption constant. We explicitly note the frequency dependence of the various quantities. The terms on the left can be written as the product of the sum and difference of n_r and n_l and the product of the sum and difference of $\varkappa_r$ and $\varkappa_l$. Thus in regions of low absorption, when $\varkappa \ll n$, since Faraday rotation per unit length $\theta = \omega(n_r - n_l)/2c$, the dispersion relation can be written

$$\bar{n}(\omega)\theta(\omega) = \frac{\omega}{2c} [n_r(\omega) + n_l(\omega)][n_r(\omega) - n_l(\omega)]$$

$$= \frac{1}{2\pi} \int_0^\infty \frac{\omega'[n_r(\omega')\alpha_r(\omega') - n_l(\omega')\alpha_l(\omega')]d\omega'}{\omega'^2 - \omega^2} \tag{16}$$

where $\bar{n}(\omega) = [n_r(\omega) + n_l(\omega)]/2$. This average index of refraction is approximately equal to the zero-field index except under very high field conditions.

Equation (16) indicates that Faraday rotation is closely related to dichroism. Indeed, for color centers in alkali halide crystals ([58,59]) it has been demonstrated that the measured Faraday rotation through the region of absorption can be related to the circular dichroism observed in the Zeeman effect. A similar relation has also been demonstrated for the case of oscillatory interband magnetoabsorption in Ge ([60]), where the interband rotation above the energy gap associated with one of the Landau transitions was related with the corresponding ellipticity, which is a measure of the dichroism.

We have used the dispersion relation Eq. (16) to calculate $n\theta/B$ below the energy gap as shown in Figs. 11 and 12. The process requires a numerical integration. The reversal in rotation is clearly shown, as is to be expected, since $n_r\alpha_r - n_l\alpha_l$ reversed sign in Figs. 11 and 12. In these calculations it is important to include the temperature dependence of the Fermi level.

A physical description of the effect may be given in terms of the dispersion effect associated with the absorption of the r.c.p. and l.c.p. radiation. Consider broad r.c.p. and broad l.c.p. lines of equal intensity, with the r.c.p. line displaced to slightly lower frequency. The dispersion at low frequencies due to these absorption lines will be dominated by the r.c.p. dispersion, which produces a net positive Faraday rotation in pure PbS. When free carriers are added the r.c.p. absorption line is weakened with respect to the l.c.p. absorption line because of the selective population of the final state of the r.c.p. transition. It is then possible for the l.c.p. dispersion tail to dominate, producing negative rotation.

Comparison With Experiment: PbS

This rigid band model has worked well to account quantitatively for the experimental results shown in Fig. 9. In Fig. 11 we have already calculated $n_r\alpha_r - n_l\alpha_l$, and $n\theta$ as a function of temperature. In Fig. 13 the fit to the interband rotation at 42 °K is shown. Here the free-carrier rotation has been calculated from Eq. (7) and subtracted from the raw data. The effective mass used was 0.103 m_0, which was determined by taking into consideration the temperature variation of mass from room temperature to liquid helium temperature. With band parameters determined from all our other studies of PbS, including $|P_{cv}|^2/2m_0 = 0.52$ eV as obtained from a fit to the zero-field absorption edge, the calculated interband rotation shows good agreement with experiment.

The low-frequency tail of $n\theta$ due to free carriers measured in a transparent region below the cyclotron frequency is proportional to the DC Hall conductivity [see Eq. (26)]. Our data imply a low-frequency tail to $n\theta$ due to interband effects also. However, since interband effects are not expected to contribute to the zero-frequency Hall conductivity, this suggests that the interband terms we consider are cancelled by other terms which originate from nonparabolic corrections to the cyclotron resonances. To avoid discussing this term in detail, we restrict the frequency range to be

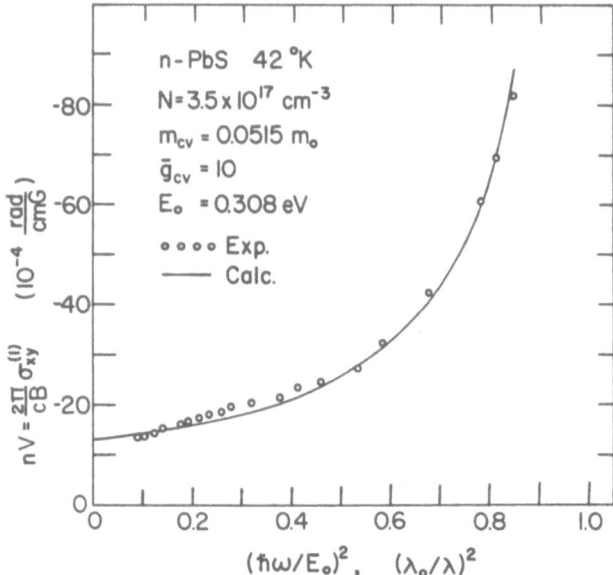

Fig. 13. Calculated and measured interband rotation in n-PbS at low temperature. The calculated free-carrier rotation has been subtracted from the raw data as discussed in the text to obtain the interband data shown here. Note the finite value for rotation in the "low-frequency limit." The energy E_0 represents the center of the strongly peaked second term in Eq. (14), which was utilized in a numerical integration of the dispersion relation Eq. (16). The wavelength λ_0 is hc/E_0, and $\bar{g}_{cv}$ is one half the g-factor combination appearing in the second term of Eq. (14).

high compared with the free-carrier resonance frequencies but still less than interband frequencies.

The experimental and calculated $n\theta(0)$ in the low-frequency limit are plotted in Fig. 14 as function of temperatures. The curve is of limited accuracy above 150 °K due to neglect of the first term in Eq. (14). The magnitude and sign of the low-frequency interband tail has been estimated by calculation of θ_{FC} due to free carriers, where the total rotation $\theta = \theta_{FC} + \theta_{IB} = 0$. It was necessary to use the proper total index of refraction in Eq. (7) for each sample for each temperature [33]. The interband tail shows the expected $1/T$ dependence similar to ordinary paramagnetic effects due to population difference of spin states.

In Fig. 15 the dependence of the $n\theta$ tail on carrier concentration has also been calculated for a temperature of 8 °K, where degeneracy persists. Although not obvious in Eq. (14), the rotation tail varies as the Fermi energy

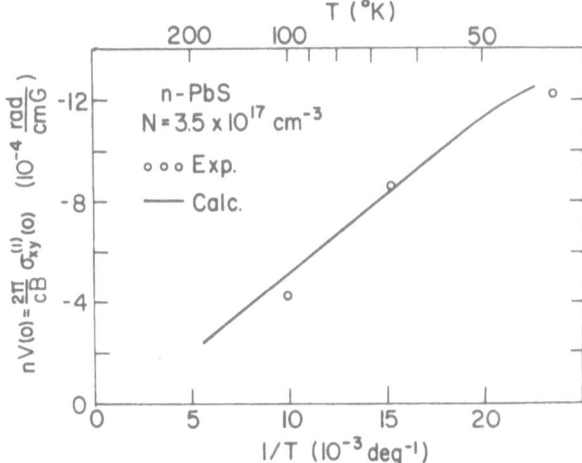

Fig. 14. Calculated and measured plots of the temperature dependence of the interband rotation in the low-frequency limit. The calculated curve includes the temperature variation of the Fermi level and band parameters but ignores the band edge contribution given by the first term of Eq. (14). Note that the rotation saturates at low temperature when the condition of strict degeneracy is attained.

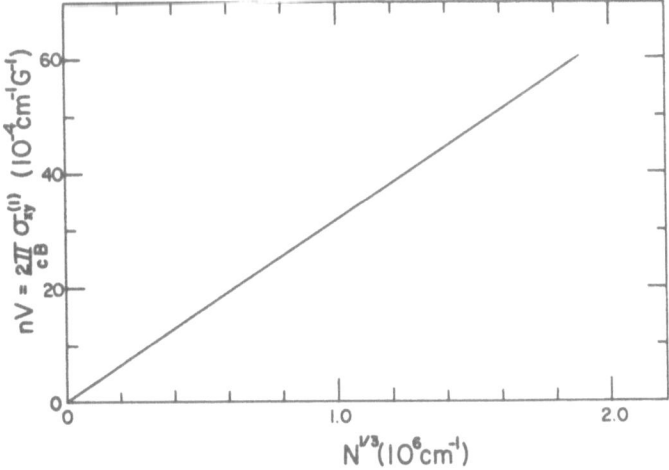

Fig. 15. Calculated plot of the carrier concentration dependence of the intraband rotation in the low-frequency, low-temperature limit.

to the one-half power, so that the carrier concentration dependence will be to the one-third power. This may be seen in Eq. (38), which is obtained in the quantum limit from a conductivity model.

p-PbS

We have measured Faraday rotation in a sample of p-PbS with $N = 8.7 \times 10^8$ cm^{-3} at $T \approx 85$ °K. In this case no reversal in interband rotation was observed, and the low-frequency tail was positive instead of negative. This behavior is in agreement with the prediction of Eq. (14). This may be seen in Fig. 3b, where now with holes populating the valence band down to a Fermi level the l.c.p. interband transitions are blocked more than the r.c.p. transitions; this means the interband rotation will not reverse sign and the low-frequency tail will remain positive. Since the valence and conduction bands are "mirror-like," the magnitude of this positive tail for a given hole concentration should be the same as the magnitude of the negative tail for the same electron concentration.

PbSe and PbTe

The rigid spherical band model works well for PbS, where $K = m_L/m_T \approx 1.3$, and should work for PbSe, where $K \approx 1.8$. However, for PbTe with $K \approx 10$ this simple model must be modified and then will only be applicable for a special case. The anisotropy introduces multiple sets of Landau levels for a magnetic field in an arbitrary direction. Furthermore, the selection rules are not separable for l.c.p. and r.c.p., as they are in the case for spherical bands. For instance, with the magnetic field in the [001] direction all ellipsoids are equivalent, so that the energy levels separate as in the simple case. However, the selection rules give both l.c.p. and r.c.p. components for a given transition, and thus the absorption edges are not separate as for the simple case. The simple model, however, still applies, and the band-population terms may be calculated by rederiving Eq. (14) with the proper selection rules and density of states for anisotropic bands. For a magnetic field in other directions the model does not apply, since not all band extrema are equivalent and there will be "spilling" of carriers from the light to heavy mass band edges.

Examination of the available data for PbSe and PbTe indicate that band population effects are also present. For n-PbTe at 300 °K the interband rotation is negative in the carrier concentration range 3–20×10^{17} cm^{-3}, while the persistent tail may be slightly negative. For p-PbTe at 300 °K the interband rotation is positive in the carrier concentration range 4–40

$\times\ 10^{17}$ cm^{-3}, with a possible positive persistent tail ([21,22,25]). This is analogous to PbS. At $\sim$77 °K the interband rotation is still positive, with a definite positive tail ([61]). Only room-temperature rotation has been measured in PbSe. For n-PbSe ([21,22,25]) with 2×10^{17} carriers/cm^3 and p-PbSe ([25]) with 2×10^{18} carriers/cm^3 interband rotation is positive, with no pronounced tail.

Band population effects should of course be present in all semiconductors with free carriers. A hint of the magnitude may be obtained from observation that the linear free-carrier Faraday rotation curve does not extrapolate to zero as $\lambda \to 0$. This implies a persistent interband tail. Such effects have been observed in GaSb ([62]). From the simple model it is evident that the magnitude and sign of $n_r\alpha_r - n_l\alpha_l$ in a pure material and a material with large carrier concentration depends on $g_c - g_v$ and $g_v + (m_c/m_v)g_c$, respectively. PbS is ideal because $g_v \approx -g_c$, which makes $n_r\alpha_r - n_l\alpha_l$, and therefore $n\theta$ below the gap small in pure material but large and of opposite sign in heavily doped material. In III–V compound semiconductors the case is more complicated because of degenerate valence bands and has not been calculated.

Interestingly, band population effects are not large in zero-magnetic-field optical experiments on PbS. Blocked interband transitions should still affect the index of refraction at all frequencies, appearing as a carrier concentration-dependence of n. Such a dependence of ε_∞ on N has been seen in plasma-reflection experiments on Pbs ([33]), but a dispersion relation analysis indicates that the decrease in absorption in the interband region is so small compared with all the interband absorption processes going on that band population cannot account for the observed change in n. In a magnetic field, however, the magnetic change in n is essentially all from the fundamental band gap absorption just above E_g, and thus is sensitive to population affects.

QUANTUM LIMIT—CONDUCTIVITY MODEL

Relations among Conductivity, Dielectric Constant, and Optical Constants

The quantum-mechanical formulation of the magnetoconductivity, dispersion relations, and the associated magnetooptical effects are thoroughly discussed in the literature ([56,57,63]). We outline the main features of this development, since it has not penetrated to the textbooks.

The plane-wave solutions to Maxwell's equations in a uniform dielectric medium may be obtained from the eigenvalue equation [56]

$$(\omega^2/c^2)\varepsilon\mathbf{E} = k^2\mathbf{E} - \mathbf{k}(\mathbf{k} \cdot \mathbf{E}) \tag{17}$$

where ε is the frequency-dependent dielectric tensor, $\mathbf{k}$ is the wave vector of a propagating mode, and $\mathbf{E}$ is the electric field associated with the wave. Unrationalized cgs units are used. This derivation assumes that the magnetic permeability is the same as for free space. The magnetic interactions are usually quite small at optical frequencies and will be neglected, although in general they must be included.

For a magnetic field in the $+z$ direction the dielectric tensor has the form [56,57]

$$\varepsilon = \begin{pmatrix} \varepsilon_{xx} & \varepsilon_{xy} & 0 \\ -\varepsilon_{xy} & \varepsilon_{xx} & 0 \\ 0 & 0 & \varepsilon_{zz} \end{pmatrix} \tag{18}$$

where the crystal is assumed to have rotational symmetry around the z axis. The components of the dielectric tensor are complex, and thus include both dispersive and dissipative effects. The diagonal components are even functions of the magnetic field, and thus, are the same as zero field if we restrict the considerations to magnetooptical effects linear in H [56,57]. The off-diagonal components are odd functions of the magnetic field, and thus generate the magneto effects we are considering.

For the symmetric case the plane-wave solutions to Eq. (17) propagating in the $+z$ direction are obtained by the substitution

$$\mathbf{E}_{r,l} = \mathrm{Re}(\mathbf{x} \pm i\mathbf{y}) \exp i(\omega t - k_{r,l}z) \tag{19}$$

where $\mathbf{x}$ and $\mathbf{y}$ are unit vectors, the subscripts r, l refer to r.c.p. and l.c.p., respectively, and Re signifies the real part. The solutions of Eq. (17) are

$$(c^2/\omega^2)k_{r,l}^2 = \varepsilon_{r,l}(\omega) = \varepsilon_{xx}(\omega) \pm i\varepsilon_{xy}(\omega) \tag{20}$$

which specify the wave vector in terms of a scalar dielectric constant $\varepsilon_{r,l}(\omega)$, which, in turn, is a function of the components of the dielectric tensor. The optical constants are obtained from

$$[n_{r,l}(\omega) - i\varkappa_{r,l}(\omega)]^2 = \varepsilon_{r,l}(\omega) = \varepsilon_{r,l}^{(1)} + i\varepsilon_{r,l}^{(2)} \tag{21}$$

The conductivity tensor σ is related to the dielectric tensor ε by

$$\varepsilon = 1 - i(4\pi/\omega)\sigma \tag{22}$$

where I is the unit tensor. Thus we can relate the optical constants to either the conductivity or the dielectric constant

$$n_{r,l}^2(\omega) - \varkappa_{r,l}^2(\omega) = \varepsilon_{r,l}^{(1)} = \varepsilon_{xx}^{(1)}(\omega) \mp \varepsilon_{xy}^{(2)}(\omega) = 1 + (4\pi/\omega)[\sigma_{xx}^{(2)}(\omega) \pm \sigma_{xy}^{(1)}(\omega)]$$
$$(23)$$
$$-2n_{r,l}(\omega)\varkappa_{r,l}(\omega) = \varepsilon_{r,l}^{(2)} = \varepsilon_{xx}^{(2)}(\omega) \pm \varepsilon_{xy}^{(1)}(\omega) = -(4\pi/\omega)[\sigma_{xx}^{(1)}(\omega) \mp \sigma_{xy}^{(2)}(\omega)]$$

where the superscripts (1) and (2) refer to real and imaginary parts, respectively.

In zero magnetic field the off-diagonal components (σ_{xy} or ε_{xy}) are zero, so that only the diagonal terms (σ_{xx} or ε_{xx}) appear in the relationship for the optical constants. In this case the dispersion relations may be applied directly. The dispersion relations relate the real and imaginary parts of the individual components of the dielectric or conductivity tensor. The general dispersion relations are ([56,57])

$$\varepsilon_{ij}^{(1)}(\omega) - \delta_{ij} = -\frac{2}{\pi} P \int_0^\infty \frac{\omega' \varepsilon_{ij}^{(2)}(\omega')\, d\omega'}{\omega'^2 - \omega^2}$$
$$(24)$$
$$\varepsilon_{ij}^{(2)}(\omega) = \frac{2\omega}{\pi} P \int_0^\infty \frac{\varepsilon_{ij}^{(1)}(\omega')\, d\omega'}{\omega'^2 - \omega^2}$$

where P indicates that the principle value is taken. The same equations apply to the conductivity components, except δ_{ij} is eliminated.

In the presence of a magnetic field the dispersion relations cannot be applied directly, since the real part of the diagonal component ($\varepsilon_{xx}^{(1)}$ or $\sigma_{xx}^{(1)}$) appears with the imaginary part of the off-diagonal component ($\varepsilon_{xy}^{(2)}$ or $\sigma_{xy}^{(2)}$). However, from Eq. (23) the differences

$$[n_r^2(\omega) - n_l^2(\omega)] - [\varkappa_r^2(\omega) - \varkappa_l^2(\omega)] = -2\varepsilon_{xy}^{(2)}(\omega) = (8\pi/\omega)\sigma_{xy}^{(1)}(\omega)$$
$$(25)$$
$$2[n_r(\omega)\varkappa_r(\omega) - n_l(\omega)\varkappa_l(\omega)] = -2e_{xy}^{(1)}(\omega) = -(8\pi/\omega)\sigma_{xy}^{(2)}(\omega)$$

depend only on the off-diagonal component. Thus in weakly absorbing regions we may apply the discussion of Eq. (16) to obtain

$$n(\omega)\theta(\omega) = (2\pi/c)\sigma_{xy}^{(1)}(\omega) \qquad (26)$$

where the Faraday rotation $\theta(\omega)$ is proportional to the frequency-dependent Hall conductivity $\sigma_{xy}^{(1)}(\omega)$ due to both free-carrier and interband processes. In insulating solids $\sigma_{xy}^{(1)}$ approaches zero at zero frequency. In conducting solids it approaches the DC Hall conductivity $\sigma_H = Nec/B$ due to free carriers only, as the interband contribution also approaches zero at zero

frequency. The dispersion relation Eq. (15) is obtained by the insertion of
Eq. (25) into Eq. (24).

The ellipticity in weakly absorbing regions depends on the absorption
difference $\varkappa_r(\omega) - \varkappa_l(\omega)$. The amplitude ellipticity for propagation a distance d is given by [60]

$$\xi(\omega) = \tanh\{(\omega d/2c)[\varkappa_l(\omega) - \varkappa_r(\omega)]\} \tag{27}$$

Thus the ellipticity and Faraday rotation are related by the dispersion
relations under certain approximations.

Conductivity Formulation

The quantum-mechanical magnetooptical effects may be developed in
two stages: First, by calculation of the components of the magnetoconductivity or magnetopolarizability by standard quantum-mechanical techniques;
second, by relating these to the optical constants $n_{r,l}(\omega)$ and $\varkappa_{r,l}(\omega)$. When
free carriers are present it is preferable to use the conductivity formulation,
since it avoids certain difficulties at zero frequency [23,57]. The basic quantum-mechanical formulation for the components of the magneto-conduc-
tivity tensor has been given by Roth [63] and Bennet and Stern [57]. We
follows the latter formulation with slightly modified notation.

A general expression for the frequency-dependent Hall conductivity
obtained by use of semiclassical perturbation theory is given by

$$\sigma_{xy}^{(1)} = \frac{Me^2}{\hbar m_0^2 V} \sum_{i<f} \left\{ \frac{|\langle f| \pi^-|i\rangle|^2}{\omega_{fi}^2 - \omega^2} - \frac{|\langle f| \pi^+|i\rangle|^2}{\omega_{fi}^2 - \omega^2} \right\} F_{if} \tag{28}$$

where M is the number of equivalent bands (four for PbS), V is a volume,
$F_{if} = F_i - F_f$ is the difference in occupation between the initial and final
states given by the Fermi function, and $\pi^\pm = (\pi_x \pm i\pi_y)/\sqrt{2}$ is the kinetic
momentum operator [64] including the spin-orbit term. In the sum over
states the initial states are restricted to lie at lower energy than the final
states $(\hbar\omega_f - \hbar\omega_i = \hbar\omega_{fi} > 0)$.

We restrict the calculation to populated bands which are spherical,
nondegenerate, and parabolic. This approximation is well satisfied for
n- and p-type PbS, for which a theoretical spherical approximation has been
developed for band extrema at the L point [41]. The calculation for anisotropic or degenerate bands is more complicated, since the separation of the
transitions into two separate groups (r.c.p. to spin-down levels, l.c.p. to
spin-up levels) is not possible. Also, the fortuitous occurrence of nearly

equal g-factors for valence and conduction bands in PbS will not occur in general, so that the sum indicated by Eq. (28) will have contributions from transitions at all frequencies, whereas for PbS the transitions in the vicinity of the Fermi level dominate.

The intraband (free-carrier cyclotron resonance) terms in Eq. (28) are evaluated using the selection rule

$$\langle f \mid \pi^+ \mid i \rangle = \hbar(m_0/m_c)[(l + 1)s]^{1/2}\delta_{l'+1,l}\delta(k_z' - k_z) \qquad (29)$$

where the conduction band is specified as the partially occupied band and $s = eB/\hbar c$. Other quantum numbers play no role in this matrix element and are neglected. For the free-carrier case (FC) we obtain

$$(FC)\sigma_{xy}^{(1)} = \frac{Me^2}{\hbar s V} \frac{\omega_c^2}{(\omega_c^2 - \omega^2)} \sum_{i<f} - (l + 1)(F_{if}^\alpha + F_{if}^\beta)$$

$$= \frac{-Me^2}{\hbar s} \frac{\omega_c^2}{(\omega_c^2 - \omega^2)} \frac{(N_\alpha + N_\beta)}{V} \qquad (30)$$

where the total number of free carriers $N_\alpha + N_\beta$ is the sum of carriers in the Kramers subbands α and β. The Kramers labels α, β have the same properties as the spin labels $\uparrow$, $\downarrow$, which, however, are no longer adequate due to spin-orbit mixing of bands. The intraband contribution to $\sigma_{xy}^{(1)}$ gives the usual expression for free carrier rotation, Eq. (7), when used in Eq. (26). It should be noted that there are no dHvA-type terms in this approximation, since the total number of carriers in the band is constant.

The eigenvalues for the energy levels in the presence of a magnetic field are given by

$$E_j(n, k_z) = E_j(0) + (l + \tfrac{1}{2})\hbar\omega_j + (\hbar^2 k_z^2/2m_j) \pm \tfrac{1}{2}g_j\beta H \qquad (31)$$

where j is the band index c or v. The effective mass m_j, g factor g_j, and cyclotron frequency ω_j are normally positive for conduction bands and negative for valence bands. Note that this is a change in notation from the previous section, where the effective masses $m_{c,v}^*$ were considered positive. The selection rules for interband transitions give the matrix element

$$\langle f \mid \pi^\pm \mid i \rangle = \pi_{j',j}^0 \delta_{l',l}\delta_{\sigma'+1,\sigma}\delta(k_z' - k_z) \qquad (32)$$

where σ is the z component of the total angular momentum, which transforms like the states at a given band edge in the absence of a magnetic field. Similarly, $\pi_{j',j}^0$ is evaluated at the band edge in zero field. The magnetic-field-dependent corrections to $\pi_{cv}^{\pm}$ are negligible for the case we consider (large g factors).

The interband (IB) contribution to the conductivity is given by

$$(IB)\sigma_{xy}^{(1)}(\omega) = \frac{M\hbar e^2 |\pi_{cv}^0|^2}{m_0{}^2} \sum_{n,k_z} \left\{ \frac{(1 - F_c^{\beta})}{(E_{cv}^-)^2 - (\hbar\omega)^2} - \frac{(1 - F_c^{\alpha})}{(E_{cv}^+)^2 - (\hbar\omega)^2} \right\} \quad (33)$$

where $F_c^{\alpha,\beta}$ is the Fermi occupation function for the Kramers α, β subbands and the energy differenc $E_{cv}^{\pm}$ is

$$E_{cv}^{\pm} = E_{cv}^0 + \left(l + \frac{1}{2}\right)\hbar(\omega_c - \omega_v) + \frac{\hbar^2 k_z{}^2}{2}\left(\frac{1}{m_c} - \frac{1}{m_v}\right) \pm \frac{(g_c + g_1)\beta H}{2} \quad (34)$$

The sum over k_z is replaced by a sum over the energy by relating the interband energy scale Eq. (34) and the conduction band energy scale Eq. (31). If $g_c\beta H/2 \ll E_F$, then Eq. (33) can be rewritten as

$$(IB)\sigma_{xy}^{(1)} = \frac{M\hbar e^2 |\pi_{cv}^0|^2}{m_0{}^2 V} \sum_{n,E} \left\{ \frac{F_+^{\alpha} - F_-^{\beta}}{[E^2 - (\hbar\omega)^2]} - \frac{[2(g_c + g_v)\beta BE](F^0 - 1)}{[E^2 - (\hbar\omega)^2]^2} \right\} \quad (35)$$

where E is given by Eq. (34) omitting the last term. The Fermi occupation functions $F_{\pm}^{\alpha,\beta}$ are functions of the variable $[(E - E_{cv}^0 - E'_{F\pm})/kT']$, where

$$E'_{F\pm} = \left(1 - \frac{m_c}{m_v}\right)E_F{}^0 \pm \left(g_v + \frac{m_c}{m_v}g_c\right)\frac{\beta H}{2} \quad (36)$$

$T' = [1 - (m_c/m_v)]T$, and $E_F{}^0$ is the Fermi level in zero-field measured relative to the conduction band edge. The first term of Eq. (35) gives contributions to the conductivity only in the vicinity of the Burstein–Moss energy $E_{cv}^0 = E_{cv}^0 + [1 - (m_c/m_v)]E_F{}^0 = E_{cv}^0 + E_F{}^{0'}$. The second term gives the contributions to the conductivity which correspond to the usual interband conductivity reduced by the band population blockage.

The major dHvA-type oscillatory terms occur in the first term of Eq. (35), which we evaluate in a form which emphasizes their physical origin. With the approximation $kT \ll E_F$ we get

$$(IB)\sigma_{xy}^{(1)} = \frac{M\hbar e^2 |\pi_{cv}^0|^2}{m_0{}^2[(E_{cv}^{0'})^2 - (\hbar\omega)^2]} \frac{N_{\alpha} - N_{\beta}}{V} + II \quad (37)$$

where II represents the second term of Eq. (35), which is not considered further. The effects we are considering thus depend directly on the difference in population between the two Kramers states of the populated band.

The evaluation ([65]) of Eq. (37) may finally be written as

$$(\text{IB})\sigma_{xy}^{(1)}(\omega) = \frac{\sqrt{2}M}{2} \frac{e^2(\mu_{cv})^{3/2} \mid \pi_{cv}^0 \mid^2}{\hbar\sqrt{m_0}[(E_{cv}^{0'})^2 - (\hbar\omega)^2]}$$

$$\times \left\{ (E_F^{0'})^{1/2}\left(g_v + \frac{m_c}{m_v}g_c\right)\beta B + \frac{2\pi(\beta B)^{1/2}m_c kT}{m_0(\mu_{cv})^{3/2}} \right.$$

$$\left. \times \sum_{r=1}^{\infty} \frac{(-1)^{r+1}\sin\{\frac{1}{2}r\pi[g_v + (m_c/m_v)g_c]\mu_{cv}\}\sin[(2\pi r/\hbar\omega_c)E_F^0 + \frac{1}{4}\pi]}{\sqrt{r}\sinh(2\pi^2 rkT/\hbar\omega_c)} \right\} \quad (38)$$

This equation has been compared with Eq. (14) for the simple rigid band model. The second term of Eq. (14) corresponds to the first term of Eq. (38) The first term of Eq. (14) corresponds to the term II in Eq. (37), which is neglected in this calculation. The oscillatory terms do not appear in Eq. (14). As mentioned above, the low-frequency tail $n\theta(0)$ is seen to vary as $N^{1/3}$ in Eq. (38). In Eq. (38) $(\mu_{cv})^{-1} = (m_c/m_0)^{-1} - (m_v/m_0)^{-1}$.

Oscillations in Interband Rotation

The nonoscillatory part of Eq. (38) shows the reversal in rotation and the low-frequency tail. In addition, the oscillatory part of Eq. (38) is observable in the Faraday rotation below the energy gap, as shown in Fig. 16. Rotation was measured for two n-type samples of PbS at $T \approx 40\,°\text{K}$ at various fixed frequencies below E_g. Measurements from 4 to 8 μ indicated that the period of the oscillations are frequency independent but carrier concentration dependent. Since only one period was observed at 40 °K, the maxima, minima and crossover points were plotted versus $1/B$ in the usual way for a dHvA experiment. The results are shown in Fig. 17. From the period of oscillations $P = (e/\pi ch)(8\pi M/3N)^{2/3}$ (one complete cycle in θ) the carrier concentrations were determined and found to agree with those obtained by Hall measurements to within $\pm 20\%$. At lower temperatures, where $kT \ll \hbar\omega_c$, the amplitude of the oscillations should increase significantly.

A study of Fig. 5 indicates that for pairs of r.c.p. and l.c.p. transitions the l.c.p. transition dominates the r.c.p. from the moment the final level of the l.c.p. transition pops through the Fermi level until the final level of the r.c.p. transition pops through the Fermi level. From then on the absorption lines are assumed to be equally strong but still slightly separated with the r.c.p. line to lower energy. Then, for example, for the transitions $(1, \downarrow) \rightarrow (1, \uparrow)$ and $(1, \uparrow) \rightarrow (1, \downarrow)$ one can expect an additional negative

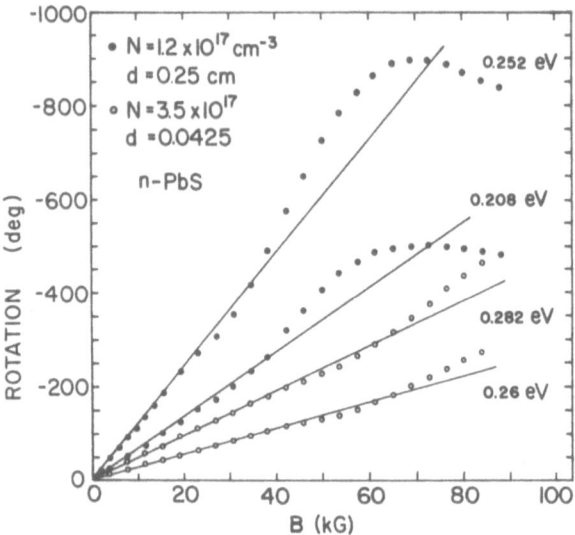

Fig. 16. Faraday rotation in two samples of *n*-PbS at ∼40°K as a function of magnetic field at several photon energies below the band gap. The period of the oscillatory part is frequency independent over the measured range 0.3–0.16 eV.

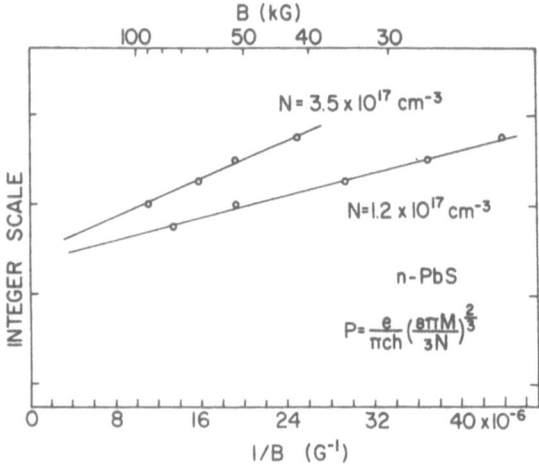

Fig. 17. The magnetic field positions of the maxima, minima, and crossover points in the oscillatory Faraday rotation of Fig. 16 plotted versus an integer scale. The period of the oscillations yielded the carrier concentrations of the two samples.

contribution to the general interband rotation to begin when the $(1, \uparrow)$ conduction level pops through and an additional positive contribution to begin when the $(1, \downarrow)$ level pops through. For 40 °K the spread in kT is about equal to the level spacing, as indicated by the T scale at the lower right in Fig. 5, so that it is difficult to ascertain when the dispersions cancel to give a crossover point in Fig. 17, or for that matter when the maxima and minima should occur. However, it is evident that the additional positive rotation starting at $\sim$73 kG for $N = 1.2 \times 10^{17}$ cm^{-3} must be due to the $(0, \uparrow) \rightarrow (0, \downarrow)$ transition, while the additional negative rotation starting at $\sim$36 kG is due to the $(0, \downarrow) \rightarrow (0, \uparrow)$ transition becoming strong, while the additional positive rotation starting at $\sim$23 kG is due to the $(1, \uparrow)$ $\rightarrow (1, \downarrow)$ transition. Similar qualitative results can be concluded for the $N = 3.5 \times 10^{17}$ cm^{-3} sample.

Magnetoplasma

The plasma reflection edge in zero magnetic field occurs at the zero in the real part of the total dielectric constant. In a magnetic field magnetoplasma edges for l.c.p. and r.c.p. radiation occur at the two zeros in the real part of the total dielectric constant $\varepsilon_{r,l}^{(1)}(\omega) = 1 + (4\pi/\omega)\sigma_{r,l}^{(2)}(\omega)$, where

$$\sigma_{r,l} = \sigma_{r,l}^{(1)} + i\sigma_{r,l}^{(2)} = (\sigma_{xx}^{(1)} \mp \sigma_{yy}^{(2)}) + i(\sigma_{xx}^{(2)} \pm \sigma_{xy}^{(1)})$$

The intraband and interband contributions to $\sigma_{xx}^{(2)}$ have been treated in detail ([66]), but neglecting possible spin-dependent population effects. We have repeated the calculation for $\sigma_{xx}^{(2)}$ with the spin and spin-orbit interaction. The details are similar to those for the calculation of $\sigma_{xy}^{(1)}$, with the result

$$\sigma_{xx}^{(2)}(\omega) = \frac{Me^2\omega}{m_0^2 V} \left\{ \frac{m_0^2(N_\alpha + N_\beta)}{m_c(\omega_c^2 - \omega^2)} - \left[\sum_{n,E} \frac{|\pi_{cv}^0|^2}{E[E^2 - (\hbar\omega)^2]} \right] \right.$$
$$\left. + \frac{|\pi_{cv}^0|^2(N_\alpha + N_\beta)}{E_{cv}^{0'}[(E_{cn}^{0'})^2 - (\hbar\omega)^2]} \right\} \tag{39}$$

where the first term is the dispersive part due to intraband transitions, the second is the dispersive part due to interband transitions in the pure crystal, and the third part is the decrease in the dispersive part due to blockage of interband transitions. It is worth noting that no dHvA-type terms appear in the expression, and to first order the expression is the same as for zero magnetic field.

Since we are interested in the frequency where the total real dielectric constant is zero, and we note that free-carrier dispersion subtracts from the interband dispersion, we may write

$$\varepsilon_{r,l}^{(1)} = 0 = 1 + (4\pi/\omega)(\sigma_{r,l_{FC}}^{(2)} + \sigma_{r,l_{IB}}^{(2)})$$

or

$$1 + (4\pi/\omega)(\sigma_{xx}^{(2)} \pm \sigma_{xy}^{(1)})_{IB} = -(4\pi/\omega)(\sigma_{xx}^{(2)} \pm \sigma_{xy}^{(1)})_{FC} \tag{40}$$

The interband term is usually replaced by the zero-magnetic-field dielectric constant $[\varepsilon^{(1)}(\omega)]_{IB}$ for the pure crystal. Dresselhaus and Dresselhaus [66] discuss the shifts of the free-carrier plasma reflection edges due to band population contributions to $\sigma_{xx}^{(2)}$, but in most cases these are small. The band population contributions to the plasma edge shifts through $[\sigma_{xy}^{(1)})_{IB}]$ are large, and in addition contain dHvA-type terms. Replacing the IB term of Eq. (40) by $[\varepsilon_{r,l}^{(1)}(\omega)]_{IB}$ and using Eqs. (30) and (39) to determine the FC term, we may write

$$[\varepsilon_{(r,l}^{(1)}\omega)]_{IB} = \Omega_p^2(\omega \mp \omega_c)/\omega(\omega^2 - \omega_c^2) \tag{41}$$

where $\Omega_p^2 = 4\pi N e^2/m_c$ and

$$[\varepsilon_{r,l}^{(1)}(\omega)]_{IB} = [\varepsilon^{(1)}(\omega)]_{IB} \pm (4\pi/\omega)[\sigma_{xy}^{(1)}(\omega)]_{IB}$$

Thus the proper high-frequency dielectric constant is not the same for r.c.p. and l.c.p. as is commonly assumed. The magnetoplasma edges $\omega_{pr,l}$ obtained from Eq. (41) are given by

$$\omega_{pr,l} = \omega_p \mp [2\pi(\omega_p/\Omega_p)^2(\sigma_{xy}^{(1)})_{IB} + \omega_c/2] \tag{42}$$

where $\omega_p^2 = \Omega_p^2[\varepsilon^{(1)}(\omega)]_{IB}^{-1}$, and the approximation $[4\pi\sigma_{xy}^{(1)}/\omega]_{IB} \ll [\varepsilon^{(1)}(\omega)]_{IB}$ has been made.

Figure 18 shows the free-carrier $\varepsilon_{r,l}$ and the saturation values for the interband $\varepsilon_{r,l}$. The plasma edges occur when the negative $[\varepsilon_{r,l}]_{FC}$ due to free carriers is equal to the positive $[\varepsilon_{r,l}]_{IB}$ from interband transitions. The population terms alter $[\varepsilon_{r,l}]_{IB}$ from its usual value ε_∞ and thus decrease the plasma edge splitting from the value ω_c. It is worth noting that the frequency difference for the two free-carrier branches is identically equal to ω_c if evaluated at the same value of ε. The vertical splitting of the two interband branches is equal to $[2n\theta\lambda B/\pi]_{IB}$ when evaluated at a fixed wavelength. This difference between ε_r and ε_l can become quite large at low frequencies or large magnetic fields.

In the usual magnetoplasma experiment with $\sigma_{xy}^{(1)}$ neglected the r.c.p. and l.c.p. plasma reflection edges move in opposite directions from ω_p by

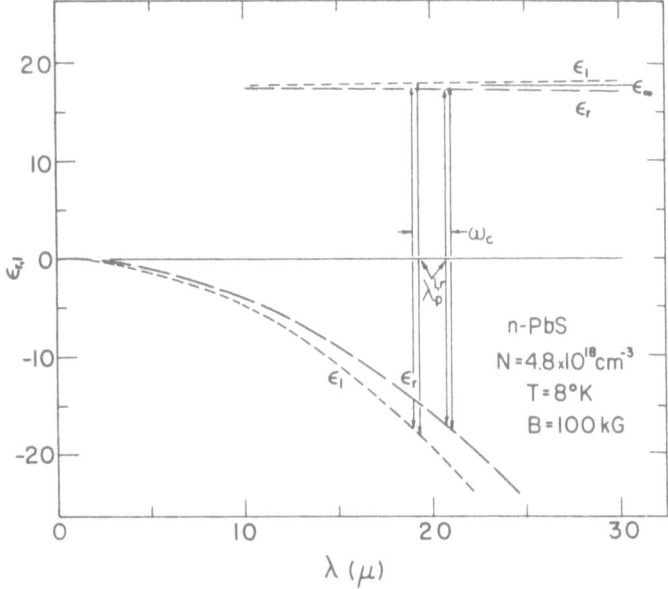

Fig. 18. Calculated interband and free-carrier contributions to the dielectric constant indicating the effect of band population on the r.c.p. and l.c.p. splitting of the plasma reflection edge.

$\pm\omega_c/2$, so that the two edges are separated by ω_c. With band population effects added this separation will be smaller than ω_c in PbS. We plot in Fig. 19 the measured magnetoplasma edge splitting in n-type PbS at ~ 20 °K together with the splittings calculated without and with the interband correction term. The dashed curve is the cyclotron frequency calculated with the effective mass $0.11 \, m_0$ for this carrier concentration range at this low temperature. When the contribution of $\sigma_{xy}^{(1)}(\omega)$ is included the difference $\omega_{pl} - \omega_{pr}$ is no longer ω_c, but is given by the solid line. Thus from the experimental splitting one would conclude an effective mass considerably too heavy ($0.16 \, m_0$) if the band population effect is ignored.

We have also measured the magnetoplasma reflection in p-PbTe at a low temperature near 20 °K. For this carrier concentration, $N = 4 \times 10^{18}$ cm^{-3}, the suceptability mass $m_s = \frac{1}{3}[(2/m_T) + (1/m_L)]$ has been determined from zero-field plasma reflectivity measurements to be $0.055 \, m_0$ [34]. For the magnetoplasma experiment in the absence of population effects $\omega_{pl} - \omega_{pr} = \omega_c$ obtains, where ω_c contains m_s. The dashed curve in Fig. 19 represents this case. The experimental splitting, given by the open circles, indicates a "heavier" effective mass. The magnitude and sign of the plasma

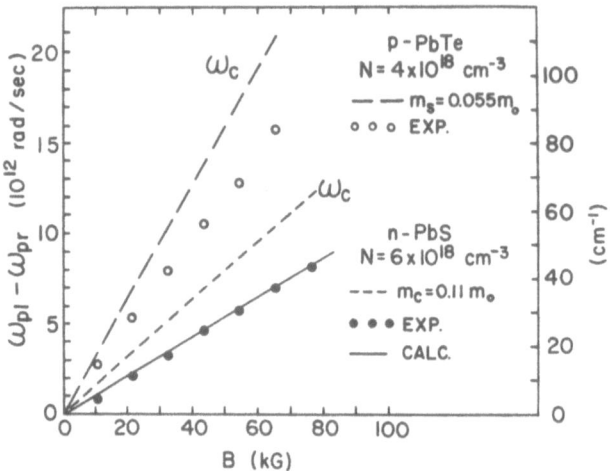

Fig. 19. Observed and calculated splitting of the r.c.p. and l.c.p. plasma reflection edges for *n*-PbS and *p*-PbTe at $T \approx 20^{\circ}$K. The dashed lines are the cyclotron frequencies calculated for the appropriate effective masses indicated. The solid line for PbS is the splitting calculated with interband population effects included. For *p*-PbTe the $\omega_{pl} - \omega_{pr}$ scale should be negative.

edge shift in *p*-PbTe are in qualitative agreement with the spherical band model for band population effects. Quantitative comparison requires an evaluation of Eq. (35) for the anisotropic bands. This calculation is straightforward but tedious. The matrix elements and Landau levels are calculable, and a calculation including anisotropy and nonparabolicity is planned for the future.

The oscillatory terms in $\sigma_{xy}^{(1)}$ should also give rise to dHvA-type oscillations of the plasma reflection edges. These have not been experimentally identified in the lead salts as yet, although such effects have been seen in antimony ([67]) but were ascribed to another mechanism. The oscillatory and nonoscillatory band population effects discussed here should be manifest in all other physical processes which depend on the background dielectric constant. We have only discussed the optical effects in the Faraday configuration, but, presumably, similar effects should occur for the Voigt configuration where the interband Voigt effect is measured. Large effects are only expected in small-band-gap materials with large spin-orbit interaction. These conditions are necessary to obtain the large *g* factors and consequent large population differences which are required. Indium antimonide, the mercury chalcogenides, and the group V semimetals are likely candidates, although as yet the effect has only been identified in the lead chalcogenides.

In this chapter we have considered the influence of free carriers on interband magnetooptical effects. The inclusion of spin and spin-orbit interaction also modified the dynamics of the free carriers themselves. These magnetoeffects are discussed by Stern ([68]).

REFERENCES

1. W. W. Scanlon, in: *Solid State Physics, Vol. 9* (F. Seitz and D. Turnbull, eds.), Academic Press, New York, 1959, p. 83.
2. Y. V. Ilisavskii, *Soviet Phys.–Solid State (English Transl.)* **4**: 674 (1962); Y. V. Ilisavskii and E. Z. Yakhind, *Soviet Phys.–Solid State (English Transl.)* **4**: 1447 (1963).
3. R. F. Burke, *Phys. Rev.* **160**: 636 (1967).
4. P. J. Stiles, E. Burstein, and D. N. Langenberg, *J. Appl. Phys. Suppl.* **32**: 2174 (1961).
5. R. S. Allgaier, *J. Appl. Phys. Suppl.* **32**: 2185 (1961).
6. R. S. Allgaier, *Phys. Rev.* **119**: 554 (1960).
7. K. F. Cuff, M. R. Ellett, and C. D. Kuglin, *J. Appl. Phys. Suppl.* **32**: 2179 (1961).
8. K. F. Cuff, M. R. Ellett, and C. D. Kuglin, in: *Report International Conference on the Physics of Semiconductors, Exeter, 1962*, The Institute of Physics and the Physical Society, London, 1962, p. 316.
9. R. S. Allgaier, B. B. Houston, R. F. Bis, J. Babiskin, and P. G. Siebenmann, in: *Proceedings of the International Conference on the Physics of Semiconductors, Paris, 1964*, Dunod, Paris, 1964, p. 659.
10. K. F. Cuff, M. R. Ellett, C. D. Kuglin, and L. R. Williams, in: *Proceedings of the International Conference on the Physics of Semiconductors, Paris, 1964*, Dunod, Paris, 1964, p. 677.
11. R. Nii, Japan, *J. Appl. Phys.* **18**: 456 (1963).
12. R. Nii, *J. Phys. Soc. Japan* **19**: 48 (1964).
13. E. Burstein, P. J. Stiles, D. N. Langenberg, and R. F. Wallis, *Phys. Rev. Letters* **9**: 260 (1962).
14. P. J. Stiles, E. Burstein, and D. N. Langenberg, *Phys. Rev. Letters* **9**: 257 (1962).
15. R. Nii, A. Kobayaski, H. Numato, and Y. Uemura, in: *Proceedings of the International Conference on the Physics of Semiconductors, Paris, 1964*, Dunod, Paris, 1964, p. 65.
16. D. L. Mitchell, E. D. Palik, J. D. Jensen, R. B. Schoolar, and J. N. Zemel, *Phys. Letters* **4**: 262 (1963).
17. D. L. Mitchell, E. D. Palik, and J. N. Zemel, in: *Proceedings of the International Conference on the Physics of Semiconductors, Paris, 1964*, Dunod, Paris, 1964, p. 325.
18. E. D. Palik, D. L. Mitchell, and J. N. Zemel, *Phys. Rev.* **135**: A763 (1964).
19. D. L. Mitchell, in: *The Use of Thin Films in Physical Investigations*, (J. C. Anderson, ed.), Academic Press, London, 1966, p. 363.
20. E. D. Palik, S. Teitler, B. W. Henvis, and R. F. Wallis, in: *Report International Conference on the Physics of Semiconductors, Exeter, 1962*, The Institute of Physics and the Physical Society, London, 1962, p. 288.

21. A. K. Walton, T. S. Moss, and B. Ellis, *Proc. Phys. Soc. (London)* **79**: 1065 (1962).
22. A. K. Walton and T. S. Moss, *Proc. Phys. Soc. (London)* **81**: 509 (1963).
23. D. L. Mitchell, E. D. Palik, and R. F. Wallis, *Phys. Rev. Letters* **14**: 827 (1965).
24. D. L. Mitchell, E. D. Palik, and R. F. Wallis, in: "Proceedings of the International Conference on the Physics of Semiconductors, Kyoto, 1966", *J. Phys. Soc. Japan Suppl.* **21**: 197 (1966).
25. S. Kurita, I. Nagasawa, K. Tanaka, Y. Nishina, and T. Fukuroi, Sc. Rep. Tohoku Univ. **A17**: 37 (1965).
26. W. W. Scanlon, *J. Phys. Chem. Solids* **8**: 423 (1959).
27. H. R. Riedl and R. B. Schoolar, *Phys. Rev.* **131**: 2082 (1963).
28. R. B. Schoolar and J. R. Dixon, *Phys. Rev.* **137**: A667 (1965).
29. R. B. Schoolar and J. N. Zemel, *J. Appl. Phys.* **35**: 1848 (1964).
30. J. N. Zemel, J. D. Jenson, and R. B. Schoolar, *Phys. Rev.* **140**: A330 (1964).
31. Yu. V. Mal'tsev, E. D. Nensberg, A. V. Petrov, S. A. Semiletov, and Yu. I. Ukhanov, *Soviet Phys.–Solid State (English Transl.)* **8**: 1713 (1967).
32. J. R. Dixon and H. R. Riedl, *Report International Conference on the Physics of Semiconductors, Exeter, 1962*, The Institute of Physics and the Physical Society, London, 1962, p. 179.
33. J. R. Dixon and H. R. Riedl, *Phys. Rev.* **140**: A1283 (1965).
34. J. R. Dixon and H. R. Riedl, *Phys. Rev.* **138**: A873 (1965).
35. H. A. Lyden, *Phys. Rev.* **135**: A514 (1964).
36. G. L. Bir and G. E. Pikus, *Soviet Phys.–Solid State (English Transl.)* **4**: 1640 (1963).
37. G. E. Pikus and G. L. Bir, *Soviet Phys.–Solid State (English Transl.)* **4**: 1530 (1963).
38. J. O. Dimmock and G. B. Wright, *Phys. Rev.* **135**: A821 (1964).
39. P. J. Lin and L. Kleinman, *Phys. Rev.* **142**: 478 (1966).
40. J. B. Conklin, L. E. Johnson, and G. W. Pratt, *Phys. Rev.* **137**: A1282 (1965).
41. D. L. Mitchell and R. F. Wallis, *Phys. Rev.* **151**: 581 (1966).
42. R. Geick, Phys. Letters **10**: 51 (1964).
43. J. N. Zemel, *Proceedings of the International Conference on the Physics of Semiconductors, Paris, 1964*, Dunod, Paris, 1964, p. 1061.
44. E. G. Bylander and M. Hass, *Solid State Commun.* **4**: 51 (1966).
45. R. Kaplan, E. D. Palik, R. F. Wallis, S. Iwasa, E. Burstein, and Y. Sawada, *Phys. Rev. Letters* **18**: 159 (1967).
46. E. Burstein, *Phys. Rev.* **93**: 632 (1954).
47. T. S. Moss, *Proc. Phys. Soc. (London)* **67B**: 775 (1954).
48. J. S. Blakemore, *Semiconductor Statistics*, Pergamon Press, New York, 1962, p. 75.
49. E. R. Washwell and K. F. Cuff, *Radiative Recombination in Semiconductors, Paris, 1964*, Dunod, Paris, 1965, p. 11.
50. J. F. Butler and A. R. Calawa, *Physics of Quantum Electronics, Puerto Rico, 1965*, McGraw-Hill Book Co., New York, 1966, p. 458.
51. M. J. Stephen and A. B. Lidiard, *J. Phys. Chem. Solids* **9**: 43 (1958).
52. R. J. Elliot, T. P. McLean, and G. G. Macfarlane, *Proc. Phys. Soc. (London)* **72**: 553 (1958).
53. J. Halpern, B. Lax, and Y. Nishina, *Phys. Rev.* **134**: A140 (1964).
54. H. S. Bennett and E. A. Stern, Univ. of Maryland Technical Report No. 197 (1960).
55. I. M. Boswarva, Ph. D. dissertation, University of Reading, 1963.
56. I. M. Boswarva, R. E. Howard, and A. B. Lidiard, *Proc. Roy. Soc. (London)* **269**: 125 (1962).

57. H. S. Bennett and E. A. Stern, *Phys. Rev.* **137**: A448 (1965).
58. F. C. Brown and G. Laramore, *Appl. Opt.* **6**: 669 (1967).
59. J. Gareyte and Y. Merle d'Aubigné, *Compt. Rend.* **258**: 6393 (1964).
60. D. L. Mitchell and R. F. Wallis, *Phys. Rev.* **131**: 1965 (1963).
61. E. D. Palik, unpublished data.
62. H. Piller, in: *Proceedings of the International Conference on the Physics of Semiconductors, Paris, 1964*, Dunod, Paris, 1964, p. 297.
63. L. Roth, *Phys. Rev.* **133**: A542 (1964).
64. Y. Yafet, in: *Solid State Physics*, Vol. 14 (F. Seitz and D. Turnbull, eds.), Academic Press, New York, 1963, p. 1.
65. A. H. Kahn and H. P. R. Fredrikse, in: *Solid State Physics*, Vol. 9 (F. Seitz and D. Turnbull, eds.), Academic Press, New York, 1965, p. 262.
66. M. S. Dresselhaus and G. Dresselhaus, *Phys. Rev.* **125**: 499 (1962).
67. M. S. Dresselhaus and J. G. Mavroides, *Solid State Commun.* **2**: 299 (1964).
68. E. A. Stern, *Phys. Rev. Letters* **15**: 62 (1965).

Chapter 4

THE COUPLING OF COLLECTIVE CYCLOTRON EXCITATIONS AND LONGITUDINAL OPTIC PHONONS IN POLAR SEMICONDUCTORS*

Sato Iwasa

Physics Department
Massachusetts Institute of Technology
Cambridge, Massachusetts

Much work has been reported recently on the coupling of longitudinal optic phonons and free-carrier collective excitations in degenerate polar semiconductors in the absence ([1-6,12]) as well as in the presence ([7-11]) of a magnetic field. The coupling takes place via "macroscopic" longitudinal electric fields, producing the combined normal modes of the system which are strong admixtures of the two excitations. We shall briefly review the simpler case of no magnetic field, for which theories and experiments are more extensive, in order to illustrate the general feature of the problem. The theory of the coupling is based upon the assumption that the polarizability of the free carriers in the random phase approximation and that of the polar ions contribute additively to make up the total dielectric constant. The $\mathbf{q}$-dependent complex dielectric function derived by Lindhard ([13]), or its approximation ([2]), is invoked to describe the former, and the polar lattice dielectric constant derived by Born and Huang ([14]) the latter. Free longitudinal oscillations will then occur in the system whenever the conditions are such that the total dielectric constant equals zero. The above procedure is particularly simple in the long-wavelength limit $\mathbf{q} \approx 0$. Assuming a polar lattice having n free carriers with one isotropic mass m^*, we have

$$\varepsilon_{\text{tot}}(0, \omega) = \varepsilon_\infty \left(1 + \frac{\omega_L - \omega_T{}^2}{\omega_T{}^2 - \omega^2} - \frac{\omega_p{}^2}{\omega^2} \right) \tag{1}$$

* Work supported in part by the U. S. Office of Naval Research.

where ε_∞ is the high-frequency dielectric constant, ω_T and ω_L are, respectively, the transverse optic (TO) and longitudinal optic (LO) phonon frequencies, and $\omega_p{}^2 = 4\pi n e^2/\varepsilon_\infty m^*$ is the $\mathbf{q} \approx 0$ plasmon frequency. By putting $\varepsilon_{tot}(0, \omega) = 0$, two resonance frequencies are found from the following quartic equation, in which ω_+ and ω_- correspond, respectively, to the high- and low-frequency normal modes:

$$\omega_\pm{}^4 - (\omega_p{}^2 + \omega_L{}^2)\omega_\pm{}^2 + \omega_p{}^2\omega_T{}^2 = 0 \tag{2}$$

In the ω_+ mode the lattice and electronic polarizations point in the same direction, whereas in the ω_- mode they point in opposite directions. For low carrier concentrations ($\omega_p \ll \omega_T, \omega_L$) the ω_+ mode is phonon-like in character, whereas the ω_- mode is plasmon-like, and for high carrier concentrations ($\omega_p \gg \omega_T, \omega_L$) the reverse is true.

The following two types of optical experiment can be performed. The two normal modes, resulting from the coupling of $\mathbf{q} \approx 0$ plasmons as the free-carrier collective excitations and $\mathbf{q} \approx 0$ LO phonons, are purely longitudinal in character and, as such, cannot interact with transverse electromagnetic (EM) radiation in bulk samples or in thin films under normal incidence conditions. An interaction of the coupled modes with EM radiation can, however, be observed in thin-film, oblique-incidence transmission experiments (Fig. 1) [9]. The absorption near the plasma frequency ω

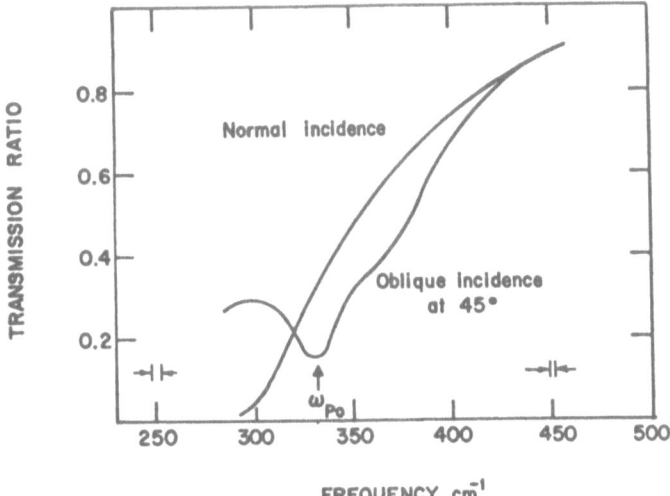

Fig. 1. Transmission spectra of single-crystal film of doped n-GaAs at normal incidence and at oblique incidence of $45°$. From Iwasa *et al.*[9].

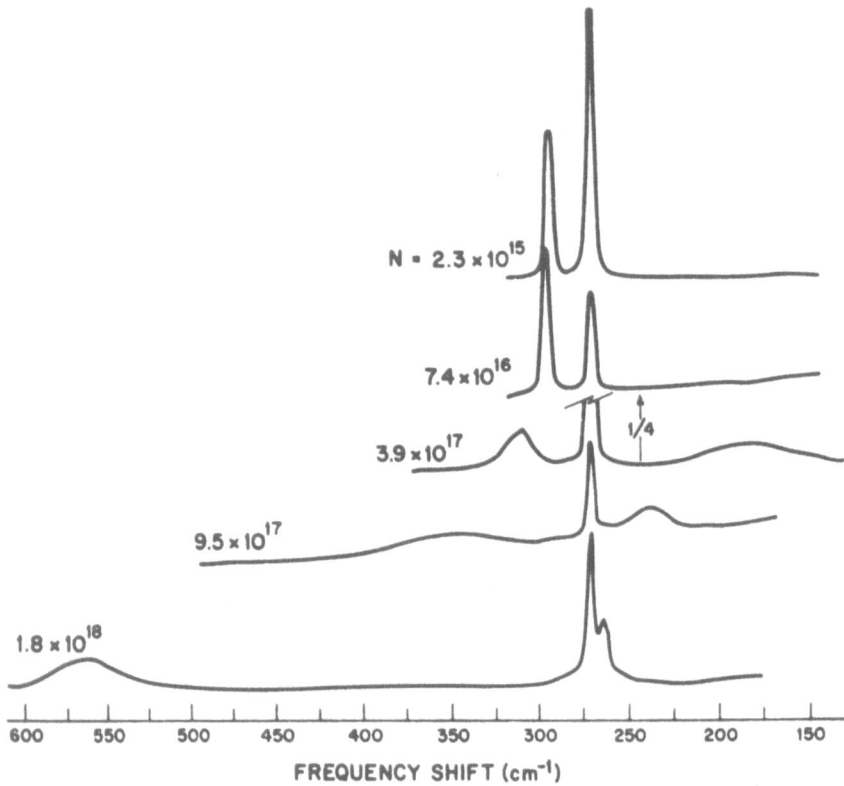

Fig. 2. The anti-Stokes Raman spectrum of *n*-type GaAs. From Mooradian and Wrigth([6]).

$= \omega_{+} \approx \omega_{p}$ is present for oblique incidence with radiation polarized in the plane of incidence, but is absent when the radiation is polarized perpendicular to the plane of incidence or is normally incident on the film. The low-frequency mode at $\omega = \omega_{-}$ is not resolved in this experiment. The coupled modes are also observable in Raman scattering experiments on semiconductors lacking a center of inversion (Fig. 2) ([6]). Three peaks are seen, one corresponding to the TO phonon and two to the LO normal modes, and the frequency dependence of the latter on the carrier concentration has shown good agreement with theory [Eq. (2)]. On the other hand, the measurement of the relative strengths and linewidths of the three Raman peaks and their theoretical interpretation have not yet been fully established. It is not only a question of the interplay of the phonon and plasmon content of a given mode, but somewhat different Raman scattering mechanisms may have to be invoked for the TO phonons and the LO normal modes in a non-centrosymmetric crystal ([15–17]). One theory on the Raman scattering efficiency by the LO normal modes, which takes into account the additional contribution due to the electrooptical effect of the longitudinal macroscopic depolarization field, has been reported ([12]).

The dispersion of the coupled modes presents yet another important problem. The **q**-dependence arises mainly from the electronic dielectric response, which is a sensitive function of the wave vector. A few numerical calculations have been undertaken ([1,2,5]). For a high carrier concentration the upper mode, $\omega_+(\mathbf{q})$, exhibits a near-quadratic dependence, as to be expected from that of the pure plasmon, whereas the lower mode, $\omega_-(\mathbf{q})$, shows something more complex—it increases monotonically from slightly below ω_T at $\mathbf{q} = 0$, until it asymptotically approaches ω_L as $\mathbf{q}$ becomes comparable to $\mathbf{q}_F$, the Fermi momentum of the electrons (Fig. 3) ([2]). Since the momentum of photons is very small, optical experiments cannot

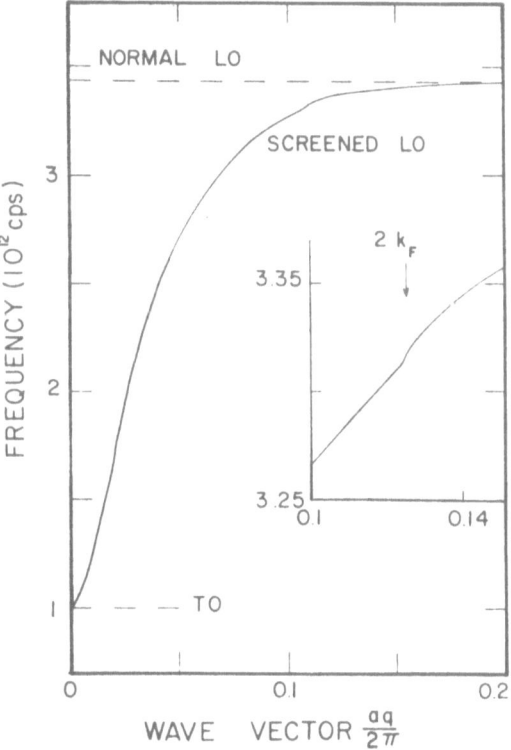

Fig. 3. The calculated LO dispersion curve in doped PbTe. The inset shows the discontinuity in the slope of the curve when the wave vector is equal to the Fermi surface diameter. From Cowley and Dolling ([2]).

probe the dispersion relation. A neutron diffraction experiment has confirmed the qualitative feature of Fig. 3, but by no means exactly [2]. No experimental information is available on the $\omega_+(\mathbf{q})$ dispersion, however.

We now introduce a static magnetic field H_0 and examine the dielectric response of the plasma alone while letting the polar phonons remain in the background. For the sake of simplicity, an isotropic plasma confined in a slab-shaped space is considered. In the Voigt (transverse) configuration, in which H_0 is parallel to the slab, the motion of each electron gives rise to a macroscopic depolarization field perpendicular to the surface, which then exerts force on all the other electrons. Hence all the electrons move "collectively." Upon closer inspection it is found that given n free electrons per unit volume, there exist altogether n normal modes ([18,19]): $n - 1$ modes in which the phase of the electron motion is related by $r_{k+1}/r_k = \exp(2\pi l_i/n)$ $(k, l = 1, 2, \ldots, n - 1)$, have their resonance at the cyclotron frequency $\omega_c = eH_0/m^*c$, and the remaining one mode in which all the n electrons move in phase has a resonance frequency given by $(\omega_c^2 + \omega_p^2)^{1/2}$. The first $n - 1$ modes produce zero net electric moment $(\sum_1^n e r_k = 0)$ and therefore do not interact with a time-varying electric field. The one mode at $(\omega_c^2 + \omega_p^2)^{1/2}$, called the collective cyclotron resonance mode, produces the net polarization of $\mathbf{p} = ne\mathbf{r}$ and will interact with electric fields (that are not parallel to H_0), and by the same token to the polar LO phonon, as will be discussed subsequently. Indeed, normal incidence infrared transmission experiments in the Voigt configuration carried out on a thin film of heavily doped n-type InSb $(\omega_p \gg \omega_T, \omega_L)$, show a resonance absorption band at the frequency given by $(\omega_c^2 + \omega_p^2)^{1/2}$ for EM radiation polarized $E \perp H_0$ [9]. No resonance absorption is seen for $E \parallel H_0$. In the Faraday (longitudinal) configuration, in which H_0 is perpendicular to the slab, the electron motion does not produce the depolarization field. The electrons are decoupled and move "singly." Hence there is no coupling to the LO phonon. The similar transmission experiments show an absorption band at ω_c for the left circularly polarized radiation but none for the right.

The calculation of the coupling between the collective cyclotron excitation and the polar LO phonon in the Voigt configuration depends, as in the zero field case, on two major assumptions, (1) that the lattice and electronic polarizations contribute additively to make up the macroscopic internal polarization P_{int}, and (2) that both the lattice and electrons are driven by the same effective macroscopic field given by $E_{eff} = E_{ext} - LP_{int}$, L being the depolarization factor, which is 4π for the thickness direction and 0 for directions parallel to the surface. There are two approaches equally feasible. One is to express the internal polarization P_{int} in terms of the externally applied field E_{ext} by solving a set of classical equations of motion for the phonon and electron amplitudes. The proportionality factor between P_{int} and E_{ext} is a dielectric "response" constant and is dependent on the actual size and shape of a specimen used in an experiment ([20]). This approach allows more physical insight, since E_{ext} is (together with H_0) the only parameter accessible to the outside. The result is particularly useful in analyzing thin-film transmission data. The other approach* is to work within the specimen and study the relationship between P_{int} and the local internal field E_{int}. The proportionality factor is a dielectric constant and is independent of the geometry of the experiment. One advantage here is the ease with which one can calculate the reflection spectra from a bulk sample or the transmission spectra from a slab of finite thickness.

We first consider the dielectric response constant. The equations of motion for the (collective) electron displacement $\mathbf{r}$ and the polar phonon amplitude $\mathbf{u}$ are given by

$$\frac{d^2\mathbf{r}}{dt^2} + \gamma \frac{d\mathbf{r}}{dt} - \omega_c\left(\frac{d\mathbf{r}}{dt} \times \frac{\mathbf{H}_0}{H_0}\right) = \frac{e}{m^*}\mathbf{E}_{\text{eff}} \tag{3a}$$

and

$$\frac{d^2\mathbf{u}}{dt^2} + \Gamma\frac{d\mathbf{u}}{dt} + \omega_T^2\mathbf{u} = \frac{e^*}{M}\mathbf{E}_{\text{eff}} \tag{3b}$$

where γ and Γ are, respectively, the collision frequencies of electrons and phonons. We shall omit the calculational detail and merely quote the expression for the transverse electric-susceptibility response constant, which gives the net internal polarization parallel to the surface for a unit external electric field applied parallel to the surface ([9]):

$$\chi_T^r(\omega) = \frac{\varepsilon_\infty - 1}{4\pi} +$$

$$\frac{\varepsilon_\infty}{4\pi}\left[\frac{\omega_L^2 - \omega_T^2}{\omega_T^2 - \omega^2 - i\Gamma\omega} - \frac{\omega_p^2}{\omega^2 + i\gamma\omega} + \frac{\omega_c^2\omega_p^2(\omega_L^2 - \omega^2 - i\Gamma\omega)}{(\omega + i\gamma)^2 D}\right] \tag{4}$$

where

$$D = \{\omega_p^2 - \omega^2 - i\gamma\omega + [\omega\omega_c^2/(\omega+i\gamma)]\}(\omega_L^2 - \omega^2 - i\Gamma\omega) - \omega_p^2(\omega_L^2 - \omega_T^2)$$

The second term in $\chi_T^r(\omega)$ has a singularity at $\omega = \omega_T$ and the last term at frequencies such that $D = 0$. The latter correspond to the two coupled normal modes which, when neglecting losses, are given by

---

* If one takes the trouble of relating E_{int} to E_{ext} for a given geometry, the dielectric constant results are seen to reduce to the previous dielectric response constant results.

$$2\omega_{\pm}^2 = \omega_p^2 + \omega_c^2 + \omega_L^2 \pm [(\omega_p^2 + \omega_c^2 + \omega_L^2)^2 - 4(\omega_p^2\omega_T^2 + \omega_c^2\omega_L^2)]^{1/2} \quad (5)$$

Using the above $\chi_T^r(\omega)$, the normal incidence transmission of a thin film for which $2\pi d/\lambda_0 \ll 1$, where d is the thickness and λ_0 is the vacuum wavelength of the radiation, is given by [9,20]

$$T = 1 - (2\pi d/\lambda_0)\,\mathrm{Im}[\chi_T^r(\omega)] \quad (6)$$

Normal-incidence transmission measurements in the Voigt configuration carried out on thin films of n-type InSb having concentrations such that ω_p is 18, 135, and 205 cm^{-1} (cf. $\omega_T = 184$ cm^{-1}), exhibit three resonance absorption bands at the above mentioned frequencies. The dependence of ω_+ and ω_- on the magnetic field is seen to agree well with Eq. (5). One may note with interest that the ratio of the phonon and electron content of a given normal mode can be calculated from the last term in Eq. (4), for instance, as a function of H_0 at a fixed frequency or as a function of ω at a fixed value of H_0. It is not surprising then that a phonon-rich mode is associated with the phonon damping constant Γ, whereas an electron-rich mode is associated with the electron damping constant γ. Since Γ and γ are normally different in magnitude, this leads to the field-dependent narrowing (or broadening) of the resonance absorption bands, as with the field-dependent absorption intensities. These effects are clearly observed in Fig. 4 and the dependence on H_0 of the apparent width and the depth of the upper and lower modes is in qualitative agreement with the theoretical prediction based on Eqs. (4) and (6).

Let us now return to the orthodoxy of the dielectric constant. In the Faraday configuration the total dielectric constant is given by

$$\varepsilon_{r,l} = \varepsilon_\infty\left[1 + \frac{\omega_L^2 - \omega_T^2}{\omega_T^2 - i\Gamma\omega_T - \omega^2} - \frac{\omega_p^2}{\omega(\omega \pm \omega_c + i\gamma)}\right] \quad (7)$$

where r and l refer to the right and left circularly polarized radiation, respectively. That there is no coupling can be inferred from the nonmixing singularities at ω_T for ε_r and at ω_T and at ω_c for ε_l. The bulk reflectivity is calculated from Eq. (7) for the following typical choice of the parameters: $\omega_T = 184$ cm^{-1}, $\omega_L = 196$ cm^{-1}, $\Gamma = 1.3$ cm^{-1}, $\omega_p = 135$ cm^{-1}, $\gamma = 4.5$ cm^{-1}, and $\omega_c = 75$ cm^{-1}, simulating a lightly doped (5.5×10^{16}) n-InSb in the field of 13.6 kG (Fig. 5). The dielectric constant in the Voigt configuration is expressed in terms of ε_r and ε_l as follows:

$$\varepsilon_{\parallel} = \varepsilon_\infty\left(1 + \frac{\omega_L^2 - \omega_T^2}{\omega_T^2 - \omega^2 - i\Gamma\omega} - \frac{\omega_p^2}{\omega(\omega \pm \omega_c + i\gamma)}\right)$$

$$\varepsilon_{\perp} = 2\varepsilon_r\varepsilon_l/(\varepsilon_r + \varepsilon_l) \quad (8)$$

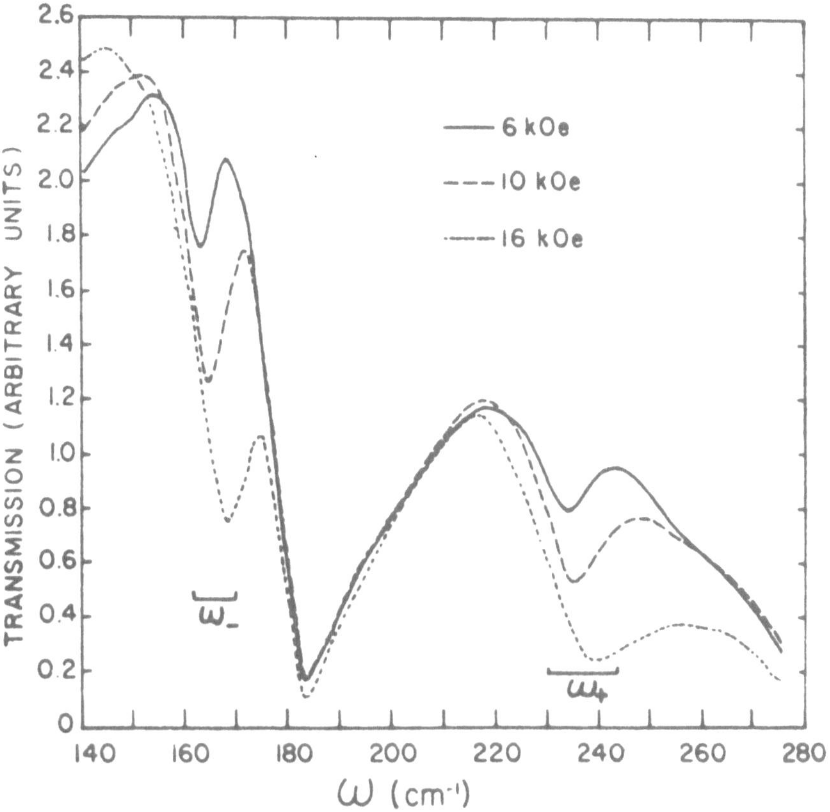

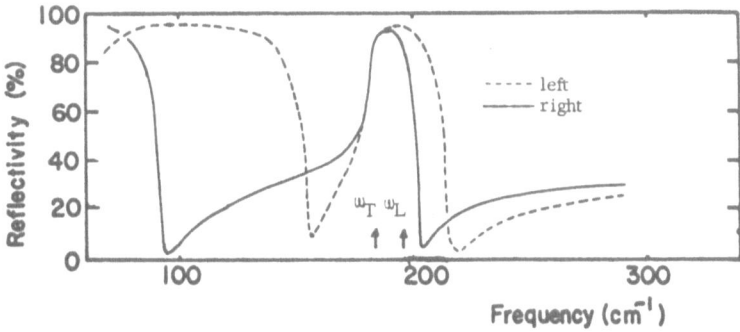

Fig. 4. Transmission spectra of n-InSb with $N = 1.4 \times 10^{17}\,\mathrm{cm}^{-3}$ for three values of magnetic field. The regions of coupled-mode absorption are indicated by brackets. From Kaplan *et al.* [11].

Fig. 5. The calculated reflection spectra in the Faraday configuration for the right and left circularly polarized light. The parameters used for n-InSb ($n = 5.5 \times 10^{16}\,\mathrm{cm}^{-3}$) with $H_0 = 13.6\,\mathrm{kG}$: $\omega_T = 184\,\mathrm{cm}^{-1}$, $\omega_L = 196\,\mathrm{cm}^{-1}$, $\Gamma = 1.3\,\mathrm{cm}^{-1}$, $\omega_p = 135\,\mathrm{cm}^{-1}$, $\gamma = 4.5\,\mathrm{cm}^{-1}$, $\omega_c = 75\,\mathrm{cm}^{-1}$, and $\varepsilon_\infty = 16$.

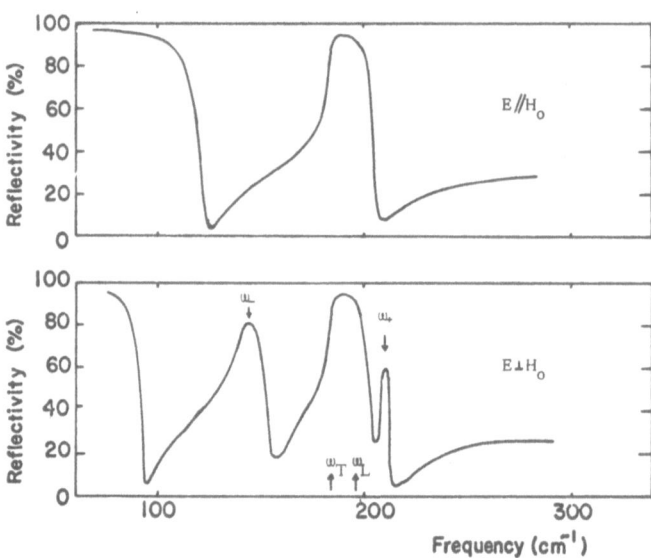

Fig. 6. The calculated reflection spectra in the Voigt configuration for the light polarized parallel and perpendicular to H_0: the frequencies of the coupled modes are indicated by arrows. The same set of parameters is used as in Fig. 5.

where $\varepsilon_{\parallel}$ and $\varepsilon_{\perp}$ are the dielectric constant for the ordinary wave ($E \parallel H_0$) and for the extraordinary wave ($E \perp H_0$), respectively. The ordinary wave shows no resonance, and it is in fact identical to the case of no magnetic field [Eq. (1)]. The reduction of $\varepsilon_{\perp}$ is a rather cumbersome affair, but the result indicates that $\varepsilon_{\perp}$ contains singularities at the identical frequencies $\omega_{\pm}$ as given in Eq. (5). One can extract from $\varepsilon_{\perp}$ various optical parameters, from which the reflection spectra of a bulk specimen and the transmission spectra of a slab of finite thickness may be obtained. The absorption coefficient arising from $\varepsilon_{\perp}$ for the lossless case has been reported ([10]). We shall similarly compute the bulk reflectivity in this configuration for the same set of parameters (Fig. 6). One may note the structures appearing at the normal mode frequencies as indicated by the arrows.

In summary, one is led to conclude that the problem of coupled modes is still at an initial stage of development. Only the simplest case of one-component isotropic plasma has been treated so far, and even then we lack (1) definitive magneto-reflection measurements, (2) magneto-Raman experiments and theory, and (3) the dispersion in the presence of magnetic field. With regard to (3), the dispersion relation for the plasma alone derived to date ([21]) is quite complex and defies easy application to the solution of the coupling problem. The problem assumes an interesting dimension when, as in the polariton problem, the mixing of the transverse photon field is considered ([22]), or, as in lead salts, multi-component anisotropic plasmas need be considered. Some new results are no doubt forthcoming in the near future.

I am grateful to Professor E. Burstein for his guidance throughout my involvement with this problem.

REFERENCES

1. B. B. Varga, *Phys. Rev.* **137**: A1896 (1965).
2. R. A. Cowley and G. Dolling, *Phys. Rev. Letters* **14**: 549 (1966).
3. R. Tsu and D. L. White, *Ann. Phys. (N.Y.)* **32**: 100 (1965).
4. Y. C. Lee and N. Tzoar, *Phys. Rev.* **140**: A396 (1965).
5. K. S. Singwi and M. P. Tosi, *Phys. Rev.* **147**: 658 (1966).
6. A. Mooradian and G. B. Wright, *Phys. Rev. Letters* **16**: 999 (1966).
7. T. N. Casselman and H. N. Spector, *Phys. Condensed Matter* **4**: 179 (1965).
8. M. P. Greene, A. Houghton, and J. J. Quinn, *Phys. Letters* **20**: 238 (1966).
9. S. Iwasa, Y. Sawada, E. Burstein, and E. D. Palik, *J. Phys. Soc. Japan Suppl.* **21**: 742 (1966).
10. I. Yokota, *J. Phys. Soc. Japan Suppl.* **21**: 738 (1966).
11. R. Kaplan, E. D. Palik, R. F. Wallis, S. Iwasa, E. Burstein, and Y. Sawada, *Phys. Rev. Letters* **18**: 159 (1967).
12. E. Burstein, A. Pinzczuk, and S. Iwasa, *Phys. Rev.* **157**: 611, (1967).
13. J. Lindhard, *Kgl. Danske Videnskab. Selskab, Mat.-Fys. Medd.* **28**: 8 (1954).
14. M. Born and K. Huang, *The Dynamical Theory of Crystal Lattices*, Clarendon Press, Oxford, 1956.
15. H. Poulet, *Compt. Rend.* **238**: 70 (1954); *Ann. Phys. (Paris)* **10**: 908 (1955).
16. R. Loudon, *Proc. Roy. Soc. (London)* **A265**: 218 (1963).
17. J. L. Birman and A. K. Ganguly, *Phys. Rev. Letters* **17**: 647 (1966).
18. R. F. Wallis, unpublished calculation (1964).
19. B. Rosenblum, *Bull. Am. Phys. Soc.* **9**: 59 (1964).
20. E. Burstein, S. Iwasa, and Y. Sawada, *Nuovo Cimento, Suppl. to Vol. 32* (1965).
21. N. J. Horing, *Plasma Effects in Solids*, Dunod, Paris, 1964, p. 107.
22. J. J. Hopfield, *J. Phys. Soc. Japan Suppl.* **21**: 77 (1966).

Chapter 5

APPLICATION OF GAS LASERS TO HIGH-RESOLUTION MAGNETOSPECTROSCOPY*

Sato Iwasa

Physics Department
Massachusetts Institute of Technology
Cambridge, Massachusetts

INTRODUCTION

The study of oscillatory interband magnetoreflection spectra has yielded much information on the energy band structure of semiconductors and semimetals. The oscillations in reflectivity, normally measured as a function of magnetic field at constant photon energies in the infrared, are associated with direct transitions from Landau levels in the valence band to Landau levels in the conduction band. In the limit of very small electronic losses both the top of valence Landau levels and the bottom of conduction Landau levels have singularly high densities of states, and the interband transition has a sharp cutoff at a critical magnetic field at which the minimum separation between the paired levels equals the photon energy. The cutoff causes a discontinuity in the reflectivity, thus making the so-called oscillatory spectra a series of reflectivity breaks occurring at each critical field for different pairs of states ([1]). Indeed, one observes such spectra on recently available high-purity single crystals of bismuth at liquid helium temperature. The spectroscopic measurements of this resonant phenomenon will require, in addition to homogeneous magnetic fields, a high degree of monochromaticity in the infrared, which may exceed the conventional capacity of dispersive spectrometers. Less critical than the monochromaticity requirement is the frequent need of pure polarization, sometimes linear and

* This research is supported by the National Aeronautics and Space Administration and Air Force Cambridge Research Laboratories.

sometimes circular, in determining the true reflectivity line shape related to transitions with simple selection rules.

Such considerations have motivated the construction of an infrared laser spectrometer using Ne–He and Xe–He gas mixtures for the purpose of high-resolution magnetospectroscopy of semimetals. In this chapter I describe recent progress in the investigation of single crystals of graphite, bismuth, and arsenic and pyrolitic graphite crystals. The research is a collaborative effort of Mr. P. R. Schroeder and Professor A. Javan of Massachusetts Institute of Technology, who had initiated the project a few years ago, and myself, who joined the group one year ago.

TABLE I

Available Laser Lines and their Relative Intensities

Neon laser lines (Ne 30 μ, He 40 μ)		Xenon laser lines (Xe 0.5 μ, He 80 μ)	
λ_{air} (μ)	Relative intensity*	λ_{air} (μ)	Relative intensity*
3.3913	0.24	3.1069	0.05
5.4033	0.030	3.3667	0.08
7.4779	0.009	3.5070	0.66
7.6142	0.009	3.9955	0.08
7.6489	0.30	5.5738	1.00
7.6994	1.00	7.3147	0.10
7.7634	0.25	9.0040	1.00
8.0066	0.18	11.290	0.11
8.0599	0.44	12.913	0.80
10.978	0.008	18.501	0.50
11.898	0.006		
12.831	0.10		
16.664	0.006		
16.889	0.006		
16.943	0.013		
17.153	0.014		
17.800	0.004		
17.837	0.004		
17.884	0.010		
18.392	0.003		
21.746	0.006		

* Measured by a copper-doped germanium detector cooled by liquid helium after the laser beam traveled through the KBr window, a KRS-5 window, and a diffraction grating 40 lines/mm blased at 22.5 μ.

LASER SPECTROMETER

Figure 1 describes the experimental arrangement. The laser, which is 2.5 cm in diameter and 4 meters long, operates continuously at some 30 wavelengths in the range $3.1069\mu \leq \lambda \leq 21.746\mu$ [2] and provides a light beam of about 1 cm in diameter, almost 100% plane polarized, by means of CsI Brewster window coupling (see Table I). When necessary the combination of an image-tilting device and a CsI Fresnel rhomb produces nearly perfect right or left circular polarization. The laser intensity is monitored either by a Au-doped Ge detector cooled by liquid N_2, or by a Cu-doped Ge detector cooled by liquid helium, at a chopping frequency of 720 cycles/ sec. The amplitude of any one of the listed lines, where the spectral separation is achieved by the use of a Bausch and Lomb 40 lines/mm diffraction grating blazed at 22.5μ, can be stabilized to better than 0.5% in level fluctuations by regulating the discharge current through an electronic feedback system. The experiment consists of observing oscillations in the reflectivity while the magnetic field is varied at a fixed photon energy. The high degree of monochromaticity of a laser line makes it an ideal light source for this type of measurement. In order to allow small fluctuations in the laser amplitude, the reflected signal is compared with the laser level either by a narrow-band differential amplifier or by ratio-taking electronics. The signal to noise ratio is further improved by the use of an electronic data processing device, the Enhancetron, which averages the signal over several magnetic field sweeps. A small reflectivity change of $\Delta R/R = 0.001$ has been resolved. A Bitter-type magnet at the National Magnet Laboratory provides the field up to 100 kG.

GRAPHITE

The two types of oscillatory interband magnetoreflection spectra [3] have been obtained for both pyrolitic and single crystals in the Faraday configuration, in which both the magnetic field H and the laser Poynting vector S are perpendicular to the c face of graphite. The measurements on a miniscual single-crystal disk of 3 mm in diameter carried out at He temperature in the wavelength range of $5.4\mu \leq \lambda \leq 21.8\mu$ constitute the first observation of the kind and clearly demonstrate an important advantage resulting from fine focusibility of the laser beam (Fig. 2). For one series of oscillations, which arises from interband transitions between the two E_3 bands near point K in the Brillouin zone, reflectivity peaks for a given photon energy occur at slightly different magnetic fields for left and right circular

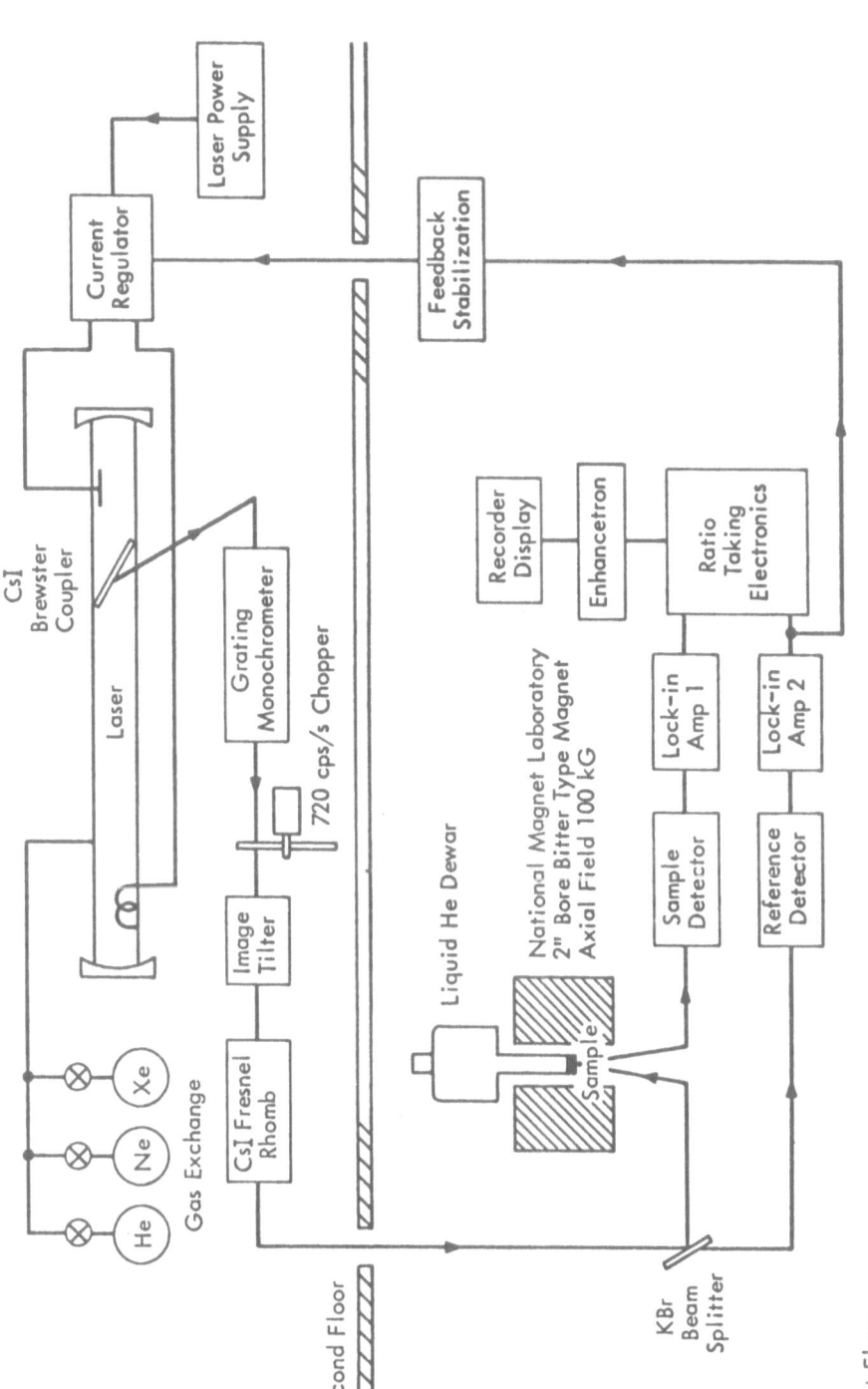

Fig. 1. Experimental arrangement of the infrared laser spectrometer with 100 kG axial magnet.

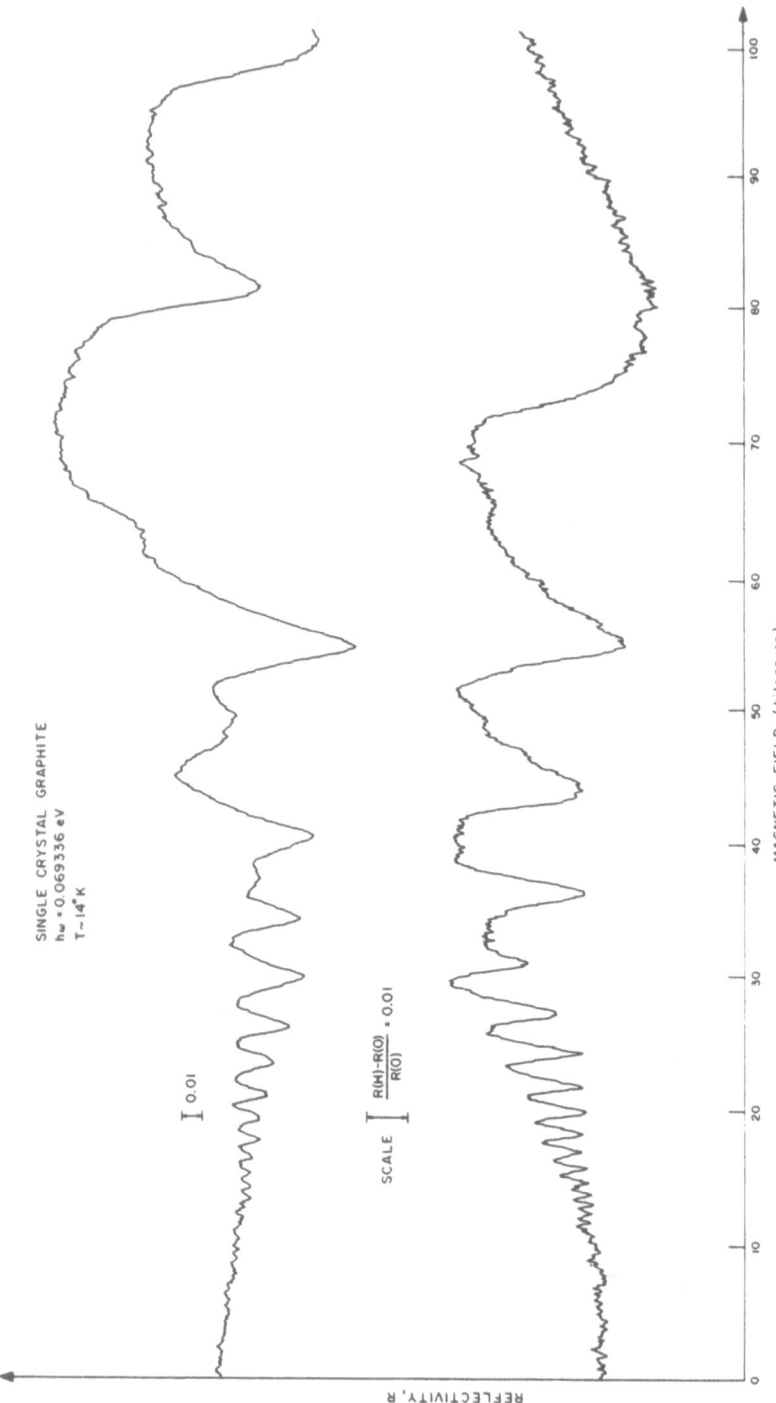

Fig. 2. Experimental traces of the oscillatory reflectivity of a single-crystal graphite sample in a magnetic field. Upper trace: for left circular polarization; lower trace: for right circular polarization. The oscillations arise from transitions about point K of the graphite Brillouin zone. Photon energy 0.069336 eV; $S \parallel H \parallel c$ axis.

polarization. This effect is a direct consequence of the small difference in the band curvature for the conduction and valence Landau levels and is brought about through the optical transition selection rule of $\Delta n = \pm 1$ for the corresponding sense of circular polarization. The difference has been accurately determined for various photon energies. Asymmetric line shapes have been observed and are fitted reasonably well by an initial line shape calculation ([4]). For the other series, arising from transitions between the degenerate E_3 bands and the degenerate E_1 and E_2 bands near H point in the Brillouin zone, the measured dependence of the Landau levels on magnetic field is identical for single and pyrolitic crystals. This is contradictory to the de Haas–von Alphen data and requires further clarification. A number of band parameters for both types of crystal can be deduced via detailed analysis of such measurements, which Mr. Schroeder is currently engaged in for his doctoral thesis. We have the collaboration of Dr. M. S. Dresselhaus of Lincoln Laboratory, M.I.T. on the theoretical aspect of the problem.

BISMUTH

The interband reflection spectra in the Faraday configuration have been obtained in two crystallographic orientations; one in which the field and the Poynting vector are parallel to the binary axis, and the other in which they are parallel to the bisectrix axis. The large oscillations shown in Fig. 3 are associated with the two electron ellipsoids, which are equivalent for the field parallel to the binary axis ([5]). The line shapes, particularly of lower quantum transitions, are more complex because of the simultaneous presence of the cyclotron resonance and the intraband magnetoplasma effect, and therefore are of theoretical interest. Here again monochromaticity of the laser spectrometer enables an unambiguous (not limited by the spectral resolution of a conventional optical system) determination of the highly asymmetric line shapes. There is a tentative indication that the line shape of higher quantum transitions undergoes a systematic change as the Landau bands of more than 0.1 eV above and below the Fermi surface are invoked in the optical process. This effect, if confirmed, is of considerable interest, since optical measurements alone can probe the band structure away from the Fermi surface, whereas in transport measurements the information is confined to within the surface. Fine structures riding on the higher-field tail of the main oscillations, which may be identified as spin nonflip transitions with the selection rule $\Delta n = 0$, $\Delta m = 0$, are under investigation.

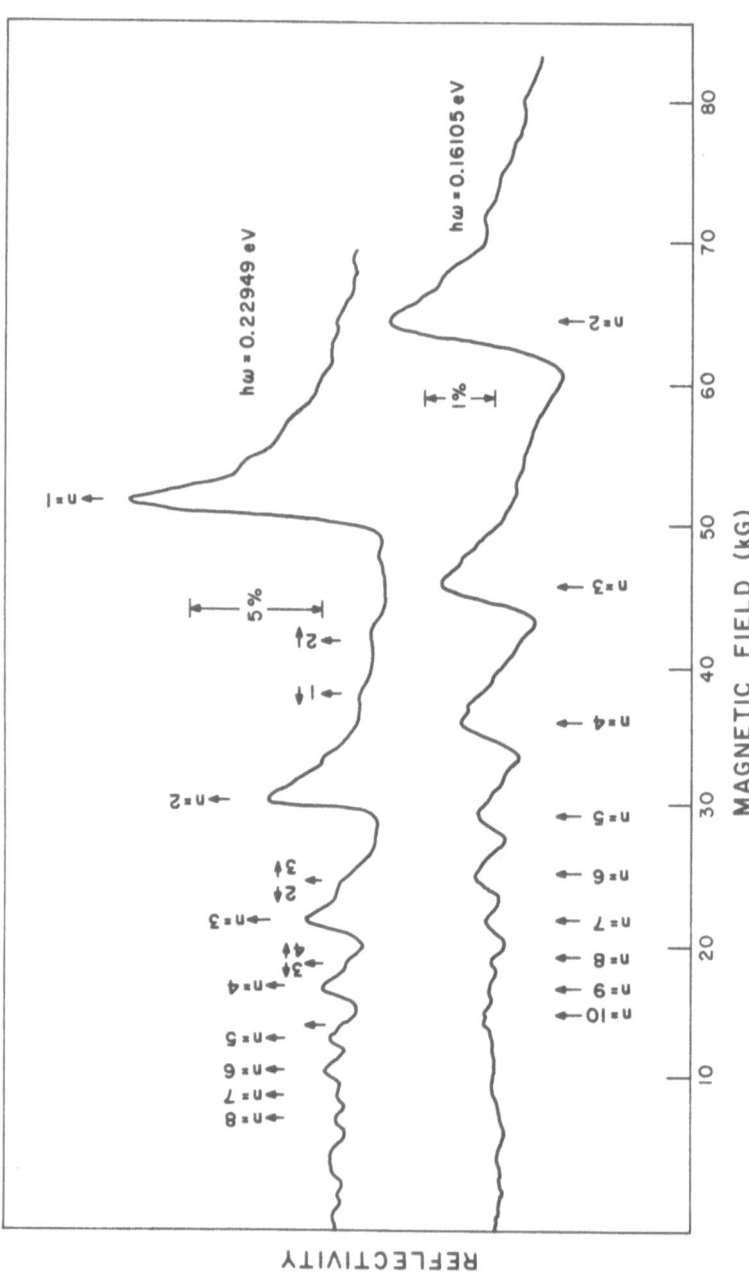

Fig. 3. Experimental traces of magnetoreflection of bismuth, where the left circularly polarized beam is incident on a binary face; S ‖ H ‖ binary axis.

ARSENIC

Recent magnetoreflection measurements using a conventional optical system have yielded small (amplitude) but well-defined oscillations for the case of S ∥ binary and S ∥ bisectrix, and unusually large and complex oscillations for S ∥ trigonal ([6]) (Fig. 4). For the latter case laser spectroscopy has separated out the presence of two sets of oscillations, one independent

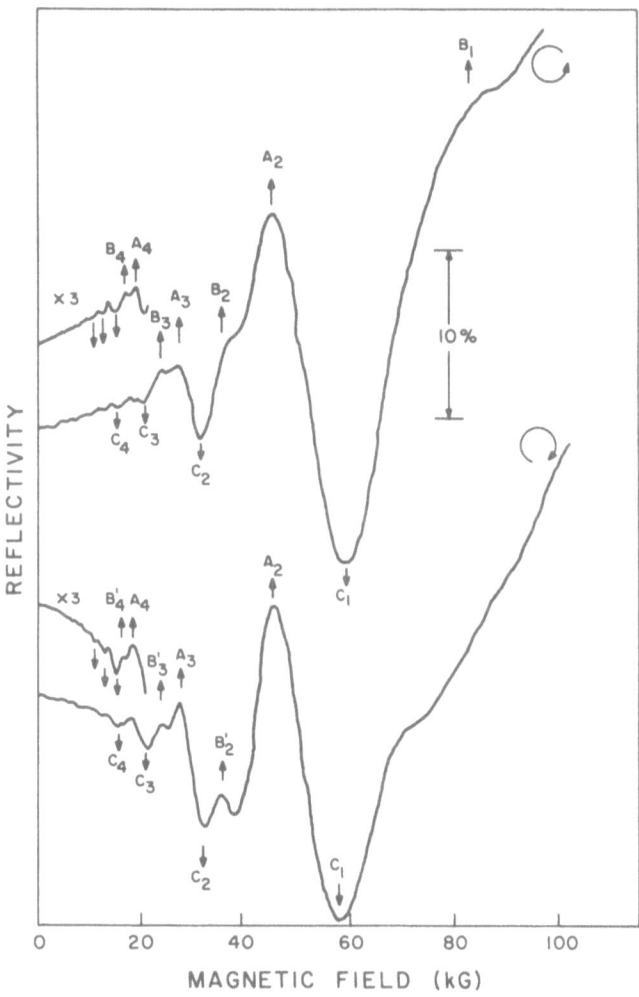

Fig. 4. Experimental traces of magnetoreflection of arsenic where the left and right circularly polarized light is incident on a trigonal face; S ∥ H ∥ trigonal axis.

of the sense of circular polarization and the other dependent on it. It is also found that the background reflectivity upon which the oscillations are superimposed increases with field for the left circular polarization, while it decreases for the right, suggesting the proximity of a classical plasma edge to photon energies of the interband transitions in question. The early stage of theory on the arsenic band structure is such that the origin of the bisectrix and binary oscillations is yet unknown and the trigonal oscillations are tentatively assigned to a part of the hole carrier surface ([7]). A trial theory is being developed in which the spectra result from the interplay of the intra- and interband contributions to the total conductivity. To help clarify the problem, reflection spectra are currently measured where S is obliquely incident by a gradually increasing angle to the trigonal axis.

In summary, the technique for high-resolution laser magnetospectroscopy has been developed for straight reflection measurements and has been successfully applied to the investigation of interband transitions in a few semimetals. Versatility of the spectrometer will be enhanced greatly by incorporating various synchronous detection methods which recently have become available in conjunction with conventional spectroscopy, such as magnetic field modulation, electroreflectance modulation, and piezoelectric stress modulation. The natural extension of the present work into high-resolution measurements of the inter- and intraband magnetoabsorption, the Faraday rotation, and the Voigt effect in solids is being contemplated.

REFERENCES

1. M. S. Dresselhaus and G. Dresselhaus, *Phys. Rev.* **125**: 449 (1962).
2. W. R. Bennett, Jr., *Appl. Opt. Suppl.* **2**: 3 (1965).
3. M. S. Dresselhaus and J. G. Mavroides, *Carbon* **1**: 263 (1964).
4. M. S. Dresselhaus, G. Dresselhaus, J. G. Mavroides, and P. R. Schroeder, *Bull. Am. Phys. Soc.* **11**: 256 (1966).
5. R. N. Brown, J. G. Mavroides, and B. Lax, *Phys. Rev.* **129**: 2055 (1963).
6. M. Maltz and M. S. Dresselhaus, to be published.
7. P. J. Lin and L. M. Falicov, *Phys. Rev.* **142**: 441 (1966).

Chapter 6

QUANTUM MAGNETOOPTICS AT HIGH FIELDS

Benjamin Lax* and Kenneth J. Button

Francis Bitter National Magnet Laboratory†
Massachusetts Institute of Technology
Cambridge, Massachusetts

INTRODUCTION

Magnetooptics dates back to the discovery by Michael Faraday in 1845 of what is now known as the Faraday effect. Near the turn of the century the more important Zeeman effect was discovered. It was subsequently explained by Lorentz, who considered a classical treatment of a bound electron in a magnetic field. Later, when quantum mechanics was developed, these phenomena were treated more rigorously and the Zeeman effect was then used to elucidate the structure of the atom. The Faraday rotation was treated quantum mechanically by Rosenfeld ([1]). This treatment only applied to atoms or to discrete impurity states in a host crystal. No significant progress in the treatment of magnetooptical phenomena in solids took place, however, until the 1950's. Rapid progress then occurred because of a series of timely coincidences in the advancement of technology and the development of theoretical concepts. Technical breakthroughs included the synthesis of pure single crystals of high quality and perfection, the widespread use of low-temperature techniques in the laboratory, the duplication of high-intensity magnetic field facilities, advances in microwave, millimeter wave, and infrared techniques, and, most recently, the introduction of the laser. On the theoretical side the most important development was the recognition of the quantum properties of electrons in a solid in the presence of a magnetic field.

* Also Physics Department, Massachusetts Institute of Technology.
† Supported by the U. S. Air Force Office of Scientific Research.

The Zeeman effect had been an important tool for atomic spectroscopy because it involved the study of resonance spectra of transitions between discrete states where the degeneracies had been removed by the presence of the magnetic field. From the theoretical point of view the application of the magnetic field lowers the symmetry of the system and permits the analysis and identification of the spectral structure. Until about 1953 transitions between discrete states could not be observed in solids such as metals and semiconductors because the energy levels merged into complicated energy bands. The detailed nature of these bands could not be identified from the scant quantitative information available from microwave, infrared, and optical spectroscopy. The infrared or optical spectroscopic studies gave information only about the location of the band edges on an energy scale, but little could be learned of the band curvatures, symmetries, or locations within the Brillouin zone. It was the magnetic field, however, which served to requantize the energy levels in the bands. These quasidiscrete levels in two dimensions led once again to resonance spectroscopy analogous to that carried out earlier for the study of atoms. Now the energy bands in solids could be studied in detail by means of the new quantum magnetooptical phenomena ([2]) discovered in the 1950's. The application of the magnetic field again lowered the symmetry of the system and permitted the examination of detailed resonance spectra which were interpreted by means of the new theoretical modes of the energy bands.

The basic concepts responsible for the new phenomena of quantum magnetooptics reside in the solution of the effective-mass Hamiltonian of an electron in the presence of a magnetic field.

$$\mathscr{H} = \frac{1}{2m^*} \left(\mathbf{p} - \frac{e\mathbf{A}}{c} \right)^2 \tag{1}$$

where $\mathbf{A}$ is the magnetic vector potential and m^* is the effective mass of the electron.

The solution of this results in the following eigenvalues:

$$\mathscr{E}_n = (n + \tfrac{1}{2})\hbar\omega_c + (p_z^2/2m^*) \tag{2}$$

where the magnetic field is taken along the z axis, $\omega_c = e\hbar/m^*c$ is the cyclotron frequency, and n is the magnetic quantum number. This result states that the magnetic levels of both the conduction and valence bands in semiconductors or metals are quantized in two dimensions and form one-dimensional bands as shown in Fig. 1. This figure illustrates the various quantum phenomena which have been studied in recent years and which are

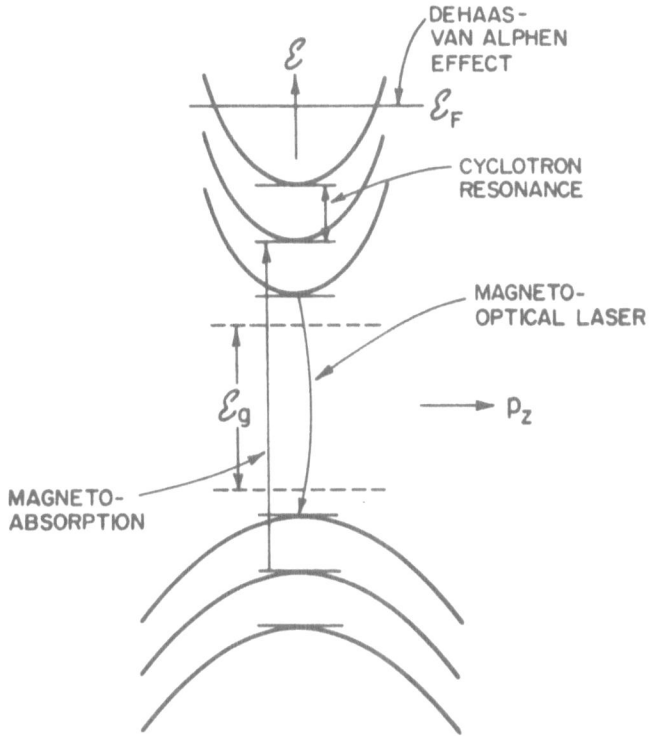

Fig. 1. Energy-momentum diagram showing the quantized subbands in the presence of a magnetic field. Several quantum phenomena have been indicated.

described in many of the chapters of this volume. For example, cyclotron resonance is represented as a transition between neighboring magnetic levels. The cyclotron resonance transition obeys the selection rule $\Delta n = \pm 1$. Classically, this corresponds to many transitions between a large number of equally spaced levels in which the distribution of electrons favors the lower states more heavily. Hence the net effect is that the induced absorption exceeds the stimulated emission and one sees cyclotron resonance absorption. The reason that this appears as a resonance peak is that the density of states at the bottom of the band exhibits a singularity; hence the absorption is peaked as if the bottom of the band represented a discrete level, as indicated schematically in Fig. 1.

Figure 1 also shows interband transitions across the energy gap. These are observed most easily at low temperature in a semiconductor when a photon induces the transition of an electron from a magnetic level in the valence band to a corresponding level in the conduction band. This is the phenomenon of magnetoabsorption, and obeys the selection rule $\Delta n = 0$. In a practical case the energy bands are somewhat more complicated, leading to violation of the simple selection rule which we shall discuss later.

Intense optical or electrical excitation of electrons to the conduction band can cause an inversion of population of the lowest level in the conduction band relative to the upper level in the valence band. This is a required condition for the operation of the interband magnetooptical laser, which can be tuned in frequency in accordance with the simple relation

$$\hbar\omega = \mathscr{E}_g + \tfrac{1}{2}\hbar\omega_c^*$$ (3)

where ω_c^* represents the cyclotron resonance frequency associated with the reduced effective mass μ,

$$1/\mu = (1/m_c) + (1/m_v)$$ (4)

Finally, Fig. 1 shows another quantum magnetic phenomenon closely related to an earlier oscillatory transport effect observed in the 1930's both in magnetoresistance and in diamagnetic susceptibility at low temperature. These are called, respectively, the de Haas–Shubnikov and the de Haas–van Alphen effects. These oscillations occur at low temperature due to the redistribution of electron population as an increasing magnetic field moves the bottom of one of the conduction bands through the Fermi energy. As the conduction band approaches coincidence with the Fermi energy and passes through it, electrons are emptied to the other magnetic levels of lower quantum number whose minima are below the Fermi energy. This fluctuation in the population as the magnetic field is increased gives rise to these oscillatory effects in the magnetoresistance and the susceptibility as well as to other related phenomena such as giant quantum oscillations of ultrasonic attenuation and magnetothermal oscillations.

CYCLOTRON RESONANCE

It is possible to derive the cyclotron resonance condition for the simple case of an electron in a magnetic field by using the classical equation of motion. This description will be quite adequate for most cases, although not for those involving complicated energy bands, such as, for example, the "quantum effects" near the degeneracy of valence bands. These exceptions to the classical treatment will be described presently.

Classical Treatment

The equation of motion of an electron in a steady magnetic field **H** and driven by an electromagnetic field $\mathbf{E}e^{i\omega t}$ is given by

$$m\frac{d\mathbf{v}}{dt} = -e\mathbf{E}e^{i\omega t} - e\frac{\mathbf{v}\times\mathbf{H}}{c} - \frac{m\mathbf{v}}{\tau} \tag{5}$$

where **H** is taken along the z axis and **E** is considered to be in the xy plane. The scattering time τ is a measure of the mean free path of the electron in its orbit. The current density is given in terms of the velocity **v** and the number of electrons N by

$$\mathbf{J} = -Ne\mathbf{v} = \sigma \cdot \mathbf{E} \tag{6}$$

and the conductivity tensor has the following simple form in our choice of an orthogonal coordinate system:

$$\sigma = \begin{pmatrix} \sigma_{xx} & \sigma_{xy} & 0 \\ \sigma_{yx} & \sigma_{yy} & 0 \\ 0 & 0 & \sigma_{zz} \end{pmatrix} \tag{7}$$

where

$$\sigma_{xx} = \sigma_{yy} = \frac{\sigma_0(1 + i\omega\tau)}{(1 + i\omega\tau)^2 + \omega_c^2\tau^2}, \qquad \sigma_{zz} = \frac{\sigma_0}{1 + i\omega\tau}$$

$$\sigma_{yx} = -\sigma_{xy} = \frac{\sigma_0\omega_c\tau}{(1 + i\omega\tau)^2 + \omega_c^2\tau^2}, \qquad \sigma_0 = \frac{Ne^2\tau}{m} = ne\mu$$

The power absorption is given by

$$\frac{P}{2P_0} = \frac{P}{\sigma_0 E^2} = \frac{1 + (\omega^2 + \omega_c^2)\tau^2}{[1 + (\omega^2 - \omega_c^2)\tau^2]^2 + 4\omega^2\tau^2} \tag{8}$$

which was obtained from the fundamental relation

$$P = \tfrac{1}{2}\,\mathrm{Re}[\mathbf{J}\cdot\mathbf{E}^*] \tag{9}$$

The resonance absorption of a linearly polarized electromagnetic wave is shown in Fig. 2, where the maximum absorption occurs as a sharp peak at the point $\omega_c = \omega$. The peak is clearly sharper for larger values of the parameter $\omega\tau$. For values of $\omega\tau < 1$ we simply have a magnetoresistance effect.

A better picture of the effect of the parameter $\omega\tau$ on the resonance linewidth is given if we consider the absorption of a circularly polarized electromagnetic wave given by

$$\mathbf{E}_\pm = E_0(\mathbf{x} \pm i\mathbf{y})$$

where **x** and **y** are unit vectors. The power absorbed for the positive sense of circular polarization is given by

$$\frac{P_+}{P_0} = \mathrm{Re}\left[\frac{\sigma_+}{\sigma_-}\right] = \mathrm{Re}\left[\frac{1}{1 + i(\omega - \omega_c)\tau}\right] = \frac{1}{1 + (\omega - \omega_c)^2\tau^2} \tag{10}$$

and a plot of this equation for several values of $\omega\tau$ is shown in Fig. 3. These absorption lines are sharper and better resolved, so that one can determine the value of the scattering time τ from a measured cyclotron resonance linewidth. The measurement of the linewidth is usually taken at the half-power point and is defined as

$$\frac{P_{1/2}}{P_0} = \frac{1}{2} = \frac{1}{1 + (\omega - \omega_c)^2\tau^2} \tag{11}$$

Then,

$$\tau = 2/\Delta\omega_c = (1.14 \times 10^{-8})/\Delta H \tag{12}$$

These classical results have shown that in order for cyclotron resonance to be observed as a sharp absorption line it is necessary for $\omega\tau$ to be greater than unity. In addition, it is obvious that $\omega\tau$ must be much greater than unity if higher resolution is needed in order to distinguish closely spaced lines or structure. This has been achieved even at relatively low microwave frequencies. The value of τ was increased by using pure, perfect crystals and by reducing the temperature to a few degrees Kelvin, where the electron–phonon interactions are then reasonably small. The lowering of the temper-

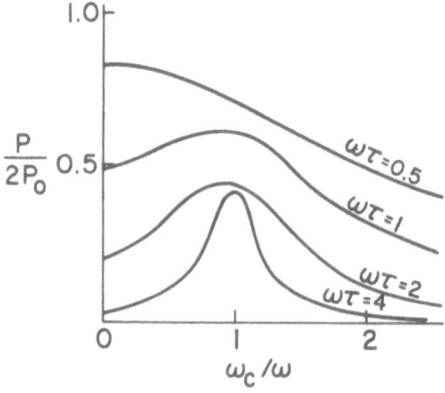

Fig. 2. Resonant absorption of linearly polarized radiation takes place in a semiconductor at the cyclotron resonance frequency for values of $\omega\tau > 1$.

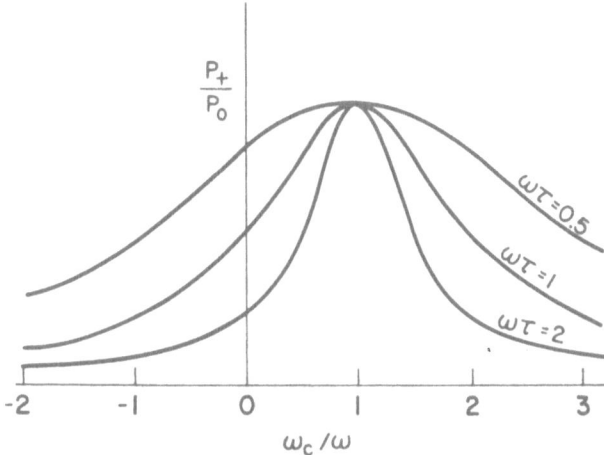

Fig. 3. Cyclotron resonance absorption in a semiconductor for
the appropriate sense of circular polarized radiation.

ature reduces the scattering of the electrons, but if the crystal is not pure
and perfect, the value of τ cannot be increased by further reducing the
temperature. Therefore to increase $\omega\tau$ it becomes necessary to increase the
frequency into the submillimeter and far infrared ranges of the spectrum.
Higher frequencies require the use of higher magnetic fields. This is the
rationale for the development of submillimeter high-field spectroscopy for
resonance.

Cyclotron Resonance in Indium Antimonide

One of the first experiments to be carried out in high-intensity magnetic
fields was that of the cyclotron resonance of electrons in indium antimonide.
It was known from microwave experiments (^{3}) that the effective mass was
very small, of the order of 1% of the free-electron mass, so that even at an
infrared wavelength of 20 μ magnetic fields from pulsed magnets would
be sufficiently large. At a longer infrared wavelength of 40 μ the magnetic
field from water-cooled solenoids would be large enough to permit observation of infrared cyclotron resonance. Such experiments were carried out
with both pulse (^{4}) and steady-field magnets (^{5}). The cyclotron resonance
data from the pulsed magnets was extended to fields above 300,000 G and
exhibited unique results which could only be explained by a quantum mechanical treatment of cyclotron resonance. Both the reflection from and
the transmission through thin samples is shown in Fig. 4. As data such as
these were taken at different wavelengths and different values of magnetic

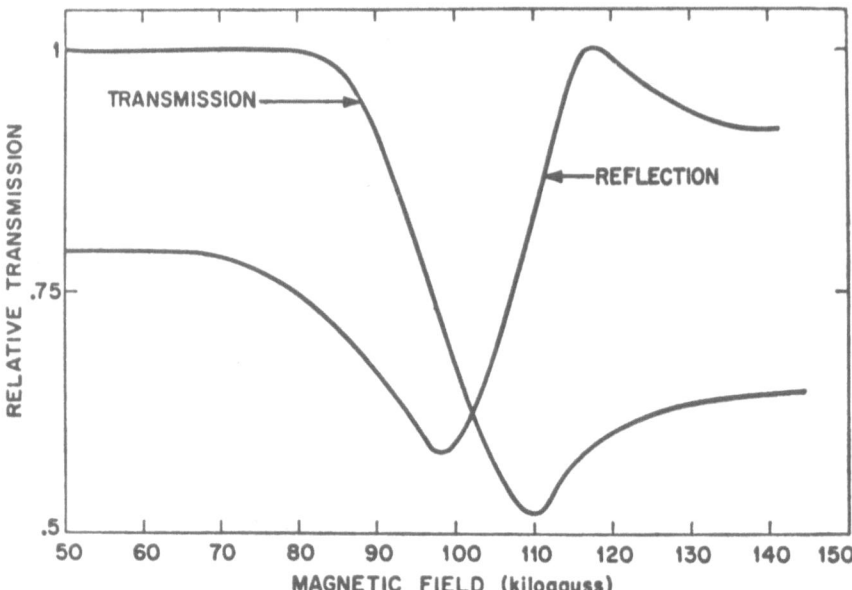

Fig. 4. High field transmission and reflection in indium antimonide showing the electron
cyclotron frequency occurring at 110 kG. From R. J. Keyes, private communication.

field intensity, it was shown that if the effective mass is interpreted by the
classical resonance formula from the measured magnetic field corresponding
to the transmission minimum, the effective mass depended on the magnetic
field intensity. The plot of the apparent effective mass as a function of the
applied magnetic field intensity is shown in Fig. 5. This field dependence can
be readily explained by taking the $\mathscr{E}$–p relation obtained by Kane [6]

$$\mathscr{E} = \left[\frac{\mathscr{E}_g^2}{4} + \mathscr{E}_g\left(\frac{p^2}{2m^*}\right)\right]^{1/2} - \frac{\mathscr{E}_g}{2} \tag{13}$$

and inserting the magnetic field by replacing p with $\mathbf{P} = \mathbf{p} - (e\mathbf{A}/c)$ where
$\mathbf{A}$ is the magnetic vector potential. When spin is also included the result
becomes [(7), p. 363]:

$$\mathscr{E}_n = \left\{\frac{\mathscr{E}_g^2}{4} + \mathscr{E}_g\left[\left(n+\frac{1}{2}\right)\hbar\omega_c \pm \frac{g\mu_B H}{2} + \frac{p_z^2}{2m^*}\right]\right\}^{1/2} - \frac{\mathscr{E}_g}{2} \tag{14}$$

where $\mu_B = e\hbar/2m_c$ is the Bohr magneton and the expression for the effective g-factor is given below. From this it is possible to calculate the
apparent effective mass from the relation $\hbar\omega_c^* = \mathscr{E}_{n+1} - \mathscr{E}_n$. When this

is done and the appropriate approximations are carried out mathematically it can be shown that

$$m^* \approx m_0^* \left(1 + \frac{4n\hbar\omega_c}{\mathscr{E}_g} \right)^{1/2} \tag{15}$$

where ω_c is the cyclotron resonance frequency corresponding to the effective mass at the bottom of the band. We can see from the above equation that the effective mass depends on the magnetic field intensity and on quantum number n. For this particular experiment, however, the transition corresponds to $n = 1$, and the theoretical result is shown by the solid line in Fig. 5 to be in good agreement with the experiment. In this same way, one can obtain an effective g-factor which decreases with magnetic field intensity and is given by

$$g = g_0^* \left[1 + \frac{2(2n + 1)\hbar\omega_c}{\mathscr{E}_g} \right]^{-1/2} \tag{16}$$

where g_0^* is the effective g-factor at the bottom of the band and is given by [8]

$$g_0^* \approx -2 \frac{m}{m_0^*} \frac{2\varDelta}{3\mathscr{E}_g + 2\varDelta} \tag{17}$$

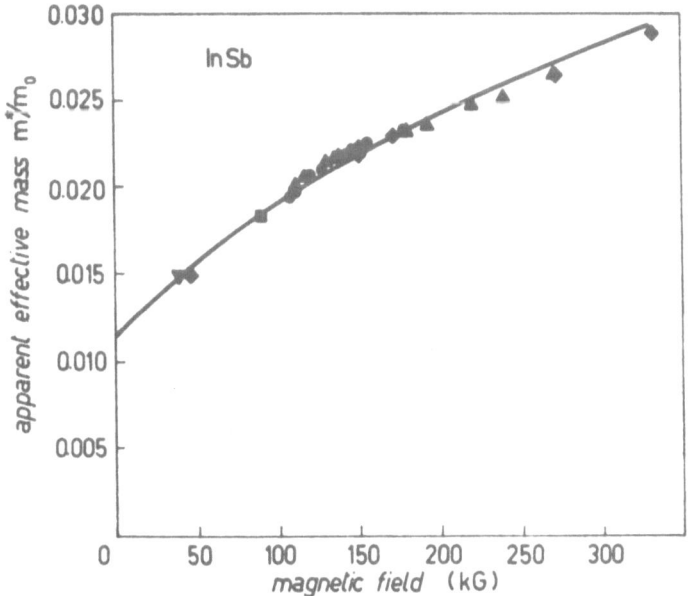

Fig. 5. Variation of apparent effective mass with magnetic field for the electron in indium antimonide. From Lax et al.[40].

where Δ is the spin-orbit splitting. The value of g_0^* is approximately equal to 50 at low temperatures in indium antimonide. These results for the spin variation have been confirmed by magnetooptical experiments [9] and spin resonance. Thus we see that high-field cyclotron resonance indeed shows very important and interesting quantum effects in low-gap, low-mass semiconductors.

Quantum Effects in Germanium and Silicon

It was pointed out by Luttinger and Kohn [10] that cyclotron resonance of holes in germanium and silicon would exhibit fine structure because the valence bands are fourfold degenerate at the center of the Brillouin zone. This structure is due to the fact that the light hole and heavy hole cannot be distinguished from each other at low values of quantum number, where their respective bands are degenerate. The magnetic level structure is a solution of a 4×4 matrix Hamiltonian which allows transitions or resonances considerably different from those corresponding simply to the light hole and heavy hole. The nature of these transitions is shown in Fig. 6, where some of the theoretically calculated transitions are shown. These

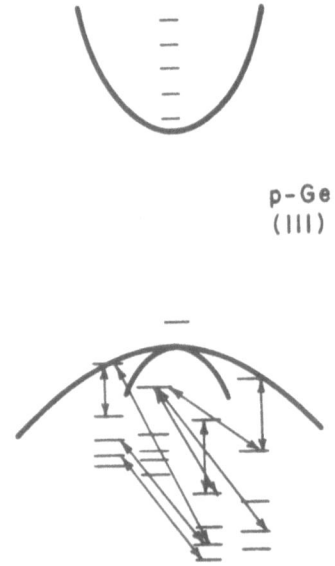

Fig. 6. Allowed transitions between quantum levels lying near the degeneracy of the valence bands of germanium. From Stickler *et al.* [41].

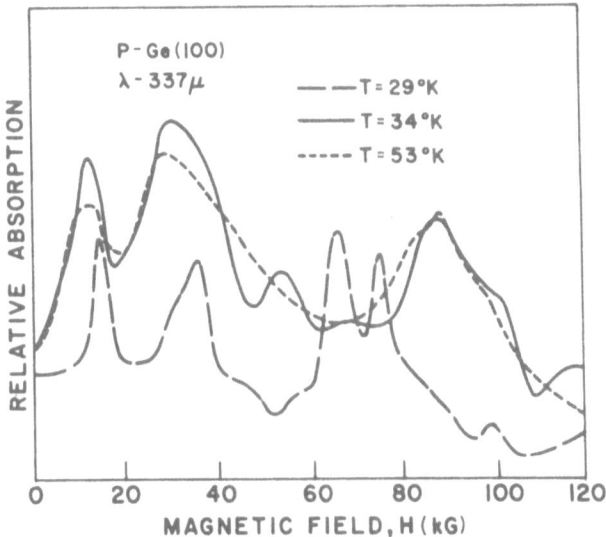

Fig. 7. Quantum effects associated with cyclotron resonance transitions near the degeneracy of the valence bands in germanium. Different transitions are observed as the change in temperature alters the population of quantum levels. From Button *et al.* ([14]).

effects were first observed experimentally at microwave frequencies in very pure crystals and at a temperature near 1 °K ([11]).

Later when millimeter waves became available these complicated transitions began to become fairly well resolved experimentally, so that a few lines could be correlated ([12]) with the theoretical interpretations. Recently the submillimeter HCN laser operating at a wavelength of $\frac{1}{3}$ millimeter has made it possible to study the quantum effects ([13]) as a function of temperature, leading to a more definite correlation between experimental and theoretical results. Some representative spectra of *p*-type Ge are shown in Fig. 7. Since the frequency and applied magnetic field are both so large, the quantum levels are widely separated, so that the energy between them is larger than the thermal energy even at temperatures up to 50 °K or more. This means that *p*-type doped specimens may be used and that the quantum levels may then be populated thermally. Formerly at low frequencies and correspondingly low temperatures the quantum levels could only be populated by exposing the crystal to white light. This did not allow one to correlate unambiguously the resonance lines with those predicted theoretically. Better identifications can now be made because one can see the dif-

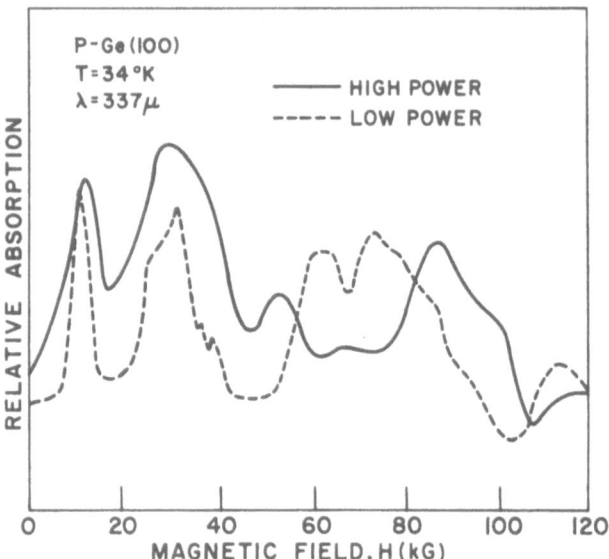

Fig. 8. Some cyclotron resonance transitions are broadened
or obliterated by the high-peak pulse power of the HCN
laser. From Button et al. ([14]).

ferent resonances come and go in Fig. 7 as the temperature is raised and the
thermal populations of initial and final states are changed. Care must still
be taken not to saturate the transitions by populating the final states by
using excessively high laser power, as demonstrated in Fig. 8. This is easily
done at microwave frequencies, where the holes are excited by white light,
because initial and final states are equally populated, but thermal population, in this present case, nearly fills the lower states and provides only a
Boltzmann population of the final states. Therefore only the high-power
pulsed HCN laser was capable of illustrating the saturation effect shown
in Fig. 8. The ordinary continuous wave laser lacks the power to saturate
the quantum effects in germanium.

INTERBAND PHENOMENA

The interband phenomena involving the transitions across the semiconductor energy gap between quantized magnetic levels was discovered
in 1957 ([2]). Since then it has been applied with significant success to a large
variety of materials including a number of semimetals. The observation of
the effect has been used along with new and elegant techniques as well as

sophisticated theoretical analysis to determine a large number of important parameters of the energy bands which could not otherwise have been measured. Some of these new techniques have been developed with the aid of large electric fields, electrooptical and piezooptical modulation, and, most recently, lasers as spectroscopic sources. These have permitted improved sensitivity and greater exploration deep into the energy band structure.

Direct Interband Magnetoabsorption

When a semiconductor in a magnetic field has both its conduction and valence band quantized the energy momentum relation of each is given by

$$\mathscr{E}_c = \mathscr{E}_g + (n_c + \tfrac{1}{2})\hbar\omega_c + (p_z^2/2m_c) \tag{18}$$

$$\mathscr{E}_v = -(n_v + \tfrac{1}{2})\hbar\omega_v - (p_z^2/2m_v) \tag{19}$$

Then, as we pointed out before it is possible to induce a transition of an electron from a magnetic level of the valence band to a corresponding level of the conduction band with the selection rule $\Delta n = 0$. The transition probability and the absorption coefficient α for this process can be calculated from time-dependent perturbation theory, whereby one obtains the "golden rule" given by

$$\alpha = C \int M_{cv}^2 \delta(\mathscr{E}_c - \mathscr{E}_v - \hbar\omega)\, d^3p \tag{20}$$

where

$$M_{cv} = \int \Psi_c^* \left(\mathbf{p} + \frac{e\mathbf{A}}{c} \right) \cdot \boldsymbol{\epsilon} \Psi_v\, d^3r$$

$\boldsymbol{\epsilon}$ is the polarization vector, $\hbar\omega$ is the photon energy of the incident radiation, and M_{cv} is the matrix element between the valence and conduction band. The coefficient C involves universal constants and the density-of-states effective mass. This particular expression can be integrated with respect to momentum and gives a result for the zero-field absorption coefficient

$$\alpha(0) = A(\omega - \omega_g)^{1/2} \tag{21}$$

where

$$A = \frac{2e^2 M_0^2}{\sqrt{\eta}\, m^2 c\omega} \left(\frac{2\mu}{\hbar^2} \right)^{3/2}$$

The symbol η denotes the index of refraction, μ is the reduced mass given by Eq. (4), and the other symbols have their usual significance. When a magnetic field is included the density of states changes and the absorption coefficient becomes

$$\alpha(H) = D \sum_n \int M_{cv}^2 \frac{\omega_c \, d\mathscr{E}}{\sqrt{\mathscr{E}}} \, \delta(\mathscr{E}_n + \mathscr{E} - \hbar\omega) \tag{22}$$

which, when integrated, gives

$$\alpha(H) = \frac{A\omega_c{}^*}{2} \sum_n \frac{1}{(\hbar\omega - \mathscr{E}_n)^{1/2}} \tag{23}$$

where $\mathscr{E}_n = (n + \tfrac{1}{2})\hbar\omega_c{}^*$. In this case the cyclotron frequency is that of the reduced effective mass, since the density of states in the integral above is that of the combined density of states of the valence and conduction bands. The definition of A remains the same. The important physical result in comparing the zero magnetic field absorption and the magnetic absorption is shown in Fig. 9, where it can be seen that the absorption exhibits peaks. In this figure we have implicitly included broadening of the

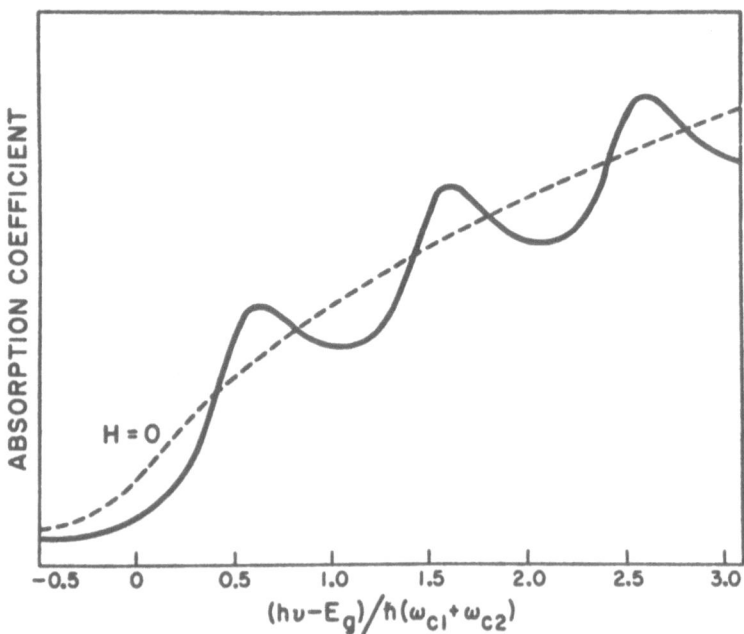

Fig. 9. Absorption coefficient as a function of energy in the presence of a magnetic field (solid line) for the direct interband transition. From Roth et al. ([8]).

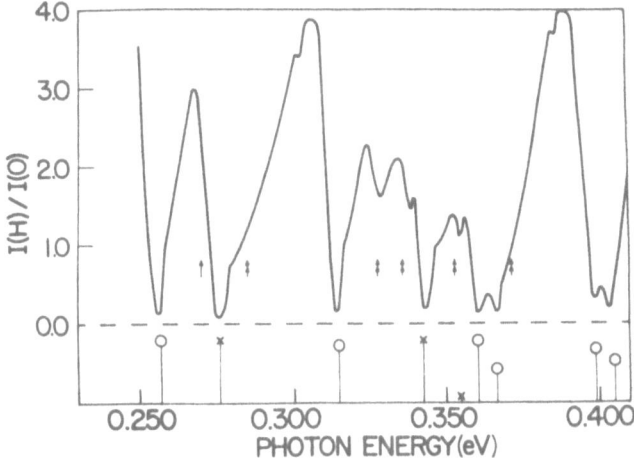

Fig. 10. Experimental observation of direct interband transitions
in the presence of a magnetic field. From Pidgeon and Brown ([9]).

peaks because the singularity in the absorption does not really exist due to
relaxation effects of both the holes and electrons. Actually, relaxation can
be introduced very readily by rewriting the absorption coefficient as

$$\alpha(H) = A'\omega_c \{\Sigma \ [\omega - \omega_n - (i/\tau_n)]^{-1/2}\} \qquad (24)$$

In an actual experiment the procedure is to apply a magnetic field to a thin
specimen at low temperature and observe the transmission of an electromagnetic wave through the specimen. As the wavelength of the monochromatic radiation is varied, the transmitted intensity is observed as shown in
Fig. 10. The complex detailed structure which is observed is due principally
to the nature of the fourfold degeneracy of valence bands as shown in Fig.
11, so that in order to interpret the details of the transitions, one has to
take this into account. The selection rules for these transitions include not
only $\Delta n = 0$, but $\Delta n = -2$. The polarization governs the selection rule
for the magnetic quantum number M_J, i.e., for the electric vector $\mathbf{E}$ of the
incident radiation parallel to the applied magnetic field, $\mathbf{E} \parallel \mathbf{H}$, $\Delta M_J = 0$
and for $\mathbf{E} \perp \mathbf{H}$, $\Delta M_J = \pm 1$. In the Faraday configuration, in which the
direction of propagation and the magnetic fields are parallel, then we
obtain either $\Delta M_J = +1$ for right-hand circular polarization or $\Delta M_J = -1$
for left-hand circular polarization. Using such polarized radiation and high-
resolution spectroscopy, the data in Fig. 10 were analyzed. This figure
shows the dominant transitions as transmission minima interpreted in

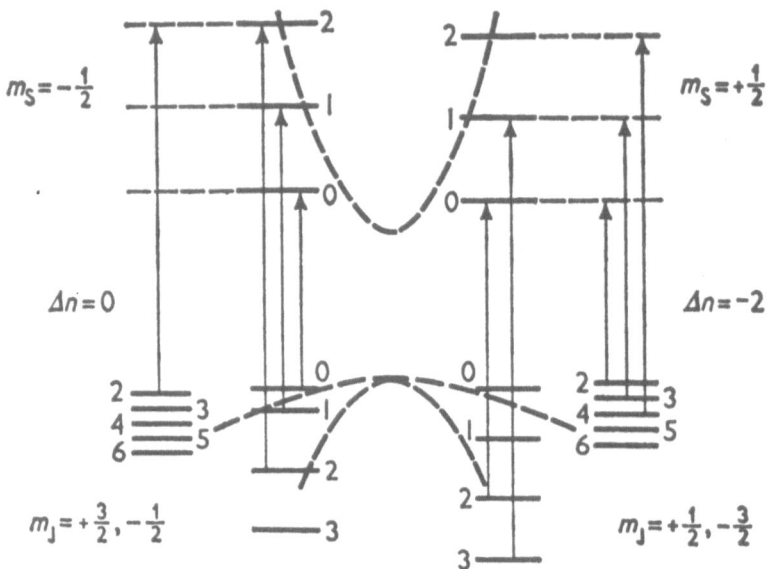

Fig. 11. Diagram of allowed direct interband transitions in germanium for the electric vector parallel to the applied magnetic field.

terms of the above selection rules. Fine structure also appears (indicated by arrows) caused by higher-order transitions associated with warping of the bands and nonparabolicity. After the identification of dominant transitions the analysis is made by fitting seven theoretical energy band parameters to the experimental data. The model of the energy bands from $\mathbf{k} \cdot \mathbf{p}$ perturbation theory introduces three of these parameters, γ_1, γ_2, and γ_3, associated with the valence band curvature. The other parameters are spin-orbit coupling, Δ, indicated in Fig. 12, which splits the $J = \frac{3}{2}$ bands from the $J = \frac{1}{2}$ band, the energy gap, $\mathscr{E}_g$, between the conduction and valence band, the effective mass of the conduction band, and the matrix elements, M_{cv}, between valence and conduction bands. A computer has been used to prepare the best fit of these seven adjustable parameters to the experimental data shown in the chart of Fig. 12. The solid lines are theoretical curves obtained when all of the parameters are mutually optimized. Actually, parameters such as the energy gap and the spin-orbit coupling are now measured quite accurately directly from such graphs constructed from electroreflection or piezoreflection data. By using suitable schemes for different portions of the chart, at high energy and low quantum numbers it is a fairly straightforward procedure to obtain the remaining five parameters almost individually.

The graph for indium antimonide in Fig. 13 illustrates how well the computer-adjusted theoretical curves can be made to fit the experimental points. It can be seen that these experiments and this procedure can now provide three-figure accuracy not only in the measurement of effective masses, but also in the determination of the g-factors of the electron and hole as well. The latter are readily deduced from the data shown in the figure. The use of this type of treatment has been extended to determine the energy band parameters for a number of materials such as InAs ([15]), GaSb, GaAs ([16]), HgTe ([17]), and most recently, gray tin ([18]). The latter two materials should be regarded as semimetals because the Fermi level lies in the conduction bands. This makes it necessary to employ magneto-reflection methods. We shall treat semimetals later.

A few comments of historical nature are in order. In the early investigation of interband transitions in germanium and indium antimonide it was

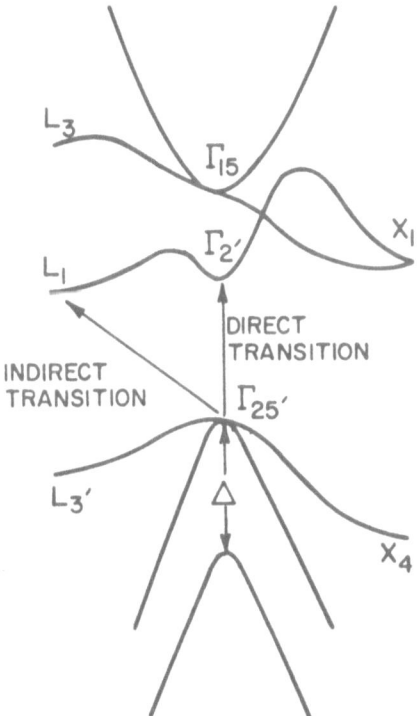

Fig. 12. Sketch of the part of the Brillouin zone in the immediate vicinity of the direct and indirect transition for germanium.

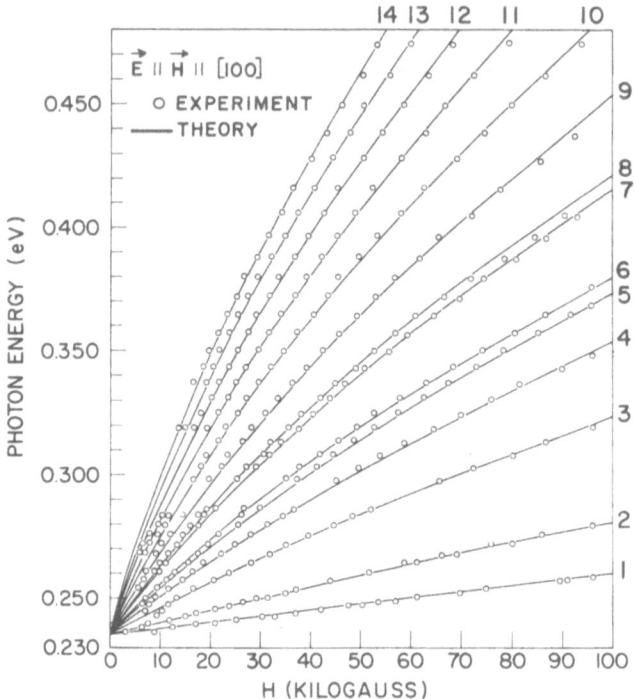

Fig. 13. Direct interband transitions in indium antimonide as a function of magnetic field intensity. From Pidgeon and Brown ([9]).

not possible to make such complete measurements as those obtained more recently. Nevertheless, unique information was gathered for the first time from the interband magnetoabsorption measurements ([19]). For example, in germanium the direct absorption occurs between energy band extrema separated by a gap that is larger in energy (see Fig. 12) than the normal (indirect) energy gap found in germanium. Hence for the first time it was possible to measure the energy gap involving a higher conduction band minimum and also to determine the effective mass and the g-factor of an electron, which could not be studied by using transport phenomena or cyclotron resonance. These latter methods could not provide the value of the anomalous g-factor in InSb, which was, however, readily determined from magnetooptical observation of the splitting of the lowest transition from the top of the valence band to the $n = 0$ level of the conduction band. The g-factor was measured to be $g \approx -50$. The negative sign actually was deduced ([8]) from theoretical work and has been given by Eq. (16).

Indirect Transition

Whereas in the previous section we discussed the phenomenon of magnetoabsorption in which a single photon alone was responsible for a transition between the valence band and conduction band at $k = 0$, another type of interband transition is possible in germanium. This one involves the simultaneous absorption of a photon as well as the absorption or creation of a phonon. This involvement of a phonon in the interband process is necessary because the electron must be transferred from the center of the Brillouin zone to the edge at the L point along the [111] direction, as indicated in Fig. 12. Momentum can be conserved only by exchanging momentum with the lattice via a phonon. Hence in order to calculate the absorption coefficient it is now necessary to utilize second-order perturbation theory. The absorption coefficient contains a product of a photon and a phonon matrix element. The former is very much like that of the direct transition, but in this indirect process the conduction band or another valence band acts as an intermediate virtual state. The phonon matrix element represents the electron–phonon interaction corresponding to translation from the zone center to the edge. The calculation of the absorption coefficient in the absence of a magnetic field yields

$$\alpha(0) = C_{\pm}(\omega - \omega_g \pm \omega_q)^2 \tag{25}$$

where

$$C_{\pm} = \pm \frac{D}{\omega} [\exp(\pm \hbar \omega_q / kT) - 1]^{-1}$$

Note that C is a coefficient which contains the phonon density population. The plus sign corresponds to phonon absorption, the negative sign denotes phonon emission, $\hbar \omega_q$ is the appropriate phonon energy at the zone edge, and $\hbar \omega_g = \mathscr{E}_g$.

The coefficient D contains the appropriate universal constants, the product of density-of-states masses for the conduction and valence bands, and the photon and phonon matrix elements. When this expression is rewritten to include the magnetic field and the appropriate integration is carried out we find

$$\alpha(H) = 2C_{\pm} \hbar^2 \omega_{c1} \omega_{c2} \sum_{n1n2} S(\omega - \omega_{1,2}) \tag{26}$$

$$\omega_{1,2} = \omega_g + (n_1 + \tfrac{1}{2})\omega_{c1} + (n_2 + \tfrac{1}{2})\omega_{c2} \pm \omega_q$$

A sketch of the "steps" exhibited at the Landau-level transitions is shown in Fig. 14 ([8]). When loss is included in the form of a phenomenological

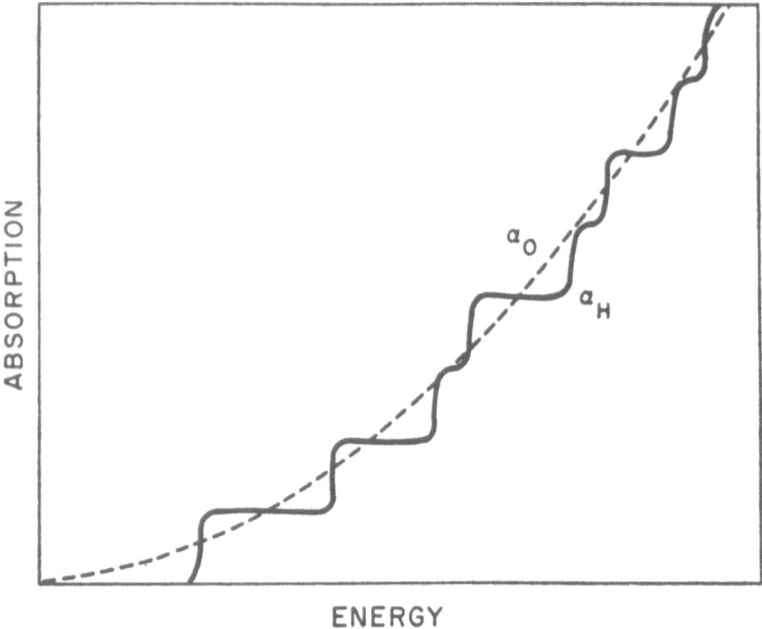

Fig. 14. Theoretical plot of the absorption coefficient for germanium in the vicinity of the indirect transition in the absence of (dashed line) and presence of (solid line) a magnetic field. From Roth *et al.* ([8]).

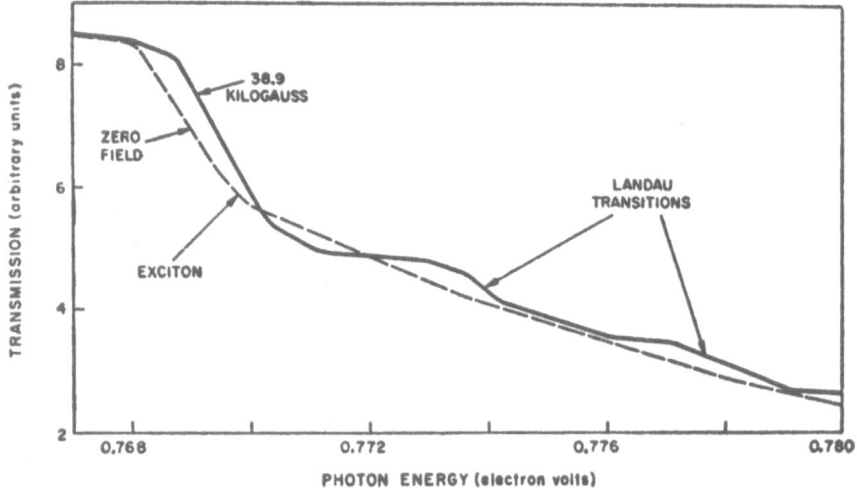

Fig. 15. Experimental observation of the indirect transition, where the first "step" at the left results from the creation of an exciton.

relaxation time τ the S-function is transformed to become $S(\omega - \omega_{1,2})$ $\rightarrow \tan^{-1}x$, where $x = (\omega - \omega_{1,2})\tau$. This causes the abrupt step to appear more like an inverted s-curve, which is shown in Fig. 15. This shows the transmission of radiation through a sample several millimeters thick. Such a large thickness implies a small value of the absorption coefficient due to the two-step photon–phonon process.

The first strong line is that associated with the indirect exciton, where the electron is captured by the hole when the hole is created by the transition. The weaker lines at higher energy, where the Coulomb attraction can be neglected, are interpreted to be, to a good approximation, transitions to free-electron Landau states. These have been analyzed ([20]) as transitions from the top of the valence band to successive levels in the conduction band. There is reasonably good quantitative agreement with the cyclotron resonance data.

The transmission technique only permitted observation and analysis of the first few transitions into the conduction band even at high magnetic

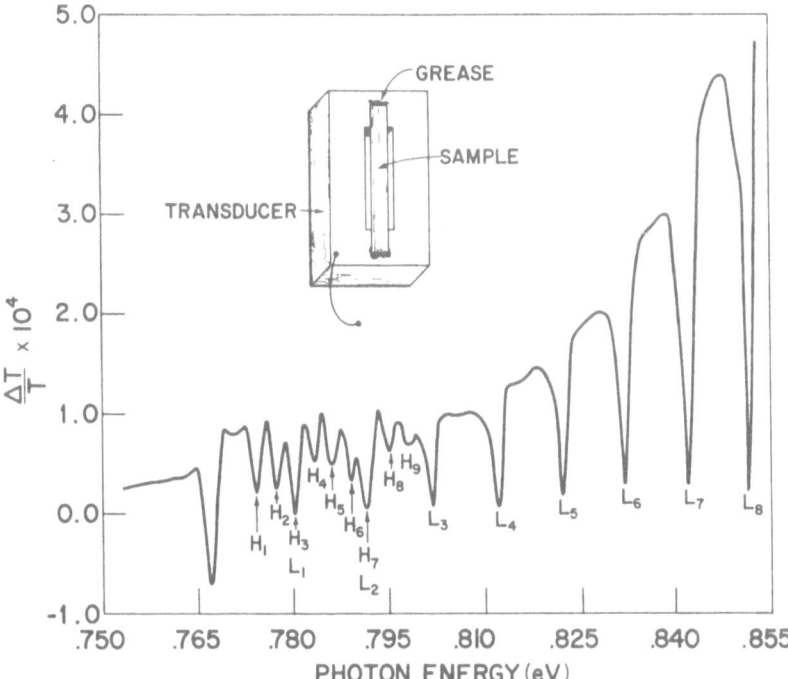

Fig. 16. Magneto-piezotransmission showing the indirect transitions in germanium, where spikes now appear in place of the "steps" shown in Fig. 15. From Aggarwal *et al.* ([21]).

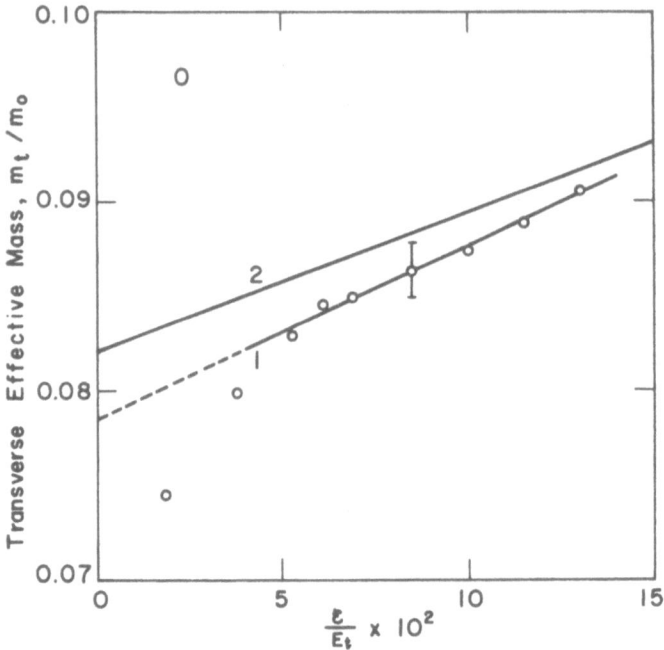

Fig. 17. Apparent effective mass of the light electron in germanium as a function of energy above the conduction band minima. From Aggarwal et al. [21].

field intensities [20]. Recently, however, it was possible to extend and improve these measurements considerably by performing a piezotransmission experiment [21,22] on a somewhat thinner specimen. It was mounted on a transducer as shown in Fig. 16. Since the modulation technique converts the display of the result to the derivative of the absorption lines, spikes appear where steps occurred previously. We can also see that the absorption line should be proportional to

$$\frac{d\alpha}{ds} = \frac{-d\omega_g/ds}{1 + (\omega - \omega_{1,2})^2\tau^2} \tag{27}$$

where s is the strain applied by the transducer. The elimination of the broad steps is helpful to the analysis of results because the position of these spikes on the energy scale can now be determined more accurately. Thus for this orientation there are transitions to two inequivalent electron levels in the conduction band. These transitions to "light" and "heavy" electron levels are designated by L and H in Fig. 16. It is visually evident that as the transitions are extended to higher energies the spacing between the light electron levels decreases. Figure 17 shows the linear variation of the light electron mass with energy above the bottom of the conduction band. The theoretical explanation for the change in effective mass emerges from the $\mathbf{k} \cdot \mathbf{p}$ perturbation treatment when the interaction between the valence and conduction bands at the L_1 and $L_{3'}$ points (see Fig. 12) at the zone edge and the resulting nonparabolicity of the conduction band are taken into account. When the theory is accordingly carried out to a suitable approximation in this energy range it is shown that the effective mass varies linearly with energy above the bottom of the conduction band. The quantitative correspondence between theory and experiment is less than perfect because the interactions with higher energy bands were not taken into account and the change in phonon momentum away from the L point has not been included. Nevertheless, the semiquantitative variation is well accounted for by this theory.

Semimetals

The techniques of magnetoabsorption used earlier for the study of semiconductors are of no value for the study of metals. Magnetoreflection techniques had to be developed ([23]). It was also found to be convenient to reflect monochromatic light from the surface of the semimetal specimen and sweep the magnetic field intensity. This avoided the necessity of correcting the observations for the frequency dependence of atmospheric absorptions. The magnetoreflection technique was first applied to bismuth ([24]). The magnetic field was applied parallel to either the bisectrix axis or the binary axis. The effective masses are small under these conditions. The reflection spectrum of Fig. 18 shows well-defined oscillatory peaks as a function of magnetic field. For the magnetic field along the bisectrix axis there are two sets of oscillations as one can see in the figure. These peaks are associated with transitions from states in the valence band well below the Fermi surface to corresponding states above the Fermi energy in the conduction band. To interpret this data quantitatively, it was necessary to develop a model of the energy bands. This was done by $\mathbf{k} \cdot \mathbf{p}$ perturbation theory assuming two interacting bands at the zone edge and two sets of ellipsoidal energy surfaces. The energy–momentum relations can be written in ellipsoidal coordinates in the nonparabolic form. When this is solved with the presence of the magnetic field included and with the zero of energy taken to be midway between these two sets of bands we obtain, to a good approxi-

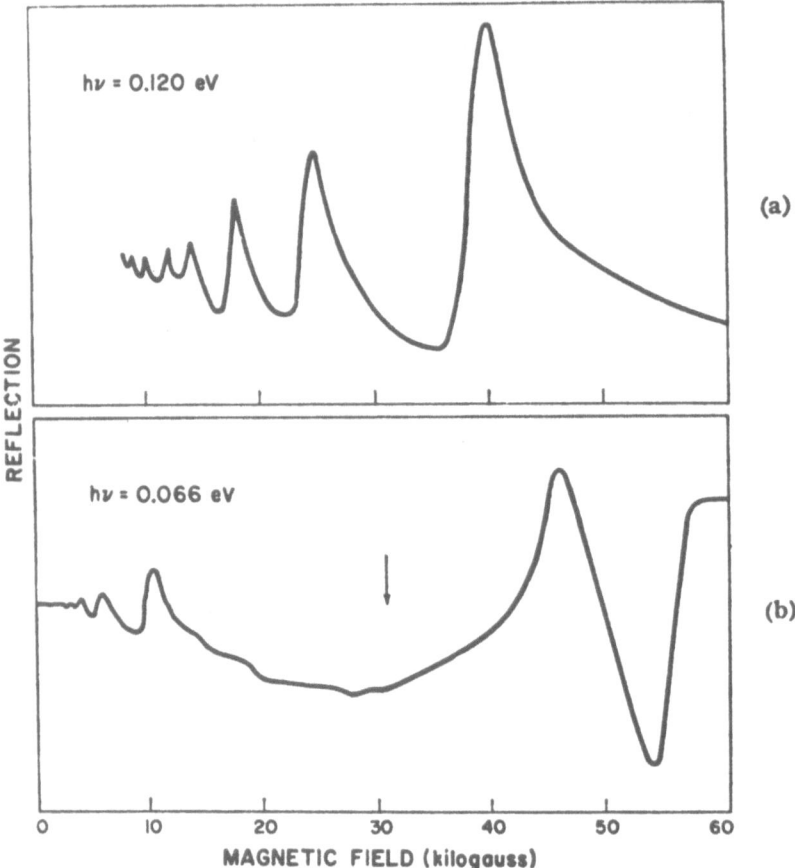

Fig. 18. Magnetoreflection oscillations showing direct transitions in bismuth as a function of magnetic field intensity. From Brown et al. ([24]).

mation, the solution for two mirror bands given by

$$\mathscr{E}_n = \pm\{\tfrac{1}{4}\mathscr{E}_g^2 + \mathscr{E}_g[(n + \tfrac{1}{2})\hbar\omega_c \pm \tfrac{1}{2}g\mu_B H]\}^{1/2} \qquad (28)$$

where $\omega_c = eH/m_0^*c$ and m_0^* is the cyclotron resonance mass at the bottom of the band. We have neglected the momentum term along the magnetic field because it is not needed to interpret energy spectra. The g-factor along the bisectrix and binary axes has been shown to be $g = 2m_0/m_0^*$, where m_0 is the free-electron mass and m_0^* is the cyclotron resonance mass at the bottom of the band, given by

$$m_0^* = [m_1 m_2 m_3/(m_1 \cos^2 \alpha + m_2 \cos^2 \beta + m_3 \cos^2 \gamma]^{1/2} \qquad (29)$$

where α, β, and γ are the angles between the magnetic field and the ellipsoidal coordinates. The effective masses m_1, m_2, and m_3 refer to the values along the principal directions. The explicit expressions for these tilted ellipsoids have been worked out [25]; [(7), p. 321]. Then the energy separation between the two bands becomes considerably simplified:

$$\Delta\mathscr{E} = (\mathscr{E}_g^2 + 4n\mathscr{E}_g\hbar\omega_c)^{1/2} \tag{30}$$

Using the selection rules $\Delta n = 0$ and $\Delta M = \pm 1$, the theoretical curves are

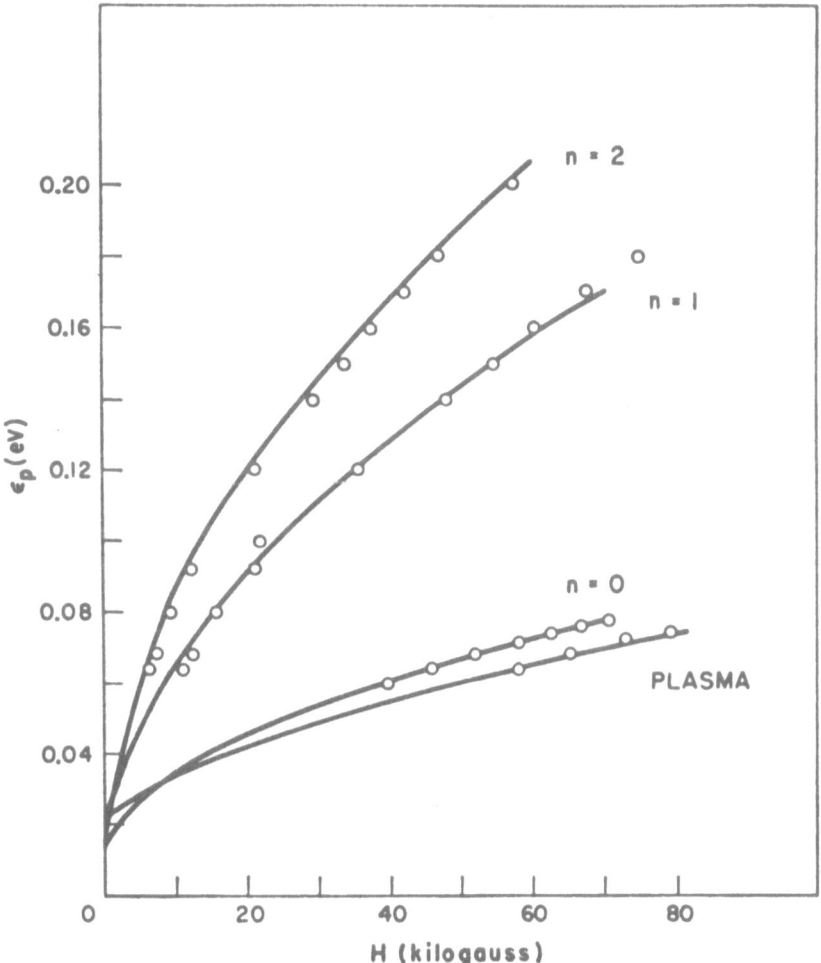

Fig. 19. Photon energies versus magnetic field for H along a binary direction in bismuth. From Brown et al. [24].

matched to the experimental data as shown in Fig. 19 by selecting the optimum value of the energy gap and the effective mass $m_0{}^*$ at the bottom of the band. This determines m_1, m_2, and m_3 when the data from both orientations are used. More recently data were also taken with the magnetic field along the trigonal axis ([26]), as shown in Fig. 20. Thus the four important parameters, $\mathscr{E}_g$, m_1, m_2, and m_3, are determined in a self-consistent way. The comparison of experiment and theory is shown beautifully by the points and curves of Fig. 21, where the magnetic field was along the binary direction.

If we look back to the Fig. 18, the vertical edge at the right-hand side of the lower curve corresponds to the cyclotron resonance transition in the conduction band. By using data from interband transitions it was found that the cyclotron resonance data was well-fitted by an expression for the effective mass similar to that given for InSb [Eq. (15)], where the apparent effective mass varies with the magnetic field. It is also important to note in Figs. 20 and 21 that the interband curves converge to a point above the origin corresponding to the energy gap. Thus in a semimetal we have been able to determine the energy gap for bands which lie beneath the

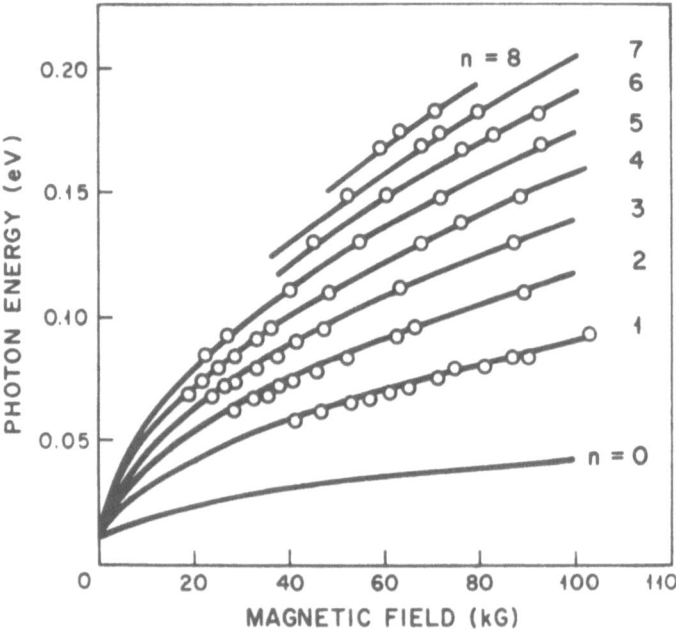

Fig. 20. Photon energies versus magnetic field for H along the trigonal direction of bismuth. From Maltz ([26]).

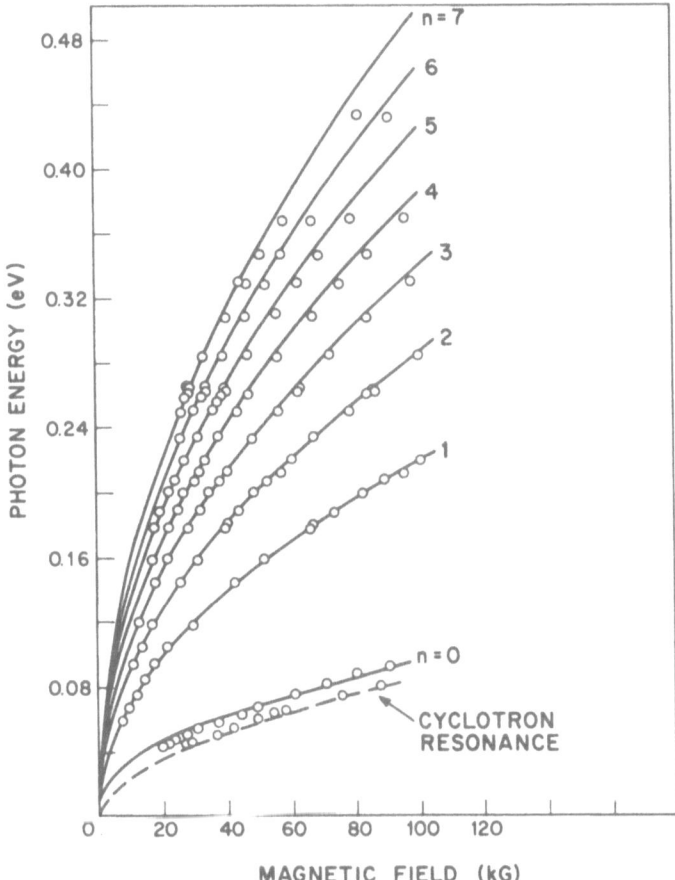

Fig. 21. Photon energies versus magnetic field with H along the binary direction in bismuth. The cyclotron resonance is shown to extrapolate to zero, but the interband transitions do not. From Maltz [26].

Fermi energy. For cyclotron resonance, however, the energy versus magnetic field curves converge at the origin in Fig. 21. One can see the highly nonparabolic nature of the bands for bismuth because the curves are not straight lines but bend considerably at high values of magnetic field.

Perhaps the most dramatic results (Fig. 22) for this type of measurement have been obtained in graphite [27], which has a hexagonal crystal structure. The hexagonally arranged atoms lie in layer planes and the magnetic field was applied along the hexagonal axis. The magnetoreflection spectrum shown in the upper part of Fig. 22 was observed in very much the same way as that described for bismuth, and again it is possible to see two

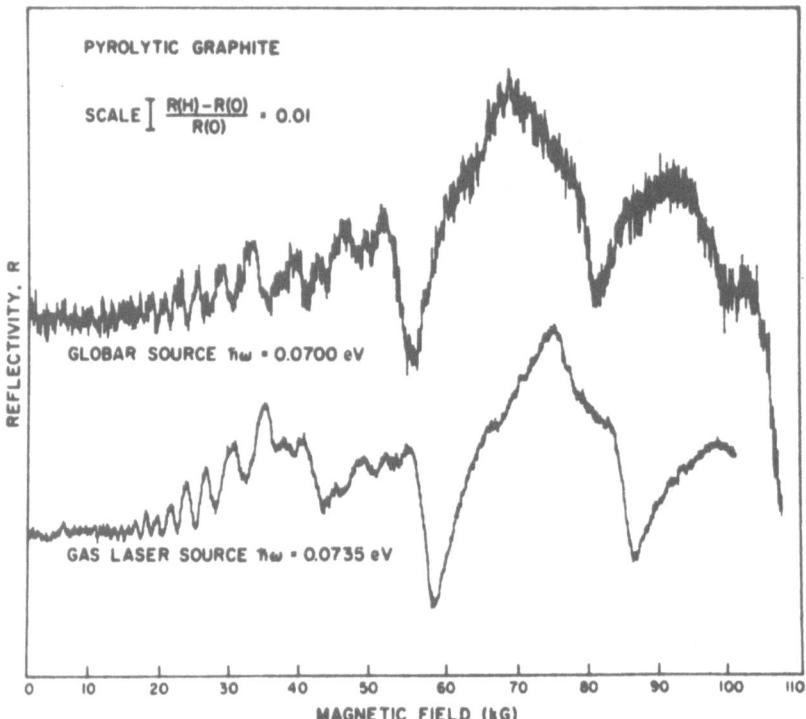

Fig. 22. Comparison of magnetoreflection spectra obtained in graphite with globar source and with laser source. The globar provides too little energy at far infrared wavelength for high-resolution spectroscopy. From Schroeder *et al.* ([28]).

sets of oscillations. Additional structure became more clearly observable, as shown in the lower part of the figure, when more intense monochromatic radiation was used from a helium–neon laser ([28]).

Two sets of data were taken in graphite, corresponding to different transitions for two sets of bands at different points at the edge of the Brillouin zone. Once again a theoretical model of the energy bands containing seven parameters is fitted optimally to the experimental data for these two transitions in the same way that was described for bismuth ([29]). Six of the seven parameters are determined from these magnetoreflection experiments and the seventh is obtained from the intraband de Haas–van Alphen or cyclotron resonance data. Indeed, this latter information could have been obtained from magnetoreflection if it had been convenient to extend the measurements to longer wavelengths.

It appears then that this one experiment has determined all of these parameters, usually to three significant figures. This represents a considerable advance in our knowledge of semimetals because in the case of graphite transport phenomena had provided us with values of only half of these parameters and they had been measured with considerably less accuracy. Furthermore, the transport phenomena such as magnetoresistance, cyclotron resonance, or de Haas–van Alphen oscillations provide information about the topology and the effective masses only at the Fermi surface. The interband magnetooptical techniques, however, provide information about the energy bands to energies above and below the Fermi surface, and in some cases considerable information. As we have seen from Fig. 21 and from the theoretical analysis, these energy bands are highly nonparabolic. Although nonparabolicity cannot be deduced from the transport phenomena, it is clearly demonstrated by the interband magnetooptical phenomena.

The numerical data obtained from these experiments are shown in Table I. These results have been used to map the Fermi surface of graphite as a function of the momentum along the hexagonal axis, as is shown in Fig. 23. The surprising feature is that the constant energy surface is highly variable along this direction because of the complicated nature of the energy bands. The diagram shows the severe warping of the Fermi surface near the lower portion, whereas the top portion has cylindrical symmetry deviating somewhat from an ellipsoidal surface where the electron and hole sur-

TABLE I

Band parameters	Dresselhaus and Mavroides determination		Other recent determinations (values, eV)
	Values (eV)	Method of determination*	
γ_0	3.21 $\pm$ 0.05	MR, high field, point H	2.8
γ_1	0.400 $\pm$ 0.005	MR, high quantum levels, point K	0.27 –0.40
γ_2	0.0185 $\pm$ 0.0005	dHvA periods	0.015–0.020
γ_4	−0.25 $\pm$ 0.02	MR, low quantum levels, point K Ratio of dHvA periods	~0.28
Δ	−0.009 $\pm$ 0.003 −0.005 $\pm$ 0.001	MR, low field, point H dHvA minority electron period	−0.02 – −0.10
E_F	0.022 $\pm$ 0.002	Number of holes = number of electrons	0.02

* MR ≡ magnetoreflection; dHvA ≡ de Haas–van Alphen.

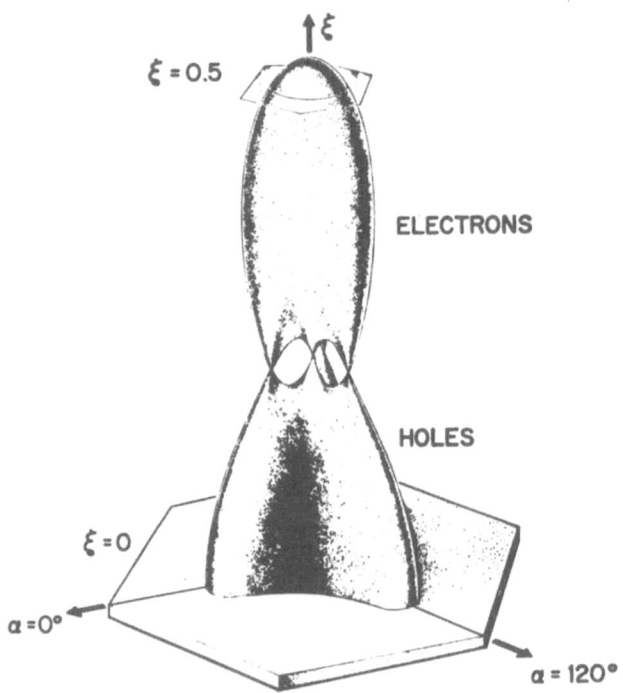

Fig. 23. Model of the graphite Fermi surface expanded along
the ξ direction by a factor of five to emphasize the trigonal
symmetry. From Dresselhaus and Mavroides ([27]).

faces touch. Recently it has been found ([30]) that the hole and electron
designation of the energy surfaces has been reversed. This was verified
from Faraday rotation and cyclotron resonance experiments using polarized
radiation. Using this model it was clear that the harmonic spectrum observed
in the cyclotron resonance was mainly due to the warped trigonal surface
at the bottom of the figure. Cyclotron resonance only measures the maximum area of the extremal surfaces. Thus, for example, the mass of the hole
is determined from the extremal cross section of the upper ellipsoidal surface.
Again we can see that the transport theory would only provide partial
information at the extremum, whereas the magnetooptical experiment
measures the parameter for the entire surface. This new understanding of
the surfaces permits previous de Haas–van Alphen and transport phenomena
to be interpreted quantitatively.

  These experiments have recently been used to study the energy bands
of antimony ([31]) and arsenic ([32]). Again accurate values of parameters

were established for those bands whose existence had been established earlier by de Haas–van Alphen experiments. New bands were also discovered by observing transitions from a full valence band below the Fermi surface to a corresponding empty conduction band above the Fermi energy, i.e., a set of bands which straddle the Fermi energy. Since these bands were full and empty, respectively, they could not be observed by using transport techniques. No electrons or holes are normally present, particularly at low temperatures. It was possible to identify the location of these bands in the Brillouin zone by using the symmetry of the polarization effects of the magnetooptical transitions and the anisotropy of the surfaces. This again is another unique feature of the magnetooptical experiments.

Multiphoton Absorption

From a physical argument we can conjecture that multiphoton absorption is essentially a time-dependent photon-assisted tunneling phenomenon. We follow the discussion by Zawadzki in Chapter 13 of tunneling in crossed electric and magnetic fields, where the model of sloping energy levels versus distance (Fig. 24) is used when a large electric field is applied. This picture is also representative of the situation when a large electric field is produced by a laser, but since this is a high-frequency electromagnetic wave, the diagram must be regarded as the instantaneous pattern. Therefore the diagram represents the situation for a half cycle of the oscillating electric field before it reverses. During this time an electron can start to tunnel across the forbidden region. Then it is almost instantaneously raised in energy by one of the many photons present in the crystal. Finally, it has time to tunnel across to the conduction band before the electromagnetic field from the laser reverses itself. Energy must be conserved, however, which means that two photons are required, because in the case we shall discuss the laser photon energy is less than the gap energy. Obviously, the argument can be extended to three- or four-photon processes.

A mathematical treatment developed by Keldysh [33] is essentially a quantitative description of the time-dependent tunneling phenomenon. We have modified this theory [34] to include a magnetic field and have worked out the results both for parallel and crossed electric and magnetic fields by the following procedure: One obtains a time-dependent wave function in the presence of an oscillating electric field and a steady magnetic field both for the valence and conduction bands. Then the matrix element for the interband transition is evaluated for these time-dependent wave functions. It is then necessary to integrate not only over space but also

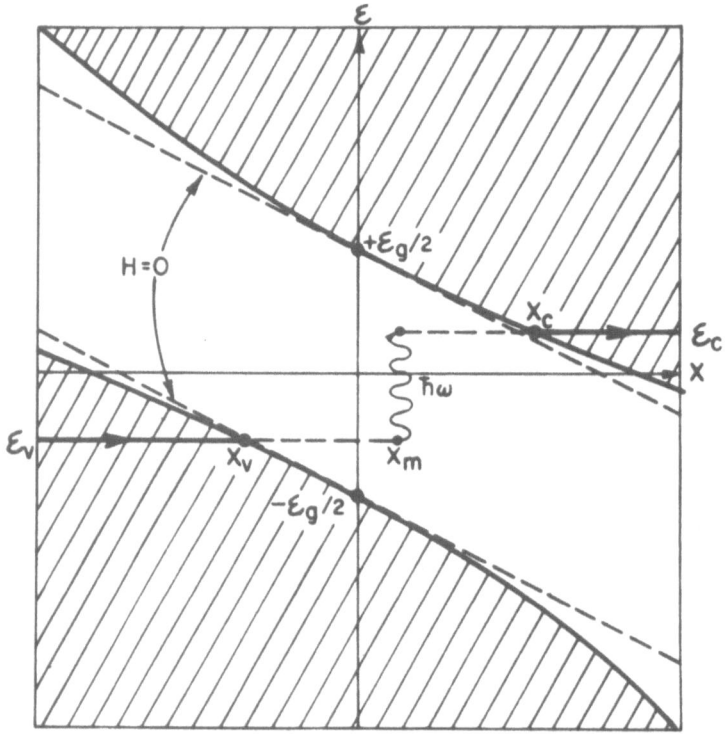

Fig. 24. Photon-assisted tunneling in crossed electric and magnetic fields. The sloping dashed lines represent the valence and conduction band edges for $H = 0$. The shaded regions represent the valence and conduction band regions with both fields present. From Weiler et al. ([42]).

over time so as to average over a period of the oscillating electric field. To do this the matrix element is expanded as a Fourier expansion in terms of the applied laser frequency and the result appears as the transition probability, which is the sum of one, two, and three or more photons. Each component term therefore represents the transition probability of an individual multiphoton process. The results of this calculation for both sets of processes are given by the following equations:

For $\mathbf{E} \parallel \mathbf{H}$:

$$W_n^{(l)} \approx \frac{\omega^2}{8l\pi^2} \left(\frac{eH}{\hbar c} \right) (2\mu)^{1/2} \left(\frac{e^2E^2}{8\mu\omega^2 \mathscr{E}_n} \right)^l \left[l\hbar\omega - \left(\mathscr{E}_n + \frac{e^2E^2}{4\mu\omega^2} \right) \right]^{-1/2}$$

$$\times \exp\left(2l - \frac{\mathscr{E}_n}{\hbar\omega} \right)$$

For $\mathbf{E} \perp \mathbf{H}$:

$$W_{n'n}^{(l)} \approx \frac{\omega^2}{4l\pi^2} \left(\frac{eH}{\hbar c} \right) \left(\frac{e^2 E^2}{8\mu\omega^2 \mathscr{E}_{n'n}} \right)^l [l\hbar\omega - (\mathscr{E}_{n'n} + \mathscr{E}_{\text{intra}})]^{-1/2}$$

$$\times \exp\left(\frac{\mathscr{E}_{n'n}}{\hbar\omega} \right) \alpha_{n'n}$$

where

$$\mathscr{E}_{\text{intra}} = \frac{eE^2}{4m_c(\omega^2 - \omega_c^2)} + \frac{eE^2}{4m_v(\omega^2 - \omega_v^2)}$$

and

$$\alpha_{n'n} = (|\, n' - n\,|!)^{-2} \left(\frac{n'!}{n!} \right)^{\pm 1} \left(\frac{\omega_\mu}{l\omega} \right)^{|n'-n|} \exp\left[\frac{2(n - n')(\omega_c - \omega_v)}{\omega} \right]$$

The cyclotron frequency ω_μ involves the reduced effective mass μ.

It is important to note that in the presence of a magnetic field, both for the parallel and transverse cases, the transition probabilities contain resonance denominators similar to that for the one-photon magnetoabsorption. Furthermore, these are weighted by the ratio of the cyclotron resonance energy to a higher power and the square of the electric field intensity, indicating that the transition probability of higher-order processes are progressively weaker, since the energy ratios in terms of the cyclotron resonance energy and the laser power to the energy gap are less than unity. Nevertheless, for a given multiphoton transition the amplitude of the individual process increases with the amplitude of the magnetic field and that of the laser power.

Another interesting feature is the intraband contribution in the resonance denominator. This contribution to the energy, which comes from the time-dependent wave function, is essentially the equivalent energy that might be absorbed by a free carrier which has a resonant denominator for both holes and electrons with $\mathbf{E} \perp \mathbf{H}$. This latter term at zero frequency is the negative energy in crossed electric and magnetic fields obtained by Aronov [35], namely, $(-c^2 E^2/H^2)(m_1 + m_2)$. This term is generalized to the oscillatory case.

Another mechanism for multiphoton processes was suggested by Braunstein and Ockman [36]; here multiphoton absorption takes place via virtual transitions between higher energy bands which act as intermediate states for the multiphoton processes. The transition probability is then calculated by higher-order perturbation theory. Such a process can take place in the presence of a magnetic field by a combination of an interband and intraband transition, as suggested by Zawadzki and co-workers

(³⁷). In this case a magnetic subband instead of a higher band acts as an intermediate state. Consequently, there is again a resonance absorption, but for this case the selection rule is $\Delta n = \pm 1$, since the laser electric field and the static magnetic field must be perpendicular to each other. For the Keldysh mechanism the selection rule is $\Delta n = 0$ for both parallel and

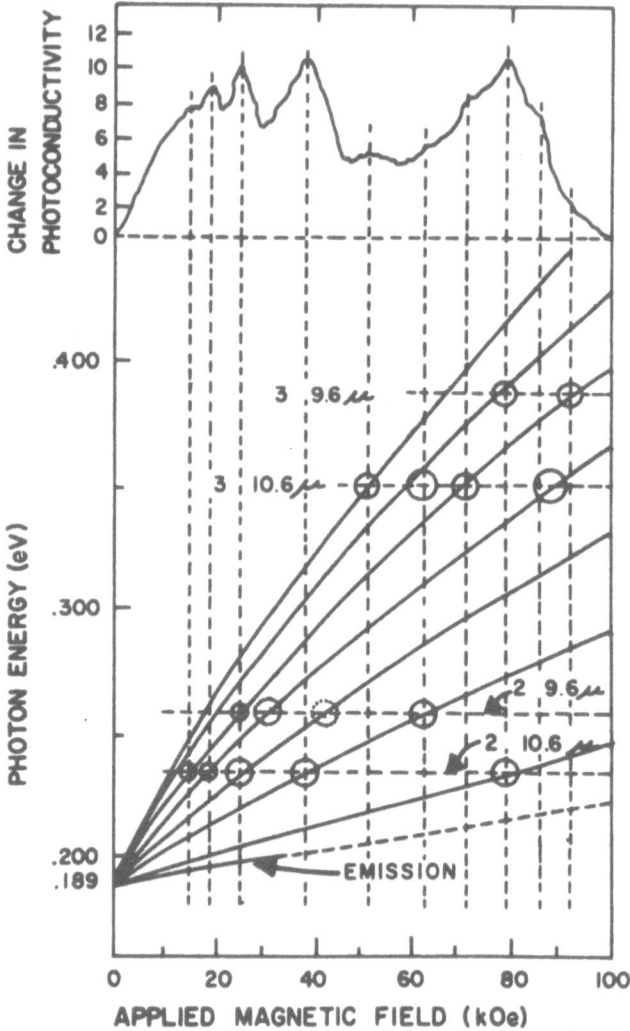

Fig. 25. The photoconductivity peaks in lead telluride (upper trace) due to multiphoton absorption compared with transitions calculated from a simple two-band nonparabolic model for selection rules $\Delta n = \pm 1$. From Button et al. (³⁸).

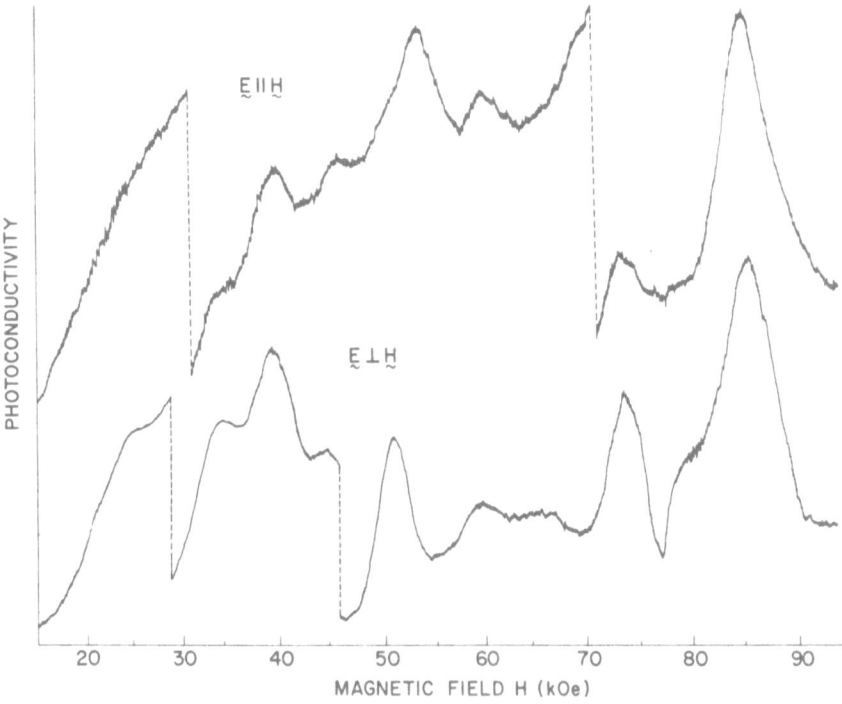

Fig. 26. Photoconductivity peaks for indium antimonide for two different orientations of the electric vector with respect to the applied magnetic field. From Weiler *et al.* ([34]).

perpendicular cases. It appears that we may have seen evidence in the experiments ([38]) for simultaneous existence of both of these phenomena.

The experimental observations of interband two-photon absorption in a steady magnetic field were made by irradiating the semiconductor specimen with the high-intensity pulses of electromagnetic radiation from a Q-switched carbon dioxide laser similar to the original technique used by Patel *et al.* ([39]). In the absence of a magnetic field, however, they could only assume that two-photon absorption was creating electrons and holes when they observed the emission of recombination radiation. When the magnetic field is applied, however, holes and electrons can be created only when the energy separation between Landau quantum levels across the gap is equal to the energy of an integral number of laser photons. Thus when the magnetic field is increased and the quantum levels are moved one expects to see a sharp increase in the photoconductivity (creation of holes and electrons) each time a pair of levels can accommodate two or more photons. This was clearly observed in InSb and PbTe, and the results for PbTe are shown in Fig. 25. This figure shows resonant peaks associated with two-photon transitions as the dominant process, which appears to have been satisfactorily interpreted in terms of the Zawadzki mechanism, which predicted the selection rule $\Delta n = \pm 1$. When $\mathbf{E} \perp \mathbf{H}$, however, resonant absorption is also seen, indicating that the Keldysh mechanism is also present, because in this configuration the perturbation theory does not account for the phenomenon. In indium antimonide it appears that the Keldysh mechanism must dominate, because in this case resonances of almost equal intensity are observed for both Faraday and Voigt configurations, as shown in Fig. 26. The data is interpreted by constructing the graph shown in Fig. 27, where the intersections of straight lines are of special significance. Horizontal lines are drawn on the energy scale corresponding to the sum of the energies of two CO_2 laser photons, namely, $2\hbar\omega$ for 10.6 μ photons, $2\hbar\omega$ for 9.6 μ, and two photons of mixed frequency. Then vertical lines are drawn corresponding to the photoconductivity peaks. Then the diagonal lines represent allowed transitions for $\Delta n = 0$, as obtained from the data of Pidgeon and Brown [9] (Fig. 13). It can be seen that the modified Keldysh theory for the magnetic case for $\Delta n = 0$ transitions fits the data quite well at the intersections of these three sets of lines.

CONCLUSION

We have briefly reviewed some of the important quantum magneto-optical phenomena that have been studied experimentally and theoretically in high magnetic fields. These phenomena can be divided into two classes: intraband and interband transitions. They have both provided much new information about the energy band parameters of a variety of semiconductors and semimetals.

We have discussed the important results obtained by the use of sub-millimeter lasers in the study of cyclotron resonance. In the near future these studies will be aided by the additional laser frequencies obtained from both the HCN and the H_2O lasers which should permit a further extension from the submillimeter range to the higher frequencies of the far infrared range.

The investigations using lasers will be augmented by the use of the Fourier transform spectroscopic technique. Although less power is available, it is convenient to be able to scan the entire wavelength range between the millimeter waves and the near infrared waves. The importance of covering a wide frequency range by either techniques will be to study relaxation times as a function of magnetic field intensity. This should give new information

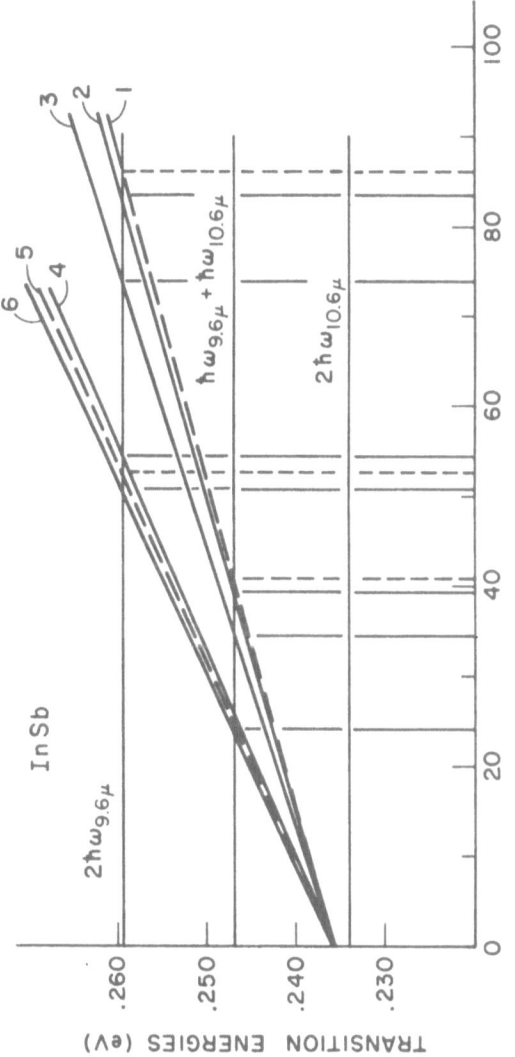

Fig. 27. Comparison of the positions of the photoconductivity peaks (vertical lines), the two-photon ener-gies supplied by the laser (horizontal lines), and the allowed transitions (sloping lines). From Weiler *et al.*[34].

about electron–phonon interactions, not only involving the acoustic phonon at long wavelengths but also the optical phonons in the far infrared region of the spectrum. Hopefully, the large-power capabilities of the lasers will eventually be adequate to enhance these electron–phonon interactions by nonlinear excitation.

The study of interband phenomena has received a significant boost since the new modulation techniques have been developed. These should prove to be even more helpful in studying new materials at high magnetic fields in the near future. Furthermore, the combination of parallel and crossed electric and magnetic fields adds another dimension, resulting in the photon-assisted tunneling phenomena. These of course have their most exciting features in the multiphoton processes whose study has just begun. In the future these nonlinear phenomena should provide information about higher bands as well as greater understanding of the mechanism of interaction of photons with electrons in a solid.

Finally, attention should be drawn to other magnetooptical phenomena involving the use of lasers in the study of solids. These include such effects as the nonparabolic nonlinearity and three-photon frequency mixing. Work is now being done on parametric effects and on particles in high magnetic fields, as well as on various scattering phenomena. The latter can simply be labeled magneto-Thomson, magneto-Raman, magneto-Rayleigh, and magneto-Compton phenomena. In addition to these single-electron excitations, collective excitation of magnetoplasmas and magneto–plasmon–phonon systems will be investigated by laser light scattering and also by magnetoreflection techniques in the far infrared. Thus it appears that there will be an exciting and fruitful period of investigation of quantum magnetooptical effects during the coming years.

REFERENCES

1. L. Rosenfeld, *Z. Physik* **57**: 835 (1930).
2. E. Burstein, G. S. Picus, and H. A. Gebbie, *Phys. Rev.* **105**: 1123 (1957); S. Zwerdling and B. Lax, *Phys. Rev.* **106**: 51 (1957).
3. G. Dresselhaus, A. F. Kip, C. Kittel, and G. Wagoner, *Phys. Rev.* **98**: 556 (1955).
4. R. J. Keyes, S. Zwerdling, S. Foner, H. H. Kolm, and B. Lax, *Phys. Rev.* **104**: 1804 (1956).
5. E. Burstein, G. S. Picus, and H. A. Gebbie, *Phys. Rev.* **103**: 825 (1956).
6. E. O. Kane, *J. Phys. Chem. Solids* **1**: 249 (1957).
7. B. Lax and J. G. Mavroides, in: *Solid States Physics* Vol. 11 (F. Seitz and D. Turnbull, eds.), Academic Press, New York, 1960.
8. L. M. Roth, B. Lax, and S. Zwerdling, *Phys. Rev.* **114**: 90 (1959).

9. C. R. Pidgeon and R. N. Brown, *Phys. Rev.* **146**: 575 (1966).
10. J. M. Luttinger and W. Kohn, *Phys. Rev.* **97**: 869 (1955); J. M. Luttinger, Phys. Rev. **102**: 1030 (1956).
11. R. C. Fletcher, W. A. Yaeger, and F. R. Merritt, *Phys. Rev.* **100**: 747 (1955).
12. J. C. Hensel, *Bull. Am. Phys. Soc.* **6**: 115 (1961); J. J. Stickler, H. J. Zeiger, and G. S. Heller, *Phys. Rev.* **127**: 1077 (1962).
13. K. J. Button, H. A. Gebbie and B. Lax, *IEEE J. Quantum Electronics* **QE2**: 202 (1966).
14. K. J. Button, B. Lax, and H. A. Gebbie, unpublished.
15. C. R. Pidgeon, S. H. Groves, and J. Feinlieb, *Solid State Commun.* **5**: 677 (1967).
16. Q. H. F. Vrehen, *J. Phys. Chem. Solids* **29**: 129 (1968).
17. S. H. Groves, R. N. Brown, and C. R. Pidgeon, *Phys. Rev.* **161**: 779 (1967).
18. S. H. Groves, C. R. Pidgeon, A. W. Ewald, and R. J. Wagner, *Bull. Am. Phys. Soc.* **13**: 410 (1968).
19. S. Zwerdling, B. Lax, L. M. Roth, and K. J. Button, *Phys. Rev.* **114**: 80 (1959).
20. J. Halpern and B. Lax, *J. Phys. Chem. Solids*, **26**: 911 (1965).
21. R. L. Aggarwal, M. D. Zuteck, and B. Lax, *Phys. Rev. Letters* **19**: 236 (1967).
22. *Phys. Rev.* (to be published).
23. R. N. Brown, J. G. Mavroides, M. S. Dresselhaus, and B. Lax, *Phys. Rev. Letters* **5**: 243 (1960).
24. R. N. Brown, J. G. Mavroides, and B. Lax, *Phys. Rev.* **129**: 2055 (1963).
25. B. Lax, K. J. Button, H. J. Zeiger, and L. M. Roth, *Phys. Rev.* **102**: 715 (1956).
26. M. Maltz, Ph. D. Thesis, MIT Dept. of Electrical Engineering (1968).
27. M. S. Dresselhaus and J. G. Mavroides, *IBM J. Res. Develop.* **8**: 262 (1964).
28. P. R. Schroeder, A. Javan, M. S. Dresselhaus, and J. G. Mavroides, *Bull. Am. Phys. Soc.* **11**: 91 (1966).
29. G. Dresselhaus and M. S. Dresselhaus, *Proceedings International School of Physics, Enrico Fermi, Varenna*, Academic Press, New York, 1966, Course 34, p. 252.
30. M. S. Dresselhaus, private communication.
31. M. S. Dresselhaus and J. G. Mavroides, *Phys. Rev. Letters* **14**: 259 (1965).
32. M. Maltz, Ph. D. Thesis, MIT Dept. of Electrical Engineering (1968); M. S. Maltz, S. Fischler, and M. S. Dresselhaus, *Bull. Am. Phys. Soc.* **11**: 917 (1966); M. S. Maltz and M. S. Dresselhaus, *Phys. Rev. Letters*, **20**: 919 (1968).
33. L. V. Keldysh, *Soviet Phys.–JETP* (*English Transl.*) **20**: 1307 (1965).
34. M. H. Weiler, M. Reine, and B. Lax, *Phys. Rev.*, **171**: 949 (1968).
35. A. G. Aronov, *Soviet Phys.–Solid State* (*English Transl.*) **5**: 402 (1963).
36. R. Braunstein and N. Ockman, *Phys. Rev.* **134**: A499 (1964).
37. W. Zawadzki, E. Hanamura, and B. Lax, *Bull. Am. Phys. Soc.* **12**: 100 (1967); W. Zawadzki, M. H. Weiler, B. Lax, E. Hanamura, K. J. Button, and M. B. Reine, *Proceedings Yerevan Conference on Nonlinear Optics, USSR, October* 1967, to be published.
38. K. J. Button, B. Lax, M. H. Weiler, and M. B. Reine, *Phys. Rev. Letters* **17**: 1005 (1966).
39. C. K. N. Patel, P. A. Fleury, R. E. Slusher, and H. L. Frisch, *Phys. Rev. Letters* **16**: 971 (1966).
40. B. Lax, J. G. Mavroides, H. J. Zeiger, and R. J. Keyes, *Phys. Rev.* **122**: 31 (1961).
41. J. J. Stickler, H. J. Zeiger, and G. S. Heller, *Phys. Rev.* **127**: 1077 (1962).
42. M. H. Weiler, W. Zawadzki, and B. Lax, *Phys. Rev.* **163**: 733 (1967).

Chapter 7

MAGNETO-QUANTUM-ELECTRIC EFFECT*

Yasuji Sawada[†]

Laboratory for Research on the Structure of Matter and Physics Department
University of Pennsylvania
Philadelphia, Pennsylvania

INTRODUCTION

In this chapter we study the motion of the center of the cyclotron orbit when the electron in the orbit absorbs photons or phonons, and we also discuss related phenomena such as the magneto-quantum-electric effect and phonon amplification. As is discussed in detail by Zawadzki in Chapter 13 of this volume, the center of the cyclotron orbit is given by $X_0 = -P_y/m\omega_c$, $Y_0 = y + (P_x/m\omega_c)$ if we choose a gauge $A = [0, H_0 x, 0]$. In this gauge the x coordinate of the center of the cyclotron orbit depends only on the "good quantum number" P_y. Therefore we should be able to shift the cyclotron orbit in the x direction by giving a momentum ΔP_y to an electron in cyclotron orbit. However, if we choose another gauge $A' = [-H_0 y, 0, 0]$, the center of cyclotron orbit is given by $X_0 = x - (P_y/m\omega_c)$, $Y_0 = P_x/m\omega_c$. In this case P_y is not a good quantum number, and it is not easy to see what will happen when the electron absorbs a momentum ΔP_y. To see what really happens to the cyclotron electron in absorbing a momentum from phonons or photons, we must carry out a gauge-independent calculation of a physical quantity such as the current associated with this shift of centers of orbits. This calculation is done in the next section. It turns out that the lateral displacement of the cyclotron orbits occurs when the momenta ΔP transfered to the electrons have components in the plane of the orbit in real space. When the electrons in a spherical energy band

* Work supported by the Advanced Research Projects Agency.
† Present address: Physics Department, Osaka University, Toyonaka, Japan.

absorb quanta in cyclotron transition ($\Delta n = \pm 1$) the centers of the cyclotron orbits are displaced in the direction perpendicular to $\Delta \mathbf{P}$ and $\mathbf{H_0}$. In the case of an ellipsoidal energy band the displacement can occur even when $\Delta \mathbf{P}$ and $\mathbf{H_0}$ are parallel to one another, provided the ellipsoid is tilted from $\mathbf{H_0}$. In this situation the absorption of quanta in either cyclotron ($\Delta n = \pm 1$) or Landau ($\Delta n = 0$) transitions leads to a displacement of the center of the cyclotron orbit which is perpendicular to $\Delta \mathbf{P}$ and to the normal to the plane of the orbit. When quanta are absorbed continuously in either situation the lateral translation of orbits constitutes a "DC current" which sets up a surface charge density Σ at the lateral boundaries and thereby a DC electric field $E = 4\pi\Sigma$.

This electric field, the "magneto-quantum-electric" (MQE) field ([1]), has been experimentally verified using a bismuth single crystal ([2]). The MQE field measured was of the order of 1 mV and was found to be able to provide us with the same amount of information about electron Fermi surfaces as the one obtained from the giant quantum attenuation experiment ([3]). The experimental arrangement and some typical data are exhibited in the section beginning on p. 193.

When the flux of phonons or photons are sufficiently high this MQE field may in turn influence the rate of absorbtion by electrons itself. In fact, it is well known that phonon flux can be amplified by applying a DC electric and magnetic field from outside the crystal [see, e.g. ([12,14])]. The relation between the MQE effect and the amplification effect is discussed in the final section.

THEORETICAL

In all the classical theories of transport phenomena in the presence of a magnetic field the shift of centers of cyclotron orbits caused by the momentum absorption has been neglected. It has been assumed that the cyclotron center is fixed during the interaction of electrons with phonons or photons. As is seen later, this orbit shift is a nonlinear effect in the sense that the "orbit-shift current" is proportional to the square of the amplitude of electromagnetic (or sound) wave. As the linear theory of the nonlocal transport phenomena in the presence of a magnetic field is complicated enough, it is easy to imagine the difficulty to be met in the nonlinear theory. On the other hand, this phenomenon can be handled very simply in quantum mechanics, as the nature of the effect is fundamentally quantum mechanical. Thus all the treatments hereafter are quantum mechanical.

Let us first consider an isotropic electron plasma in the absence of magnetic field. The Hamiltonian of an electron in the plasma is

$$\mathscr{H}_0 = (1/2m)P^2 \tag{1}$$

Constants of motion of the electron are the energy ε and the three components of momentum P_x, P_y, and P_z. They are eigenvalues of the operators $\mathscr{H}_0$ and P_x, P_y, P_z, which commute by definition with the Hamiltonian itself. We are generally not interested in measuring these quantities, and even if we are interested it is usually not easy to measure them.

We assume some perturbation applied from outside on the system such as electromagnetic radiation or sound waves. This perturbation interacts with the electron system and changes the constants of motion, making it possible for us to study the interaction of the electrons with the perturbation. Let us assume the following form for the perturbation

$$\mathscr{H}' = \mathscr{H}_1 \exp(-i\omega t) = \frac{eP_x}{mc} A' \exp(iqy - i\omega t) \tag{2}$$

This equation corresponds to an electromagnetic wave or transverse sound wave polarized in the x direction and propagating along the y direction. In the presence of this perturbation Hamiltonian the energy operator $P^2/2m$ and a momentum operator P_y do not commute with the total Hamiltonian $\mathscr{H}_T = \mathscr{H}_0 + \mathscr{H}'$. They are no longer constants of the motion, and we can calculate the rate of the change as follows

$$\dot{\mathscr{H}}_0 = \frac{1}{i\hbar} [\mathscr{H}_0, \mathscr{H}_T] = \frac{1}{i\hbar} [\mathscr{H}_0, \mathscr{H}'] = \frac{iP_y q}{m} \mathscr{H}' \tag{3}$$

$$\dot{P}_y = \frac{1}{i\hbar} [P_y, \mathscr{H}_T] = \frac{1}{i\hbar} [P_y, \mathscr{H}'] = iq\mathscr{H}' \tag{4}$$

The expectation value of the time-derivative operator $\dot{\mathscr{H}}_0$ is given by;

$$d\varepsilon/dt = \langle \dot{\mathscr{H}}_0 \rangle = \gamma \, \Delta\varepsilon \tag{5}$$

where $\Delta\varepsilon = \hbar q P_y/m$ is the change of energy of the electron by absorbing a phonon and γ is the rate of phonon absorbtion, given by

$$\gamma = \frac{\pi}{\hbar} \sum_{nn'} \delta(\varepsilon_n - \varepsilon_{n'} - \hbar\omega)\{f(\varepsilon_{n'}) - f(\varepsilon_n)\} \, |\langle n | \mathscr{H}_1 | n' \rangle|^2 \tag{6}$$

where $f(\varepsilon)$ is the Fermi distribution function. The derivation of Eq. (5)

is given in the Appendix. Similarly, the rate of change of P_y is given by

$$\frac{dP_y}{dt} = \langle \dot{P}_y \rangle = \hbar q \gamma = \gamma \, \Delta P_y \tag{7}$$

The change of energy of the system can be studied experimentally either by measuring the temperature rise of the crystal which contains the plasma or by measuring the attenuation of the incident perturbation through the plasma. The change in P_y can be studied by measuring the electric potential difference along the direction of wave propagation. This effect is known as the acoustoelectric effect (4) in the case of acoustic waves and as radiation pressure in the case of electromagnetic waves. The momentum distribution of the electrons in the plasma is homogeneous in the absence of the perturbation, and the statistical average of the momentum P_y over the electrons vanishes. However, when the interaction of the incident perturbation with electrons is turned on the momentum of the electrons along the direction of the wave propagation increases, as shown in Eq. (4). Therefore the statistical average of the momentum over the electrons no longer vanishes. This nonvanishing momentum constitutes a current and thereby a potential difference is established along the current direction.

So far we have discussed the constants of motion of electrons in the absence of an applied magnetic field H_0. A natural extension of this consideration is to ask what the constants of motion are in the presence of H_0. The Hamiltonian of an electron in this case is

$$\mathcal{H} = (1/2m)\pi^2 \tag{8}$$

where $\pi = P + (e/c)A$, with A the vector potential which satisfies $H_0 = \nabla \times A$. To look for the constants of motion, we must know the commutation relations between the components of π and the components of r. Assuming a scalar function ϕ, we obtain the commutation relations between the components of π:

$$(\pi \times \pi)\phi = \frac{e}{c}(A_0 \times P + P \times A_0)\phi = -\frac{i\hbar e}{c}\phi(\nabla \times A_0) = -\frac{i\hbar e}{c}H_0\phi$$

thus

$$\pi \times \pi = -(i\hbar e/c)H_0 \tag{9}$$

The commutation relations between the components of π and the components of r are the same as those for P and r:

$$[\pi_i, r_j] = [P_i + (e/c)A_{0i}, r_j] = [P_i, r_j] = -i\hbar\delta_{ij} \tag{10}$$

Using these commutation relations and assuming H_0 to be applied along the Z axis, we obtain the following relations:

$$\dot{\pi}_x = (1/i\hbar)[\pi_x, \mathscr{H}_0] = -\omega_c \pi_y, \qquad \dot{\pi}_y = (1/i\hbar)[\pi_y, \mathscr{H}_0] = \omega_c \pi_x$$

$$\dot{\pi}_z = (1/i\hbar)[\pi_z, \mathscr{H}_0] = 0, \qquad m\dot{\mathbf{r}} = (1/i\hbar)[\mathbf{r}, \mathscr{H}_0] = \pi \qquad (11)$$

where $\omega_c = eH_0/mc$ is the cyclotron frequency. Using these relations, we find the following operators which commute with the Hamiltonian: $X_0 = x - (\pi_y/m\omega_c)$, $Y_0 = y + (\pi_x/m\omega_c)$, $\pi_z = P_z$, and $\mathscr{H}_0$ itself. We again have the energy and the momentum components along H_0 as constants of motion. They are not affected by the application of H_0. However, the other components of momentum, P_x and P_y, are no longer constants of the motion. New constants of motion X_0 and Y_0 correspond to the x and y coordinates of the center of the cyclotron orbit.

Now we calculate the change of this center of the cyclotron orbit due to the same type of perturbation as given by Eq. (2):

$$\dot{Y}_0 = \frac{1}{i\hbar}[Y_0, \mathscr{H}_T] = \frac{1}{i\hbar}[Y_0, \mathscr{H}'] = \frac{1}{i\hbar}\left[y + \frac{\pi_x}{m\omega_c}, \mathscr{H}'\right]$$

$$= 0 \qquad (12)$$

$$\dot{X}_0 = \frac{1}{i\hbar}[X_0, \mathscr{H}_T] = \frac{1}{i\hbar}\left[x - \frac{\pi_y}{m\omega_c}, \mathscr{H}'\right]$$

$$= i\frac{cq}{eH_0}\mathscr{H}' \qquad (13)$$

One notices that these expressions are independent of the gauge chosen for A. We can rewrite these equations as follows:

$$\dot{Y}_0 = 0 \qquad (14)$$

$$\dot{X}_0 = (i/\hbar)\,\Delta X_0 \mathscr{H}' \qquad (15)$$

where $\Delta X_0 = c\hbar q/eH_0$ is the displacement of the center of the cyclotron orbit. Eqs. (14) and (15) are completely analogous to Eqs. (5) and (7). If we define an operator $\mu = e\mathbf{R}_0$, the components of the time derivative of this operator correspond to a "polarization current" for the electron. The expectation value of this current can be calculated using the method of Eq. (5):

$$\langle \dot{\mu}_y \rangle = e\langle \dot{Y}_0 \rangle = 0 \qquad (16)$$

$$\langle \dot{\mu}_x \rangle = e\langle \dot{X}_0 \rangle = e\,\Delta X_0 \gamma \qquad (17)$$

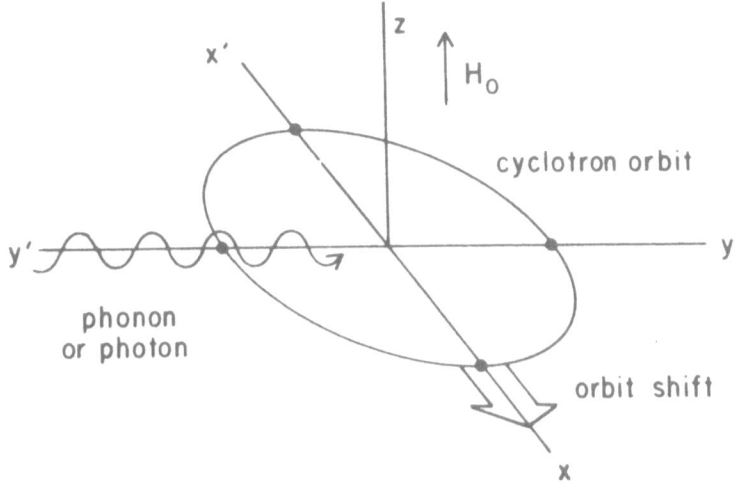

Fig. 1. When the electrons in a spherical energy band absorb quanta in cyclotron transition the centers of the cyclotron orbits are displaced in the direction perpendicular to q and H_0.

where γ is again probability of absorption of a quantum by an electron per unit time. According to Eqs. (16) and (17) the polarization current flows in the direction mutually perpendicular to the magnetic field and to the direction of wave propagation, as shown in Fig. 1. Let us consider a plasma consisting of N electrons/cm³ and a flux of phonons φ cm⁰/sec impinging upon it. For this system $\gamma = \Gamma\varphi$, where Γ is the attenuation constant (cm⁻¹), and the polarization current is given by*

$$J_p = N\langle \dot{\mu}_x \rangle = e\,\Delta X_0 \Gamma\varphi \tag{18}$$

This J_p is the origin of the MQE field which appears across the lateral boundaries only when phonons or photons are absorbed by electrons (i.e., $\gamma \neq 0$).

So far we have discussed the MQE effect for a simple isotropic plasma and magnetic field applied perpendicular to the wave propagation. In this

---

* One notices that this polarization current, J_p, is a DC current and proportional to E^2, the square of the electric field associated with the perturbation wave. The reason for this is that the probability of absorption, $\gamma = \alpha\varphi$, is proportional to the square of the matrix element $|\langle n | \varkappa_1 | n \rangle|^2$, as is given by Eq. (6), where $\varkappa_1 = (eP_x/mc)A'\exp{(iqy)}$, and the vector potential A' associated with the perturbing wave is related to the electric field by $i\omega A'/C = E$.

case only $\Delta n = \pm 1$ transitions are involved in the mechanism.* In the case of a plasma consisting of electrons in an ellipsoidal energy band whose axes are tilted from H_0 the transfer of momentum $\mathbf{q}$ along H_0 causes both the longitudinal acoustoelectric effect (or radiation pressure) and the transverse MQE effect. Let us assume the energy ellipsoid given by

$$\varepsilon = \frac{1}{2m}(\alpha_{xx}P_x{}^2 + \alpha_{yy}P_y{}^2 + \alpha_{zz}P_z{}^2 + 2\alpha_{yz}P_yP_z) \tag{19}$$

This represents an ellipsoid whose axis is tilted in the yz plane. We apply a magnetic field H_0 along the y direction, which is parallel to the direction of phonon (or photon) propagation. To simplify the discussion we choose a gauge $A = [0, 0, -H_0x]$ for the vector potential, which immediately gives us a physical insight. By performing a calculation similar to the one which we did for a spherical energy band, we find the x and z coordinates of the center of the cyclotron orbit given by

$$\mathbf{x}_0 = -\frac{c}{eH_0}\left(P_z + \frac{\alpha_{yz}}{\alpha_{zz}}P_y\right) \tag{20}$$

$$\mathbf{z}_0 = z + (c/eH_0)P_x \tag{21}$$

If we send a perturbation in the z direction, which is perpendicular to H_0, and change P_z, the center of the orbit will be shifted again in the x direction, as is the case for a spherical energy band. In the case of ellipsoidal energy band, however, the orbit shift can take place even if the momentum P_y, which is parallel to H_0, is changed by the phonons or photons. From Eq. (20) the change of P_y by $\Delta P_y = \hbar q$ leads to a change in the $\mathbf{x}_0$ given by $\Delta X_0 = -(c\hbar q/eH_0)(\alpha_{yz}/\alpha_{zz})$, where α_{yz}/α_{zz} is the term which is responsible for the tilting of the ellipsoid. This situation for the tilted ellipsoid is shown in Fig. 2. The momentum absorption along the direction of H_0 can be achieved either by a $\Delta n = 0$ transition or a $\Delta n = \pm 1$ transition. Thus longitudinal sound waves can also be utilized in the experiment. In either case the general expression (17) for the polarization current holds.

As experiments are usually performed under an open-circuit condition, we calculate the magnitude of the electric field (MQE field) which is caused by the polarization current J_p. The continuous absorption of phonons or photons builds up a surface charge Σ at the boundaries which sets up an electric field across the crystal. This electric field E induces an ohmic current

---

* For a metal where the skin depth is small transitions involving $\Delta n = \pm 2, \pm 3, \ldots$ may also be important.

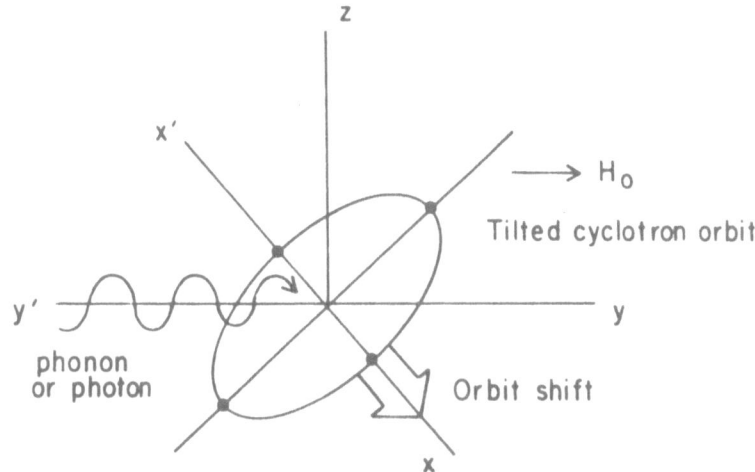

Fig. 2. In the case of an ellipsoidal energy band the displacement can occur even when q and H_0 are parallel to one another, provided the ellipsoid is tilted from H_0. In this situation the absorption of quanta leads to a displacement of the center of the cyclotron orbit which is perpendicular to q and to the normal to the plane of the orbit.

(screening current J_s) which tries to neutralize this surface charge. The continuity equation for Σ at the surface is

$$d\Sigma/dt = J_p + J_s \tag{22}$$

where*

$$J_s = E_{MQ}/\varrho_{xx}(H_0) \tag{23}$$

where $\varrho(H_0)$ is the magnetic-field-dependent resistivity tensor. In the steady state $d\Sigma/dt = 0$.

---

* Notice that $J_s \neq \sigma_{xx}(H_0)E_{MQ}$. The reason for this is that

$$J_{sx} = \sigma_{xx}E_x + \sigma_{xy}E_y + \sigma_{xz}E_z$$

However, other boundary conditions are given by

$$J_y = \sigma_{yx}E_x + \sigma_{yy}E_y + \sigma_{yz}E_z = 0$$
$$J_z = \sigma_{zx}E_x + \sigma_{zy}E_y + \sigma_{zz}E_z = 0$$

By solving these three equations we get

$$E_x = \varrho_{xx}J_{sx}, \qquad E_y = \varrho_{yx}J_{sx}, \qquad E_z = \varrho_{zx}J_{sx} \tag{24}$$

The first equation is equivalent to Eq. (23).

From Eqs. (18) and (23) we get

$$E_{MQ} = -e \, \Delta X_0 \Gamma \varphi \varrho_{xx}(H_0)$$

$$= -\Gamma S \frac{c}{v_s} \frac{\varrho_{xx}(H_0)}{H_0} \frac{\alpha_{yz}}{\alpha_{zz}} \qquad (25)$$

where $S = \varphi \hbar \omega$ is the phonon energy flux and $v_s = \omega/q$ is the sound wave velocity.

Before estimating the magnitude of this field it is helpful for the discussion below to examine the experimental conditions under which the MQE effect occurs. As for any other effect which involves Landau levels, $\omega_c \tau > 1$ must be satisfied, i.e., the Landau-level separation is greater than the collision broadening of the energy level. The second requirement is that an electron in the cyclotron orbit must experience the effect of the wave over at least one wavelength. This condition is expressed by $qR > 1$, where R is the radius of the orbit or the projection of the orbit along the direction of the wave propagation if the orbit is tilted. If these two conditions are satisfied, $ql > 1$ is automatically satisfied, as $ql = qR\omega_c\tau > 1$.[*] Under these conditions, the interaction of electrons with phonons or photons is quantum mechanical and the electrons absorb them as quanta (a quantity with a well-defined momentum and energy).

The ordinary acoustoelectric field experiment [5] has been carried out under the classical condition of $ql \ll 1$ in a material which contains an uncompensated plasma. In this case the sound waves cannot be treated as quanta, but they are treated as elastic waves, and electrons flow with this elastic wave as if they were surf-riding. In this situation holes would also be driven in the same direction simultaneously if the plasma consists of electrons and holes. Thus there is no acoustoelectric effect if the plasma is compensated. Under quantum mechanical conditions the situation is different. The energy and momentum conservation relations between electrons and phonons (photons) are satisfied only at certain definite values of the magnetic field strength, in either $\Delta n = 0$ or $\Delta n = \pm 1$ transitions [6,7], and as these values are generally not the same for the electrons and holes, the acoustoelectric field, E_{AE}, does not vanish even for the compensated plasma.

---

[*] The $ql > 1$ condition has the following interesting meaning. The uncertainty in momentum P_y is given by $\delta P_y = \hbar/l$. Therefore the uncertainty of the position of orbit center is $\delta x_0 = (c/eH_0)\delta P_y = c\hbar/eH_0 l$. This uncertainty must be smaller than the amount of orbit shift $\Delta x_0 = c\hbar q/eH_0$. The inequality $ql > 1$ is equivalent to the condition $\Delta x_0 > \delta x_0$.

Now let us compare the magnitude of the longitudinal acoustoelectric field E_{AE} and the transverse magneto-quantum-electric field E_{MQ} under quantum mechanical conditions for both compensated and uncompensated plasmas. In the acoustoelectric effect the relation between the attenuation constant and E_{AE} for a free-electron gas was given by Weinreich ([8]):

$$E_{AE} = \Gamma S / Nev_s \tag{26}$$

From Eqs. (25) and (26) we obtain the ratio of E_{MQ} to E_{AE}:

$$E_{MQ}/E_{AE} = Nec\varrho_{xx}(H_0)/H_0 \tag{27}$$

The magnitude of this ratio will depend on whether or not the plasma is compensated. For an uncompensated plasma $\varrho_{xx}(H_0)$ becomes constant at high magnetic field strength and the ratio approaches $1/\omega_c\tau$. Thus the transverse MQE field is smaller than the longitudinal AE field by the factor $\omega_c\tau$. However, in this case there will be an associated Hall electric field set up in the z direction which can be larger than the MQE field. The Hall field, arising from the effect of the magnetic field on the screening current,* is given from Eq. (24) by

$$E_H = \varrho_{zx}J_s = (\varrho_{zx}/\varrho_{xx})E_{MQ} \approx E_{AE} \tag{28}$$

Thus the Hall field is of the same order of magnitude as the AE field. For compensated plasma ϱ_{xx} varies as $H_0{}^2$. In this case the ratio of the MQE field to the AE field is greater than $\omega_c\tau$. Thus in either the compensated or the uncompensated case we can expect to observe a transverse field—either the MQE field or the associated Hall field—which is of the same order or greater than the longitudinal AE field.

EXPERIMENT

The experiment in which the quantum mechanical conditions are most clearly satisfied is the giant quantum attenuation (GQA) of sound waves (see Chapter 20 by Shapira in this volume). We looked for the MQE fields in GQA experiments on bismuth which had already been carried out at our laboratory ([3]); bismuth is one of the best known materials ([9]), and the Fermi surfaces for electrons and holes are fairly well established.

---

* The magnetic field acts only on the screening current. Therefore although the screening current cancels the polarization current in the steady state and there is a zero net current, a Hall field is set up in the direction mutually perpendicular to J_s and H_0.

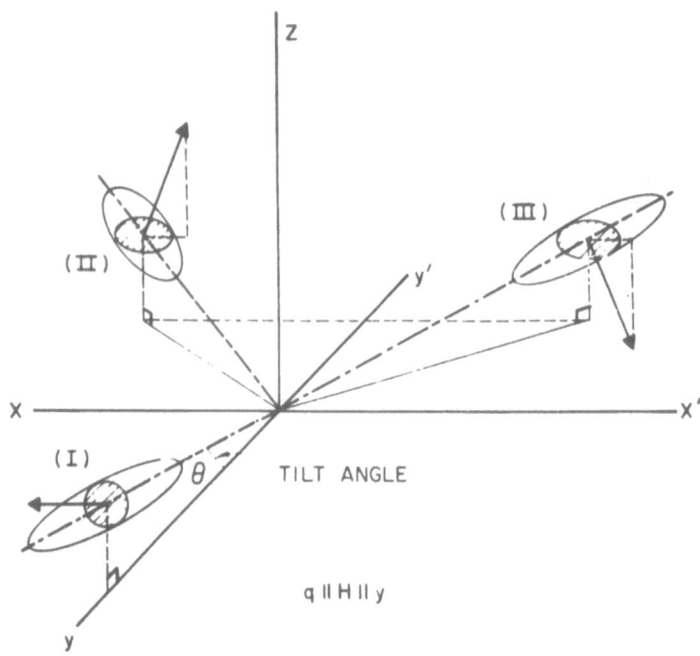

Fig. 3. Schematic picture for the electron ellipsoids in bismuth. The directions of orbit shift are shown for $\mathbf{q} \parallel \mathbf{H} \parallel \mathbf{Y}$.

The electron Fermi surface consists of three equivalent ellipsoids, labeled for convenience as I, II, and III. Ellipsoid I has its major axis in the binary (yz) plane, but it is tilted out of the xy plane by $\sim 6°$ and can be represented by [10]

$$\varepsilon[1 + (\varepsilon/\varepsilon_g)] = (1/2m)(\alpha_{xx}P_x^2 + \alpha_{yy}P_y^2 + \alpha_{zz}P_z^2 + 2\alpha_{yz}P_yP_z) \qquad (29)$$

where the α_{ij} are the components of the inverse-mass tensor α. Ellipsoids II and III are equivalent to I, but are rotated about the z axis by $\pm 120°$, respectively, as shown in Fig. 3.

As discussed in the previous section, there exists in the x direction a "polarization current," J_{MQ}, and an associated electric field E_{MQ} for the ellipsoid I, when the sound wave propagation and the applied magnetic field are along the y direction. The magnitudes are given by

$$J_{MQ} = \frac{\Gamma Sc}{v_s H} \frac{\alpha_{yz}}{\alpha_{zz}} \qquad (30)$$

$$E_{MQ} = \frac{\Gamma Sc}{v_s} \frac{\varrho_{xx}(H)}{H} \frac{\alpha_{yz}}{\alpha_{zz}} \qquad (31)$$

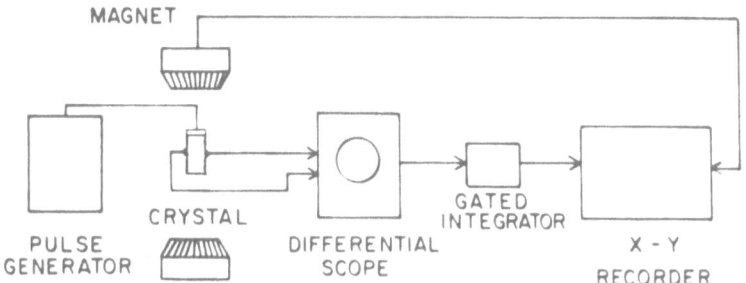

Fig. 4. A block diagram of the apparatus for measuring MQE field.

where Γ is the attenuation constant per unit length and S is the phonon energy flux. The MQE field is proportional to the off-diagonal component of the inverse effective-mass tensor, α_{yz}, which is responsible for the tilting of the ellipsoid. Therefore a rotation of the crystal by $180°$ about the z axis should result in a change of the sign of α_{yz} and thus a change of sign in the voltage.*

The bismuth single crystal used for the experiment had a resistivity ratio $\varrho_{300°K}/\varrho_{4.2°K} \approx 100$, and ql was estimated to be about 10 for the frequency of 220 MHz which was employed for the experiment. The sample, with dimensions of approximately $5 \times 5 \times 6$ mm³, was cut and polished such that its faces were normal within $\pm 2°$ to the binary, bisectrix, and trigonal axes. An x-cut quartz transducer of fundamental frequency 20 MHz was bonded to one bisectrix face of the sample and driven at its higher harmonics. The experimental set-up is shown in Fig. 4. Acoustic pulses 1-μ sec long were produced at the transducer by an rf pulse generator. Enamel-covered copper wires spark-welded to the binary faces were used to detect a voltage pulse which appeared across the sample as the acoustic pulse passed the point of contact. A differential scope was used to display the difference voltage across the sample faces. The difference voltage ΔV was taken from the vertical output of the oscilloscope and applied to a gated integrator whose gate was opened for only 0.1-μ sec while the acoustic pulse was passing the voltage leads. The integrated signal was amplified by a DC amplifier and recorded as a function of magnetic field.

Figure 5 shows typical data of ΔV versus H_0 taken for several values of θ, where θ is the angle between H_0 and the bisectrix direction. Because sound waves propagating in the y direction in bismuth couple only to

* The ordinary galvanomagnetic transport phenomena generally include the off-diagonal component of α squared, and the sign thus remains unchanged under a $180°$ rotation.

ellipsoid I (11), ΔV shows only one set of peaks periodic in $1/H_0$ for all θ. The magnitudes of the peaks in ΔV at $H = 10$ kG are about 1 mV. This agrees with the magnitude calculated using Eq. (27) for $S = 1$ W/sec-cm² and $\Gamma = 20$ dB/cm. The sign of the voltage peaks reversed when the direction of $\mathbf{H_0}$ was changed by 180°, again in agreement with the theory. Thus we believe we have observed the MQE effect associated with GQA of sound waves in bismuth.

To make sure that this voltage is not due to a pickup of any sort, we also measured the voltage across the other pair of faces (trigonal). This voltage is shown in Fig. 6. We will not go into the detailed analysis of this electric field here. It is clear, however, that the magnetic field dependence

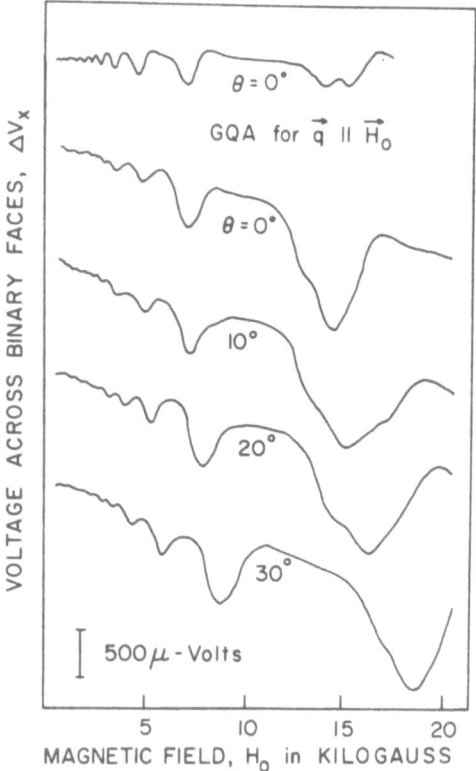

Fig. 5. Voltage across the binary (x) faces for $\mathbf{q} \parallel \mathbf{Y}$ for several values of θ, where θ is the angle between $\mathbf{H_0}$ and $\mathbf{Y}$. The ordinate zero is arbitrary for each curve. For reference GQA for $\mathbf{H_0} \parallel \mathbf{q} \parallel \mathbf{Y}$ $(\theta = 0)$ is also shown.

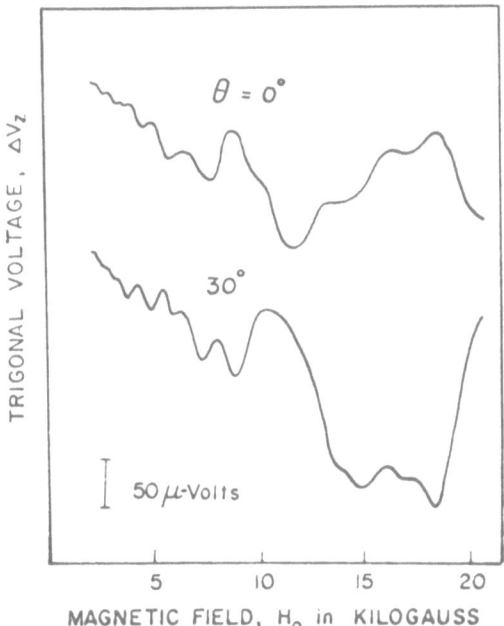

Fig. 6. Voltage across trigonal (z) faces for $\mathbf{q} \parallel \mathbf{Y}$ for two values of θ.

is entirely different from that appearing across the bisectrix plane. As a matter of fact, the voltage shown in Fig. 6 is composed of more than one period. It is necessary to take into consideration the effect produced by the acoustoelectric current flowing parallel to the sound waves to understand this magnetic field dependence ([2]).

DISCUSSION

In this section we discuss the relation between this MQE field and other effects such as sound wave amplification ([12]), the conductance anomaly ([13]), etc. In the MQE experiment we observe a DC electric field produced by phonons. On the other hand, in sound wave amplification we observe an increase in the sound wave amplitude by applying a DC electric field across the sample. The anomaly of the transverse conductance at $E = E_c$, known as "kink effect" ([13]), is also very closely related to the sound wave amplification. Before discussing these phenomena it is helpful to remember the behavior of electrons in the presence of a DC electric and a DC magnetic field. The Hamiltonian of an electron in the presence of H_0 along z and E_0 along y is given by

$$\mathscr{H} = \mathscr{H}_0 - eE_0y \qquad (32)$$

where

$$\mathscr{H}_0 = \frac{P_x{}^2}{2m} + \frac{1}{2m}\left(P_y + \frac{e}{c}H_0x\right)^2 + \frac{P_z{}^2}{2m}$$

The motion of the center of the cyclotron orbit is obtained in the same way as in the case of phonon absorption by assuming $-eE_0y$ as a perturbation:

$$\dot{X}_0 = \frac{1}{i\hbar}[X_0, -eE_0y] = \frac{1}{i\hbar}\left[-\frac{cP_y}{eH_0}, -eE_0y\right]$$

$$= \frac{cE_0}{H_0} \qquad (33)$$

$$\dot{Y}_0 = \frac{1}{i\hbar}[Y_0, -eE_0y] = \frac{1}{i\hbar}\left[y + \frac{cP_x}{eH_0}, -eE_0y\right]$$

$$= 0 \qquad (34)$$

Thus there is no current in the direction of the applied electric field, but there is a Hall current $J_H = necE_0/H_0$ in the direction mutually perpendicular to E_0 and H_0.* We consider now how the mechanism of phonon absorption is affected by the presence of DC electric field. This problem has been studied by Kazarinov and Skobov ([14]) for $\mathbf{q} \perp \mathbf{H}_0$, and by Spector ([15]) for $\mathbf{q} \parallel \mathbf{H}_0$. We follow the treatment by Kazarinov and Skobov here.

As is discussed in Chapter 13 by Zawadzki the presence of a DC electric field E_0 does not drastically alter the situation. The expression for the wave function in the presence of E_0 is identical to the one in the absence of E_0 except that the explicit form of the center of cyclotron orbit is given by

$$X_0 = -\frac{P_y}{m\omega_c} + \frac{E_0}{m\omega_c{}^2} \qquad (35)$$

The energy contains a new potential energy term due to the electric field:

$$\mathscr{E}(n, k_z, X_0) = \hbar\omega_c(n + \tfrac{1}{2}) + (\hbar^2k_z{}^2/2m) - eE_0X_0 \qquad (36)$$

Thus the energy depends on P_y through the x coordinate of the orbit center. In other words, the degneracy of energy among the states of different

---

* One must not confuse this constant drift of cyclotron orbit centers with the instantaneous shift when a phonon is absorbed (MQE). Hall current exists without phonons and MQE current exists without a DC electric field.

P_y is split by a DC electric field. The position dependence of the electron energy appears in two aspects. First, the energy conservation relation is modified. Second, the Fermi energy becomes position dependent.

When an electron which is at a state $\alpha = (n, k_z, X_0)$ absorbs a phonon propagating along the y direction and makes a transition to a new state $\beta = (n + 1, k_z, X_0 + \Delta X_0)$ the energy conservation relation is expressed as follows:

$$\mathscr{E}_\beta - \mathscr{E}_\alpha = \hbar\omega \tag{37}$$

Using Eq. (36) this can be written as

$$\hbar\omega = \hbar\omega_c - eE_0 \Delta X_0 = \hbar\omega_c + \frac{cE_0}{H_0} \hbar q$$
$$= \hbar(\omega_c + qv_d) \tag{38}$$

The second term on the right is the new term arising from the presence of E_0. This relation can be interpreted using a classical picture of cyclotron orbit drifting along y direction. The cyclotron electron drifting with a drift velocity v_d experiences sound waves varying with a frequency of $\omega - qv_d$ instead of ω. It is one kind of Doppler-shifted cyclotron resonance.*

The other effect of the applied DC electric field is that the Fermi level is now position dependent. Although by definition the kinetic energy of the electrons, $\hbar\omega_c(n + \frac{1}{2}) + (\hbar^2 k_z^2/2m)$, at the Fermi level is independent of the position of cyclotron orbit, the total energy of the electrons at the Fermi level depends on the position of the orbit center. Therefore when one of these electrons absorbs phonon energy $\hbar\omega$ this energy does not totally become kinetic energy, since the electron must suffer orbit shift and it costs the potential energy $-E_0 \Delta X_0$ to do so. Thus the gain in the kinetic energy of the electron is $\Delta\mathscr{E}_{\text{kin}} = \hbar\omega + eE_0 \Delta X_0$. This quantity must be positive for the electron to make a transition at $T = 0\,°\text{K}$, since the energy states below Fermi level are completely occupied; thus

$$\hbar\omega + eE_0 \Delta X_0 > 0 \tag{39}$$

or

$$E_c > E_0 \qquad (v_s > v_d) \tag{40}$$

Thus the absorbtion of the phonons occurs only if the drift velocity $v_d = cE_0/H_0$ is smaller than the sound velocity v_s.

---

* The ordinary Doppler-shifted cyclotron resonance experiment is performed in the configuration $\mathbf{q} \parallel \mathbf{H}$. In this case the resonance condition is given by $\hbar\omega = \hbar(\omega_c + qv_F)$, where v_F is the Fermi velocity.

When v_d exceeds v_s the process of phonon absorption is forbidden and phonon emission takes place instead. In the case of phonon emission an electron loses an amount $\hbar\omega$ of its total energy and the center of cyclotron orbit displaced by the amount ΔX_0 to the direction opposite from the one for the absorption process. The increment of the kinetic energy in this case is $\Delta\mathscr{E}_{kin} = -\hbar\omega - eE_0\,\Delta X_0$. This quantity is positive if $v_d > v_s$ is satisfied. Thus phonons are emitted instead of absorbed from electrons when v_d is greated than v_s, or $E_0 > E_c\ (= v_sH_0/c)$.

These situations are illustrated in Fig. 7. Four cases are shown (A) $E_0 = 0$, (B) $0 < E_0 < E_c$, (C) $E_0 = E_c$, and (D) $E_c < E_0$. In each case the parabola in the middle corresponds to the energy band of the electron before transition (the center of orbit is at $x = X_0$). The parabola on the left is the energy band of the electron after absorbing a phonon (the center

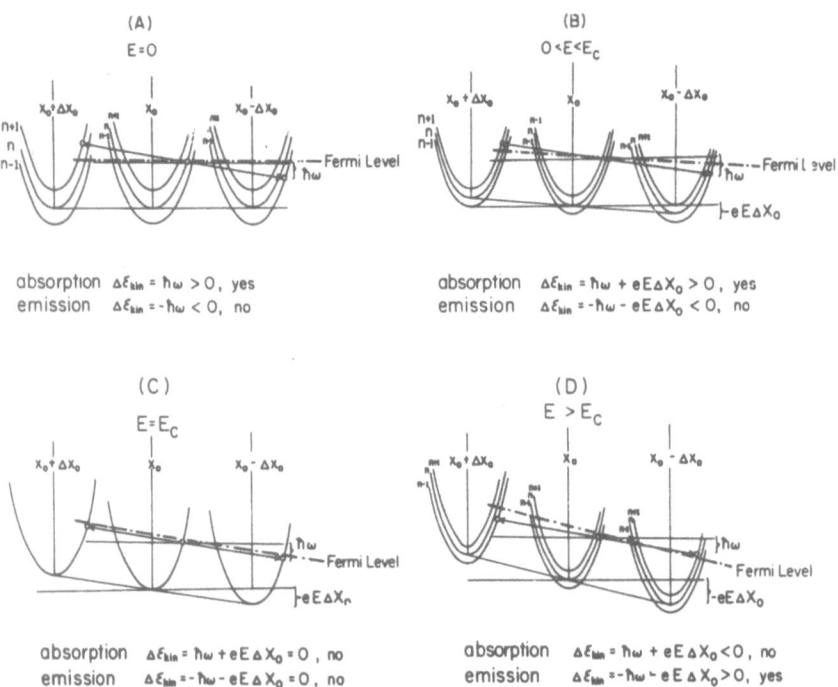

Fig. 7. In the presence of E_0 (x direction) and H_0 (z direction) the electron must suffer an orbit shift by the amount $\pm\Delta X_0$ to absorb (emit) a quanta. Accordingly the electron gains potential energy by the amount $\pm eE\,\Delta X_0$. Thus the net gain in the kinetic energy of the electron is $\Delta\mathscr{E}_{kin} = \pm(\hbar\omega + eE\,\Delta X_0) = \pm\hbar\omega[1 - (v_d/v_s)]$ for the absorption (emission) process. In the cases (A), (B), $\Delta\mathscr{E}_{kin}$ is positive for the absorption process. In case (D) $\Delta\mathscr{E}_{kin}$ is positive for the emission process. In case (C) $\Delta\mathscr{E}_{kin} = 0$ for both processes.

of orbit is at $x = X_0 + \Delta X_0$), and the parabola on the right is the energy band of the electron after emitting a phonon (the center of orbit is at $x = X_0 - \Delta X$). The broken line shows the Fermi level (a constant kinetic energy level) which is parallel to the line connecting the bottoms of the parabolas having the same quantum number n. For $E_0 = 0$ and $0 < \overset{c}{E_0} < E_c$ only the phonon absorption process is allowed at $T = 0\,°K$, as the final states for the transition are empty, whereas the final states for the phonon emission process are occupied. For $E_0 = E_c$ neither the absorption nor the emission of phonons are allowed, as the kinetic energy gain is zero and thus the final states for both of these processes are occupied. When E_0 is greater than E_c only the phonon emission process is allowed for a similar reason.

The discussion above is strict only at $T = 0\,°K$, where the Fermi level is sharply defined. When the temperature is finite the Fermi level has a width of the order of kT. Since kT is greater than $\hbar\omega$ in the ordinary experimental condition, both the phonon absorption and phonon emission can take place at either case, $E_0 < E_c$ or $E_0 > E_c$. To calculate the transition probability at a finite temperature we have to go back to Eq. (6), which includes both absorption and emission. The modification we have to make in the presence of a DC electric field is that the Fermi function is only a function of the kinetic energy and not a function of the total energy. Thus by replacing $f(\mathcal{E}_n)$ in Eq. (6) by $f(\mathcal{E}_\alpha + eE_0X_0)$ we get

$$\gamma(E_0) = (\pi/\hbar) \sum_{\alpha\beta} \delta(\mathcal{E}_\alpha - \mathcal{E}_\beta - \hbar\omega) \, | \langle \alpha \, | \mathcal{H}_1 | \, \beta \rangle \, |^2$$
$$\times \{ f(\mathcal{E}_\beta + eE_0X_0') - f(\mathcal{E}_\alpha + eE_0X_0) \} \tag{41}$$

Assuming that $kT \gg \hbar\omega$, we expand the difference of Fermi functions as follows;

$$f(\mathcal{E}_\beta + eE_0X_0') - f(\mathcal{E}_\alpha + eE_0X_0) = \{ (\mathcal{E}_\beta - \mathcal{E}_\alpha) + eE_0(X_0' - X_0) \} \frac{\partial f}{\partial \mathcal{E}} \Big|_{\mathcal{E}_F}$$
$$= -(\hbar\omega + eE\,\Delta X_0) \frac{\partial f}{\partial \mathcal{E}} \Big|_{\mathcal{E}_F}$$

Combining this approximate form with Eq. (41), we obtain

$$\gamma(E_0) = \gamma(0)\left(1 + \frac{eE_0\,\Delta X_0}{\hbar\omega} \right)$$
$$= \gamma(0)\left(1 - \frac{v_d}{v_s} \right) \tag{42}$$

where $\gamma(0)$ is the probability for the phonon absorption given by Eq. (6)

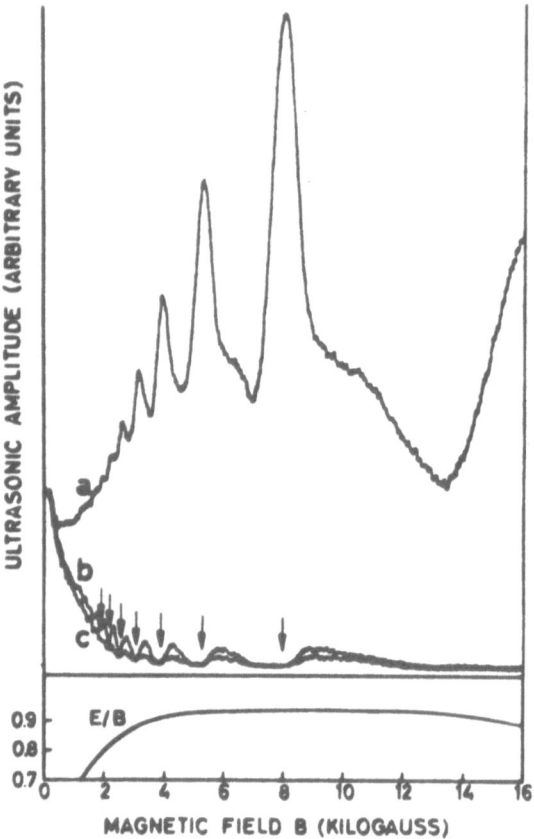

Fig. 8. Received amplitude of shear wave at 188 MHz in bismuth as a function of magnetic field strength. (*a*) **B** "normal," drift field applied; (*b*) **B** "normal," without drift field; (*c*) **B** "reversed," drift field applied. Lower curve: E/B in units of 10^5 cm/sec. From Walther ([12]).

in the absence of a DC electric field. One sees that the absorption constant becomes negative (net emmision) even at finite temperature when v_d is greater than v_s. This equation suggests that the attenuation peak which is observed, for example, in a giant quantum attenuation experiment would be reversed in the presence of a DC electric field which is greater than the critical field $E_c = H_0 v_s/c$. This has been observed experimentally in bismuth by Walther ([12]). Figure 8 shows the data he obtained. Curve (*a*) shows the magnetic field dependence of the ultrasonic amplitude in the presence of an electric field greater than the critical field. Curve (*b*) shows the ordinary giant quantum attenuation without electric field. Curve (c) shows the attenuation in the presence of an electric field and a reversed magnetic field.

Above the critical field E_c phonons can be generated spontaneously without applying sound waves from outside the crystal. This spontaneous emission of phonons leads to the conductance anomaly observed by Esaki ([13]) in bismuth. In the absence of impurity scattering at low temperature there is no current flowing along the direction of applied electric field if a magnetic field is simultaneously applied perpendicular to the electric field.* The reason for this is that, as noted above, the electrons drift in the direction mutually perpendicular to the electric field and the magnetic field. However, when the applied electric field E_0 exceeds E_c electrons emit phonons in the direction mutually perpendicular to H_0 and E_0 and displace their orbit centers to the direction along E_0. Thus a "polarization current" is created in the direction of applied field at an electric field greater than E_c. This sharp change of the conductance at $E_0 = E_c$ is known as the "kink effect" ([13]).

The MQE effect bridges these two observed phenomena, namely, here an electric field or an electric current is produced by electrons absorbing phonons. This electric field, given by Eq. (25), increases when the phonon flux increases and it effects the absorption constant γ given by Eq. (42). We have neglected the effect of the MQE field on the attenuation constant A in deriving Eq. (25). Although it was a good approximation when $E_{MQ} \ll E_c$, we may not be able to neglect this effect when the phonon flux is high. From Eq. (42) we can write attenuation constant as a function of E_{MQ}:

$$\Gamma(E_{MQ}) = \Gamma(0)\left(1 - \frac{E_{MQ}}{E_c}\right) \qquad (43)$$

Combining Eqs. (25) and (42), we obtain the following expressions for the attenuation constant and the M_{QE} field as functions of phonon flux

$$E_{MQ} = [\beta/(1 + \beta)]E_c \qquad (44)$$

$$\Gamma(E_{MQ}) = [1/(1 + \beta)]\Gamma(0) \qquad (45)$$

where $\beta = S\Gamma(0)(c/v_s)^2(\varrho_{xx}/H_0^2)$ is a nonlinear constant. Thus the MQE field cannot exceed the critical field no matter how large a value β may take, and the attenuation constant decreases with increasing phonon flux. In our experiment the value of β for bismuth at $H = 10$ kG, $T = 4.2\,°$K,

---

* The transverse conductivity $\sigma_{xx}(H) = \sigma_0/(1 + \omega_c^2\tau_c^2)$ goes to zero for infinite τ.

and $S = 1$ W/sec is estimated to be less than 10^{-3}. Thus the MQE field in this case is not large enough to modify the attenuation constant appreciably.

In summary, we proposed and calculated the MQE effect in general, presented some experimental evidence in bismuth, and discussed the relation of this effect to other phenomena which involve phonons and crossed electric and magnetic fields. The MQE effect itself will be as useful as the direct attenuation experiment to study the Fermi surface of electrons and other properties of the electron plasma. The combined study of the MQE effect and attenuation will provide us with interesting information about electron–phonon coupling.

APPENDIX

The density matrix ϱ obeys the following equation of motion:

$$i\hbar\, \partial\varrho/\partial t = [\mathscr{H}, \varrho] \tag{A1}$$

If the Hamiltonian $\mathscr{H}$ consists of an unperturbed part $\mathscr{H}_0$ and a small perturbation $\mathscr{H}_1 e^{-i\omega t}$, we can expand the density matrix as follows:

$$\varrho = \varrho_0 + \varrho_1 e^{-i\omega t} \tag{A2}$$

where ϱ_0 represents the density matrix at the equilibrium state, and satisfies the following relation for the electron gas:

$$\varrho_0 \mid n\rangle = f_0(E_n) \mid n\rangle \tag{A3}$$

where $f(E_n) = \{[\exp(E_n - \mu/kT)] + 1\}^{-1}$ is the Fermi distribution function. From Eqs. (A1) and (A2) we obtain the relation

$$\hbar\omega\varrho_1 = [\mathscr{H}_0, \varrho_1] + [\mathscr{H}_1, \varrho_0] \tag{A4}$$

If we take the matrix element of the operator relation (A4) using Eq. (A3) we get

$$\langle n \mid \varrho_1 \mid n'\rangle = \frac{\{f(E_n) - f(E_{n'})\}\langle n \mid \mathscr{H}_1 \mid n'\rangle}{E_n - E_{n'} - \hbar\omega + i\delta}$$

$$= i\pi\delta(E_n - E_{n'} - \hbar\omega)\{f(E_n) - f(E_{n'})\}\langle n \mid \mathscr{H}_1 \mid n'\rangle \tag{A5}$$

The statistical average of some physical quantity represented by an operator A is given by

$$\bar{A} = \mathrm{Tr}\{\varrho A\} = \sum_n \langle n \mid \varrho A \mid n\rangle$$

$$= \sum_{nn'} \langle n \mid \varrho \mid n'\rangle\langle n' \mid A \mid n\rangle \tag{A6}$$

Utilizing Eq. (A5), (A6) can be written as

$$\bar{A} = i\pi \sum_{nn'} \delta(E_n - E_{n'} - \hbar\omega)\{f(E_n) - f(E_{n'})\}$$
$$\times \langle n \mid H_1 \mid n'\rangle\langle n' \mid A \mid n\rangle \tag{A7}$$

Futhermore, if A has a special form given by

$$A = (i/\hbar)\,\Delta Q \mathscr{H}_1, \tag{A8}$$

then Eq. (A7) becomes $\bar{A} = \gamma\,\Delta Q$ where

$$\gamma = (\pi/\hbar) \sum_{nn'} \delta(E_n - E_{n'} - \hbar\omega)\{f(E_n) - f(E_{n'})\} \mid \langle n \mid \mathscr{H}_1 \mid n'\rangle \mid^2 \tag{A9}$$

REFERENCES

1. Y. Sawada, E. Burstein, W. Salaneck, and L. Testardi, *Phys. Rev. Letters* **18**: 776 (1967).
2. W. Salaneck, Y. Sawada, and E. Burstein, *Phys. Rev. Letters* **18**: 779 (1967).
3. Y. Sawada, E. Burstein, and L. Testardi, in: "Proceedings of the International Conference on the Physics of Semiconductors, Kyoto, 1966", *J. Phys. Soc. Japan Suppl.* **21**: 760 (1966).
4. R. H. Parmenter, *Phys. Rev.* **89**: 990 (1953).
5. G. Weinreich, T. M. Sanders, Jr., and H. White, *Phys. Rev.* **114**: 33 (1959).
6. V. L. Gurevich, V. G. Skobov, and Yu. A. Firsov, *Soviet Phys.–JETP (English Transl.)* **13**: 552 (1961).
7. D. N. Langenberg, J. J. Quinn, and S. Rodriguez, *Phys. Rev. Letters* **12**: 104 (1964).
8. G. Weinreich, *Phys. Rev.* **107**: 317 (1957).
9. Y. Sawada and E. Burstein, *Phys. Rev.* **150**: 456 (1966).
10. B. Lax, J. G. Mavroides, H. J. Zeiger, and R. J. Keyes, *Phys. Rev. Letters* **5**: 241 (1960).
11. S. Mase, Y. Fujimori, and H. Mori, *J. Phys. Soc. Japan* **21**: 1744 (1966).
12. K. Walther, *Phys. Rev. Letters* **16**: 642 (1965).
13. L. Esaki, *Phys. Rev. Letters* **8**: 4 (1962).
14. R. F. Kazarinov and V. G. Skobov, *Soviet Phys.–JETP (English Transl.)* **16**: 1057 (1963).
15. H. N. Spector, Phys. Letters **16**: 163 (1964).

Chapter 8

ELECTRON BAND STRUCTURE STUDIES USING DIFFERENTIAL OPTICAL TECHNIQUES AND HIGH MAGNETIC FIELDS

J. G. Mavroides

*Lincoln Laboratory**
Massachusetts Institute of Technology
Lexington, Massachusetts

INTRODUCTION

The recent incorporation of differential techniques in optical measurements ([1-6]) has resulted in an enhancement in sensitivity of two to three orders of magnitude over conventional methods, thereby providing a powerful tool for studying optical properties as well as the band structure of solids. In this technique the crystal under investigation is perturbed by some external force, such as an alternating electric field, stress, or temperature. This perturbation modulates the parameters of the band structure and thereby the complex dielectric constant, which determines both the absorption and the reflection. By synchronously detecting the resulting modulation of the optical properties it is possible to obtain an enhancement over the background of the structure arising from electronic transitions between critical points in the energy bands.

It is well known that the application of a magnetic field to conventional optical methods gives additional information about the energy bands. We shall show that this also is the case with differential optical techniques. What we would like to do in this chapter is first, to describe and compare the various modulation techniques, second, to discuss the effect of the application of a magnetic field, and finally, present results which have been obtained to date using magnetic fields.

* Operated with support from the U. S. Air Force.

METHODS OF DIFFERENTIAL OPTICAL MODULATION

Electric Field Modulation

It was shown theoretically by Franz [7] and Keldysh [8] that the application of an electric field to a semiconductor increases the probability of electron tunneling from the valence band to the conduction band, thereby causing an exponential absorption tail and a decreased energy gap. This phenomenon was subsequently observed and confirmed in semi-insulating GaAs by Moss [9], who found a shift of the edge of $\sim$200 μeV for an electric field of 5000 V/cm. However, it was not until electric-field modulation techniques were applied that sensitive and high-resolution measurements of critical points became possible. Frova and Handler [2] used germanium p–n junctions in reverse bias to study the electroabsorption effect in the vicinity of the direct edge (Fig. 1), while Seraphin [3] made a condenser-like configuration consisting of a transparent SnO_2 electrode which was separated from the germanium sample by a 0.01-mm Mylar film. The junction technique is limited to a small number of materials and works only at certain impurity concentrations. The Seraphin arrangement, although

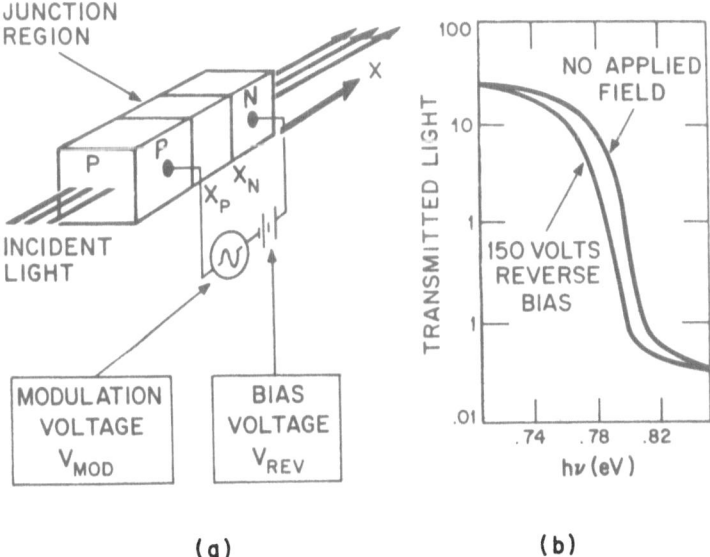

(a)   (b)

Fig. 1. (a) Schematic representation of the arrangement for electrotransmission experiment. (b) Transmitted light through the germanium diode junction for 0 and 150 V reverse bias [$F_{max} \approx 10^5$ V/cm]. From Frova and Handler [2].

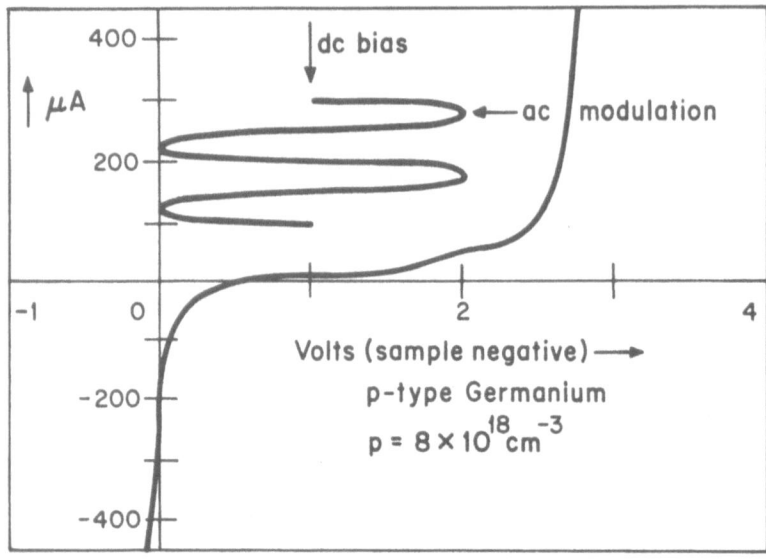

Fig. 2. The *I–V* curve of a *p*-type germanium sample with $p = 8 \times 10^{18}$ cm^{-3}.
From Cardona *et al.* ([10]).

more versatile, suffers from a number of disadvantages, such as multi-reflection and vibration complications as well as limitations in spectrum coverage due to the transmission response of the conducting electrode.

Shaklee *et al.* ([4]) eliminated some of these complications by immersing the sample in an electrolyte, biasing it in the reverse direction (Fig. 2), and looking at electroreflectance so that higher bands could be studied. A space charge layer, typically of $\sim 10^{-5}$ cm for a carrier concentration $\sim 10^{17}$ cm^{-3}, is set up on the semiconductor surface and by keeping the ionic concentration in the solution high enough so that the Gouy layer is thinner than the depletion layer most of the applied voltage will be across the semiconductor layer. Thus an applied voltage ~ 1 V produces an average field $\sim 10^5$ V/cm, which is comparable to the other modulation methods. Figure 3 gives some of Cardona's typical results in germanium ([10]); note that the direct edge is not shown, because the electrolyte was not transmitting in this energy range. Cardona and coworkers studied a large number of semiconductors with this technique, including germanium, silicon, α-Sn, GaP, GaAs, GaSb, InSb, ZnO, and HgTe. However, even this technique suffers from a long-wavelength cutoff of 1μ and a low-temperature limit of $-120\,°C$ (for an ethyl alcohol electrolyte). By mounting the sample very close to the window with only a thin film of electrolyte Groves *et al.* ([11])

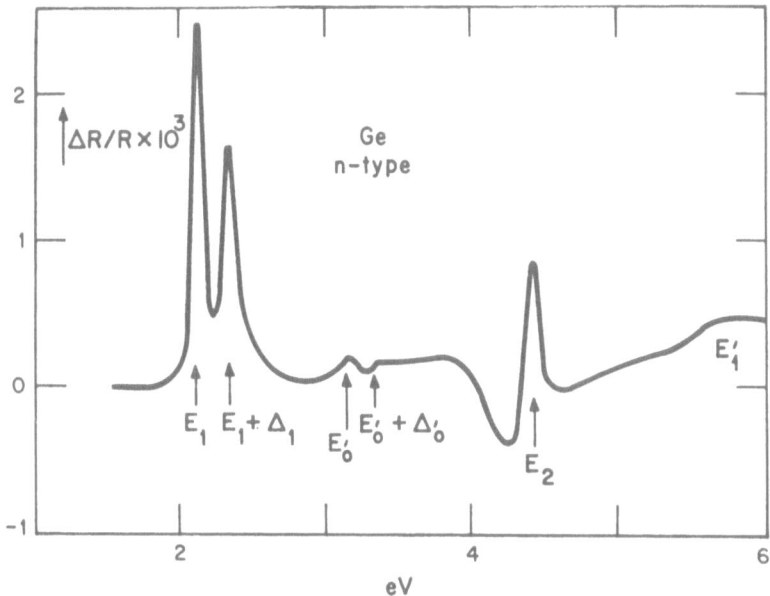

Fig. 3. Electroreflectance spectrum of n-type germanium with 5×10^{14} electrons/cm^3 at room temperature. $V_{DC} = 1.5$ V; $V_{AC} = 1$ V rms. From Cardona *et al.* [10].

were able to extend this technique to 3 μ and thereby study the low-energy interband transitions in InSb.

So far the techniques which have been discussed are limited to room temperature and slightly below. The advantages of studying solids at low temperatures are well known. The primary effect of low temperatures is to decrease the scattering and thereby sharpen the spectral lines so that fine structure and splittings may be observable. In order to extend the electric field technique to low temperatures and also to eliminate the wavelength restriction, Pidgeon *et al.* [12] developed a dry package in which a plastic insulating film is sandwiched between the crystal to be studied and a nickel electrode which is transparent at least out to 20 μ. This mechanically integrated thin-film package has been successfully operated at temperatures as low as 1.5 °K.

In summary, of the various electromodulation techniques, the liquid electrolyte technique is the simplest and most versatile for room temperature studies of energy bands with gaps as low as ～0.4 eV, while the dry package, although more complex, gives at least comparable results at room temperature and even works well at very low temperatures.

Piezomodulation

The first optical stress experiments involved the application of static stresses ([13]) and consequently were lacking in sensitivity. Stress affects the electron energy bands primarily in two general ways. In nondegenerate bands it can change the energy gaps or the spacing between bands. In degenerate bands it can lower the symmetry and thereby remove degeneracies. Two examples of the latter are illustrated by considering the semiconductor germanium or silicon. Here the appropriate stress can remove the degeneracy at $\mathbf{k} = 0$ of the $p_{3/2}$ valence bands or the equivalence of the multiellipsoidal conduction bands. Thus one of the early uses of stress

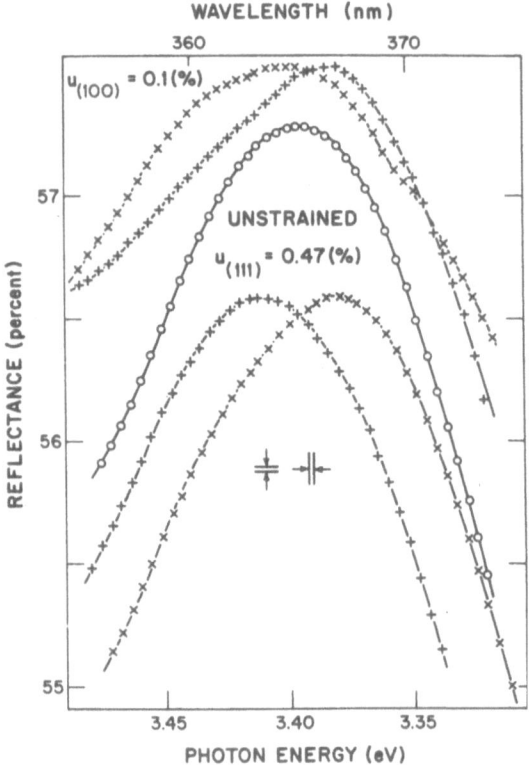

Fig. 4. Piezoreflectance of silicon near 3.4 eV for unstrained and strained samples. Here u_{100} and u_{111} are the strain components of a stress applied to the [100] and [111] axes, respectively. The reflectance scale refers to the unstrained samples. +Compression, × extension. From Gerhardt ([14]).

experiments was for identification and location of band edges in the Brillouin zone. Although piezoresistance experiments proved quite profitable for studies of the lowest or occupied band edges, the early static piezo-optical measurements with unpolarized light did not offer promise, apparently because of the breadth of the reflection peaks and also the lack of sensitivity. However, Gerhardt ([14]) did observe small shifts in the 3.4- and 4.5-eV reflectance peaks of Si. For example, in studying the 3.4-eV peak it was found that a [111] stress produced a shift of the zero-stress peak without broadening, whereas a [100] stress distorted the line (Fig. 4). This was interpreted as evidence that an essential contribution to the 3.4-eV peak is due to transitions to energy bands located along the [100] axes of the Brillouin zone, since all [100] degeneracies are affected equally by a [111] stress.

The next advance, namely, to increase the sensitivity and yet use smaller strains, was achieved in room-temperature measurements by Engeler *et al.* ([5]) and independently by Gobeli and Kane ([6]), who applied alternating stresses on the crystal and made use of phase-sensitive detection techniques. Both groups obtained the AC stress by mounting the sample on a piezoelectric transducer. Using a nonresonant system in which thin semiconductor single crystals or thin metal films were either cemented to or evaporated on top of a lead zirconate titanate transducer, Engeler *et al.* obtained strains $\sim 8 \times 10^{-5}$. Gobeli and Kane achieved about an order-of-magnitude larger strain by means of a compound resonator in which a long thin silicon rod was cemented to the end of an x-cut quartz driver with a fundamental longitudinal resonance at $f_0 \approx 130$ kHz; the silicon rod length was carefully ground and etched until the composite system had a resonance $f_0 \pm 0.5$ Hz. The nonresonant technique is more stable, less critical, easier to achieve, and requires smaller samples; however, the resonant configuration has the advantage of applying a purely longitudinal stress on the sample. Thus with this latter technique Gobeli and Kane were able to study the piezoreflectance with polarized light either parallel or perpendicular to the stress direction and thereby obtain the frequency dependence of three (for cubic materials) fundamental constants W_{ijkl} which express the relation between the variation of the dielectric constant ε_{ij} with strain u_{kl}, namely, $\Delta \varepsilon_{ij} = W_{ijkl} u_{kl}$. Typical results of $\Delta R/R$ versus photon energy are shown in Fig. 5.

The extension of the nonresonant technique to liquid helium temperatures is achieved in a fairly straightforward manner by mounting the appropriate titanate transducer and sample on the bottom of an optical Dewar. Two opposing effects come into play as far as optimum perform-

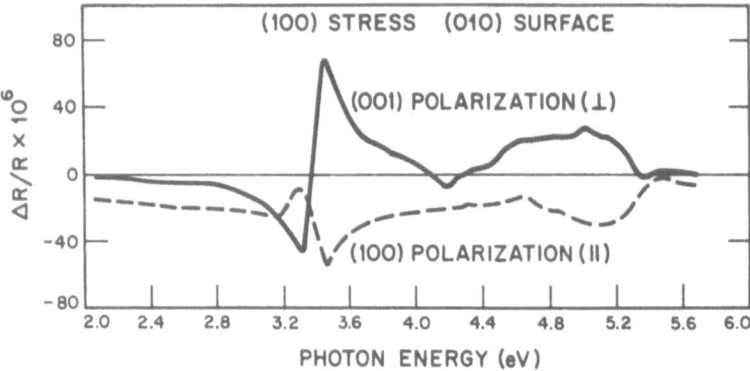

Fig. 5. Plot of $\Delta R/R$ versus photon energy in silicon for polarization modes parallel and perpendicular to stress field. From Gobeli and Kane ([6]).

ance is concerned. First, the sample is made as thin as possible in order to obtain large AC strains and thereby large modulation effects. However, if the crystal, which is cemented to the transducer, is made too thin, then it will become strained when cooled because of differential contraction between it and the transducer. This problem may be partially alleviated by using a high-viscosity silicone oil rather than Duco cement as the adhesive. The oil doesn't freeze until quite low temperatures. Typical low-temperature germanium studies, for example, utilized 20-mil samples on a 2.5 mm thick transducer.

Thermal Modulation

It is well known that the energy gap of semiconductors varies with temperature. For example, Fig. 6 shows the temperature dependence in germanium for both the direct and indirect energy gaps ([15]). Batz ([16]) utilized essentially this effect in thermal modulation studies in germanium when he passed low-frequency pulsed electric currents through thin samples and synchronously detected the resulting modulated reflectance $\Delta R/R$. This technique differs from the other two in that the modulating parameter is a scalar quantity and therefore cannot be expected to yield as much detailed information as, for example, piezomodulation. This method has several advantages over electroreflectance. First, the technique is quite simple. Second, the incident light does not pass through either an electrode or an electrolyte, so that there is no wavelength limit of the measurements or transmission loss on this account. Finally, it is fairly easy to cool the films to somewhat low temperatures, which cannot be done in the liquid

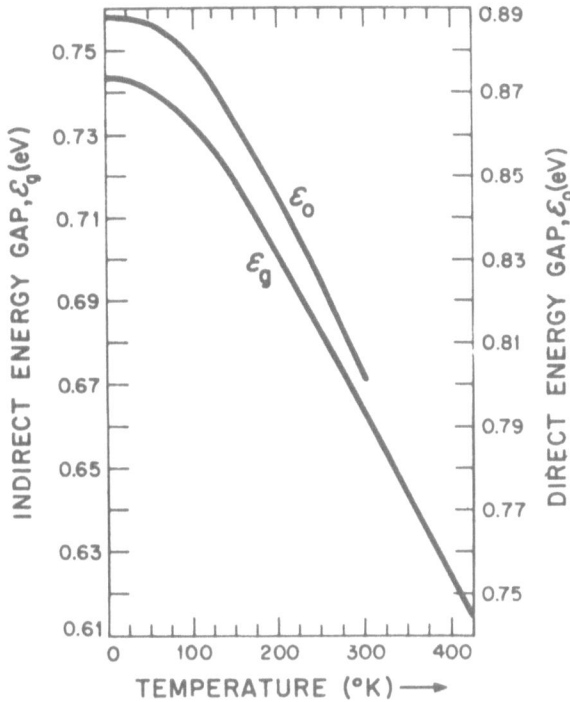

Fig. 6. The direct and indirect energy gaps $\mathscr{E}_g$ and $\mathscr{E}_0$, respectively, as functions of temperature in germanium. From McLean ([15]).

electroreflection technique. However, examination of Fig. 6 indicates that the sensitivity of this technique may be poor at very low temperatures. Except possibly for simplicity, there are no pronounced advantages over the dry electrolyte electroreflectance or the piezoreflection technique. Actually, the latter two methods both allow a much lower sample-temperature operation.

In typical low-temperature measurements in gold by Scouler ([17]) 6-A, 15-Hz current pulses, with unit duty cycle, were passed through a 2000-Å film which had been evaporated on a 0.15-mm glass substrate; the substrate in turn was attached to a nitrogen-cooled copper block by means of silicone grease which was mixed with silver powder to increase the thermal conductivity. The AC temperature modulation was probably of the order of a few degrees, whereas the DC temperature rise was ∼40° above the liquid nitrogen temperature. The enhancement of this structure in $\Delta R/R$ over that of R is shown in Fig. 7.

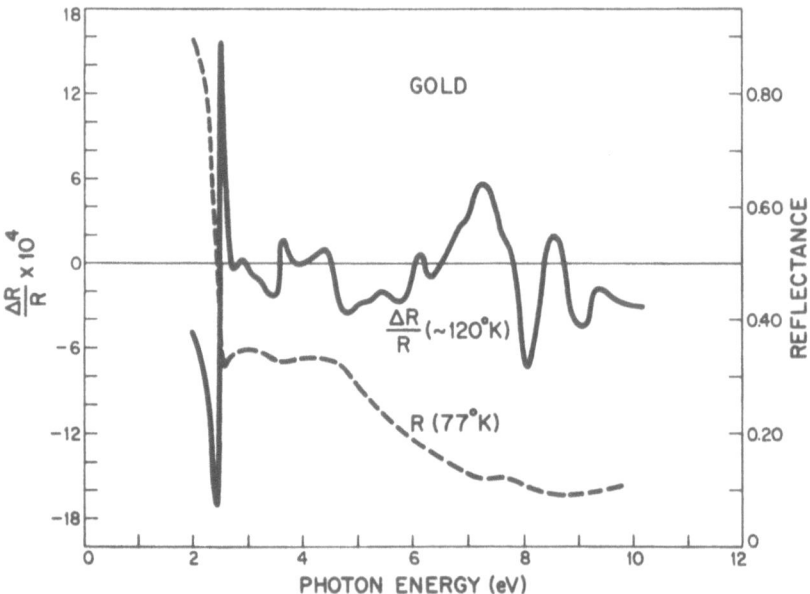

Fig. 7. Comparison between thermal reflectance, $\Delta R/R$, and reflectance, R, for a gold film. From Scouler ([17]).

EXPERIMENTAL TECHNIQUES

The experimental arrangement for the magnetodifferential optical measurements is basically the conventional magnetooptical set-up with the addition of an arrangement for modulating the optical constants of the sample. A block diagram of a typical set-up is shown in Fig. 8. The arrangement shown here is for piezomodulation; the electro and thermal modulation schemes are essentially the same. A single-frequency light beam from the monochromator is fed to the sample, which is strained periodically at ~1000 cycles by means of a transducer. The light reflected from the sample is modulated, since the strain varies the energy gap and thus the optical constants. If we disregard the 152-cycle chopper for the moment, the 1000-cycle ΔR signal is detected; preamplified; synchronously amplified, with the reference signal coming from the transducer driving circuit; and then displayed on the recorder. The 152-cycle signal is used to normalize the ΔR signal and to obtain $\Delta R/R$. This is accomplished by feeding the output from the 152-cycle lock-in amplifier to a servo system which adjusts the input potentiometer to give a constant reflectivity output.

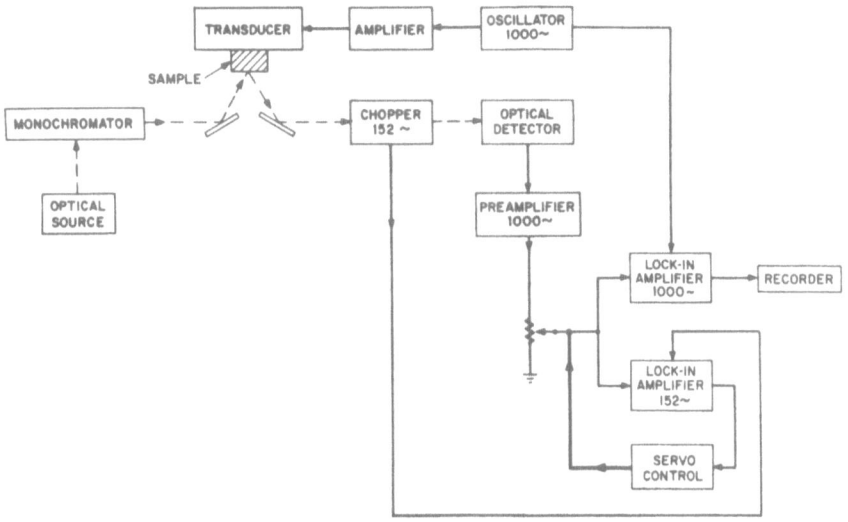

Fig. 8. Block diagram of experimental piezomodulation set-up.

THEORETICAL CONSIDERATIONS

The optical properties of a solid can be well described by means of the complex index of refraction n^*, where

$$n^{*2} = (n + ik)^2 = \varepsilon_1 + i\varepsilon_2 \tag{1}$$

It can easily be shown from Maxwell's equations that the reflectivity, R, at normal incidence is given by

$$R = \frac{(n - 1)^2 + k^2}{(n + 1)^2 + k^2} \tag{2}$$

and the absorption coefficient α

$$\alpha = 4\pi k / \lambda \tag{3}$$

Furthermore, by expansion of Eq. (1)

$$n^2 - k^2 = \varepsilon_1 \tag{4}$$

and

$$2nk = \varepsilon_2 \tag{5}$$

so that it is fairly straightforward to express both the reflectivity R and the absorption α in terms of ε_1 and ε_2.

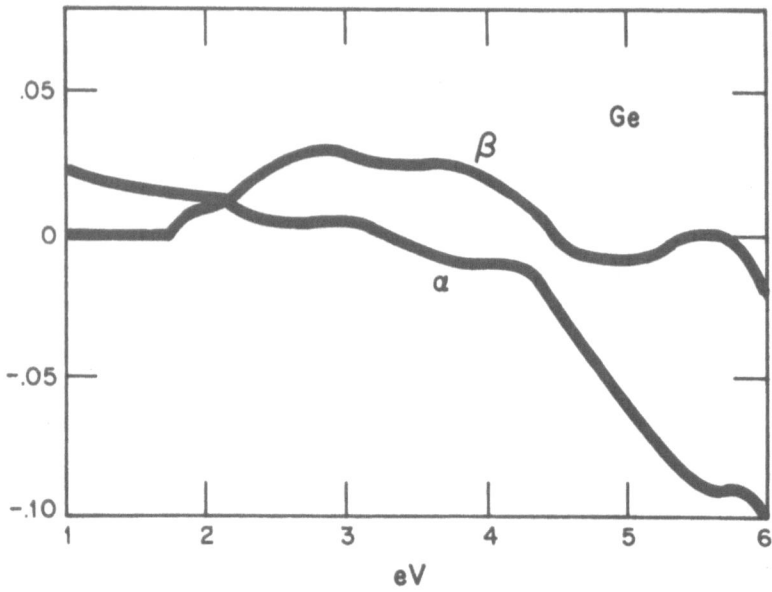

Fig. 9. Spectral dependence of the functions a and b governing the contribution of $\Delta\varepsilon_1$ and $\Delta\varepsilon_2$ to a reflectance modulation spectrum for a germanium sample immersed in water. From Cardona *et al.* ([10]).

In the experiments to be discussed it is the reflectivity $R(\omega; H)$ or changes in the reflectivity $\Delta R(\omega, H)$ which are measured. Since the reflectivity can be expressed in terms of ε_1 and ε_2, then

$$\Delta R/R = a\Delta\varepsilon_1 + b\,\Delta\varepsilon_2 \tag{6}$$

where ([10])

$$a \equiv \frac{1}{R}\,\frac{\partial R}{\partial\varepsilon_1} = \frac{\sqrt{2}\,(\varepsilon_1 + \varepsilon)^{1/2}}{\varepsilon[(\varepsilon_1 - 1)^2 + \varepsilon_2{}^2]}\,[2\varepsilon_1 - 1 - \varepsilon] \tag{7}$$

$$b \equiv \frac{1}{R}\,\frac{\partial R}{\partial\varepsilon_2} = \frac{\sqrt{2}}{\varepsilon(\varepsilon_1 + \varepsilon)^{1/2}[(\varepsilon_1 - 1)^2 + \varepsilon_2{}^2]}\,[2\varepsilon_1 - 1 + \varepsilon] \tag{8}$$

and

$$\varepsilon^2 \equiv \varepsilon_1{}^2 + \varepsilon_2{}^2 \tag{9}$$

Thus in general $\Delta R/R$ depends on ε_1, ε_2, $\Delta\varepsilon_1$, and also $\Delta\varepsilon_2$ in a rather complex manner. However, the coefficients a and b may be calculated if ε_1 and ε_2 are known, and given the specific modulation technique, $\Delta\varepsilon_1$ and $\Delta\varepsilon_2$ may be estimated. Usually, the change due to $\Delta\varepsilon_1$ is not equal and opposite to that arising from $\Delta\varepsilon_2$, so that these two terms do not cancel one another.

The relative importance of a and b varies with frequency. For example, Cardona et al. [10] has calculated these coefficients in the case of germanium in an aqueous electrolyte, and the results are shown in Fig. 9. We see that it is $\Delta\varepsilon_1$ which is significant at low frequencies or energies near the direct gap; at higher energies $\Delta\varepsilon_1$ and $\Delta\varepsilon_2$ contribute equally, and at still higher energies $\Delta\varepsilon_1$ becomes dominant again. In view of this complexity we shall not pursue this general problem any further but shall make a simplification. Although $R(\omega)$ is a function of both $\varepsilon_1(\omega)$ and $\varepsilon_2(\omega)$, it is $\varepsilon_2(\omega)$ which is usually considered the more fundamental one. The real part of the dielectric constant, $\varepsilon_1(\omega)$, depends most strongly on the scattering processes and has contributions present even at energies below an interband transition. However, for small damping the imaginary part $\varepsilon_2(\omega)$ is zero until the frequency ω exceeds a threshold, at which an interband transition can take place; then there is a contribution due to the interband transition. Thus the principal features of $R(\omega)$ are approximately reproduced at the same energies in $\varepsilon_2(\omega)$. Experimentally, this is borne out in germanium and silicon by an examination of the reflectivity results of Philipp and Taft [18,19]. Figure 10 gives the comparison for germanium.

Cohen [20] has shown that the contribution to the imaginary part of the dielectric constant of an interband transition from a simple valence band to a simple conduction band can be expressed as

$$\varepsilon_2(\omega) = \frac{4\pi^2 e^2 \hbar}{3m^2\omega^2} \sum_{v,c} \int_{BZ} \frac{2}{(2\pi)^3} \, \delta[\omega_c(\mathbf{k}) - \omega_v(\mathbf{k}) - \omega] \, | \, M_{c,v}(\mathbf{k}) \, |^2 \, d^3k \quad (10)$$

Knowing $\varepsilon_2(\omega)$, it is possible to calculate $\varepsilon_1(\omega)$ by use of the Kramers–Kronig relation

$$\varepsilon_1(\omega_0) = 1 + \frac{1}{\pi} \int_{-\infty}^{\infty} \varepsilon_2(\omega) \, \frac{d\omega}{\omega - \omega_0} \quad (11)$$

Assuming that the matrix element $M_{c,v}(\mathbf{k})$ is constant, which in many cases is a fair assumption, then

$$\varepsilon_2(\omega) \sim \frac{2}{(2\pi)^3} \int_{BZ} \delta[\omega_c(\mathbf{k}) - \omega_v(\mathbf{k}) - \omega] \, d^3k \equiv J_{v,c}(\omega) \quad (12)$$

where $J_{v,c}(\omega)$ is the joint density of states, i.e., $J_{v,c}(\omega) \, \Delta\omega$ gives the number of pairs of states in the bands v and c which have an energy between $\hbar\omega$ and $\hbar(\omega + \Delta\omega)$. The volume integral of Eq. (12) can be transformed to an

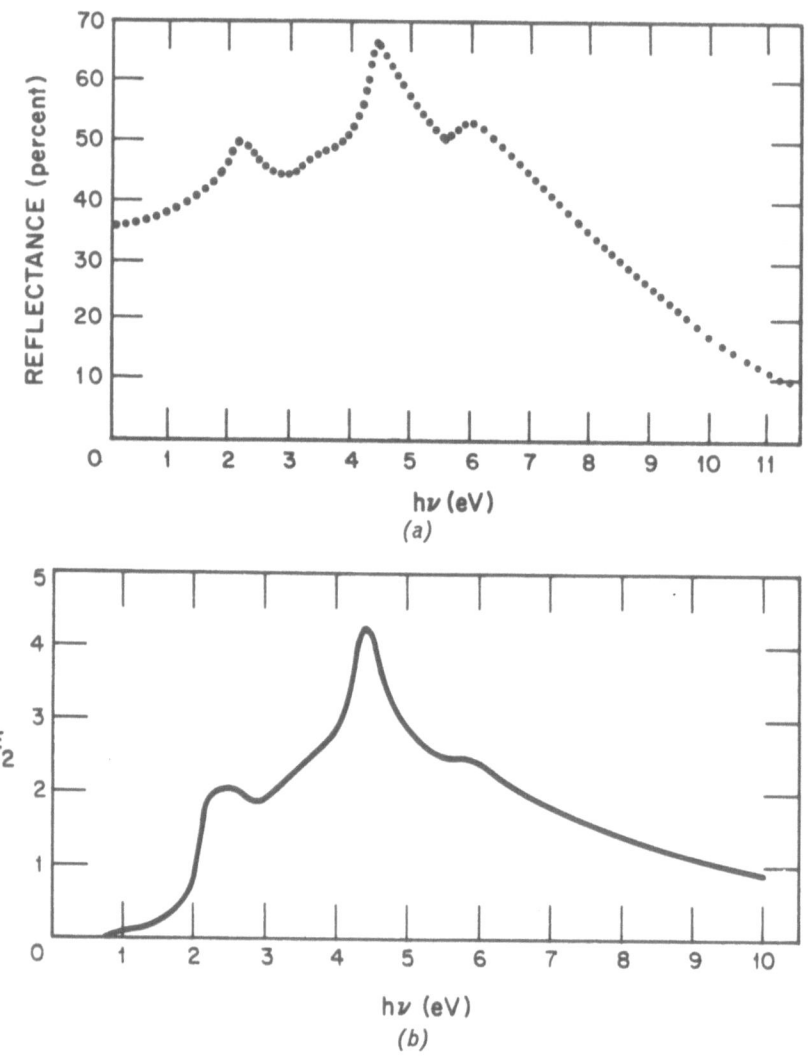

Fig. 10. (a) Spectral dependence of the reflectance of germanium. (b) Spectral dependence of the imaginary part of the dielectric constant, ε_2, of germanium. From Philipp and Taft ([18]).

integral over the surface S defined by

$$\omega_c(\mathbf{k}) - \omega_v(\mathbf{k}) = \omega_{c,v} \tag{13}$$

so that

$$J_{v,c}(\omega) = \frac{2}{(2\pi)^3} \int \frac{dS}{\nabla_k[\omega_c(\mathbf{k}) - \omega_v(\mathbf{k})]} \tag{14}$$

From Eq. (14) it is evident that the density of states and therefore ε_2 have a singularity when

$$\nabla_k[\omega_c(\mathbf{k}) - \omega_v(\mathbf{k})] = 0 \tag{15}$$

Points in $\mathbf{k}$ space which satisfy this condition are called critical points. There are two ways of obtaining this singularity. The first involves two bands with extrema at the same point in $\mathbf{k}$ space, i.e.,

$$\nabla_k\omega_c(\mathbf{k}) = \nabla_k\omega_v(\mathbf{k}) \tag{16}$$

The other case involves two bands which have parallel surfaces, i.e.,

$$\nabla_k\omega_c(\mathbf{k}) = \nabla_k\omega_v(\mathbf{k}) \neq 0 \tag{17}$$

Van Hove ([21]) has considered the expansion of the density of states about a critical point $\mathbf{k}_0$. By expanding in a Taylor series

$$\omega_{c,v}(\mathbf{k}) = \omega_{c,v}(\mathbf{k}_0) + \sum_{j=1}^{3} \frac{1}{2\mu_i} (k_i - k_{0i})^2 \tag{18}$$

where the reduced effective mass $1/\mu = (1/m_c) + (1/m_v)$ for simplest singularities of the type M_0 which can be described by Eq. (16). Substituting Eq. (18) into Eq. (14), we obtain

$$J_{v,c}(\omega) \sim A \qquad\qquad \omega < \omega_g \tag{19}$$

$$J_{v,c}(\omega) \sim A + B(\omega - \omega_g)^{1/2} \qquad \omega > \omega_g \tag{20}$$

where $\omega_g \equiv \omega_{c,v}(\mathbf{k}_0)$. Thus, considering only the nonanalytical components

$$\varepsilon_2(\omega) \approx C(\omega - \omega_g)^{1/2} \tag{21}$$

In the differential optical experiments we measure

$$\Delta\varepsilon_2(\omega) \approx -(C/2)(\omega - \omega_g)^{-1/2} \Delta\omega_g \tag{22}$$

where we have neglected the change in C, which is relatively small and nonresonant. Thus in the derivative technique critical points are enhanced and show up as singularities. Of course, in practice the lines have finite widths due to lifetime broadening effects.

This treatment can be extended to include the effects of a magnetic field. Taking the magnetic field in the z direction, the energy levels become quantized and Eq. (18) becomes

$$\omega_{c,v}(\mathbf{k}) = \omega_{c,v}(\mathbf{k}_0) + \left(n + \frac{1}{2}\right)\omega_c{}^* \pm \frac{\beta}{2\hbar}(g_c + g_v)H + \frac{1}{2\mu_z}(k_z - k_{0z})^2 \tag{23}$$

where the reduced cyclotron frequency $\omega_c{}^* = eH/\mu c$, β is the Bohr magneton, and g_c and g_v are the spectroscopic spin factors for the conduction and valence bands, respectively. Again substituting into Eq. (14), we obtain

$$J_{v,c}(\omega, H) \sim A' + B'\omega_c \sum_n [\omega - \omega_g - (n+\tfrac{1}{2})\omega_c{}^* \pm (\beta/2\hbar)(g_c+g_v)H]^{-1/2} \quad (24)$$

so that $J_{v,c}(\omega, H)$ and therefore $\varepsilon_2(\omega, H)$ exhibit a singularity or peak whenever a transition takes place between a Landau level in the valence band to a Landau level in the conduction band subject to the selection rules $\Delta n = 0$ and $\Delta k_z = 0$. In this case, again keeping only the highest-order term,

$$\Delta\varepsilon_2(\omega, H) \approx \frac{1}{2} C'\omega_c{}^* \sum_n \frac{\Delta\omega_g - (n + \tfrac{1}{2})(\omega_c{}^*/\mu)\Delta\mu \pm (\beta H/2\hbar)(\Delta g_c + \Delta g_v)}{[\omega - \omega_g - (n + \tfrac{1}{2})\omega_c{}^* \pm (\beta/2\hbar)(g_c + g_v)H]^{3/2}}$$

$$(25)$$

which indicates that with a magnetic field the differential technique gives higher-order singularities than are obtained with the conventional magneto-optical method. In a manner analogous to the case for magnetoreflection we again obtain a singularity, or a sharp line if damping is included, whenever the photon energy and magnetic field are appropriate for a transition from a Landau level in the valence band to one in the conduction band, subject to the proper selection rule.

MAGNETO-OPTICAL MODULATION EXPERIMENTS

The first magneto-optical modulation experiments were carried out at MIT in germanium ([22,23]) at room temperature using the piezomodulation technique. Although magnetoabsorption oscillations associated with the direct energy gap at $\mathbf{k} = 0$ had been previously observed in thin germanium samples at room temperature ([24]), the corresponding magnetoreflection effect had never been reported. However, by modulating the Landau levels, as has been discussed, one obtains an enhancement of the oscillations as shown in Fig. 11; here a comparison is made between the magnetic field dependences of the reflectivity and of the piezoreflectivity. The incident photon energy is at 0.826 eV, which is larger than the smallest $\mathbf{k} = 0$ direct gap. Each of the oscillations corresponds to an interband transition from a Landau level in the valence band to one in the conduction band. The frequency variation of the piezoreflectivity and the magnetopiezoreflectivity in this region of the spectrum is given in Fig. 12. This figure emphasizes the enhancement of the resonant structure by the magnetic field. Above $\sim$20 kG the amplification is linear, as one would expect from density-of-states considerations. There is also a shift in the piezoreflection peak with magnetic field. This shift corresponds to the motion of the lowest Landau level.

Since a line-shape calculation had not been worked out for piezoreflectivity, an attempt was made to obtain an experimental criterion for determining where to read the energy gap on the zero-field piezoreflection line. This criterion was found by plotting the photon energy and magnetic field dependence of the piezoreflection maximum, inflection point, and minimum corresponding to the lowest interband transition as functions of magnetic field. As may be observed in Fig. 13, all three points on the magnetic-field resonance line extrapolate to the inflection point of the zero-field piezoreflection line, suggesting that this point corresponds to the energy gap.

There was also a question concerning the resonance position within the magnetopiezoreflection linewidth. This point was resolved by comparing slope in the three lines in Fig. 13 with the $n = 0$ room-temperature magnetoabsorption data (^{24}). The best fit to these data was obtained by taking the

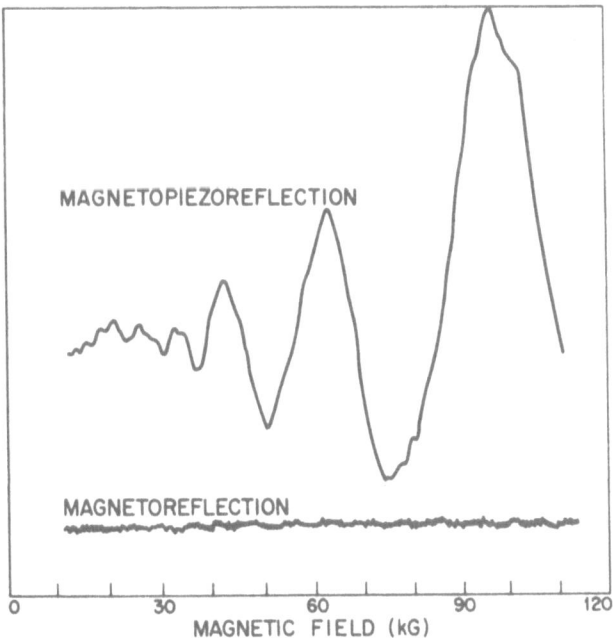

Fig. 11. Comparison between magnetopiezoreflection and magnetoreflection for a (111) face of germanium at room temperature and $\hbar\omega = 0.826$ eV. From Mavroides *et al.* (^{23}).

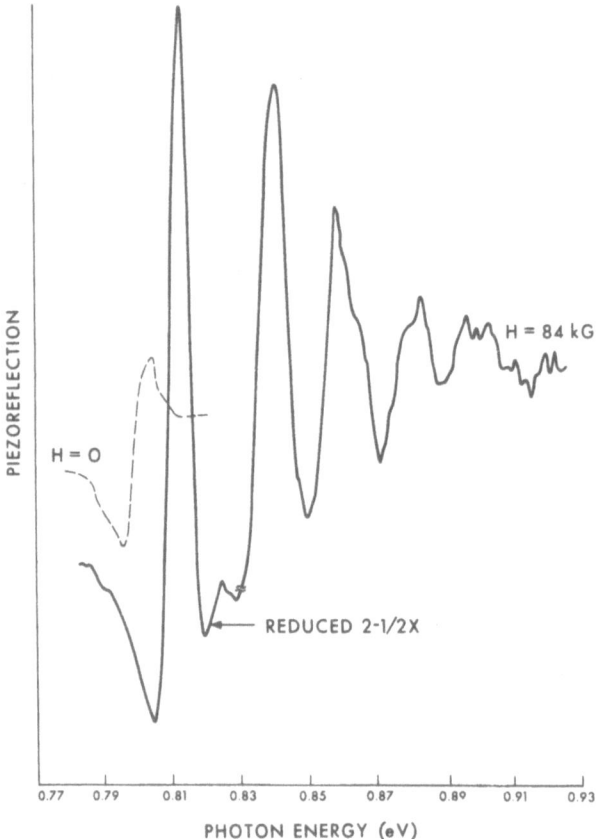

Fig. 12. Variation in germanium of the piezoreflection and magnetopiezoreflection with photon energy near the direct gap. From Mavroides *et al.* ([23]).

resonance point at the magnetopiezoreflection maximum. However, since the fit obtained by selecting the inflection point was almost as good, this criterion was not well established, although later low-temperature experiments have indicated that it is correct. Also easily observable by this technique were the interband transitions in InSb.

The success of these preliminary experiments encouraged the extension of the electroreflectance technique to magnetic field measurements. Groves *et al.* ([11]) obtained some very beautiful results in germanium at room temperature. Figure 14 gives a plot of their interband electroreflectance data in the region of photon energy from 0.75 eV to 1.20 eV. In this spectral region interband transitions are observed not only from the principal valence

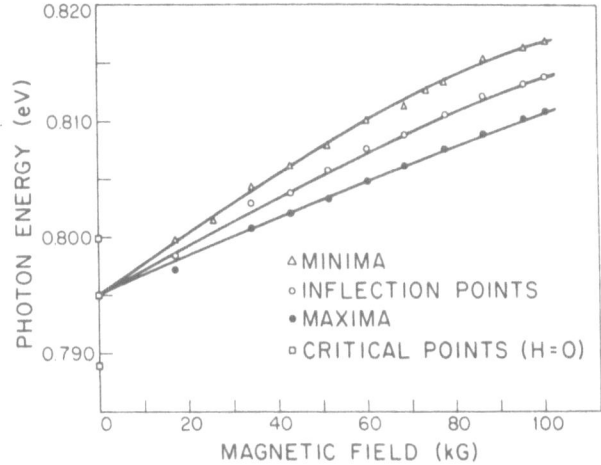

Fig. 13. Magnetic field dependence of the piezoreflection maximum, inflection point, and minimum associated with the lowest interband transition in germanium. From Mavroides *et al.* ([23]).

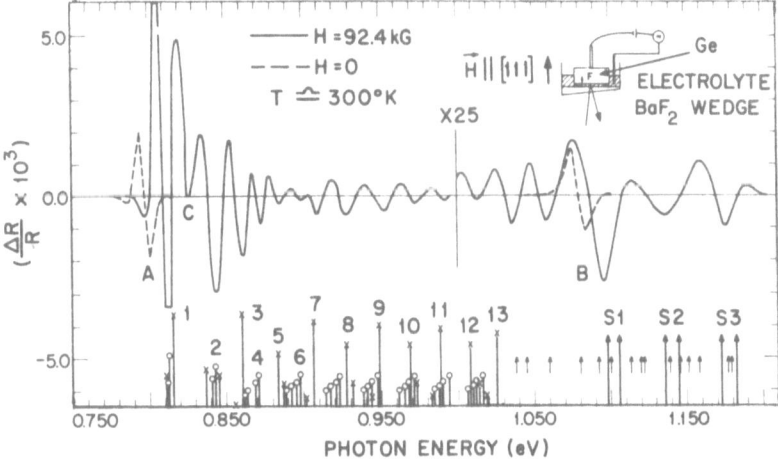

Fig. 14. Interband electroreflectance in germanium for $H = 0$ (dashed line) and $H = 92.4$ kG (solid line) in the parallel electric and magnetic field configuration at $T \approx 300\,°K$. Theoretical energies and relative strengths are shown for allowed light-hole ($\times$) and heavy-hole ($\bigcirc$) to conduction-band transitions up to 1.03 eV. Above this energy the positions of the principal valence-to-conduction-band transitions are indicated by small arrows. The relative strengths and positions of all the allowed split-off-band to conduction-band transitions are shown by heavy arrows. A schematic diagram of the experimental arrangement is shown (inset). From Groves *et al.* ([11]).

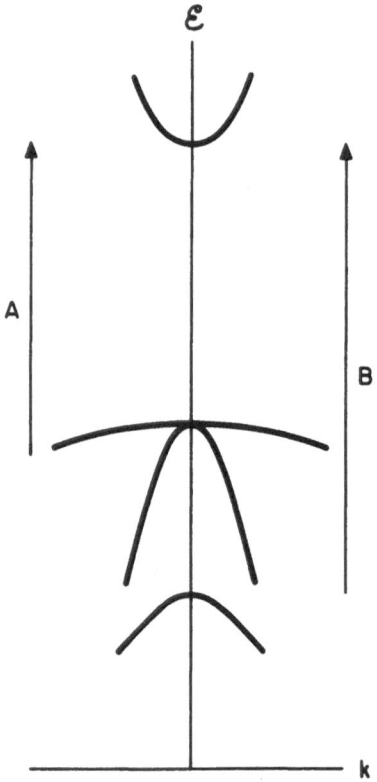

Fig. 15. Energy bands at $k = 0$ involved in
interband transitions of Fig. 14.

to the lowest $k = 0$ conduction band (A) but also from the spin-orbit split-
off valence band to the lowest $k = 0$ conduction band (B) (Fig. 15). The
dashed curves of Fig. 14 represent the zero-field transitions and the solid
lines the magnetoelectroreflectance at 92.4 kG. Theoretical energies and
relative strengths are shown below the experimental data. A summary of
the results (Fig. 16) indicates that transitions to levels quite high in the bands
can be observed and studied. The points are experimental; the solid lines
are calculated from effective-mass theory using 77 °K band parameters
and then reducing the calculated energies by 0.082 eV to make the calculated
gap agree with experiment. At the higher energies there is evidence of nonparabolicity. The parameters obtained [the energy gap, $\mathscr{E}_g = 0.801$ eV,
and the $k = 0$ conduction mass $m_c = (0.042 \pm 0.005)m$, calculated by
subtracting the computed heavy-hole energies for the heavy-hole to conduction-band transitions numbered 2 and 4] are in agreement with the

piezoreflection data. Also plotted here are the split-off valence-band to conduction-band transitions. These extrapolate at $H = 0$ to 1.083 eV, which corresponds to the energy gap. From these transitions a reduced effective mass $m_r = (0.028 \pm 0.007)m$ is obtained. The expression $m_r = m_c m_{so}/m_c m_{so}$ yields a mass for the spin-orbit split-off band $m_{so} = 0.084m$,

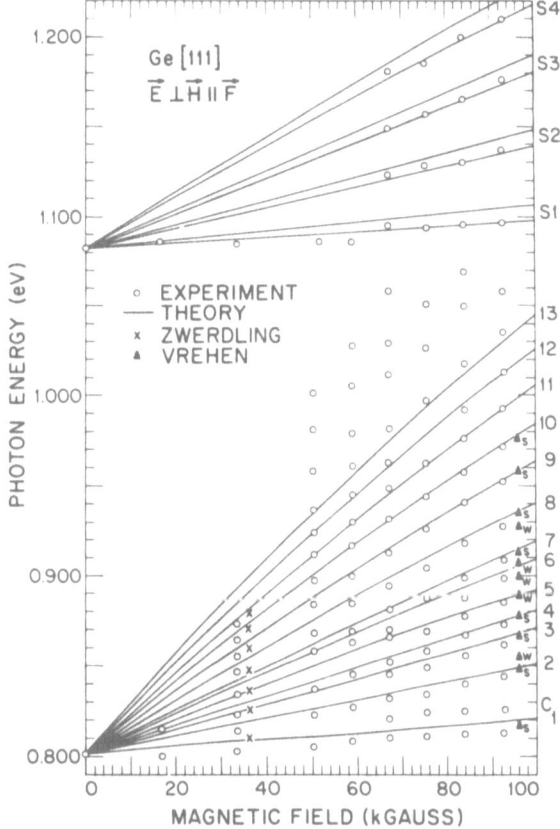

Fig. 16. Plot of the photon energy of magnetoelectrore-flectance minima in germanium as a function of magnetic field. The circles show the experimental points. The crosses give the positions of magnetoabsorption peaks (i.e., transmission minima) of Zwerdling, Roth, and Lax for $H = 35.7$ kG and the solid triangles the peaks of Vrehen for $H = 96$ kG (S indicates strong feature and W weak feature). The solid lines give the theoretical energies for the principal allowed valence- to conduction-band transitions labeled 1 to 13. The split-off-band to conduction-band transitions are labeled S1 to S4. From Groves et al. [11].

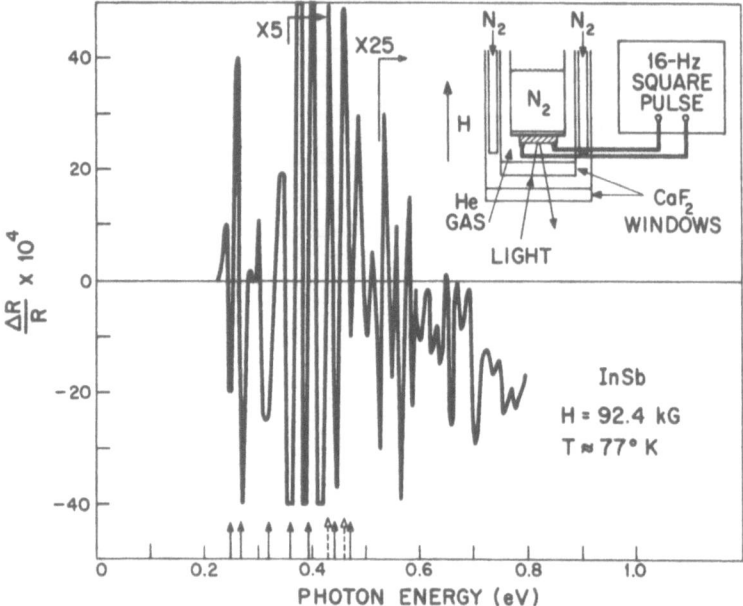

Fig. 17. Magnetothermal reflectance in InSb at ∼77°K. Inset shows the experimental arrangement. From Feinleib *et al.* ([25]).

which is in reasonable agreement with theoretical estimates and the magneto-piezoreflection data.

Other semiconductors studied by magnetoelectroreflectance were un-doped *p*-type GaSb near the direct absorption edge at ∼0.74 eV and intrinsic InSb in the region of transition from the spin-orbit split-off valence band to the lowest conduction band, ∼0.97 eV. The 0.18-eV transition from the principal valence to the conduction band could not be observed because of the low-frequency cutoff of the electrolyte. For the transition associated with spin-orbit split-off band no quantum oscillations were observed in a magnetic field, probably because of lifetime broadening of the submerged valence band. However, as we shall see, by going to low temperatures one is actually able to resolve such structure.

In connection with low-temperature measurements, the thermoreflec-tance technique ([16]) appeared somewhat promising because of its simplicity. Feinleib *et al.* ([25]) studied the magnetothermoreflectance of InSb using this method. The sample was secured to one side of a thin copper foil by means of a light coating of vacuum grease which also serves as an electrical insulator. The other side of the foil was cooled by the liquid nitrogen reservoir, and to provide further sample cooling, helium exchange gas was used. A trace of the variation of the reflectance $\Delta R/R$ with photon energy at a magnetic field of 92.4 kG is given in Fig. 17. The temperature of the sample was ∼77 °K. The agreement with previous measurements is excellent, and it is evident that this technique is quite sensitive when it can be used. However, it has the limitation that it cannot be used at really low temperatures because of the power input required for thermomodulation. Consequently, in order to study fine structure at low temperatures it became evident that the piezoreflection and dry-electrolyte electroreflection techniques offered the most promise.

In the piezoreflection technique the principal problem in going to low temperatures arose from induced strains in the thin sample due to the differential expansion coefficient between the sample and the transducer to which it was acoustically in contact. However, Aggarwal and Lax [26] were able to overcome this obstacle by optimizing the sample thickness, and they obtained beautiful results in germanium, such as shown in Fig. 18. The enhanced resolution here exhibits fine structure up to 1.3 eV, allowing the determination at ∼20 °K of the spin-orbit splitting, $\Delta = 0.297 \pm 0.065$ eV, and also the spin-orbit hole effective mass, $m^* = (0.10 \pm 0.01)m$. A similar investigation was carried out in InSb, where magnetopiezoreflec-

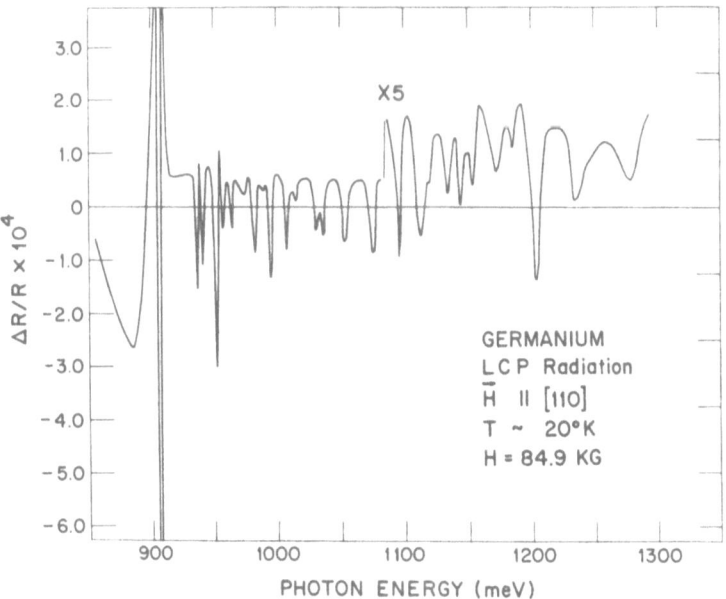

Fig. 18. Magnetopiezoreflectance at 20 °K in germanium using circularly polarized light. From Aggarwal and Lax [26].

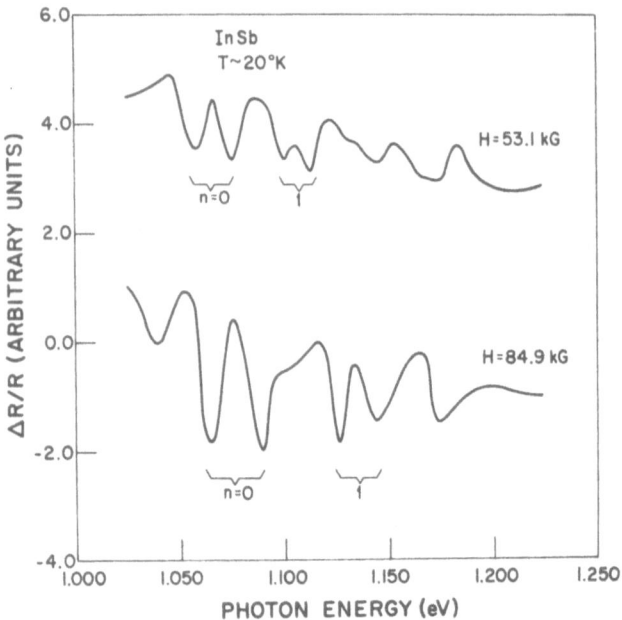

Fig. 19. Magnetopiezoreflection spectrum of the direct optical transition from the split-off valence band to conduction band in InSb for **H** ‖ [111]. From Aggarwal ([27]).

tion oscillations were observed at photon energies as high as 1.2 eV in magnetic fields up to 85 kG. From data like that of Fig. 19 Aggarwal ([27]) deduced a spin-orbit splitting, $\Delta = 0.803 \pm 0.005$ eV, and a split-off hole mass, $m^* = 0.1m$.

Similar measurements have now been obtained by using a dry electro-reflection technique. It will be recalled that the Seraphin transparent-electrode method could not be used below ∼80 °K because the optical index matching oil, which is used to eliminate interference effects arising from mechanical vibrations, freezes and thereby becomes opaque; to over-come this difficulty, Groves et al. ([12]) developed a mechanically integrated dry-electrolyte package. With this system magnetoelectroreflection measure-ments were taken as low as ∼1.5 °K and in fields as high as 84 kG. Figure 20 depicts typical traces in InSb obtained in the Faraday geometry, i.e., with the propagation in the direction of the magnetic field, which in this case is in the [111] direction. The oscillations here correspond to transitions from the magnetic levels of the complex valence bands to the $n = 0$ con-duction band levels. In addition to the allowed transitions, labeled I, III, IV, and V in Fig. 20, two extra lines are observed (II and VI) with comparable strengths to the allowed ones. These extra transitions are present when the magnetic field is in the [111] direction and absent for H ‖ [100] or [110].

An interpretation of these extra lines can be made by an examination of the Landau levels near the band edge, which are given in Fig. 21. From a consideration of the energy and polarization dependence of these additional lines it has been concluded by Groves and Pidgeon that they arise from terms in the magnetic Hamiltonian which were omitted by Luttinger ([28]) to make the problem exactly solvable. These terms connect the closely lying states $a^-(2)$ and $b^+(0)$ shown by the heavy arrow in Fig. 21. Since these states are nearly degenerate, their wave functions are sufficiently admixed to give transitions of the size of II and VI.

There are two terms which connect the desired levels and which have the correct angular dependence. One term is proportional to the warping of the valence-band energy surfaces, which can be measured from cyclotron resonance ([29]); the other term depends on the spin-orbit splitting and is smaller than the warping effect. The observed strength of the extra transitions can be accounted for entirely by the warping term. Both terms are

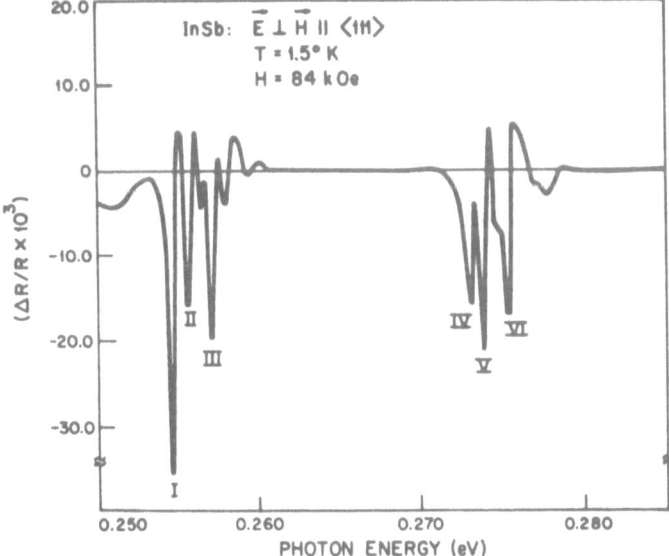

Fig. 20. Lowest transitions in the InSb magnetoelectroreflectance spectrum for both σ_r and σ_l Faraday configurations. Transitions I, III, IV, and V are "allowed" and are seen for H along all directions. Transitions II and VI are identified as "warping-induced" transitions and appear for H along [111] but not for H along [110] or [100]. From Groves and Pidgeon ([30]).

needed to explain the relative energies and strengths of the transitions in Fig 20.

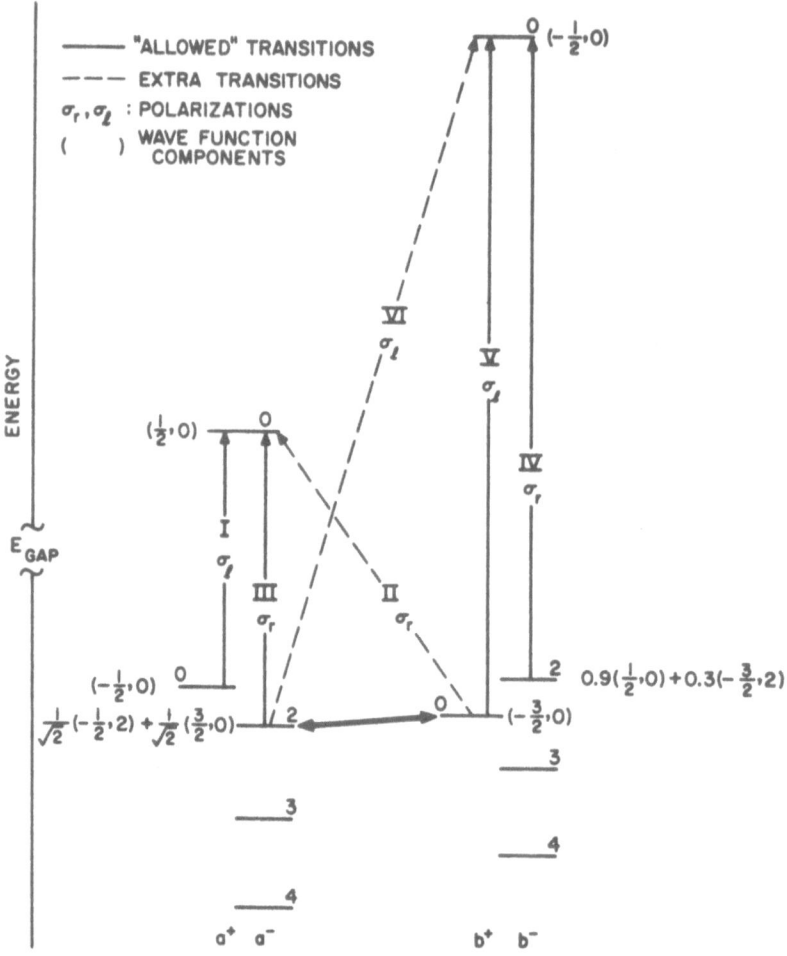

Fig. 21. Schematic representation of uppermost valence-band and lowermost conduction-band Landau levels in InSb. The labeling of the levels and ladders follows the notation of Luttinger ([28]). The Roman numerals identify the transitions with those shown in Fig. 20, with the "allowed" transitions drawn as solid lines and the extra transitions as dashed lines. Here σ_r and σ_l denote the two polarizations. The parentheses give the components of the wave function. For example, $(1/\sqrt{2})(-\tfrac{1}{2}, 2) + (1/\sqrt{2})(\tfrac{3}{2}, 0)$ denotes the wave function $\psi = (1/\sqrt{2})u_{-1/2}\phi_2 + (1/\sqrt{2})u_{3/2}\phi_0$, where u_m is a band edge Bloch function of M units of angular momentum along H and ϕ_n is the one-dimensional harmonic oscillator function of order n. From Groves and Pidgeon ([30]).

CONCLUSION

In conclusion, the recently developed optical modulation and magneto-modulation techniques with their accompanying two to three orders of improvement over conventional methods represent a major advance in the field of experimental optics as well as in electron band structure studies. Already these techniques have made possible the observation of many transitions which were previously unobservable. One profitable area which has not been discussed involves the combination of these methods with a static uniaxial stress. Further polarization magneto-optical modulation experiments will also be very fruitful. So far the work in this field has been primarily experimental. The understanding of the effects themselves and the subtle differences in the information obtained by the different techniques presents a challenge to the theoretician and should yield a wealth of new information.

ACKNOWLEDGMENTS

It is a pleasure to acknowledge helpful comments and discussions on this chapter by Dr. M. S. Dresselhaus. The author also wishes to express his gratitude to his other colleagues for discussions of their work, particularly to Drs. S. H. Groves, R. L. Aggarwal, G. F. Dresselhaus, J. Feinleib, and W. J. Scouler.

REFERENCES

1. Q. H. F. Vrehen and B. Lax, *Phys. Rev. Letters* **12**: 471 (1964); M. Chester and P. H. Wendland, *Phys. Rev. Letters* **13**: 193 (1964); Y. Yacoby, *Phys. Rev.* **142**: 445 (1965).
2. A. Frova and P. Handler, in: *Proceedings of the 7th International Conference on the Physics of Semiconductors, Paris, 1964*, Dunod, Paris, 1964, p. 157.
3. B. O. Seraphin and R. B. Hess, *Phys. Rev. Letters* **14**: 138 (1965); B. O. Seraphin, in: *Proceedings of the 7th International Conference on the Physics of Semiconductors, Paris, 1964*, Dunod, Paris, 1964, p. 165.
4. K. L. Shaklee, F. H. Pollak, and M. Cardona, *Phys. Rev. Letters* **15**: 883 (1965).
5. W. E. Engeler, H. Fritzsche, M. Garfinkel, and. J. J. Tiemann, *Phys. Rev. Letters* **14**: 1069 (1965).
6. G. W. Gobeli and E. O. Kane, *Phys. Rev. Letters* **15**: 142 (1965).
7. W. Franz, *Z. Naturforsch* **13a**: 484 (1958).
8. L. V. Keldysh, *Soviet Phys.—JETP* (*English Transl.*) **34**: 788 (1958).
9. T. S. Moss, *J. Appl. Phys. Suppl.* **32**: 2136 (1961).
10. M. Cardona, K. L. Shaklee, and F. H. Pollak, *Phys. Rev.* **154**: 696 (1967).

11. S. H. Groves, C. R. Pidgeon, and J. Feinleib, *Phys. Rev. Letters* **17**: 643 (1966).
12. C. R. Pidgeon, S. H. Groves, and J. Feinleib, *Solid State Commun.* **5**: 677 (1967).
13. H. R. Philipp and E. A. Taft, *Phys. Rev.* **120**: 37 (1960).
14. U. Gerhardt, *Phys. Letters* **9**: 117 (1964).
15. T. P. McLean, in: *Progress in Semiconductors, Vol. V*, John Wiley and Sons, New York, 1960, p. 53.
16. B. Batz, *Solid State Commun.* **4**: 241 (1966).
17. W. J. Scouler, *Phys. Rev. Letters* **18**: 445 (1967).
18. H. R. Philipp and E. A. Taft, *Phys. Rev.* **113**: 1002 (1959).
19. H. R. Philipp and E. A. Taft, *Phys. Rev.* **120**: 37 (1960).
20. M. H. Cohen, *Phil. Mag.* **3**: 762 (1958).
21. L. Van Hove, *Phys. Rev.* **89**: 1189 (1953).
22. R. L. Aggarwal, L. Rubin, and B. Lax, *Phys. Rev. Letters* **17**: 8 (1966).
23. J. G. Mavroides, M. S. Dresselhaus, R. L. Aggarwal, and G. F. Dresselhaus, in: "Proceedings of the International Conference on the Physics of Semiconductors, Kyoto, 1966", *J. Phys. Soc. Japan Suppl.* **21**: 184 (1966).
24. B. Lax and S. Zwerdling, in: *Progress in Semiconductors, Vol. V*, John Wiley and Sons, New York, 1960, p. 221.
25. J. Feinleib, C. R. Pidgeon, and S. H. Groves, *Bull. Am. Phys. Soc.* **11**: 828 (1966).
26. R. L. Aggarwal and B. Lax, *Bull. Am. Phys. Soc.* **11**: 828 (1966).
27. R. L. Aggarwal, *Bull. Am. Phys. Soc.* **12**: 100 (1967).
28. J. M. Luttinger, *Phys. Rev.* **125**: 1869 (1962).
29. D. M. Bagguley, M. L. A. Robinson, and R. A. Stradling, *Phys. Letters* **6**: 143 (1963).
30. S. H. Groves and C. R. Pidgeon, *Bull. Am. Phys. Soc.* **13**: 428 (1968); C. R. Pidgeon and S. H. Groves, *Phys. Rev. Letters* **20**: 1003 (1968).

Chapter 9

HIGH FIELD MAGNETOSPECTROSCOPY AND BAND STRUCTURE OF Cd$_3$As$_2$

E. D. Haidemenakis

Ecole Normale Supérieure
Paris, France

PROPERTIES AND PREPARATION

Cadmium arsenide has a tetragonal structure with unit cell dimensions $a = b = 12.65$ Å and $c = 25.45$ Å. The preparation of monocrystals is rather difficult due to a solid transformation point at 578 °C. The density of cadmium arsenide is 6.21 gm/cm^3, its fusion temperature is 721 °C, and its thermal conductivity is 0.014 W/cm-°K at room temperature. The crystals used had a carrier concentration between 2×10^{18} and 10^{19} cm^{-3}. Their mobilities were about 10^4 cm^2/V-sec at 300 °K and 10^5 cm^2/V-sec at 4 °K.

The crystals were grown in four steps ([1]): (1) Melting pure arsenic and cadmium stoichiometrically. (2) Distilling through hydrogen and condensation. (3) Melting the monocrystals in needle form and condensing with rotation. (4) Baking at 500 °C for 8 days.

EXPERIMENTAL

The zero-field reflection, magnetoplasma reflection, and interband magnetoreflection and magnetoabsorption measurements were taken using three different experimental set-ups. The low-field measurements were done at the Ecole Normale Supérieure, and the high-field measurements, at both the National Magnet Laboratory of MIT and the Naval Research Laboratory in Washington, D.C.

The ENS measurements were prformed using the magnetooptical system shown in Fig. 1, with standard Perkin-Elmer and Leeds and Northrup

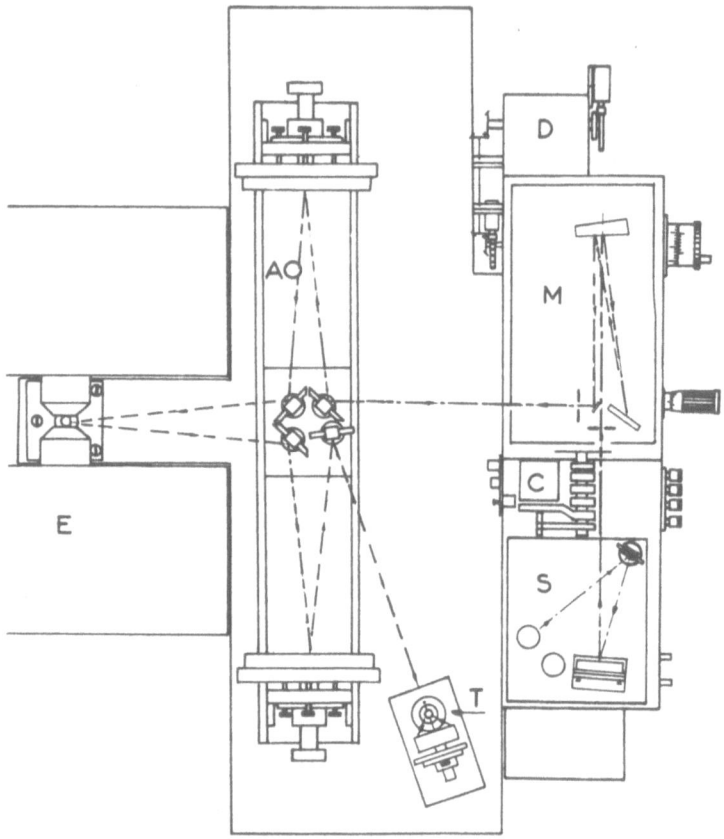

Fig. 1. Top view of the ENS magnetooptical system: (S) source; (C) chopper; (M) monochromator; (D) wavelength drive; (T) thermocouple; (AO) after-optics; (E) electromagnet with sample.

optical and detection and amplification equipment. Various automatic raising–lowering mechanisms were constructed for the room-temperature measurements for both transmission and reflection.

For the 9–14-μ range a 60 lines/mm grating blazed at 12 μ and an InSb transmission filter to suppress higher orders were used. For measurements at higher wavelengths a number of LiF and CaF_2 crystals served as choppers and as restrahlen plates.

For the low-temperature measurements a liquid helium cryostat with a tail width of 11 cm was constructed and was slid vertically to bring the sample and mirror into focus by means of compressed air in a special sample–mirror interchange mechanism.

The magnetoreflection measurements at higher fields were made using the high-field facilities at the National Magnet Laboratory, MIT; the apparatus is described in ([2,3]). The radiation was parallel to the magnetic field, which was swept from zero to a maximum of 120 kG. The sample was mounted on a copper block in contact with liquid helium. A bucking system allowed the recording of the slightest signals, which were detected by a thermocouple kept away from the magnet in order to shield against pick-up.

Most measurements, however, were taken at the magnet facilities of the Naval Research Laboratory ([4,5]). The magnets were Bitter-type, air-core solenoidal, with an aperture of 2.5 in.; they produced a maximum field of 112 kG. The optical system was composed of a prism monochromator, a high-power globar source, a 13-cps thermocouple, and a number of polarizers and flat and spherical mirrors, which directed the radiation to the surface of the sample from below through the magnet's aperture. Both longitudinal and transverse orientations were used. In the Voigt configuration the beam was reflected by one or two small flat mirrors located in the middle of the cyclindrical magnet. The electronics included standard Perkin-Elmer equipment and a Reeder thermocouple detector. The output of the Princeton Applied Research Corporation lock-in amplifier was fed into a Brown recorder and presented as a function of wavelength. The magnetic field was either swept to 112 kG at fixed photon energy or was kept fixed while the photon energy was varied.

Various prisms were used to cover the range between 5 and 35 μ, and certain reflection and transmission filters were placed near focal points for different spectral regions. Both linear and circular polarization had to be used. The former was obtained from a AgCl polarizer near the entrance slit of the monochromator. A NaCl Fresnel rhomb together with a linear polarizer produced circular polarization of the radiation in the longitudinal orientation case.

The data were taken at 300 °K and 4 °K. For the low-temperature free-carrier experiments the samples were polished mechanically and glued with silicone cement directly onto the tailpiece of the liquid helium cryostat. The helium Dewar was designed so that the optical beam could enter the CsBr windows either through the side or through the bottom of its tailpiece; thus, the Poynting vectors S of the radiation field were either parallel or perpendicular to the magnetic field. For the transmission measurements the samples were glued with infrared transparent cellulose caprate thermoplastic cement to a CsI crystal and polished to a thickness of between 50 and 100 μ. The supporting CsI crystal was then embedded in a specially constructed tailpiece and surrounded by proper radiation shields. To prevent sample heating an Eastman Kodak filter which was transparent beyond $3\,\mu$ was used. The entire optical system was enclosed in a box made from plexiglass, and the air was dried with a molecular sieve.

ZERO-FIELD REFLECTION

The reflection coefficient R for normal incidence on a semi-infinite conducting slab is given by

$$R = \frac{(n-1)^2 + k^2}{(n+1)^2 + k^2} \tag{1}$$

In addition to the complex dielectric constant $K^* = K' - iK''$ associated with the intrinsic dielectric properties of a crystal one should, for a semi-conductor with free carriers, include a complex conductivity $\sigma^* = \sigma' - i\sigma''$ associated with its free carriers. These free carriers produce abrupt transitions in the reflectivity spectra in a region where the medium is essentially dispersive. Zener [6] was the first to study these effects based on the Drude [7] theory for an electron in a region where $\omega\tau \gg 1$. Spitzer and Fan [8] have used the minima of the reflectivity in semiconductors to determine effective masses after having previously measured the carrier concentration by Hall effect methods.

If ε_0 is the permittivity of free space, the index n of refraction and the extinction coefficient k are related by

$$n^2 - k^2 = K' - (\sigma''/\omega\varepsilon_0), \qquad 2nk = K'' + (\sigma'/\omega\varepsilon_0) \tag{2}$$

For the case where there are no losses, i.e., for $\tau \to \infty$, these equations become

$$n^2 - k^2 = K' - (\omega_p^2/\omega^2), \qquad 2nk = 0 \tag{3}$$

with the plasma frequency ω_p given by

$$\omega_p^2 = (Ne^2/m^* K' \varepsilon_0) \tag{4}$$

Since n and k are real and neither is equal to zero, for $\lambda \leq \lambda_p$ the solutions to these equations are

$$n^2 = K'\left(1 - \frac{\lambda^2}{\lambda_p^2}\right), \qquad k = 0, \qquad R = \frac{(n-1)^2}{(n+1)^2} \tag{5}$$

and for $\lambda \geq \lambda_p$,

$$n = 0, \qquad k^2 = -K'\left(1 - \frac{\lambda^2}{\lambda_p^2}\right), \qquad R = 1 \qquad (6)$$

In Fig. 2 n^2, k^2, and R are plotted against the wavelength squared [9,10]. In plotting the above parameters we assumed that K' is a constant. The free-carrier susceptibility subtracts from the dielectric constant as the wavelength increases. When n becomes 0 the reflectivity R falls to zero and then rises to unity at λ_p, after which it remains constant. When the reflectivity is zero, or if the dielectric constant is equal to 1, the wavelength of the reflectivity minimum is given by

$$\lambda_{min}^2 = \left(\frac{K' - 1}{K'}\right)\lambda_p^2 \qquad (7)$$

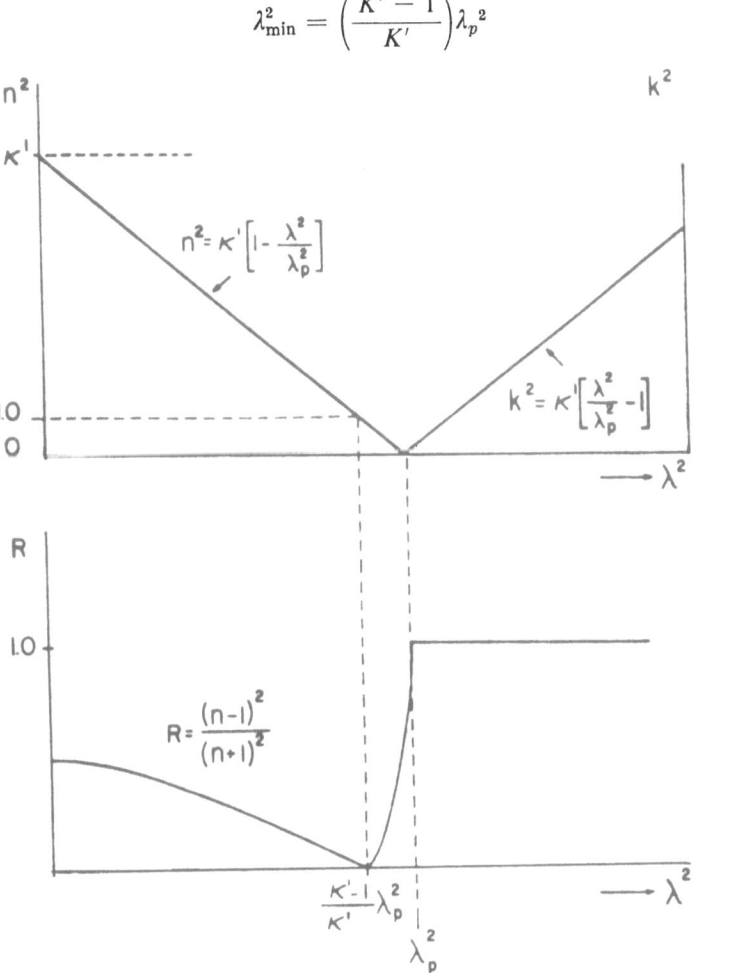

Fig. 2. A plot of n^2, k^2, and R versus wavelength squared.

from which

$$K' = \frac{1}{1 - (\lambda_{min}/\lambda_p)^2} \qquad (8)$$

The reflectivity rises to unity when the index of refraction goes to zero because of a further decrease in the dielectric constant. As the wavelength increases further the index of refraction becomes imaginary and is plotted as k^2.

If losses are not negligible, i.e., if the relaxation time $\tau \ll \infty$, or when K'' or σ' in Eq. (2) are finite, the sharp corners in the curves of Fig. 2 become rounded.

For k negligible and for $\lambda < \lambda_p$, n is given by

$$n = (1 + \sqrt{R})/(1 - \sqrt{R})$$

Thus a plot of the experimental value of n^2 versus the square of the wavelength should yield a straight line whose slope is given by

$$\frac{\Delta \lambda^2}{\Delta n^2} = - \frac{m^*}{m} \frac{1}{N} 10^{-22} \frac{1}{8.97}$$

where the wavelength is in microns and N is in cm^{-3}. The slope ceases to be a straight line when $k^2 \geq 1$. In a similar manner one can determine the wavelength of the reflectivity minimum for k^2 negligible as a function of K', m^*/m, and N.

Figure 3 shows zero-field reflection data taken at different intervals. The curves indicate a change in carrier concentration with time and after heat treatment. When the sample was placed on a hot plate its carrier concentration doubled, as can be seen from the shift of the plasma edge to a higher frequency. Small oscillations near the reflection minimum which disappear after the heat treatment cannot be explained. These could be due to constraints or traps in the bulk, interference from a surface layer, underlying interband effects, or multiple plasma edges.

The plot of the index of refraction squared versus the wavelength squared was found to be a straight line. Its intercept determined the high-frequency dielectric constant, which was equal to 16.3, following the relation

$$\varepsilon = n^2 = \varepsilon_\infty \left(1 - \frac{\omega_p^2}{\omega^2}\right) = \varepsilon_\infty - \frac{e^2 N \lambda^2}{\pi c^2 m^*}$$

Values of N/m^* are obtained from the slope of the straight line. The susceptibility mass m^* was determined from Hall measurements for two different

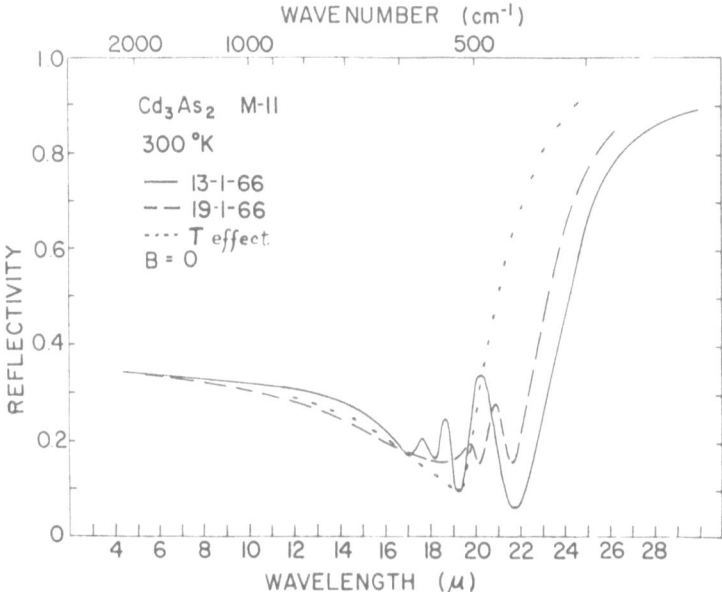

Fig. 3. Typical reflectivity curves of Cd_3As_2 crystals showing time and temperature effects on carrier concentration.

samples of known carrier concentration, and was found to be equal to $0.05m$.

In the presence of a magnetic field the plasma reflection edge splits in two edges and the separation between the two minima increases linearly with field. This separation is equal to the cyclotron frequency. The application of the magnetic field B changes the index of refraction. In the longitudinal case the index of refraction is given by [11]

$$n_{\pm}^2 = K'\left(1 - \frac{\omega_p^2}{\omega(\omega \pm \omega_c)}\right)$$

where the plus and minus sings refer to left and right circularly polarized radiation, respectively. In the transverse case and for $E \perp B$ the index of refraction becomes

$$n^2 = \left[1 - \left(\frac{\omega_p^2}{\omega^2}\right)\left(\frac{\omega^2 - \omega_p^2}{\omega^2 - \omega_p^2 - \omega_c^2}\right)\right]$$

Effective masses may be obtained both from the separation of the two minima and from a plot of the plasma edge shifts at an isoreflection point

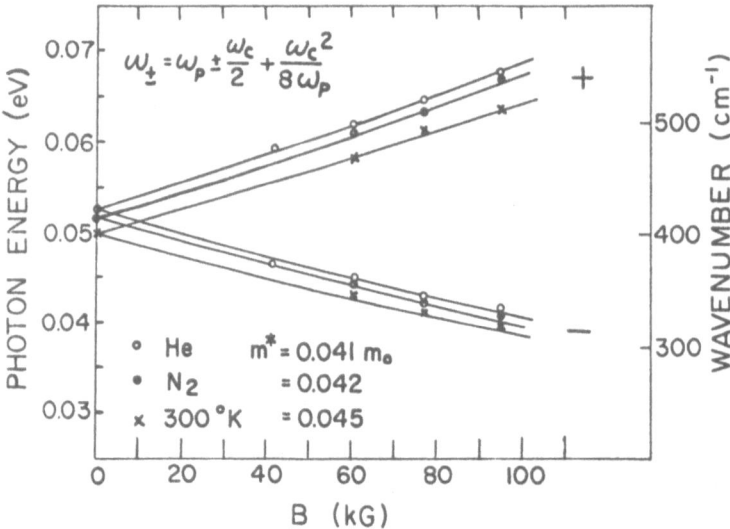

Fig. 4. Isoreflection points at 0.55 magnetoplasma reflectivity for left (+) and right (−) circular polarization at the Faraday orientation.

as a function of magnetic field. Figure 4 shows the results of magnetoplasma reflectivity measurements in the Faraday configuration for fields up to 95 kG using left (+) and right (−) circular polarization. The data were taken along the 0.55-isoreflection line on the steeply rising reflection edge. This point is very near the plasma frequency, for which according to classical theory the formula

$$\omega_{\pm} = \omega_p \pm \frac{\omega_c}{2} + \frac{\omega_c^2}{8\omega_p}$$

holds to a good approximation.

The left circular edge isoreflection points move to high energies with increasing magnetic field, whereas the (−) ones move to lower energies. The solid lines were drawn using the values for the effective masses as shown. The experimental points at zero field were used in place of the plasma frequency in the above equation. The last term $\omega_c^2/8\omega_p$ of this equation brings an asymmetry in the reflection edge splitting which always remains equal to the cyclotron frequency. From the values $m/m_0 = 0.041$, 0.042, and 0.045, for $T = 4\,°K$, 77 °K, and 300 °K, respectively, one can see that the effective masses decrease slightly with decreasing temperature, similar to the case for lead salts, implying nonparabolicity in the conduction band.

Similar curves were obtained in the Voigt configuration. A zero-field reflection minimum at 0.049 eV splits into two minima whose separation is equal to the cyclotron frequency ω_c for $\mathbf{E} \perp \mathbf{B}$. For parallel polarization a slight effect was observed, indicating some mass anisotropy, since in a spherical band material the parallel magnetoplasma effect is absent.

The subsidiary reflection maxima and minima which were observed near the main minimum at zero magnetic field seem to follow in both the Faraday and Voigt configurations and to parallel the two magnetoplasma minima; they disappear only after heat treatments. This treatment also produced a shift of the minimum in the transverse case from 0.05 eV to 0.08 eV. The new effective mass ratio became 0.054 and 0.045 at 300 °K and at 4 °K, respectively, which implied nonparabolicity in the conduction band.

INTERBAND MAGNETOABSORPTION AND MAGNETOREFLECTION

The first experiments in interband magnetoabsorption were carried out by Burstein and Picus in 1957 on InSb [12] and were soon followed by the Lincoln Laboratory group of Lax and co-workers [13,14] on germanium. A number of experiments followed, not only on semiconductors but on semimetals such as bismuth and antimony [15,16]; more recently similar measurements were done on lead salts at the U.S. Naval Research Laboratory [2,17-20].

In the presence of a magnetic field the Schrödinger equation in the effective-mass approximation for an electron is given by

$$\frac{1}{2m_c}\left(\mathbf{P} + \frac{e}{c}\,\mathbf{A}\right)^2 \psi = \mathscr{E}\psi$$

where m^* is the effective mass of the electron, $\mathbf{p}$ is the momentum operator and $\mathbf{A} \cdot \frac{1}{2}(\mathbf{r} \times \mathbf{H})$ is the vector potential. The solutions of the above equation are the equations of the energy levels for the conduction and valence bands for a set of simple idealized parabolic bands of a semiconductor:

$$\mathscr{E}_c = (n + \tfrac{1}{2})\hbar\omega_c + (\hbar^2 k_z^2/2m_c)$$

and

$$\mathscr{E}_v = -\mathscr{E}_g - (n' + \tfrac{1}{2})\hbar\omega_v - (\hbar^2 k_z^2/2m_v)$$

where c and v refer to the conduction and to the valence band, respectively, the integer n is the magnetic quantum number, and ω_c is the cyclotron frequency. The energy levels for a simple case of spherical energy bands are called Landau levels, and were first obtained by Landau in 1930 [21].

For zero magnetic field the absorption constant for transitions between valence and conduction bands is given by ([22])

$$\alpha = \frac{Ke^2(2\mu)^{3/2} \, |\, \mathbf{P}_{cv} \,|}{\omega 16\pi\varepsilon m^2\hbar^{5/2}} \, (\omega - \omega_g)^{1/2}$$

where $\mu = m_c m_v/(m_c + m_v)$. In the presence of a magnetic field an oscillatory effect is observed which is due to allowed transitions between Landau levels. The above constant now becomes ([23,24])

$$\alpha(H) = \frac{Ke^2(2\mu)^{3/2} \, |\, \mathbf{P}_{cv} \,|}{2\omega 16\pi\varepsilon m^2\hbar^{5/2}} \, \omega_{cv} \sum_n \frac{1}{(\omega - \omega_n)^{1/2}}$$

where

$$\omega_{cv} = eH/\mu c, \qquad \omega_n = \omega_g + (n + \tfrac{1}{2})\omega_{cv}$$

This equation demonstrates the existence of singularities or peaks for transitions between Landau levels. These are direct transitions and the selection rules are $\Delta n = 0$ and $\Delta k_z = 0$. They occur between the valence and conduction bands when the levels have the same Landau quantum number n. The density of states is infinite for each value of n at $k_z = 0$ and decreases with increasing k_z, resulting in a change in the absorption and the appearance of absorption lines.

At the maxima of the absorption constant the photon energy is equal to

$$\hbar\omega = \mathscr{E}_g + \hbar\omega_c(n + \tfrac{1}{2}) + \hbar\omega_v(n + \tfrac{1}{2})$$

where ω_c and ω_v are the cyclotron frequencies of the conduction and valence bands, respectively. The energy gap $\mathscr{E}_g$ and the reduced effective mass $\mu = m_c m_v/(m_c + m_v)$ can be deduced from the maxima of the absorption constant by measuring the positions of the absorption lines as a function of magnetic fields for different values of the quantum number n.

We can also plot the absorption coefficient as a function of energy for indirect transitions. These refer to the case where we have more than one photon, such as the emission or the absorption of a phonon of energy $\hbar\omega_{ph}$. The application of a magnetic field in this case produces the well-known staircase curve of the absorption coefficient. Since there are no phonon transitions here, there are no selection rules for the Landau quantum numbers.

For a degenerate semiconductor the number of transitions possible from valence to conduction band is reduced due to the Pauli principle, and some of the available energy states in the conduction band are occupied by the added carriers. The semiconductor becomes more transparent around the band edge, and the lines in this region of small absorption are relatively weak. To detect them experimentally, it is simpler to fix the frequency and sweep the magnetic field instead of fixing the field and varying the photon energy. As the magnetic field increases oscillations are observed when each Landau line passes a fixed photon energy. The peaks of the absorption are due to the large density of states at $k_z = 0$.

Well-developed oscillations were observed as the magnetic field was swept at liquid nitrogen and helium temperatures for energies between 0.10 and 0.18 eV. Measurements were taken in the transverse configuration for parallel polarizations. Near the Burstein–Moss edge the oscillations were the strongest, and decreased at lower photon energies, where the absorption was due to free carriers rather than interband transitions. These oscillations were periodic in $1/H$ and are not of the de Haas–van Alphen type (variation of the Fermi level) because they are not independent of the photon energy.

One should also note that transmission experiments (^{3}) on interband transitions usually show larger bands than reflection experiments. In addition, the forbidden energy gap is found to be smaller in transmission experiments as compared to experiments in reflection. This observation is also valid in the case of bismuth, as observed by Engeler in magnetotransmission experiments and by Brown et al. (^{2}) in magnetoreflection.

Figure 5 shows data obtained in magnetotransmission experiments on two samples in the Voigt configuration for energies down to 0.13 eV. Straight-line extrapolation gives an energy gap of 0.025 eV. With lower carrier concentrations one would obtain additional points below 0.13 eV. Purer samples would undoubtedly show nonparabolic effects at lower photon energies, and would verify whether or not the bands are quadratic in k for the magnetic fields used. Recent results with $Cd_xZn_{3-x}As_2$ alloys have shown substantial nonparabolicity and an energy gap close to 0 near $x = 2.6$. Spin-splitting of the Landau levels due to the effective g-factor should become evident, especially for a material with a small effective mass, such as cadmium arsenide.

There was no clear temperature dependence of the gap; the transmission, however, increased to higher photon energy as the sample was cooled, implying that the Fermi-level edge was sharpening up. A reduced effective mass ratio was deduced from the energy separations between adjacent lines at fixed magnetic fields, and was found to be equal to 0.031. From the conduction band mass already obtained and the expression for the reduced effective mass ratio a valence band mass was calculated to be equal to 0.12.

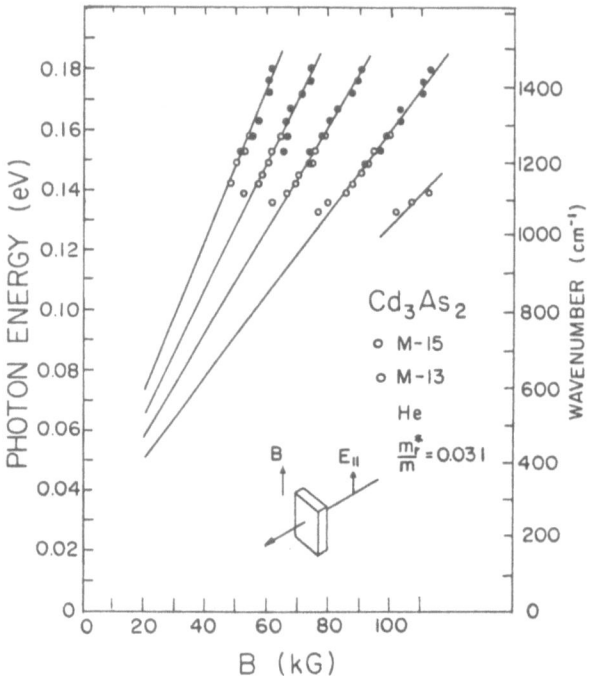

Fig. 5. Variation of the photon energy of the interband magnetoabsorption lines with magnetic field in the Voigt orientation.

The transitions appear to originate from a heavy-hole valence band. The quantum numbers 3, 4, 5, 6, and 7 were assigned to the levels observed. It was not possible to determine whether the transitions were direct or indirect; from line shape at fixed magnetic field direct transitions produce an oscillatory effect in the absorption constant, while indirect transitions show stairsteps.

The Fermi level was calculated to be about 0.12–0.14 eV above the band edge assuming all 2×10^{18} carriers are in a single band. The Burstein–Moss edge is thus found to be in the right range, assuming that the energy gap lies between 0 and 0.03 eV, and that the transitions are coming from a heavy-hole valence band.

REFERENCES

1. B. Koltirine and M. Chaumereuil, *Phys. Status Solidi* **13**: K1 (1966).
2. R. N. Brown, J. G. Mavroides, B. Lax, *Phys. Rev.* **129**: 2055 (1963).

3. E. D. Haidemenakis, J. G. Mavroides, M. S. Dresselhaus, and D. F. Kolesar, *Solid State Commun.* **4**: 65 (1966).
4. E. D. Palik, D. L. Mitchell, and J. N. Zemel, *Phys. Rev.* **135**: 3A (1964).
5. E. D. Haidemenakis, M. Balkanski, E. D. Palik, and J. Tavernier, *J. Phys. Soc.* **21** (1966).
6. C. Zener, *Nature* **132**: 968 (1933).
7. P. Drude, *The Theory of Optics*, Dover, New York, 1959.
8. W. G. Spitzer and H. Y. Fan, *Phys. Rev.* **106**: 882 (1957).
9. F. Galeener, Master's Thesis, MIT, 1961.
10. G. Wright, Thesis, MIT, 1960.
11. G. B. Wright and B. Lax, *Bull. Am. Phys. Soc.* **6**: 18 (1960).
12. E. Burstein and G. S. Picus, *Phys. Rev.* **105**: 1123 (L) (1957).
13. S. Zwerdling and B. Lax, *Phys. Rev.* **106**: 51 (1957).
14. S. Zwerdling, B. Lax, and L. M. Roth, *Phys. Rev.* **108**: 1402 (1957).
15. R. N. Brown, J. G. Mavroides, M. S. Dresselhaus, and B. Lax, *Phys. Rev. Letters* **5**: 506 (1960).
16. L. Hebel and G. E. Smith, *Phys. Letters* **10**: 273 (1964).
17. D. L. Mitchell, E. D. Palik, J. D. Jensen, R. B. Schoolar, and J. N. Zemel, *Bull. Am. Phys. Soc.* **8**: 309 (1963).
18. D. L. Mitchell, E. D. Palik, J. D. Jensen, R. B. Schoolar, and J. N. Zemel, *Phys. Letters* **4**: 262 (1963).
19. D. L. Mitchell, E. D. Palik, J. D. Jensen, R. B. Schoolar, and J. N. Zemel, *Bull. Am. Phys. Soc.* **9**: 292 (1964).
20. D. L. Mitchell, E. D. Palik, and J. N. Zemel, in: *Proceeding of the 7th International Conference on the Physics of Semiconductors*, Dunod, Paris, 1964, 325.
21. L. D. Landau, *Z. Physik* **64**: 629 (1930).
22. L. H. Hall, J. Bardeen, and F. J. Blatt, *Proceedings of the Conference on Photoconductivity*, Atlantic City, 1956, p. 146.
23. L. M. Roth, B. Lax, and S. Zwerdling, **114**: 90 (1959).
24. B. Lax, L. M. Roth, and S. Zwerdling, *J. Phys. Chem. Solids* **8**: 311 (1959).
25. R. J. Wagner, E. D. Palik, and E. M. Swiggard, *Bull. Am. Phys. Soc. Sec. 11*, Vol. 13, p. 28 (1968).

Chapter 10

EFFECTS OF HIGH MAGNETIC FIELDS ON ELECTRONIC STATES IN SEMICONDUCTORS—THE RYDBERG SERIES AND THE LANDAU LEVELS

H. Hasegawa

Department of Physics
Kyoto University
Kyoto, Japan

INTRODUCTION

The first theoretical prediction that shallow impurity states in semiconductors should be greatly influenced by an external static magnetic field of technically attainable intensity was given by Yafet, Keyes, and Adams (YKA) in 1956 ([1]). In this chapter we will describe an outgrowth of the theory originating from this work.

Our subject here is the effect of *competition* of the external magnetic force with the internal Coulomb one, both acting on a small number of carriers in the conduction or valence band of a hypothetical semiconductor. The Coulomb force is assumed to arise from a charged impurity center. Its characteristic strength is given by

$$R_y = m^* e^4 / 2\hbar^2 \varkappa^2 \tag{1}$$

i.e., the Rydberg energy of a hydrogenic atom with an effective electron mass m^* and the screened nuclear charge $e/\varkappa$. The theory may equally apply, with a certain generalization, to a more fundamental object in semiconductors, namely, the exciton, where the Coulomb force shielded by the same dielectric constant $\varkappa$ acts between an optically produced electron–hole pair. The strength of a magnetic field H, on the other hand, may be characterized by the shift of the band edge due to the field, i.e. the zero-point energy of the lowest Landau level, given by

$$\tfrac{1}{2}\hbar\omega_c = (e\hbar/2m^*c)H \tag{2}$$

The comparison of (1) and (2) can also be interpreted in terms of the two kinds of orbital radius, i.e., the effective Bohr radius

$$a^* = \hbar^2\varkappa/m^*e^2 \tag{3}$$

and the cyclotron radius

$$l = (\hbar c/eH)^{1/2} \tag{4}$$

respectively, since the following relations hold:

$$R_y = (\hbar^2/2m^*)a^{*^{-2}}, \qquad \tfrac{1}{2}\hbar\omega_c = (\hbar^2/2m^*)l^{-2} \tag{5}$$

The work of Yafet et al. ([1]) showed that when the magnetic field is strong enough so that $\tfrac{1}{2}\hbar\omega_c$ is comparable to or larger than R_y a considerable compression of the electronic wave function of the ground state occurs because its orbital radius tends to become smaller and smaller in accordance with (4) as the field is increased. This shrinkage of the wave function in turn causes the ground-state electron to be affected by a stronger binding of the attractive Coulomb potential, and thus results in an increase of the ionization energy.

The above authors pointed out that the effect could be observed as a decrease in number of conduction carriers due to a thermal deionization against an increase of the field H, and was subsequently confirmed by the Hall measurements carried out in n-type InSb ([2]). The effect of the carrier redistribution over the energy band and the impurity states in InSb, named the "magnetic freeze-out effect," has received interest in connection with consideration of the impurity band problem ([3]).

We plan to treat the "freeze-out effect" separately on another occasion* by investigating the nature of the Landau subband edge with many impurities. Here we confine ourselves to the problem of the energy spectrum of an isolated hydrogen atom in a magnetic field. This is a specialized, but conceptually interesting subject of basic quantum mechanics,[†] which aims to clarify how the whole spectrum of such a well-defined system changes with the variable parameter, the magnetic field. It involves a special mathematical complexity owing to the presence of groups of discrete and continuous spectra, and thus may be considered as an example of the "coexistence" problem of localized and band characters in solids ([5]).

---

* See Note 1 added in proof.
[†] The first systematic study was given by Schiff and Snyder ([4]).

The magnetooptic spectroscopy in semiconductor physics has necessi-
tated the study of the above subject, and although the final goal remains
to be reached, a considerable progress has already been achieved, demonstrating a general feature of coexistence of the two types of spectra, namely
the Rydberg series and the Landau levels.

PRELIMINARIES

The magnetic field dependence of the ionization energy of the hydrogen
ground state calculated by Yafet et al. [1] is shown in Fig. 1. It is based on a
variational method: The quantum expectation value of the Hamiltonian

$$
\mathscr{H} = \frac{1}{2m^*} \left(\mathbf{p} + \frac{e}{2c} \mathbf{H} \times \mathbf{r} \right)^2 - \frac{e^2}{\varkappa r}
$$

$$
= \frac{p^2}{2m^*} + \frac{eH}{2m^*c} (x p_y - y p_x) + \frac{e^2 H^2}{8m^*c^2} (x^2 + y^2) - \frac{e^2}{\varkappa r}, \qquad (\mathbf{H} \parallel z)
$$
(6)

is minimized with a trial wave function of the form

$$
\psi(\mathbf{r}) = \frac{1}{[(2\pi)^{3/2} a_\perp^2 a_\parallel^2]^{1/2}} \exp\left(-\frac{x^2 + y^2}{4a_\perp^2} - \frac{z^2}{4a_\parallel^2} \right)
$$
(7)

where the two radii $a_\perp$ and $a_\parallel$ are the variation parameters. The ionization
energy of this state may be defined by the difference between the energy
of the lowest free Landau level, $\frac{1}{2}\hbar\omega_c$, and the minimized energy. To understand that this is the correct definition of the ionization energy we give the
YKA diagram [1] of the free Landau levels in Fig. 2, which illustrates the

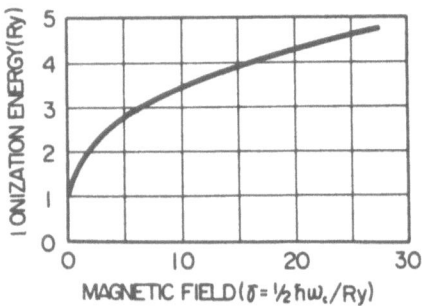

Fig. 1. The theoretical ionization energy of
the ground-state hydrogen atom versus an
increase of a static magnetic field H. From
Yafet et al. [1].

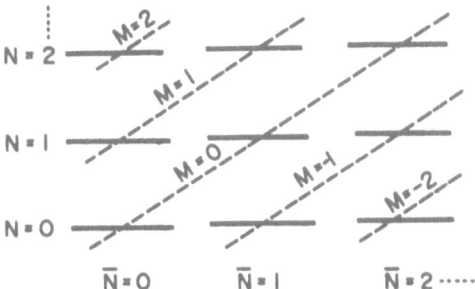

Fig. 2. An energy diagram of the free Landau levels for the symmetric gauge of the magnetic field, $A = (\frac{1}{2}Hy, -\frac{1}{2}Hx, 0)$, where the angular momentum M along the field direction is a constant of motion.

infinitely degenerate structure. This degeneracy of each energy level ($N = 0$, 1, 2, ...) is of course a consequence of the Landau coalescence, and reflects the fact that for any cyclotron orbit there exist other orbits with the same energy but of different locations of the orbit center which is "quantized," as shall be seen later. In the YKA diagram this situation is indicated by the dotted lines labeled by M, the angular momentum quantum number, showing that the Landau level N is degenerate for the states $M = N, N - 1$, $\cdots \rightarrow -\infty$. The ionization process of the ground-state electron of the form (7) is the one in which the electron jumps into the $N = 0$, $M = -\infty$ level whose energy is nearly equal to $\frac{1}{2}\hbar\omega_c$, as Yafet et al. discussed. This is because the state with a large $|M|$ value corresponds to the orbit whose center is located far from the center of the Coulomb force due to the centrifugal potential and hence with a negligible binding effect. The angular momentum quantum number M is an exact quantum number for the Hamiltonian (6), the one-impurity problem with central symmetry, so that the above statement is meaningful.

On the other hand, it is noted that the "orbit center" and the Landau quantum number N are of exact meaning only for the free Landau levels, i.e., when the static potential is absent. In particular, the "orbit center" is a hidden variable which does not enter the kinetic energy part,

$$\mathscr{H}_0 = \frac{1}{2m^*} \left(\mathbf{p} + \frac{e}{2c} \mathbf{H} \times \mathbf{r} \right)^2$$

thus yielding the degeneracy. Its significance was first noted by Johnson and Lippmann ([6]). Later, Kubo et al. ([7]) used the concept of the orbit-center coordinates on a rigorous basis to formulate the transport theory at high magnetic fields. Following the latter theory one can write the transverse components of the electron coordinate vector **r** as

$$x = \xi + X, \qquad y = \eta + Y \tag{8}$$

Here (ξ, η) and (X, Y) are the relative and the center coordinates of the cyclotron motion, respectively, and the former are related to the canonical momentum, $\mathbf{p} + (e/2c)\mathbf{H} \times \mathbf{r}$, by

$$\xi = \frac{c}{eH}\left(p_y + \frac{e}{2c} Hx\right), \qquad \eta = -\frac{c}{eH}\left(p_x - \frac{e}{2c} Hy\right) \tag{9}$$

Accordingly, the center coordinates may be written as

$$X = -\frac{c}{eH}\left(p_y - \frac{e}{2c} Hx\right), \qquad Y = \frac{c}{eH}\left(p_x + \frac{e}{2c} Hy\right) \tag{10}$$

which together with (ξ, η) form a complete set of the canonical variables in the transverse motion. In quantum terminology these satisfy the commutation relations

$$\begin{aligned}
[\xi, \eta] &= l^2/i, \qquad [X, Y] = il^2 \\
[\xi, X] &= [\xi, Y] = [\eta, Y] = [\eta, X] = 0
\end{aligned} \tag{11}$$

where l is the elementary radius defined by Eq. (4). The hidden character of the center coordinates (X, Y) is that

$$\begin{aligned}
\mathcal{H}_0 &= (\hbar\omega_c/2l^2)(\xi^2 + \eta^2) \\
&= \hbar\omega_c(N + \tfrac{1}{2})
\end{aligned} \tag{12}$$

and therefore

$$\dot{X} = \dot{Y} = 0 \tag{13}$$

The angular momentum along the **H** direction, M, is given in terms of these operators as

$$\begin{aligned}
M &= (1/2l^2)(\xi^2 + \eta^2) - (1/2l^2)(X^2 + Y^2) \\
&= N - \bar{N},
\end{aligned} \tag{14}$$

where N and $\bar{N}$ are the number operators of the two harmonic oscillators associated with (ξ, η) and (X, Y), respectively. For low-lying Landau levels, therefore, the states with large negative values of M correspond to those with large magnitudes of $X^2 + Y^2$, the distantly located orbits.

SPECTRA IN THE HIGH FIELD LIMIT

The free Landau levels are a good starting basis for obtaining the energy spectra of the Hamiltonian (6) when $\frac{1}{2}\hbar\omega_c \gg R_y$. In fact, the result of the variational calculation demonstrates that when $\frac{1}{2}\hbar\omega_c$ exceeds about 30 R_y the transverse radius $a_\perp$ is nearly equal to the cyclotron radius l ([1]), so that the groundstate wave packet of the transverse motion in such a high field region is quite similar to that of the particular Landau level, $N = M = 0$. In this limit the effect of the Coulomb potential or any static potentials may be treated naturally as perturbations acting on free Landau states, which will cause shifts and splitting of the degenerate levels. These features are important in transport phenomena ([7]), and also in magnetooptics— for example, one observes a weak, low-frequency optical line corresponding to the *same* Landau subband. The perturbational treatment is, however, generally not simple, because the unperturbed states involve a continuous spectrum due to the electron plane-wave motion along the field **H**. Thus in general one must solve a one-dimensional Schrödinger equation of the form

$$\left(-\frac{\hbar^2}{2m^*} \frac{d^2}{dz^2} + V_{NM}(z) - \varepsilon_{NM} \right) F = 0, \qquad (15)$$

where the effective potential $V_{NM}(z)$ is given by

$$V_{NM}(z) = \int dx\, dy\, |\phi_{NM}(xy)|^2 \frac{-e^2}{\varkappa(x^2 + y^2 + z^2)^{1/2}} \qquad (16)$$

i.e., the Coulomb potential averaged over the free packet of the transverse motion (N, M). In this formulation the energy eigenvalue ε_{NM} represents the shift measured from the unperturbed energy, $(N + \frac{1}{2})\hbar\omega_c$, and therefore its lowest value gives the ionization energy directly. The Gaussian form of the trial solution for Eq. (15) as assumed in (7) does not give a very good result, nor does it give a systematic method of obtaining all other solutions. Attempts to improve this situation were made by Elliott and Loudon ([8]), Loudon ([9]), and Hasegawa and Howard ([10]). In particular, the latter authors obtained a logically consistent method of solving the necessary boundary value problem which secures the exact field dependence of the eigenvalue spectrum as the field **H** tends to infinity.

The effective potential defined in (16) is one-dimensional Coulomb-like, but distorted by a smearing effect of the wave packet circulating about the origin whose center is located at the distance $l(2\bar{N} + 1)^{1/2}$. Thus the magnitude of $V_{NM}(z)$ at the origin is finite, and for large values of z,

$|z| \gg l(2\bar{N}+1)^{1/2}$, it can be approximated by the pure Coulomb potential. These properties can be summarized by writing

$$V_{NM}(z) = -(e^2/\varkappa l)v_{NM}(z/l) \tag{17}$$

and showing the behavior of the function $v(t)$ as

$$v_{NM}(t) = \frac{1}{t} + O\left(\frac{1}{t^3}\right) \qquad\qquad t \to \infty \tag{18a}$$

$$\begin{aligned} v_{NM}(t) &= v_{NM} + O(t) & (M = 0) \\ &= O(t^2) & (M \neq 0) \end{aligned} \qquad t \to 0 \tag{18b}$$

The finite value of the potential at the origin in the three important cases $N = M = 0$, $N = M = 1$, and $N = 0$, $M = -1$ are given by

$$v_{0,0} = (\pi/2)^{1/2}, \qquad v_{1,1} = v_{0,-1} = \tfrac{1}{2}(\pi/2)^{1/2} \tag{19}$$

The principle of the quantum-defect theory ([11]) is well adapted to this type of potential problems, where the form of the eigenfunction for the outer region is known to be the Whittaker function.* The quantum condition for both the discrete and continuous spectra may be set up by a connection of such an outer solution to the inner solution, which should be solved by taking full account of the distortion of the potential from the pure Coulomb. The point of the connection must be suitably chosen. For the present problem it was proved that the following choice of the point of connection, z_0 on the z axis, suffices ([10]):

$$z_0 = Cl^\varrho = C(c\hbar/eH)^{\varrho/2}, \qquad 0 < \varrho < \tfrac{2}{3} \tag{20}$$

where C is an arbitrary constant. It can then be shown that both the outer and the inner solutions may have their *logarithmic derivatives* asymptotically represented as

$$\eta(z_0,\varepsilon) = \frac{\partial}{\partial z_0}\log F(z_0,\varepsilon) \tag{21a}$$

$$= -\frac{2}{a}\log z_0 + \text{const (independent of } z_0) \tag{21b}$$

the rest of the terms vanishing in powers of either z_0 or $lz_0^{-1} \propto l^{1-\varrho}$ as H tends to infinity. The required connection may be established by equating

* A mathematical formalism is given in ([12]).

the two expressions of $\eta(z_0, \varepsilon)$ for the outer and the inner solutions, the resulting quantum condition being expressed by an equality involving the two constant terms in Eq. (21b) which are *independent of z_0*. Therefore this procedure avoids any ambiguity which might arise if the result were not well-fixed because of the arbitrariness in the choice of z_0 in (20).

The quantum condition so derived for the bound solutions of Eq. (15) is explicitly written as

$$f(\varepsilon) \equiv (-\varepsilon)^{1/2} + \log(-2\varepsilon/\hbar\omega_c) + 2\psi\left[\frac{1}{(-\varepsilon)^{1/2}}\right] + 2\pi \cot \frac{\pi}{(-\varepsilon)^{1/2}} + \beta_{NM}$$
$$= 0 \tag{22}$$

where the unit of energy ε is the Rydberg, $R_y = (\hbar^2/2m^*)a^{*-2}$, and $(-\varepsilon)^{1/2}$ is defined in such a way that it is a positive quantity for $\varepsilon < 0$. The quantity $\psi(x)$ is the so-called digamma function, defined by $\psi(x) = (d/dx) \log \Gamma(x)$, and β_{NM} is a fraction of order 1 depending on (N, M). For example, in terms of Euler's constant γ,

$$\beta_{00} = \log 2 + 3\gamma = 2.4247\ldots$$
$$\beta_{11} = \beta_{00} + 1 = 3.4247\ldots \tag{23}$$
$$= \beta_{0-1}$$

There is one and only one bound solution whose ionization energy ε tends to infinity as $\hbar\omega_c \to \infty$, the behavior of which can be seen quantitatively from Eq. (22). This ionization energy versus $\frac{1}{2}\hbar\omega_c$ computed from (22) for the lowest Landau state $N = M = 0$ is plotted in Fig. 3 and compared with the previous calculations.* It is not strange that the present result gives about a 10% greater value of the ionization energy than that provided from the YKA-type variational method in the range $30R_y \lesssim \frac{1}{2}\hbar\omega_c \lesssim 100R_y$. We estimate an error involved in our method to be about 5% in the same range. The predicted H-dependence of the ionization energy is weaker than any power law, and is proportional to $(\log H)^2$ for $H \to \infty$. More recently Larsen [13] has made a revised variational calculation assuming an improved trial function of the form $\psi(\mathbf{r}) = N \exp[-(x^2+y^2/4a_\perp^2) - \varkappa(x^2+y^2+\alpha z^2)^{1/2}]$. His result is included in Fig. 3. This has narrowed the gap between the two types of approximation.

Our quantum condition (22) also gives the spectrum of the excited bound solutions of Eq. (15) which correspond to the even parity, i.e., those eigenfunctions of (15) which satisfy $F(-z) = F(z)$. There must be

---

* The results of R. W. Keyes and R. J. Elliott cited in Fig. 6 of (⁸).

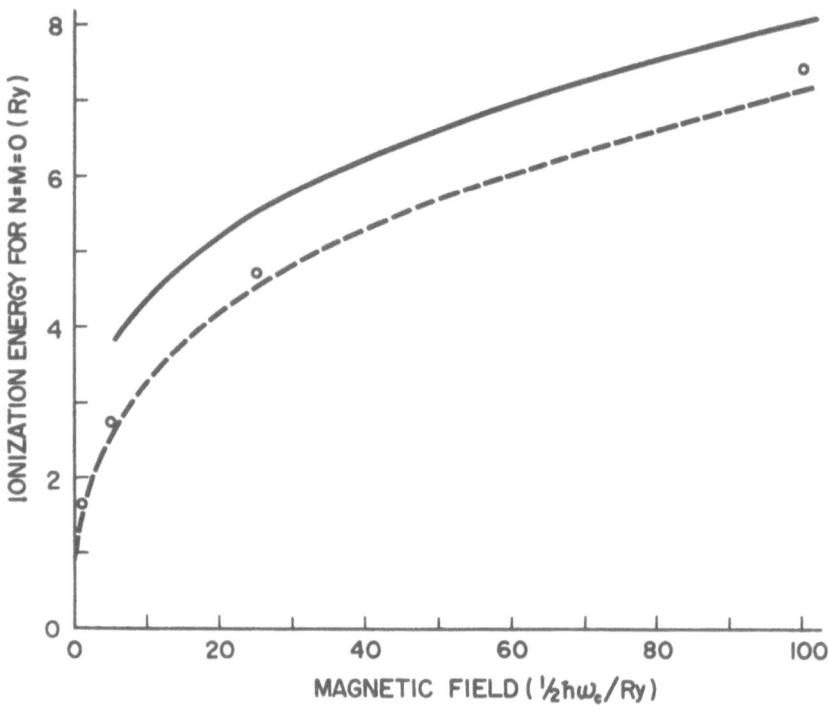

Fig. 3. A revised theoretical ionization energy for the ground-state hydrogen atom from Eq. (22), which is exact for $\frac{1}{2}\hbar\omega_c \gtrsim 30\,R_y$. The units are the same as in Fig. 1. The dotted line corresponds to the previous result as cited in ([8]). Circles indicate a revised variational result given by Larsen ([13]).

another sequence of solutions corresponding to the odd parity, $F(-z) = -F(z)$. It can be shown that for $H \to \infty$ both these sequences converge to a spectrum of the inverse-square law, i.e., a "one-dimensional" Rydberg series, but each in different manner. It is convenient to express the energy terms in the way suggested from the quantum defect theory:

$$\varepsilon_{\nu\pm} = -\frac{1}{(\nu + \delta_\nu{}^\pm)^2}, \qquad \nu = 1, 2, \ldots \tag{24}$$

where the $+$ and $-$ signs denote the even- and odd-parity solutions, respectively. Then the quantum defects of the even-parity spectrum tend, as $H \to \infty$, to zero like $(\log H + \text{const})^{-1}$, while those of the odd-parity spectrum like AH^{-1} ($A < 0$), thus they tend to zero from opposite directions, with the latter faster than the former. ([10]) The structure is shown schematically in Fig. 4. (See Note 2 added in proof.)

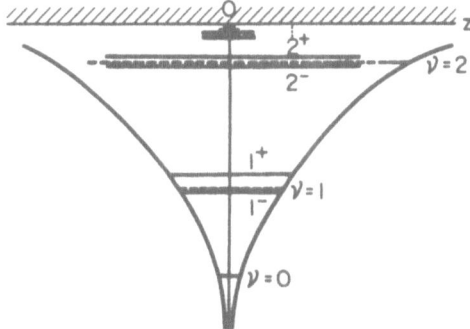

Fig. 4. A schematic diagram of the energy spectrum of the one-dimensional Schrödinger equation (15) with a distorted Coulomb potential. Here $\{0, 1^+, 2^+, \ldots\}$ represents the even-parity series and $\{1^-, 2^-, \ldots\}$ the odd-parity one.

We turn to the continuous spectrum of the solutions of Eq. (15) corresponding to the positive energy, $\varepsilon > 0$. The quantum condition for the continuous spectrum is the condition which determines the phase shift of the scattered wave. Because of the long-range character of the potential function (17) the asymptotic solution for $|z| \to \infty$ is not a simple plane wave but is modified as

$$\exp \pm i[kz + (1/ka) \log 2kz] \tag{25}$$

where k is related to the energy ε through

$$\varepsilon = (\hbar^2/2m^*)k^2 \tag{26}$$

Such an asymptotic solution may be provided again in terms of the Whittaker function (with an imaginary parameter). Let an incident wave on the positive side $z > 0$ be given by

$$F(z, \varepsilon) = \frac{e^{-\pi/2\sqrt{\varepsilon}}}{\sqrt{2\pi}}\, W_{-i/\sqrt{\varepsilon},\,1/2}(i2\sqrt{\varepsilon}\, z)$$

$$\xrightarrow{z \to \infty} \frac{1}{\sqrt{2\pi}} \exp\left(-i\sqrt{\varepsilon}\, z - \frac{i}{\sqrt{\varepsilon}} \log \sqrt{\varepsilon}\, z\right) \tag{27}$$

Then the phase shift of the scattered wave on the same side (the reflected wave) may be defined by

$$F(z, \varepsilon) = \frac{e^{-\pi/2\sqrt{\varepsilon}}}{\sqrt{2\pi}} \, W_{i/\sqrt{\varepsilon},1/2}(-i2\sqrt{\varepsilon}\,z)e^{i\delta}$$

$$\xrightarrow{z\to\infty} \frac{1}{\sqrt{2\pi}} \exp\left(i\sqrt{\varepsilon}\,z + \frac{1}{\sqrt{\varepsilon}} \log\sqrt{\varepsilon}\,z + i\delta\right) \qquad (28)$$

The quantum condition to determine the phase shift $e^{i\delta}$ is closely related to the corresponding bound-state condition (22), and is written as

$$e^{i\delta} = \frac{f^*(\varepsilon)}{f(\varepsilon)} \frac{\Gamma(-i/\sqrt{\varepsilon})}{\Gamma(i/\sqrt{\varepsilon})} \qquad (29)$$

where $f(\varepsilon)$ is given by Eq. (22), in which $(-\varepsilon)^{1/2}$ for a positive value of ε is defined to take the branch $-i\sqrt{\varepsilon}$ ($\sqrt{\varepsilon} > 0$). The solution to be obtained from this phase shifts corresponds to an even-parity solution, which is only a special solution of the continuous spectrum: The most general form of the solutions of the scattering problem must be a linear combination of both parity eigenfunctions; we will not go into the detail here.

The importance of the phase shift obtained in (29) is that it modifies the amplitude of the solution $F(z, \varepsilon)$ near the Coulomb center $z = 0$ from the simple plane wave, especially for slow incident electrons ($ka \ll 1$). Thus

$$| F(0, \varepsilon) |^2 = A\sqrt{\varepsilon}, \qquad \varepsilon \to 0 \qquad (30)$$

the constant A vanishing as the magnetic field H tends to infinity [10]. This is essentially due to the long-range nature of the Coulomb potential, and is caused by the characteristic factor [14]

$$\frac{\sqrt{\varepsilon}}{2\pi} (1 - e^{-2\pi/\sqrt{\varepsilon}}) \; *$$

From this fact one is led to the conclusion that the optical density at the continuum edge of the Landau subband has no significant peak in spite of the $1/\sqrt{\varepsilon}$ singularity of the density of states of the one-dimensional continuous spectrum [8,10]. Note that the density-of-state function itself

---

* In the present problem of one-dimensional modified Coulomb scattering this factor appears in the numerator of the scattered amplitude rather than in the denominator as is the case in the pure three-dimensional Coulomb scattering [14].

for such a single-center problem is the same as that of the free-electron spectrum, given simply from (26) by

$$\frac{dk}{d\varepsilon} = \left(\frac{m^*}{2\hbar^2} \right)^{1/2} \frac{1}{\sqrt{\varepsilon}} \tag{31}$$

CONTINUATION BETWEEN THE RYDBERG AND LANDAU SPECTRA

Donor States in InSb

The discussion so far furnishes in principle the energy spectra of a hydrogen atom in a field of extremely high intensity, $\frac{1}{2}\hbar\omega_c \gg R_y$. For experiments this is of special interest for semiconductors with a small effective mass and a large dielectric constant, in particular, for InSb. One of the interesting points in optical studies of donors in InSb is impurity-induced transitions between those levels which belong to the same Landau subband $N = 0$. This was first considered by Wallis and Bowlden [15] and has recently been studied experimentally by Kaplan [16].

From our standpoint let us ask what the limiting absorption spectrum due to dipole excitations from the ground state is. Here the ground state means the state whose ionization energy is calculated in Fig. 3 and is denoted by $(N = M = 0, v = 0)$. We may then ask further about the continuation of the spectrum as a function of the field H to its lower values.

We answer the first question by dividing the whole spectra into three groups: (A) the intensity tends to a finite nonvanishing value, (B) the intensity vanishes like $(\log H + \text{const})^{-p}$, $p > 0$, and (C) the intensity vanishes like $H^{-r}(\log H + \text{const})^{p'}$, $r > 1$. Thus in the actual limit $H \to \infty$ only the line that belongs to (A) survives. This is the transition to $N = M = 1$, $v = 0$, the lowest bound state of Eq. (15) associated with the *second* Landau level, and is allowed for the electric vector $\mathbf{E}$ left-circularly polarized perpendicular to $\mathbf{H}$. All other lines or absorption bands are considered to be satellites of this main line. In particular, weakly allowed lines which belong to (B) are the transitions from $N = M = 0$, $v = 0$ to $N = M = 0$, v^- that are allowed for $\mathbf{E} \parallel \mathbf{H}$, and those that belong to (C) the transitions to $N = 0$, $M = -1$, v^+ for $\mathbf{E} \perp \mathbf{H}$ right-circularly polarized, their absorption frequencies being considerably lower than $\hbar\omega_c$, the fundamental cyclotron frequency.

Using a photoconductivity detection technique Kaplan observed in addition to the main peak (A) some transitions corresponding to (B) and

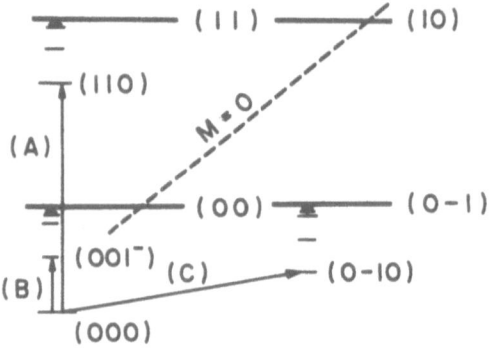

Fig. 5. The three most important dipole excitations from the ground-state hydrogen atom in the high field limit are indicated by the arrows. These have been identified in Kaplan's experiment in n-InSb ([17]).

(C) in a submillimeter region in InSb samples with donor concentrations of about 6×10^{13} cm^{-3}. The significant peaks that Kaplan assigned are: (A), transition from (000) to (110); (B), from (000) to (001$^-$); and (C), from (000) to (0 − 10); these are indicated in Fig. 5. Note that the excited spectrum (11$\nu^\pm$) is exactly identical with (0 − 1$\nu^\pm$) due to a dual symmetry between the two oscillators, N and $\bar{N}$, and hence to the identiy $V_{NM}(z) = V_{N-M,-M}(z)$ ([10]). Note also that the oscillator strengths of the two transitions (A) and (C) may be written in terms of the matrix elements of the high-frequency coordinate $\xi + i\eta$ and the low-frequency coordinate $X - iY$, respectively ([10]), indicating that *the latter transition is the one in which the electron changes its center coordinate*. This is particularly interesting in connection with transport theory at the high-field quantum limit ([7]).

We now come to our chief question: What is the *original* spectrum of the hydrogen atom which tends to these three peaks when the field H varies from zero to infinity? A variational calculation of YKA-type suggested that the three most elementary excitations, $1s \rightarrow 2p$ ($m = \pm 1, 0$), correspond to these peaks ([15]). The question may be generalized to ask about the continuation of the whole energy spectrum of a hydrogen atom from the one limit $H = 0$ to the other $H = \infty$.

This is a difficult question, and in fact has not been answered convincingly. However, a suggestive prediction has been given by Kleiner ([17]) and by Elliott and Loudon ([8]) based on an argument of the "nodal-surface conservation." Let (n, l, m) be the set of quantum numbers for a bound eigenstate in the Rydberg series. Then the prediction says that this eigenfunction varies into a strong-field one $(N, M, \nu^{\pm})$ without changing the *number* of nodal surfaces of the wave function, and consequently the following equalities hold:

$$m = M, \qquad n - l - 1 = N - \tfrac{1}{2}(M + \mid M \mid),$$

$$l - \mid m \mid = 2\nu \qquad (\nu = \nu^{+} \quad \text{even parity}) \qquad (32)$$

$$= 2\nu - 1 \qquad (\nu = \nu^{-} \quad \text{odd parity})$$

Here the first equality just represents the angular momentum conservation, the second the equality between the number of *nodal spheres* $(H \to 0)$ and of *nodal cylinders* $(H \to \infty)$, and the third the equality between the number of *nodal cones with the z axis* $(H \to 0)$ and of the *nodal planes perpendicular to the z axis* $(H \to \infty)$, on the left and the right sides, respectively. If this is applied, it is easy to see that the three $2p$ states $2m = 1$, -1, and 0 continues to the three high-field ones (110), $(0-10)$ and (001^{-}), respectively, in agreement with the variational result. The question of how the energy of each individual level changes quantitatively with the field H has not been settled yet except for such low-lying states.

Another interesting point in Kaplan's results is the absence a of photoionization peak at the continuum edge, which gives a direct verification of the theory of the optical density mentioned in the last section. It is noted that optical studies of the magnetic-field-induced donor levels in InSb as explored by Kaplan involve many subjects needing further consideration which should be of importance for the future development of far-infrared and submillimeter optics.

Exciton Spectra in Layer-Type Semiconductors

The continuation of the Rydberg spectra of a hydrogen atom to the spectra of the Landau ladders especially in the transition region $\tfrac{1}{2}\hbar\omega_c \lesssim R_y$ has been an important subject in studies of excitons since the Rochester Conference in 1958 [18].* In these years a number of magnetooptical spectra have accumulated on numerous semiconductors, and it is quite common now to work with oscillatory peaks in an interband continuous background on which some exciton structures usually appear to superpose. An obvious question in analysing such spectra is how to denote each peak in referrence to excitons or Landau subbands; these two concepts are inevitably redundant.

* Also see the papers and discussions of the Session on Magnetooptical Effects in [19].

A good example of the question of continuation which has been examined explicitly is magnetooptical spectra (both the absorption and Faraday rotation) in a layer-type semiconductor, GaSe ([20],[21]). In this compound one observes a well-resolved first exciton peak about 0.01 eV below the edge of the fundamental absorption continuum in the absence of a magnetic field. When the field **H** is applied along the c axis several "Landau-like" oscillatory peaks appear above the interband threshold, while the lowest exciton peak remains almost unchanged up to $H = 200$ kG, for which $\frac{1}{2}\hbar\omega_c$ is estimated to be only about one-half of the effective Rydberg, R_y (the binding energy of the exciton).

One should not hasten to assign these quasi-Landau peaks by labeling them from the lower energy one simply as $N = 0, 1, 2, \ldots$, since in such a "low"-field region the approximation in the last section just does not apply. Instead, for this special example of the layer structure Shinada and Sugano ([22]) proposed another approximation, i.e., a "two-dimensional" approach, where one tries to solve an eigenvalue problem of the Hamiltonian similar to (6) without restricting the discussion to the high-field limit:

$$\left\{ \frac{1}{2m_\perp^*} (p_x^2 + p_y^2) + \frac{eH\alpha}{2m_\perp^* c} (xp_y - yp_x) + \frac{e^2 H^2}{8m_\perp^* c^2} (x^2 + y^2) \right.$$

$$\left. - \frac{e^2}{\varkappa(x^2 + y^2)^{1/2}} \right\}\psi = E\psi \tag{33}$$

Here for the exciton purpose m^* is the reduced mass of an electron–hole pair, α a certain constant, and the equation describes the relative motion of the pair. The approximation is expected to be best, of course, for vanishingly small values of the mass ratio $m_\perp^*/m_\parallel^*$. This may not actually be the case, and then one must make a suitable correction due to the presence of the freedom of the third dimension parallel to **H**. Nevertheless, the solutions of Eq. (33) give a useful starting point.

The program for solving the above eigenvalue equation has been carried out in detail by Akimoto and Hasegawa using the WKB method ([23]) and by Shinada and Tanaka using a method of numerical integration. ([24]) It has turned out that the WKB or quasiclassical approach, which was also proposed by Zhilich and Monozon ([25]) provides an excellent tool in getting an understanding of the continuation problem between the two opposite limits, the Rydberg and the Landau limits, and, furthermore, provides good approximations both for the energies and eigenfunctions below the transition range of the field, $\frac{1}{2}\hbar\omega_c \lesssim R_y$. In this approach the energy term may be obtained from the well-known formula of the semiclassical quantization:

$$\int_{r_1}^{r_2} \left[\frac{2m^*}{\hbar^2} \left(E - \frac{\hbar \omega_c}{2} M - \frac{m^* \omega_c^2}{8} r^2 - \frac{\hbar^2}{2m^*} \frac{M^2}{r^2} + \frac{e^2}{\varkappa r} \right) \right]^{1/2} dr$$

$$= (\bar{n} + \tfrac{1}{2})\pi, \qquad M = 0, 1, \ldots, \bar{n} = 0, 1, 2, \ldots \qquad (34)$$

The interval of the integration is indicated in Fig. 6, where the shape of the effective potential that binds our two-dimensional particle is also shown. (For simplicity, the formula is written not for the exciton but for a particle in the fixed Coulomb field.)

It can be proved easily that in the two limiting cases (1) $\omega_c = 0$, and (2) $\omega_c \to \infty$ (equivalent to dropping the Coulomb potential in (34) to zeroth approximation) the semiclassical quantization rule yields the exact results:

$$E_{\bar{n}M} = -\frac{R_y}{(\bar{n} + |M| + \tfrac{1}{2})^2}, \qquad \omega_c = 0 \qquad \text{(Rydberg)} \qquad (35)$$

$$E_{\bar{n}M} = (\bar{n} + \tfrac{1}{2}(M + |M| + 1))\hbar \omega_c, \qquad \omega_c \to \infty \qquad \text{(Landau)} \qquad (36)$$

Note that the half-odd integer values appear on the right-hand side of the quantum condition (34), which is proper for the two-dimensional particle in the Coulomb field (plus any additional field with a normal behavior at the origin) ([23]). It is therefore reasonable to expect that the quantum number $\bar{n}$ introduced in (34) may be an adequate one for the assigning purpose. This appears to be correct, and it can in fact be proved that for a fixed value

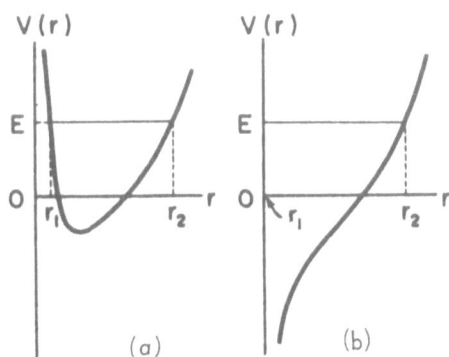

Fig. 6. The potential function $V(r)$ consisting of the Coulomb, $-e^2/\varkappa r$, the magnetic-field, $(e^2 H^2/8m^* c^2)r^2$, and the centrifugal $(\hbar^2/2m^*)$ (M^2/r^2) forces acting on a two-dimensional particle. (a) $M \neq 0$ (b) $M = 0$.

of M the energy $E_{\bar{n}}$ at the Rydberg limit $\omega_c = 0$ varies, *without intersecting other values for different $\bar{n}$'s*, to change into the energy of the Landau limit with the same $\bar{n}$. As can be seen from the general WKB procedures, this quantum number $\bar{n}$ is indeed the number of nodal surfaces (nodal cyclinders for this case) of the eigenfunction and thus Kleiner and Elliott–Loudon prediction is explicitly verified in this special case. In terms of the usual principal quantum number n for the Rydberg case and of the Landau quantum number N, this is expressed as

$$\bar{n} = n - |M| - 1 = N - \tfrac{1}{2}(M + |M|) \qquad (37)$$

in agreement with Eqs. (32), where the third quantum number ν is set equal to zero.

Akimoto and Hasegawa [23] computed the energy eigenvalues presrcibed by (34) for the special case $M = 0$, which corresponds to the "allowed" exciton and is actually the case in GaSe [22]. An intermediate low-lying excited spectrum, neither exciton-like nor pure Landau, has indeed come out near the transition range, as shown in Fig. 7. They also presented a method of correcting for the third dimension, using a perturbation technique for considering the effective potential $V(z)$ similar to (16). Here the concept of the effective potential is equivalent to the usual adiabatic approximation by which the motion along the z direction can be separated, and is not necessarily restricted to the high-field limit. The effect of this potential is to increase the energy of the purely two-dimensional spectrum calculated in Fig. 7, and also to give rise to a sequence of excited states denoted by $\nu^\pm$. The perturbational treatment involves an expansion of $V(z)$ near the origin and thus merely gives an estimate of the zero-point energy of the lowest state $\nu = 0$. To determine a reliable theoretical spectrum laborious, numerical computations are necessary to determine the correct behavior of $V(z)$, and these are now under way, since magnetooptical experiments on GaSe have recently provided excellent fine structures [26].

Here we discuss a question of practical importance concerning the continuation problem. Let us ask whether the observed first quasi-Landau peak continues eventually to the high-field state in the $(N, M, \nu^\pm)$ notation (100) or to the state denoted by (001$^+$).* (That is to say, the lowest level $\nu = 0$ associated with the *second* Landau subband, or the second excited level $\nu^+ = 1^+$ associated with the *first* Landau subband.) The answer depends partly on the experimental accuracy and partly on the validity of the pre-

---

* This assignment has been given by Halpern [21].

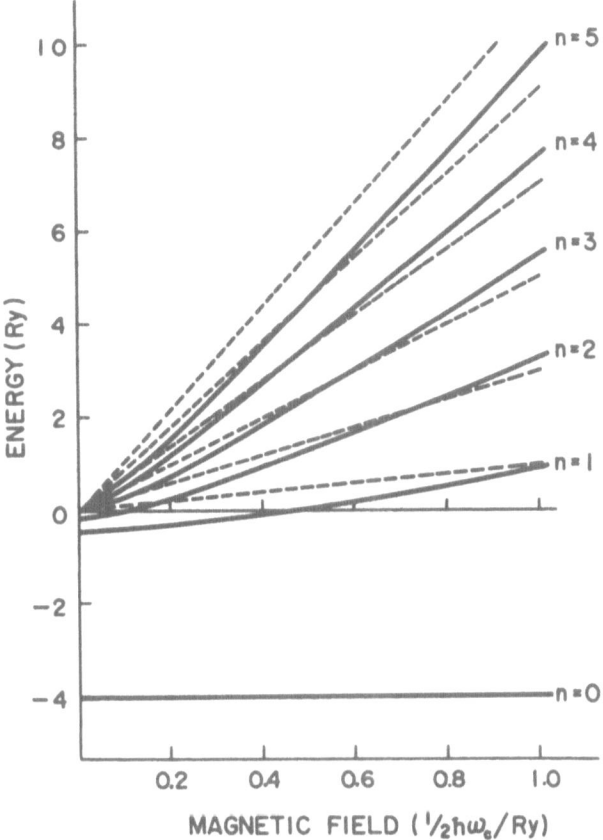

Fig. 7. The energy versus $\frac{1}{2}\hbar\omega_c$ for several low-lying levels calculated from the semiclassical quantum condition (34) with
$M = 0$ [23].

diction (32). Assume that Eqs. (32) are valid. Then we expect the two
continuations:

$$
\begin{array}{cccccc}
n & l & m & N & M & \nu^{\pm} \\
(2 & 0 & 0) & \to & (1 & 0 & 0) \\
(3 & 2 & 0) & \to & (0 & 0 & 1^{+})
\end{array}
\tag{38}
$$

But the experimental data indicates, within a limited accuracy, that the peak
originates from the "$2s$" ($\bar{n} = 1$, $l = 0$) state at $H = 0$. Therefore we have
a strong reason to favor the former assignment (100) as correct.* Questions
remain, however, about the instrumental resolution and the validity of

* See Note 3 added in proof.

(38): These continuity relations imply that intersections of some energy levels are inevitable when the third dimension is included. At the same time the Landau-limit spectra contain continuous subbands due to this freedom, whose formation should be closely related to the occurrence of the intersections.

While an ultimate understanding of the continuation problem remains open at this point, much has been learned about the nature of the quasi-Landau peaks by means of the two-dimensional model. One thing for which a clearcut understanding has been achieved is the intermediate nature of these peaks. The magnetic field dependence of the peak positions for $\bar{n} \geq 1$ is quite "Landau"-like: They rise rapidly, intersecting the $E = 0$ line (the continuum edge at $H = 0$) at values of H much below the critical value $\frac{1}{2}\hbar\omega_c = R_y$. The result of the WKB calculation gives for this case

$$\tfrac{1}{2}\hbar\omega_c(E = 0) = 1.56 R_y/(\bar{n} + \tfrac{1}{2})^3 \tag{39}$$

which means that the first quasi-Landau peak ($\bar{n} = 1$) intersects the $E = 0$ line at $0.46\,R_y$, while the lowest exciton peak intersects at $12.5\,R_y$. On the other hand, the theoretically predicted intensities of these peaks show significant deviations from the pure Landau limit for which all the allowed peaks are to have the same intensity proportional to H.* For convenience we summarize the theoretical intensities of the allowed case ($M = 0$), the square of the amplitude of the wave function at the origin in the relative coordinate space, $\psi^2(0)$, for both limits.

Rydberg Limit

$$\frac{2}{(\bar{n} + \tfrac{1}{2})^3} \frac{1}{a^{*2}} = \frac{2m^*}{\hbar^2} \frac{dE}{d\bar{n}} \tag{40a}$$

Landau Limit

$$\frac{1}{l^2} = \frac{m^*}{\hbar^2} \frac{dE}{d\bar{n}} \tag{40b}$$

Interpolations have also been carried out by the WKB method and by numerical integration ([23,24]). Shinada and Tanaka ([24]) have demonstrated from their numerical solutions how each individual eigenfunction is squeezed toward the origin by an increase of the field H. In the region $\frac{1}{2}\hbar\omega_c \lesssim R_y$ the square of the amplitude $\psi^2(0)$ exhibits a strong n-dependence for several low-lying numbers, and thus the bound nature of these eigenfunctions is

---

* See the intensity formula for interband magnetooptical effects cited in, for example, ([27]). For the two-dimensional case the state density due to the free motion along the z direction is replaced by δ-functions, giving a discrete spectrum of equal weights.

shown to persist. Such behavior has actually been observed experimentally, and has been analyzed thanks to Faraday rotation measurements ([28]) that are more accurate than absorption for the intensity purpose (see Fig. 8).

Let us recall the spirit of the orthogonalized-plane-wave theory [e.g., ([29])] in order to get further insight on the above results. We assume that the ground-state wave function of Eq. (34) is known. Let it be ψ_g, and let the energy be E_g. To describe excited states we prepare the set of the Landau base $\{\phi_N, N = 0, 1, 2, \ldots\}$. Since the latter is a complete orthogonal set, it is sufficient to write the secular equation taking into account the Coulomb potential and solve it, but such a complication is not desirable. To avoid difficulty we make use of our knowledge about the ground state and set upt the equation on the "orthogonalized Landau base," $\{\tilde{\phi}_N, N = 0, 1, 2, \ldots\}$, where

$$\tilde{\phi}_N(\mathbf{r}) = \phi_N(\mathbf{r}) - \psi_g(\mathbf{r})S_N, \qquad S_N = (\psi_g, \phi_N) \tag{41}$$

A variational principle may be used to derive the secular equation for the energy, thereby restricting the basis function only to a few $\tilde{\phi}$'s. For example, from a single $\tilde{\phi}_N$ one obtains

$$E_N = E_N{}^0 + \frac{S_N{}^2}{1 - S_N{}^2}(E_N - E_g) \tag{42}$$

where

$$E_N{}^0 = (N + \tfrac{1}{2})\hbar\omega_c - \left(N \left| \frac{e^2}{\varkappa r} \right| N\right) \tag{43}$$

is the diagonal energy for the pure Landau state ϕ_N. It is greatly an *overestimate* of the Coulomb binding effect in the low-field case ($\tfrac{1}{2}\hbar\omega_c \lesssim R_y$ for the energy of the quasi-Landau levels), and is partly cancelled by the second term in (42), an increase of the energy due to the orthogonalization correction. At the same time the expression (41) indicates that the amplitude of $\tilde{\phi}_N$ at the origin must be modified from $\phi_N(0)$ to be reduced. This is what one ascribes, at least qualitatively, to the nature of the quasi-Landau levels discussed above, conforming to the general result of the OPW or pseudopotential theory.

EXCITON BAND IN A MAGNETIC FIELD

As a final topic we briefly discuss how the external magnetic field influences an exciton band.* Here the exciton band means a continuous spectrum of the excited states of an intrinsic semiconductor which corre-

---

* See Note 4 added in proof.

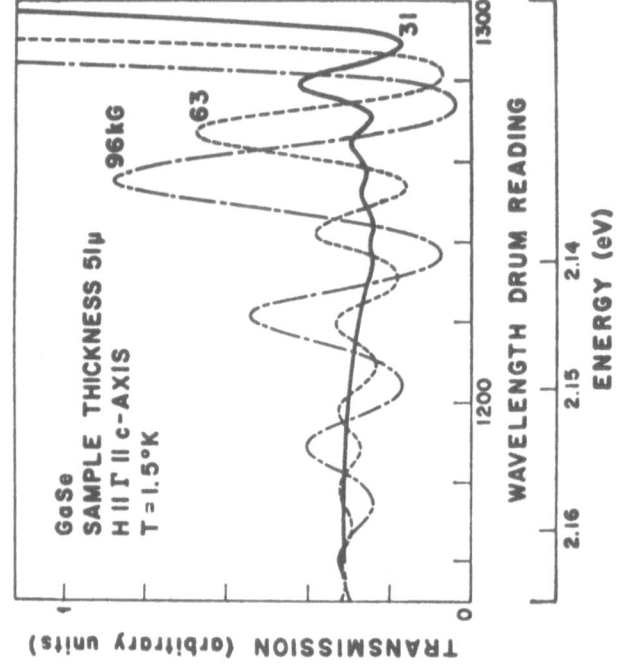

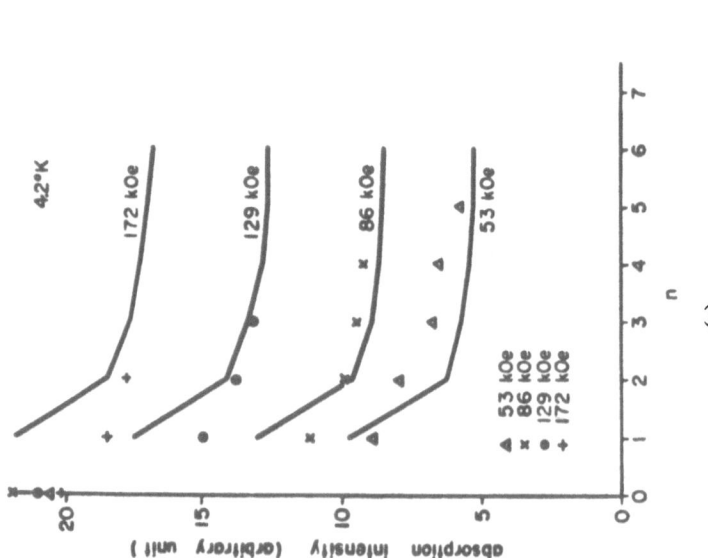

Fig. 8. (a) Comparison between the relative intensities of the low-lying quasi-Landau peaks measured in GaSe by Faraday rotation and those calculated by the WKB method [20]. (b) Experimentally observed oscillatory absorption peaks in GaSe [from Halpern] [21].

sponds to the translational motion of one electron–hole pair, and may be denoted by $\mathscr{E}(\mathbf{P})$, where $\mathbf{P}$ is the momentum of the exciton. It is well known that for an electron and a hole moving in simple parabolic bands interacting with each other through a static potential depending only on the relative coordinate of the two particles, the motion can be separated into the relative or internal motion and the motion of their center of gravity [see, e.g., ([30])]. The energy of this latter is expressed for such a simple case to be (M is the mass of the center of gravity, $m_e + m_h$)

$$\mathscr{E}(\mathbf{P}) = (1/2M)\mathbf{P}^2 \tag{44}$$

i.e., the exciton band; this expression is the same form for all the energy levels of the internal motion in this case. For a more complicated band structure the exciton band can still be constructed from the Wannier point of view, as discussed by Dresselhaus ([31]), although the exciton band may then be different for different bound states of the internal motion.

When a static magnetic field is applied a coupling arises between the two dynamical systems, i.e., between the internal motion and the translational motion of the exciton. The first experiment demonstrating this coupling was the magneto-Stark effect experiment of Thomas and Hopfield on CdS ([32]) which showed that for the exciton with a finite momentum the coupling is effectively such as to act on the bound exciton state as an electric field of

$$\mathbf{E} = (1/Mc)\mathbf{P} \times \mathbf{H} \tag{45}$$

We point out that in addition to the Stark shift of each hydrogenic exciton level this coupling gives rise to a change of the exciton band associated with that level. To lowest order the effect is a change of the exciton band mass in the direction transverse to the magnetic field, and can be written as

$$\mathscr{E}(\mathbf{P}, \mathbf{H}) = \frac{1}{2M}\left(1 - \frac{\alpha_i H^2}{Mc^2}\right)(P_x{}^2 + P_y{}^2) + \frac{P_z{}^2}{2M} \tag{46}$$

where α_i is a static polarizability associated with ith exciton level. For example, the $1s$ hydrogenic exciton has a polarizability given by

$$\alpha_{1s} = \tfrac{9}{2}a^{*3} \tag{47}$$

where a^* is the effective Bohr radius defined in (3). If α_i is written in terms of a dimensionless constant χ_i as

$$\alpha_i = \chi_i a^{*3} \tag{48}$$

The change of the inverse mass of the exciton band is given by

$$\Delta\left(\frac{1}{M}\right) = -\chi_i \frac{m_e m_h}{(m_e + m_h)^3 \varkappa} \left(\frac{a^*}{l}\right)^4 \tag{49}$$

For a hydrogen atom χ_i is a positive and rapidly increasing function of the principal quantum number n (on the order of n^6) [33]. One sees that the static magnetic field is quite effective in reducing the curvature of the exciton band, so that when the cyclotron radius l is comparable to the radius of the exciton orbit na^* of the principal quantum number n the picture of the exciton band in the zero-field sense would break down. One is then concerned with the exciton band from the high-field point of view.

To be precise and more general, we consider an effective-mass Hamiltonian for an electron–hole pair of the form

$$\mathcal{H} = E_c\left(\mathbf{p}_e + \frac{e}{2c}\mathbf{H} \times \mathbf{r}_e\right) - E_v\left(\mathbf{p}_h - \frac{e}{2c}\mathbf{H} \times \mathbf{r}_h\right) + V(\mathbf{r}_e - \mathbf{r}_h) \tag{50}$$

and summarize the general nature of the exciton band in the magnetic field **H** as follows:

1. A vector $\bar{\mathbf{P}}$ defined by

$$\bar{\mathbf{P}} = \mathbf{p}_e + \mathbf{p}_h - \frac{e}{2c}\mathbf{H} \times (\mathbf{r}_e - \mathbf{r}_h) \tag{51}$$

is the constant of motion of the Hamiltonian (50), and each energy eigenvalue of it is a function of $\bar{\mathbf{P}}$, which yields a dispersion of the exciton in a magnetic field. The transverse components of $\bar{\mathbf{P}}$ can be related to the cyclotron center coordinates in Eq. (10) of the electron and the hole as

$$\Delta X = X_e - X_h = -\frac{c}{eH}\bar{P}_y, \qquad \Delta Y = Y_e - Y_h = \frac{c}{eH}\bar{P}_x \tag{52}$$

Unlike Eq. (11), these two components satisfy $[\Delta X, \Delta Y] = 0$, and therefore may be two independent constants.

2. For vanishing magnetic field (the Rydberg limit) $\bar{\mathbf{P}}$ reduces to the exciton momentum **P**, and the dispersion is identical with the *ordinary* exciton band. The effect of the magnetic field is in general to reduce the curvature of the exciton band perpendicular to the field, and in the high-field limit (the Landau limit) the curvature vanishes, so that each exciton level is degenerate with respect to ΔX and ΔY.

These considerations are important in a line-shape study of the quasi-Landau peaks, since the exciton absorption line shape was shown to be sensitive to the mass of the exciton band ([34]).

ACKNOWLEDGMENTS

Part of this chapter was written when I was staying at the Institute of Solid State Physics, University of Tokyo in the summer of 1967. I wish to thank Professors Toyozawa and Sugano for making this stay possible. Thanks are also due to Drs. Shinada and Tanaka and Mr. Akimoto for comments on several points in the manuscript.

REFERENCES

1. Y. Yafet, R. W. Keyes, and E. N. Adams, *J. Phys. Chem. Solids* **1**: 137 (1956).
2. R. W. Keyes and R. J. Sladek, *J. Phys. Chem. Solids* **1**: 143 (1956); R. J. Sladek, *J. Phys. Chem. Solids* **5**: 157 (1958).
3. H. Miyazawa and H. Ikoma, *J. Phys. Soc. Japan* **23**: 290 (1967).
4. L. I. Schiff and H. Snyder, *Phys. Rev.* **55**: 59 (1939).
5. Y. Toyozawa, M. Ionoue, T. Inui, M. Okazaki, and E. Hanamura, *J. Phys. Soc. Japan* **22**: 1337 (1967).
6. M. H. Johnson and B. A. Lippmann, *Phys. Rev.* **76**: 828 (1949).
7. R. Kubo, H. Hasegawa, and N. Hashitsume, *J. Phys. Soc. Japan* **14**: 56 (1959); R. Kubo, S. Miyake, and N. Hashitsume, in: *Solid State Physics, Vol.* 17 (F. Seitz and D. Turnbull, eds.), Academic Press, New York, 1966, p. 269.
8. R. J. Elliott and R. Loudon, *J. Phys. Chem. Solids* **15**: 196 (1960).
9. R. Loudon, *Am. J. Phys.* **27**: 649 (1959).
10. H. Hasegawa and R. E. Howard, *J. Phys. Chem. Solids* **21**: 179 (1961).
11. F. S. Ham, *Solid State Physics*, Vol. 1 (F. Seitz and D. Turnbull, eds.), Academic Press, New York, 1955, p. 127.
12. D. R. Hartree, *Proc. Cambridge Phil. Soc.* **24**: 426 (1928).
13. D. M. Larsen, private communication.
14. L. D. Landau and E. M. Lifshitz, *Quantum Mechanics*, Pergamon Press, London, 1958, p. 522.
15. R. F. Wallis and H. J. Bowlden, *J. Phys. Chem. Solids* **7**: 78 (1958).
16. R. Kaplan, *J. Phys. Soc. Japan Suppl.* **21**: 249 (1966).
17. W. H. Kleiner, Lincoln Lab. Progr. Rep., Feb. 1958.
18. E. F. Gross, *J. Phys. Chem. Solids* **8**: 172 (1959).
19. "Proceedings of the International Conference on Semiconductors", *J. Phys. Chem. Solids* **8**: 305 (1959).
20. S. Sugano *et al.*, *J. Phys. Soc. Japan Suppl.* **21**: 174 (1966).
21. J. Halpern, *J. Phys. Soc. Japan Suppl.* **21**: 180 (1966).
22. M. Shinada and S. Sugano, *J. Phys. Soc. Japan* **21**: 1936 (1966).
23. O. Akimoto and H. Hasegawa, *J. Phys. Soc. Japan* **22**: 181 (1967).

24. M. Shinada and K. Tanaka, unpublished.
25. A. G. Zhilich and B. S. Monozon, *Soviet Phys.—Solid State* (*English Transl.*) **8**: 2846 (1967).
26. E. Mooser, private communication.
27. R. J. Elliott nd R. Loudon, *J. Phys. Chem. Solids* **8**: 382 (1959).
28. Y. Nishina, S. Kurita, and S. Sugano, *J. Phys. Soc. Japan* **21**: 1609 (1966).
29. P. W. Anderson, *Concepts in Solids*, W. A. Benjamin, New York, 1963, Chapter 2.
30. R. S. Knox, *Solid State Physics*, *Vol. 5* (F. Seitz and D. Turnbull, eds.), Academic Press, New York, 1963, pp. 37, 79.
31. G. Dresselhaus, *Phys. Chem. Solids* **1**: 14 (1956).
32. D. G. Thomas and J. J. Hopfield, *Phys. Rev.* **124**: 657 (1961).
33. H. A. Bethe and E. E. Salpeter, *Quantum Mechanics of One and Two-Electron Atoms*, Academic Press, New York, 1957, p. 228.
34. Y. Toyozawa, *Prog. Theor. Phys.* **20**: 53 (1958).

NOTES ADDED IN PROOF

The following recent developments are pertinent to this contribution:

1. A theory of a Landau subband edge perturbed by the presence of many impurities has been worked out: it includes the possibility of an impurity-band formation. (To be published in *J. Phys. Soc. Japan*, 1969.)

2. In the text we have stated that the odd-parity spectrum tends to the one-dimensional Rydberg series from the *lower* energy side with the quantum defects like $\delta_\nu^- = AH^{-1}$ ($A < 0$) (p. 254). This fact was not well accounted for in (^{10}), and Dr. Kaplan has inquired into this point (private communication). According to the author's check this can be shown, but in a very complicated way. Moreover, an inconvincible point has been found, so that it is difficult to give the constant A explicitly. In the lowest-order approximation one finds only $\delta_\nu^- = 0$. For most practical purposes this result may be enough.

3. A precision measurement of the exciton-magnetooptical spectra in GaSe has been reported by Brevner, Halpern, and Mooser. (Presented at the International Symposium on Anisotropy in Layer Structures, Taormina, Italy, in 1968; to be published in *Nuovo Cimento*.) The fine structure they observed seems to shed light on the problem of "continuation" between the Rydberg levels 2s, 3d (m = 0), and the high-field levels (100), (001$^+$) (see p. 263, 264). This has led Shinada to a rule about the nodal-surface conservation which is more reasonable than Kleiner's. (To be published in *Nuovo Cimento*.)

4. The nature of the exciton band has been discussed also by Gor'kov and Dzyaloshinskii, *Soviet Phys.–JETP* (*English Transl.*) **26**: 449 (1968).

Chapter 11

FARADAY ROTATION IN SOLIDS

A. Van Itterbeek*

Institute for Low Temperatures and Applied Physics
Leuven, Belgium

Faraday rotation is a nice indirect method for studying the magnetic properties of solids, gases, liquids, and solutions from room temperature down to very low temperatures.

Before treating this optical–magnetic method we will briefly discuss the magnetic properties of matter.

First we have to consider the diamagnetic substances which obey Curie's laws:

1. The magnetic susceptibility χ is proportional to H and its value is negative and very small, of the order of 10^{-6} e.m.u.

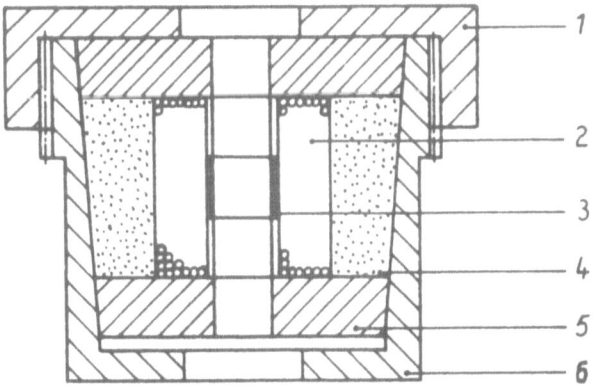

Fig. 1. Construction of a coil with stainless steel reinforcement cylinder. (1) Stainless steel cover; (2) windings; (3) pick-up coil; (4) araldite; (5) celeron flange; and (6) stainless steel cylinder.

* Deceased.

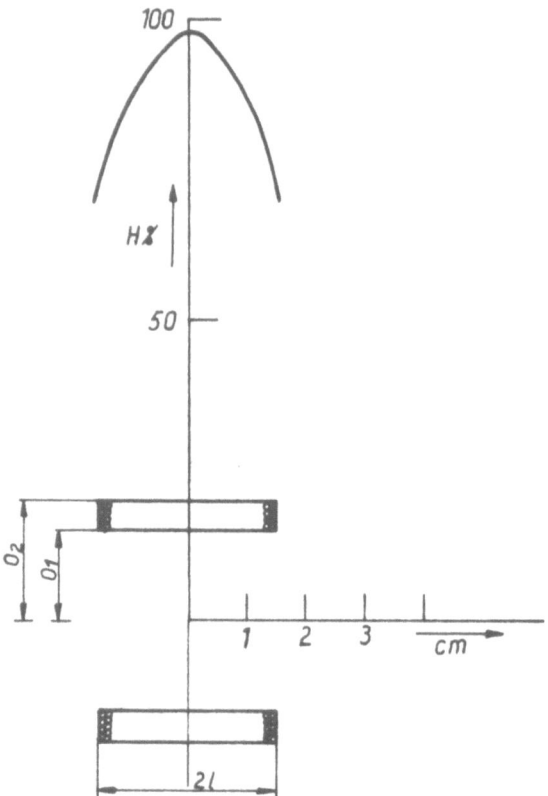

Fig. 2. Field homogeneity inside a coil as measured by
means of a second pick-up coil.

2. χ is independent of temperature over a wide temperature range, although there is small deviation from this at very low temperatures (Landau electronic paramagnetism).

3. χ is also independent of the physical state, which has been proved by the nice measurements carried out by Kamerlingh Onnes on the magnetic susceptibility of hydrogen gas and liquid.

Most of the substances (metals, organic compounds, semiconductors) are diamagnetic, as the magnetism is related to the atomic properties which appear in a magnetic field. Under the influence of the magnetic field a Lorentz force acts on the electrons circling around the nucleus and a magnetic moment is induced which is proportional to the field strength H and R^3 (R being the mean radius of the atom). One can easily see that ΔM

is independent of field direction, so that the magnetic moments are added.

One can only expect to obtain deviations when the magnetic field is very strong (approximately 700 kOe, which is the strength of the orbital magnetic field at the nucleus); thus in this case a deformation of the atom may be considered.

In the past the magnetic susceptibility was determined by using the Faraday method, in which the magnetic force is measured by a weight acting on a sample suspended in the inhomogeneous part of an electromagnet, a field of the order of 15 kOe being used.

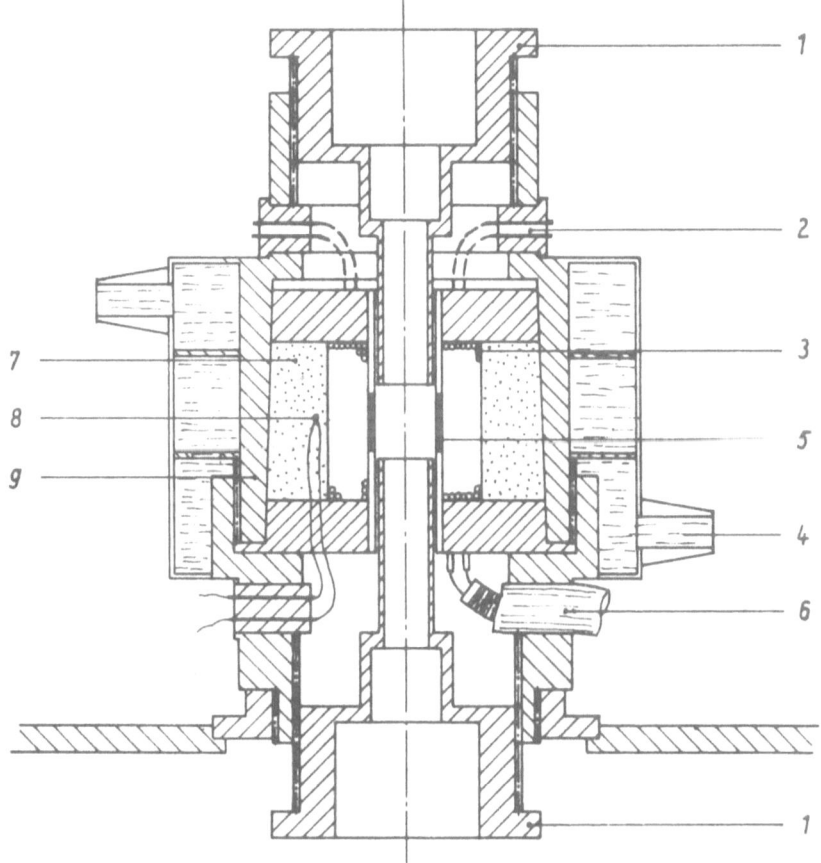

Fig. 3. Coil with cooling mantle developed for measuring the Faraday rotation at ordinary temperatures. (1) Sample holder; (2) conductors of the coil; (3) coil; (4) cooling water; (5) pick-up coil; (6) coaxial conductor for pick-up coil; (7) araldite; (8) thermocouple; and (9) reinforcing cylinder.

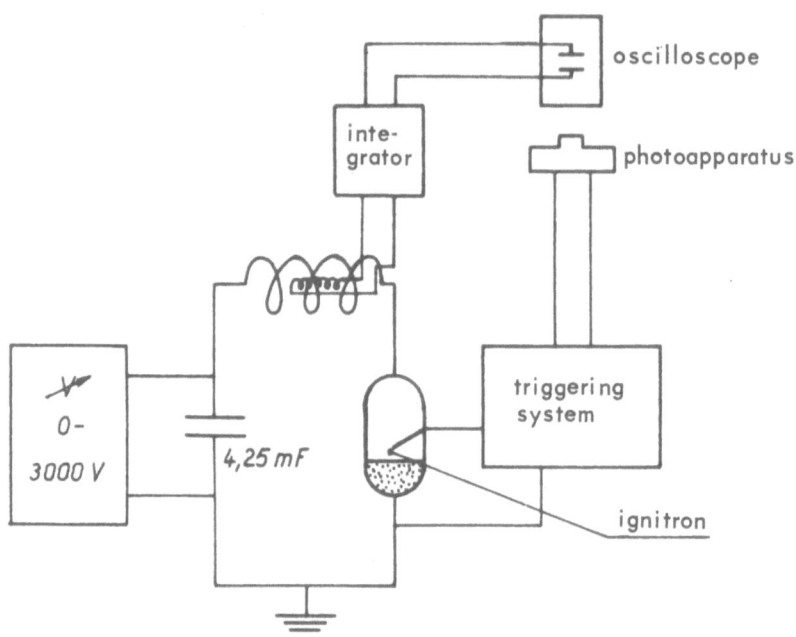

Fig. 4. Block diagram of the pulse apparatus—oscillating circuit together with an ignitron trigger system.

In general a good confirmation of Curie's law is found. While the addition rule in organic compounds is also well-satisfied, the deviation from this rule can give theoretical information on the chemical binding.

Paramagnetism is quite different, as it results from the existence of a magnetic moment μ (orbital or spin) associated with the atom or molecule. A resulting moment ΔM appears in a magnetic field produced by an orientation of the moment in the magnetic field. Here the sign of ΔM is changed as the direction of the field is changed, so that magnetic moments in the opposite direction have to be subtracted. Here the temperature comes into the phenomenon through the Boltzmann factor $\mu H/e^{kT}$ in the supposition that there is no interaction between the magnetic dipoles.

Paramagnetism obeys the Curie–Langevin law

$$\sigma = \sigma_0[(\coth a) - (1/a)]$$

where σ is the magnetization $= \chi H$, $\sigma_0 = N\mu$ (N is the number of dipoles), and the parameter $a = \mu H/kT$.

In first approximation one obtains the Curie law

$$\chi = \sigma_0\mu/3kT$$

The Curie–Langevin equation gives rise to a saturation phenomenon which can appear at low temperatures and high magnetic fields.

In this connection we recall the measurements of Kamerlingh Onnes and Pervier on gadolinium sulfate.

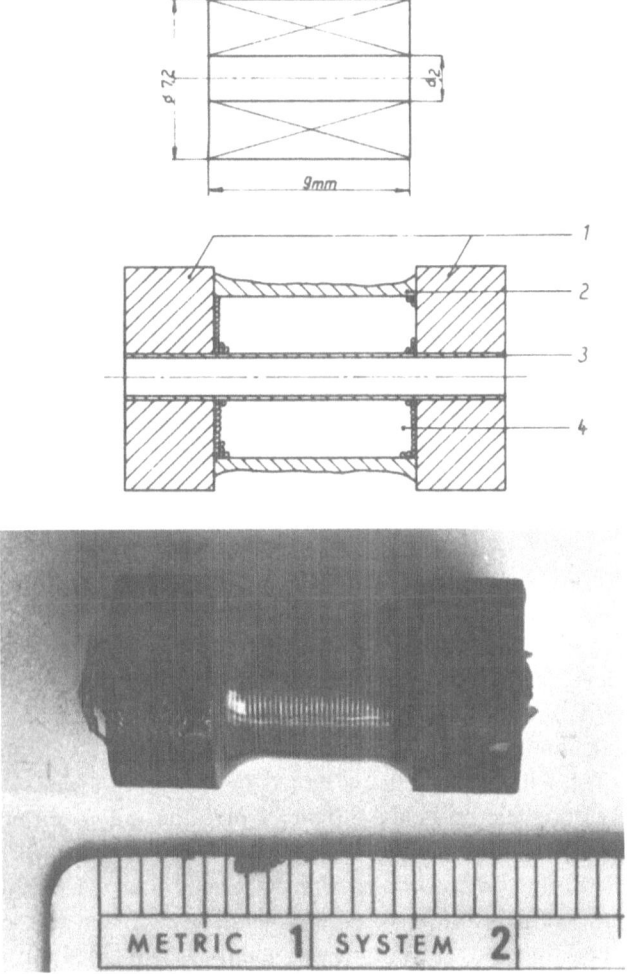

Fig. 5. Construction of a miniature coil which allows one to obtain 200 KOe; pulse time: 25 μsec. (1) Celeron; (2) cement--UHU plus; (3) glass; and (4) windings.

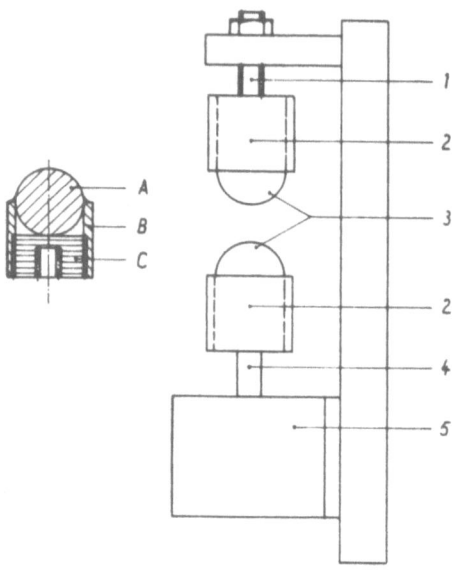

Fig. 6. Moving spark gap discharge system;
pulse time: 25 µsec; condensor bank: 1 F. (1)
Regulation of the upper ball; (2) messing
holder; (3) balls; (4) moving part of the magnet; (5) electromagnet; (A) stainless steel ball;
(B) brass; (C) plexiglass.

Before we discuss our measurements on Faraday rotation we will
describe briefly the experimental method which we use to produce intense
pulsed magnetic fields.

We developed three types of coils. The first two were used to measure
Faraday rotation with a pulse time of 15 msec. The first one (Fig. 1) is
cooled with liquid nitrogen and allows one to obtain a field maximum of
570 kOe. The field strength is measured by means of a pick-up wound in
the middle on a celeron-holder. Figure 2 gives an indication of the homogeneity inside of the coil.

Figure 3 shows a coil cooled with water and used for Faraday rotation
at ordinary temperatures. Maximum fields of 320 kOe can be obtained
with this coil, but most of the measurements were carried out at 200 kOe.
For the discharge we use a battery system of 4.43 mF and a high tension
of 3000 V. The electrical block diagram is shown in Fig. 4. The minimum
discharge time for the ignitron is 1 msec. In addition to our work on de-

veloping coils for Faraday rotation measurements last year we carried out some experimental work on the construction of coils with small dimensions (mini-coils) cooled with liquified gases. An example of one of the types developed is given in Fig. 5. Since the pulse time of this coil is much smaller, of the order of 1 μsec, the ignitron was replaced by another discharge system (see Fig. 6).

Faraday rotation was first treated by Verdet. This phenomenon consists in a change of the plane of rotation of a linear polarized light beam, the direction of the light beam being in the direction of the field. The magnitude of rotation is given by the equation $\alpha = VlH \cos\theta$, or, better, $\alpha = V \int_0^l H \, dx$, where H, the magnetic field, is not homogeneous; θ is the angle between

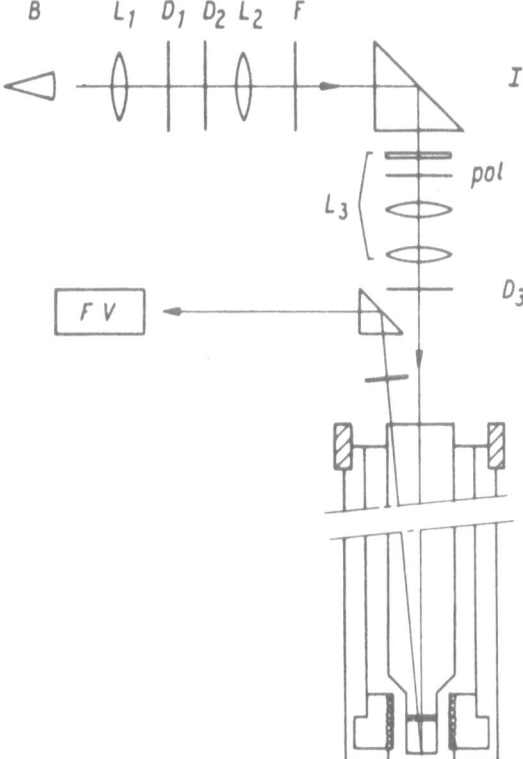

Fig. 7a. Optical system for the measurement of the Faraday rotation ($\lambda = 5460$ Å) at liquid nitrogen temperatures. (B) Light source; (L_1, L_2, L_3) lenses; (D_1, D_2, D_3) diaphragms; (F) filter; (I) prism; (pol) polarizer; and (F V) photomultiplier.

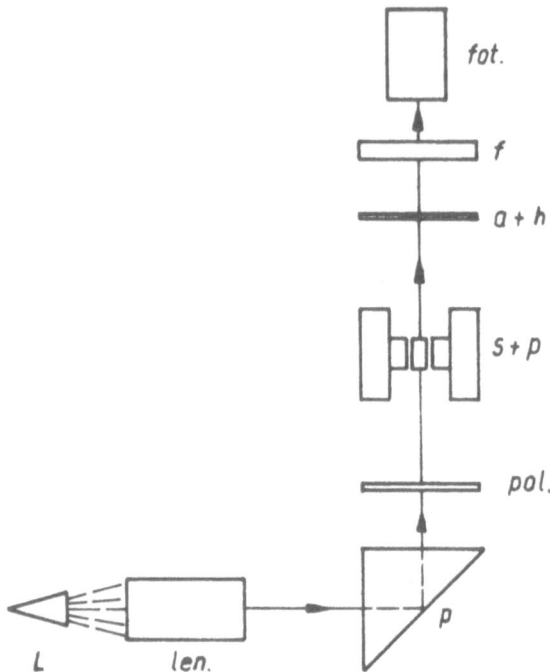

Fig. 7b. Optical system for the measurement of the Faraday rotation at room temperature. (L) light source; (len) collimating system; (p) prism; (pol) polarizer; (s + p) coil with sample; (a + h) analyzer mounted in goniometer; (f) filter—5460 Å); and (fot) photomultiplier.

the light direction and the magnetic field (in general, $\cos \theta$ is approximately equal to 1); V is the so-called Verdet constant; and l the length of the sample in which Faraday rotation is measured. The constant V is a function of temperature for paramagnetic Faraday rotation, but is nearly independent of temperature in the case of diamagnetic Faraday rotation.

On the other hand, V is a function of the concentration, in the case of mixtures, and of the wavelengths (Brulat's equation, $\alpha \propto 1/\lambda^2$). The theory has been treated successively by Fresnel, Becquerel, and Sommerfeld. Faraday rotation is a magnetooptical effect and is explained by the variation under the influence of a magnetic field of the difference in the refractive indices for the right and the left circular polarized vectors which comprise the resulting linear polarized light vector.

According to Fresnel's equation

$$\alpha = (\Delta n/\lambda)l$$

i.e., n is related to a change of the light velocity under the influence of the magnetic field, which in turn is determined by the factor $[(\mu_1 + \Delta\mu_1) \times (\varepsilon + \Delta\varepsilon)]^{1/2}$, where $\Delta\varepsilon$ is the variation of the dielectric constant and $\Delta\mu_1$ the variation of the magnetic permeability, which, also in turn, is directly related to a variation of the magnetic moment ΔM. In this manner we can see how Faraday rotation is directly connected with the diamagnetic and paramagnetic properties of material. For diamagnetic substances there is a Verdet constant which is independent of temperature; but here an influence can be expected when field strengths of the order of 500 kOe or more are used. On the other hand, paramagnetic rotation is given by the equation

$$\alpha = A \tanh (\mu H/kT) + \text{B.H.}$$

where B.H. corresponds to the universal diamagnetic properties of all

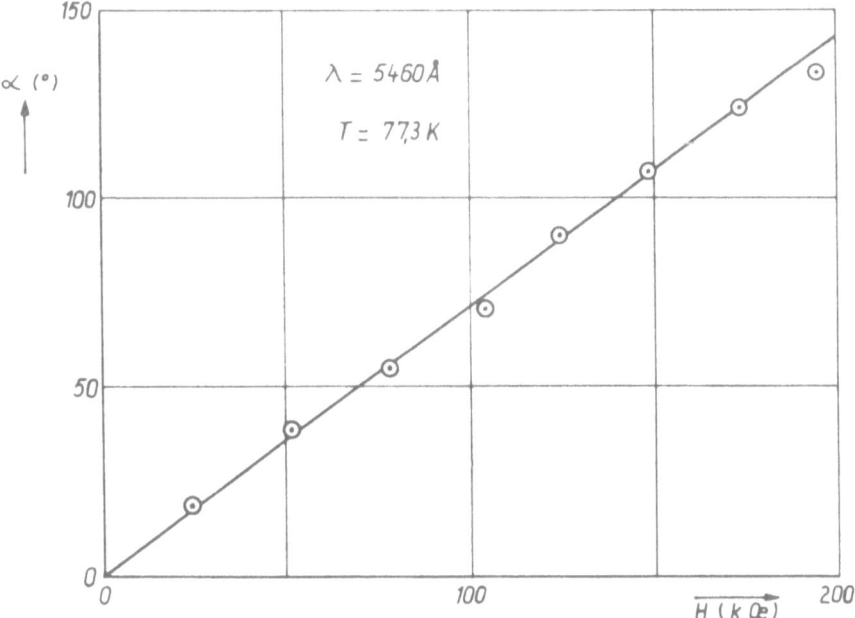

Fig. 8. Measurement of the Faraday rotation at low temperatures. Length of the sample: 0.943 cm.

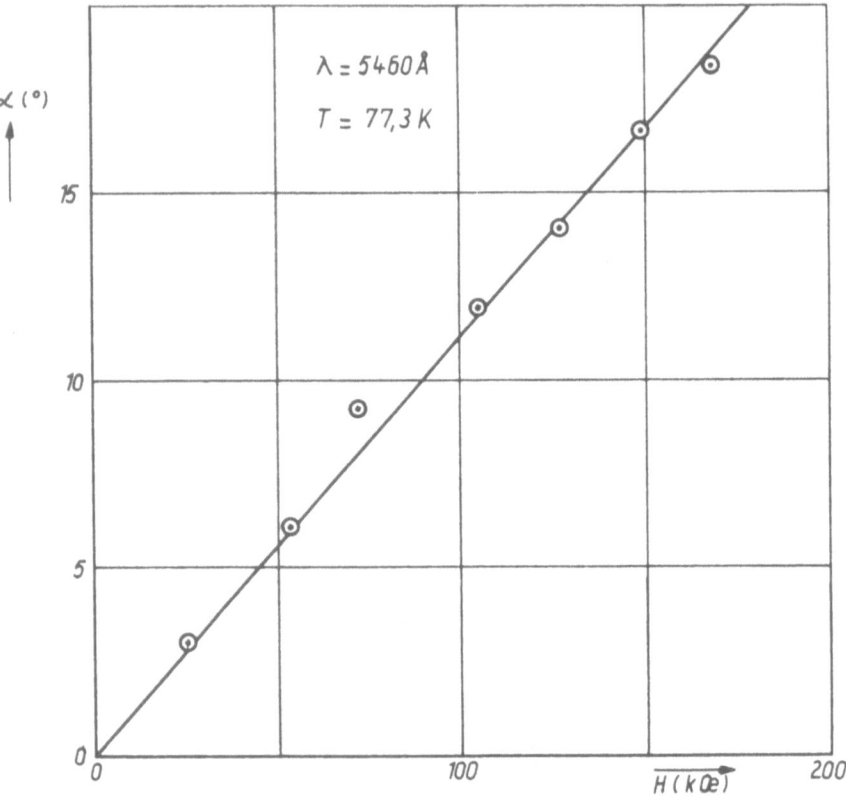

Fig. 9. Measurement of the Faraday rotation in flint glass at low temperatures. Length of the sample: 0.086 cm.

substances. Thus, in first approximation, α will be proportional to H. For all the different cases it is supposed that no magnetic interaction exists.

EXPERIMENTAL METHOD AND RESULTS OBTAINED FOR FARADAY ROTATION

For Faraday rotation measurements in diamagnetic transparent solids a reflection method using the under surface of the sample can be used, as diamagnetic rotation is independent of field direction and is, of course, additive. In this manner an amplification of the rotation can be obtained.

Figure 7 shows the experimental set-up which we used for measurements on Faraday rotation in ordinary glass, flint glass, and plexiglass. Figures 8, 9, and 10 show, respectively, the results of measurements in glass (thickness $2l = 0.943$ cm), flint glass ($2l = 0.086$ cm), and plexiglass ($2l = 1.863$ cm) at 77 °K. From these figures we see that α as a function of H is, up to 200 kOe, a straight line. We obtain the following values for the temperature-independent Verdet constants:

Mirror glass: $V = 2.16 \times 10^{-2}$ min/Oe-cm

Flint glass: $V = 4.07 \times 10^{-2}$ min/Oe-cm

Plexiglass: $V = 1.41 \times 10^{-2}$ min/Oe-cm

Paramagnetic solutions are also excellent substances for measuring paramagnetic rotation. Small vessels with plane-parallel cover glasses for microscopy were filled with solutions of $CeCl_3$ of different concentrations. The results obtained for one solution are given in Fig. 11. Here we again

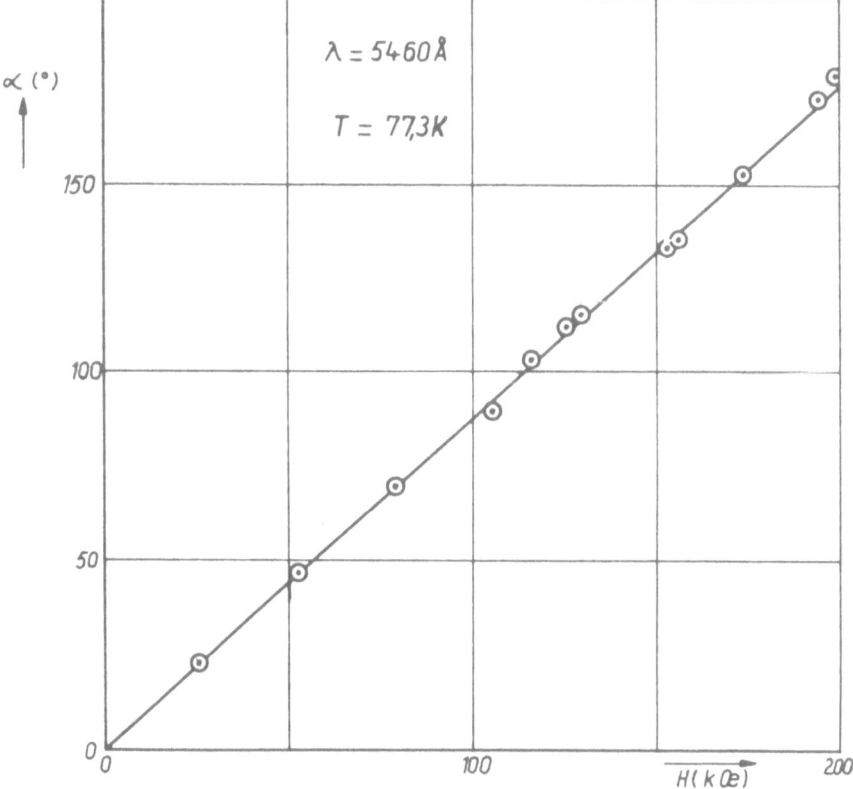

Fig. 10. Measurement of the Faraday rotation in polymethyl methacrylate at low temperature. Length of the sample: 1.86 cm.

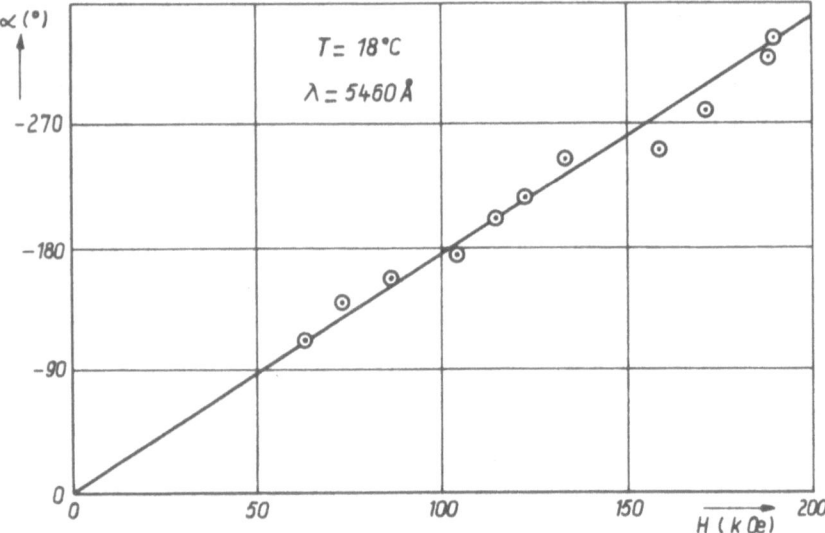

Fig. 11. Faraday rotation of a solution of CeCl$_3$ in water (4 mol). Length of the sample: 4 cm.

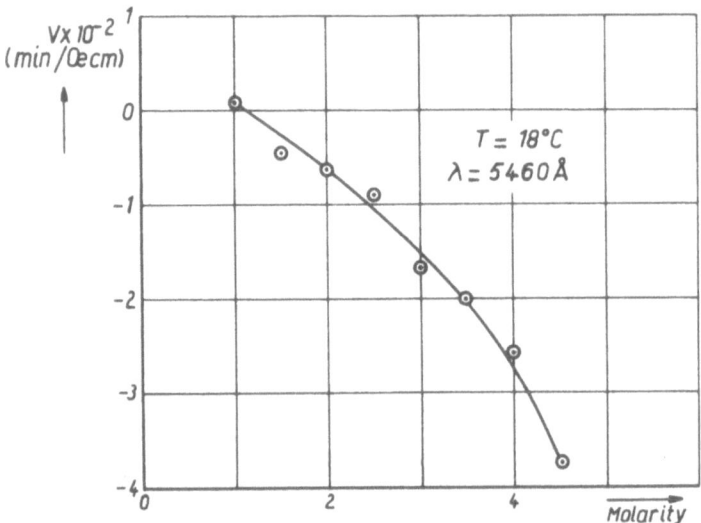

Fig. 12. Verdet constant as a function of the concentration of CeCl$_3$ in H$_2$O.

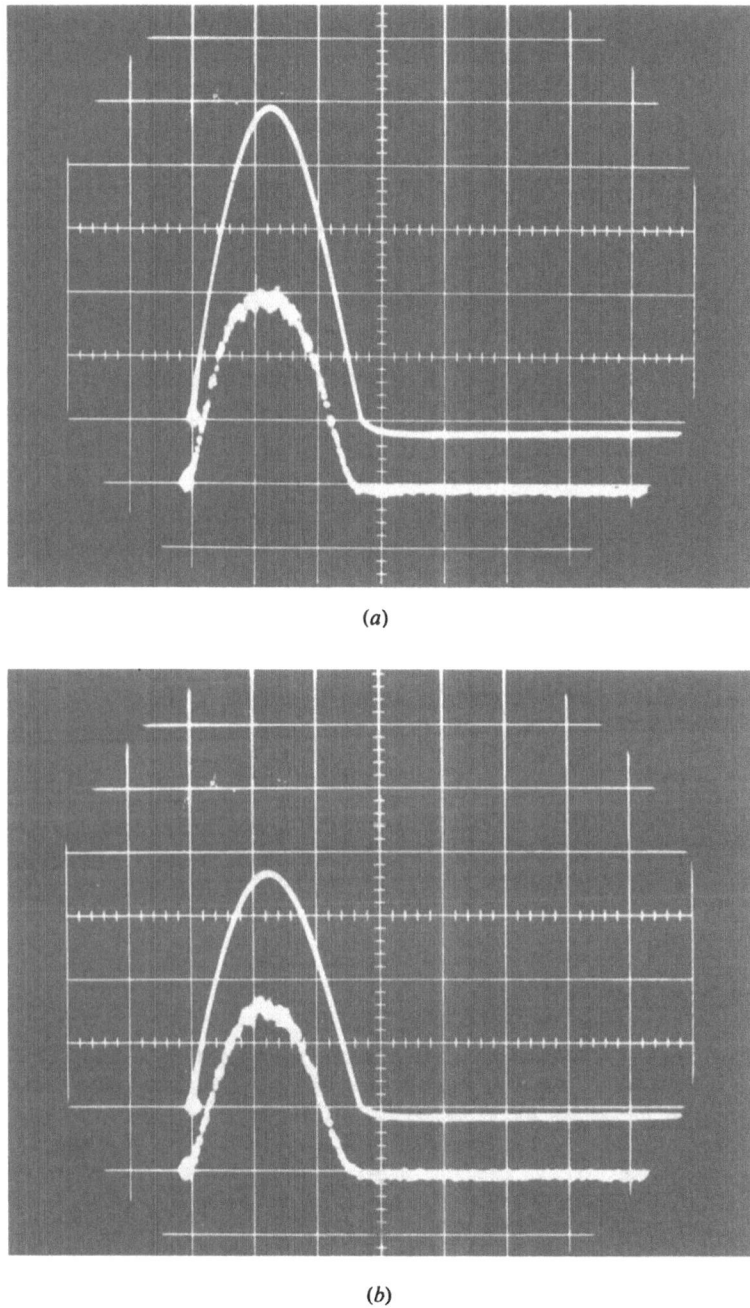

(a)

(b)

Fig. 13. Oscillograms of the magnetic field variation and the rotation during half a pulse.

see that a straight line is obtained for α as a function of H, again up to 200 kOe.

In Fig. 12 the variation of Verdet's constant in $CeCl_3$ is given as a function of concentration at room temperature. As can be seen, a curve is obtained and not the straight line one would expect from theory. This is naturally a problem of physical chemistry, which indicates an interaction between the paramagnetic ions and the water molecules.

Finally, Fig. 13 shows an oscillogram of the magnetic field variation and the rotation during half a pulse.

As a general conclusion one can say:

1. Faraday rotation is an excellent tool for measuring field intensity, and the Verdet equation is well fulfilled even for very high intense fields.

2. In order to expect more interesting phenomena one has to carry out the measurements up to higher field strengths.

3. During our investigations we have not to deal with a relaxation phenomenon. This can only be expected at very low temperatures, for paramagnetic rotation, which is then also related to paramagnetic relaxation phenomena.

ACKNOWLEDGMENTS

The work reported in this paper is the result of collaborations with H. Hendrickx, H. Myncke, G. Pitsi, W. Van Driessche, I. De Grave, E. Van Ocken, H. Peters and H. Verhaegen.

We take this opportunity to express our thanks for financial aid to the Belgian Interuniversity Institute for Nuclear Sciences.

REFERENCES

1. H. Hendricks, A. Van Itterbeek, H. Myncke, and G. Pitsi, *Comm. Flemish Acad. Sc.* Vol. 26, No 6 (1964); A. Van Itterbeek, G. Pitsi, H. Myncke, and H. Hendrickx, *Appl. Sci. Res.*, *Sect. B*, **11**: 433 (1964); H. Hendrickx, H. Myncke, and A. Van Itterbeek, (1964) *Proc. Int. Congr. Refrig.*, *11th, Munich, 1963* Vol. 1, p. 219 (1964); A. Van Itterbeek, W. Van Driessche, and H. Myncke, *Bull. Belg. Phys. Soc. Vol. IV*, 6, 389 (1965); A. Van Itterbeek, W. Van Driessche, I. De Grave, and H. Myncke, *Bull. Belg. Phys. Soc. Vol. V*, No. 2–3, 188 (1966).
2. H. Abraham and J. Lemoine, *Compt. Rend.* **30**: 429 (1900); H. Abraham and J. Lemoine, *Ann. Chim. Phys.* **20**: 264 (1900); H. Becquerel, *Compt. Rend.* **125**: 679 (1897); C. Bruhat, Optique, Paris, 1954; E. Verdet, *Ann. Chim. Phys.* **41**: 570 (1854).

Chapter 12

A TECHNIQUE FOR THE MEASUREMENT OF THE FARADAY EFFECT IN PULSED MAGNETIC FIELDS AT LOW TEMPERATURES

G. Sacerdoti

Comitato Nazionale per l'Energia Nucleare
Roma, Italy

This chapter describes a technique for measuring the spin-lattice relaxation time in paramagnetic crystal and the Faraday effect in different materials. The technique was suggested by Prof. Toraldo di Francia of the University of Florence, and an apparatus based on this technique has been developed for use in the Synchrotron Laboratory at Frascati, Italy. The sample is put in a pulsed magnetic field whose duration is comparable to (or larger than) the relaxation time of the spin-lattice of the crystal. By measuring the magnetic field H versus time and the magnetization of the specimen it is possible to get the value of τ as a function of temperature and magnetic field.

The magnetization M of the specimen is revealed by measuring the angle of Faraday rotation θ that a linearly polarized beam of light undergoes upon passing through the specimen; M is proportional to this angle of rotation at temperatures sufficiently low, on the order of 1 °K (Fig. 1).

The experimental apparatus is shown in Fig. 2. A photograph of the same apparatus is shown in Fig. 3. The light from a laser passes through an orientable glass deflecting system, having all mechanical degrees of freedom, and a diaphram to limit the intensity of the beam; the light then crosses a glass window in the Dewar and a Glan Thompson polarizer at room temperature* in the vacuum chamber of the Dewar; the angular position

---

* At low temperature both the nicols and the Glan Thomson become opaque.

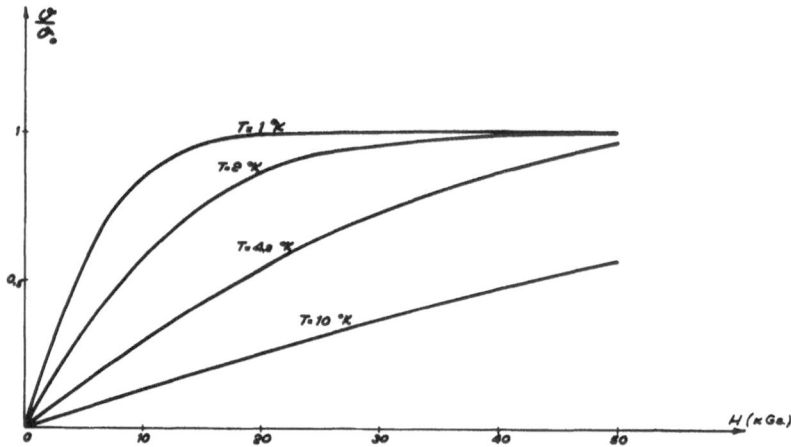

Fig. 1. Behavior of Faraday rotation ϑ versus magnetic field. Curves are relative to the ethyl sulfate hydrate of neodymium.

is easily regulated and read on a circular graduated scale. Then at the temperature of liquid N_2 there is a thermal screen with a small hole. The light enters the liquid helium container through an optically-flat glass window.

After a few millimeters of liquid He there is another window through which the light enters the glass tube containing the specimen; the glass tube is evacuated to avoid too long a path in liquid helium for reasons to be given shortly. From the container of the specimen the light passes through the analyzer (another Glan Thomson polarizer) and then through another window to the photomultiplier tube.

Around the tube that contains the specimen there is a coil of known dimensions for measuring the magnetic field. The magnetic field is obtained by discharging a condenser bank through a multilayer coil immersed in liquid N_2. The energy of the bank is 136 kJ and the maximum voltage is 9 kV. Figure 4 is a photograph of the coil with inner radius $r_i = 38$ mm and with height $h_0 = 180$ mm. The maximum field obtainable is about 100,000 G and the time of discharge is 60 msec. At the maximum field it is necessary to use a crowbar to avoid reversal recharge of condenser bank. The magnet is suspended from the upper plate of the stainless steel Dewar connectors with rubber lining to decrease mechanical vibrations of frequency higher than the frequency of discharge of the condenser bank, but this doesn't become necessary until 60% of the maximum field is reached.

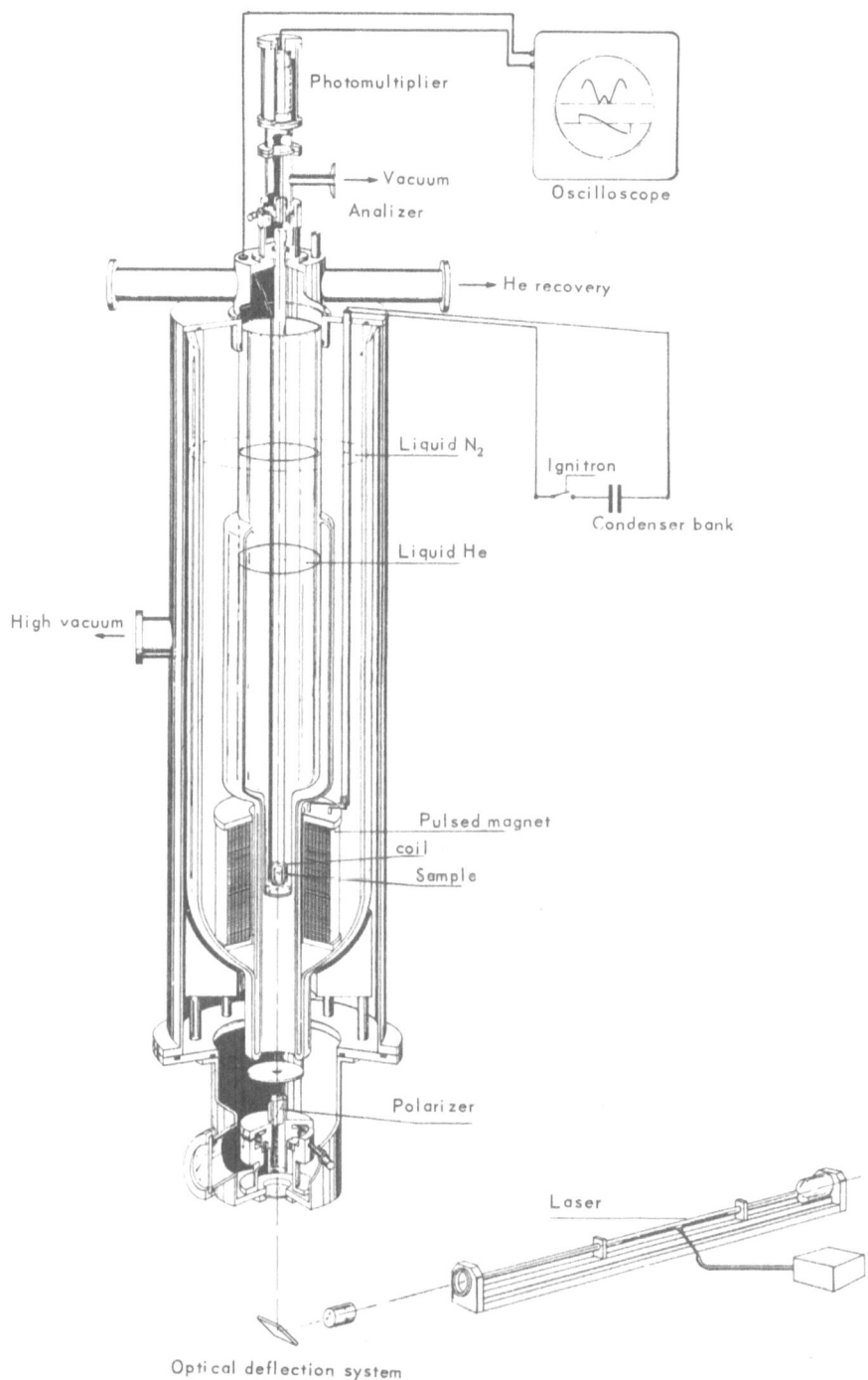

Fig. 2. Schematic of the experimental apparatus (vertical beam).

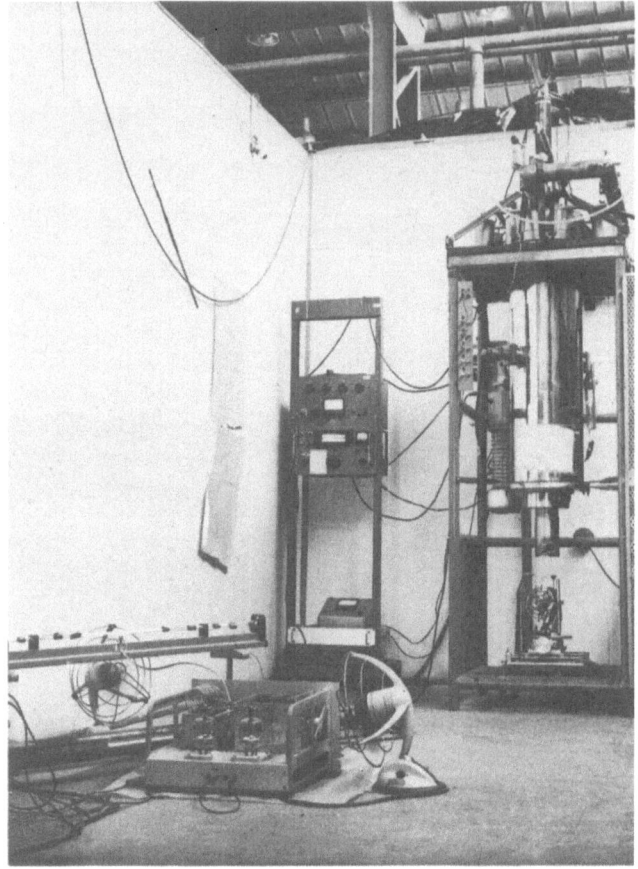

Fig. 3. Photograph of the experimental apparatus.

The following points will be discussed:

1. The reasons we have preferred to work with a vertical rather than with a horizontal light beam.

2. The reasons the specimens are not directly immersed in liquid helium.

3. The best fashion for detecting and measuring the angle of rotation as a function of time and of magnetic field.

4. The main errors in these measurements.

5. The possibility of working at higher fields with a similar apparatus.

Fig. 4. Photograph of the coil.

1. With a horizontal beam the layout of the apparatus is that shown in Fig. 5. However, the different parts of the apparatus are at different temperatures, and the resulting thermal deformations cause changes in optical alignment, particularly at different levels of liquid He.

To avoid this problem, it is necessary to use the arrangement shown in Fig. 6), where the geometrical position of the different parts of the Dewar are fixed by pyrex supports (Fig. 7); the design is complicated.

2. At first the specimen was immersed in liquid helium, but at temperatures between 4.2° and 2.2 °K (λ point of liquid He) the passage of the beam through the helium caused noise because there was a depolarization of light due to the formation of bubbles when helium passed from liquid to gas. Below $T = 2.2$ °K the noise disappeared, and it was then possible to work with the specimen immersed in liquid He.

With the actual apparatus in use we reduced the path in helium to a few millimeters, and the bubble production did not create detectable disturbances. The thermal conductivity was sufficient to guarantee that the specimen was at the temperature of liquid He.

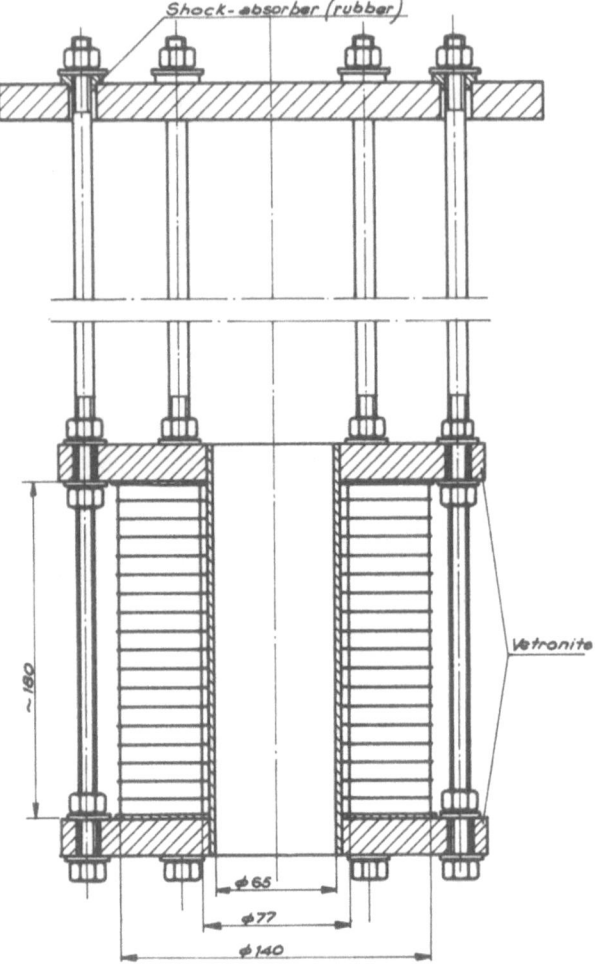

Fig. 5. Scheme of the pulsed coil.

3. Now we describe the measurement of the Faraday rotation, and thus Verdet's constant ($\tau \ll T$, the duration of the pulsed magnetic field). The output signal from a photomultiplier in the case of monocromatic light is (when $\tau_{\text{relaxation}}$ is zero and $M \equiv H$)

$$V_u(t) = K \cos^2\left(\theta_0 + V \int_0^l H \, dl\right) = K \cos^2(\theta_0 + \theta_r) \tag{1}$$

where K is a coefficient proportional to the initial beam intensity and to the sensitivity of the photomultiplier, θ_0 is the angle between the light polarizer and light analyzer, θ is the Faraday rotation angle, V is the Verdet constant, and l is the length of the specimen. When the light is not monochromatic the signal output is given by

$$V_u(t) = \int_{\lambda_{min}}^{\lambda_{max}} I(\lambda)T(\lambda)\delta(\lambda) \cos^2\left[\theta_0 + V(\lambda) \int_0^l H \, dl\right] d\lambda \qquad (2)$$

where $I(\lambda)$ is the intensity of the light source (a function of wavelength), $T(\lambda)$ is the transmissibility of the specimen (a function of wavelength), $\delta(\lambda)$ is the sensitivity of photomultiplier (a function of wavelength), l is the length of specimen, and $V(\lambda)$ is Verdet's constant (a function of wavelength).

We tried to make measurements using a light source from a sodium arc lamp ($\lambda_1 = 5890$ Å, $\lambda_2 = 5896$ Å) using an RCA 6217 photomultiplier, but the influence of the pulsed magnetic field on the source had the effect of changing the emission intensity by a factor of 30% to 50% despite screening by permalloy around the lamps. We also tried using an incandescent lamp and an RCA 6217 tube, but because of the many parameters in Eq. (2) it was very difficult to get noise-free data on the constant $V(\lambda)$. The photomultiplier used had low sensitivity in the red region, at which frequencies our specimens are transparent. As a result we had to use a helium–neon laser with a Philips-type 150CUP photomultiplier with a good sensitivity in red region.

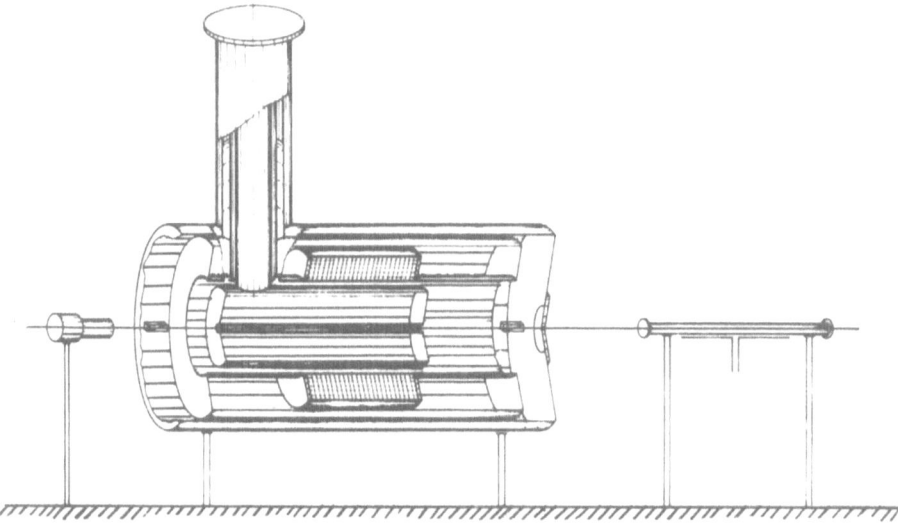

Fig. 6. Scheme of a possible experimental apparatus (horizontal beam).

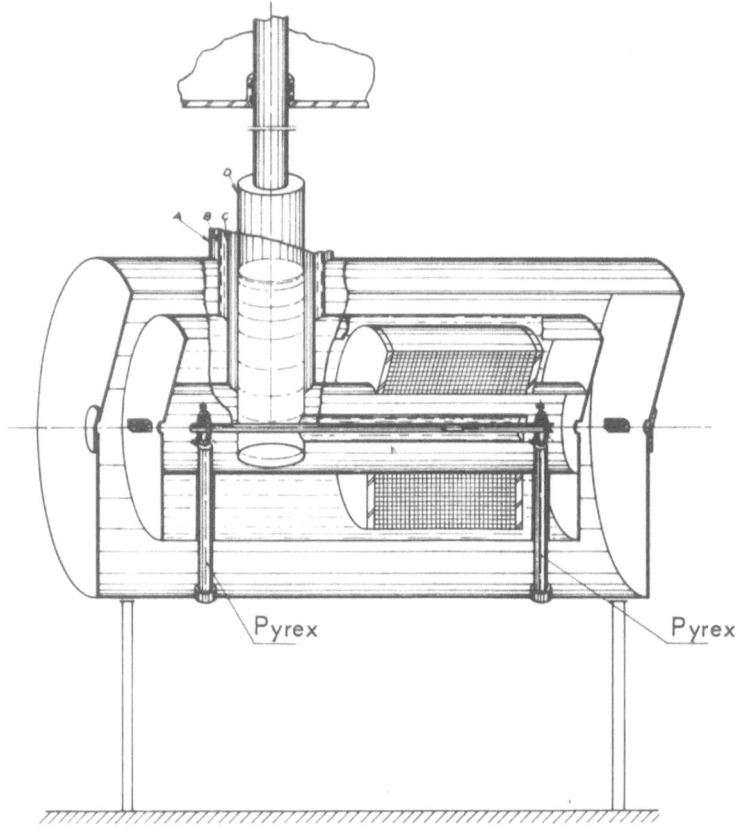

Fig. 7. Detail of the apparatus with horizontal beam.

Equation (1) was used to analyze the results of the measurements, and the coefficient we looked for was Verdet's constant V.

Different methods of data reduction can be used.

(i) We assume *a priori* that during the pulse the value of K remains constant.

The best fit of the curves $V(t) = K \cos^2(\theta_0 + V \int H \, dl)$ taken from a photograph at the oscillograph of the signal $V(t)$ (θ_0 is known) and of the signal proportional to dH/dt (from the pick-up coil of known area) enable one to obtain the values of V and K. We tried this method using a scanning machine and a PDP-8 computer to get the points of the curves $(dH/dt)(t)$ and $V_u(t)$ from the negative of the photograph; to find the best fit it is possible to use a computer or to proceed by manual calculation (Fig. 8).

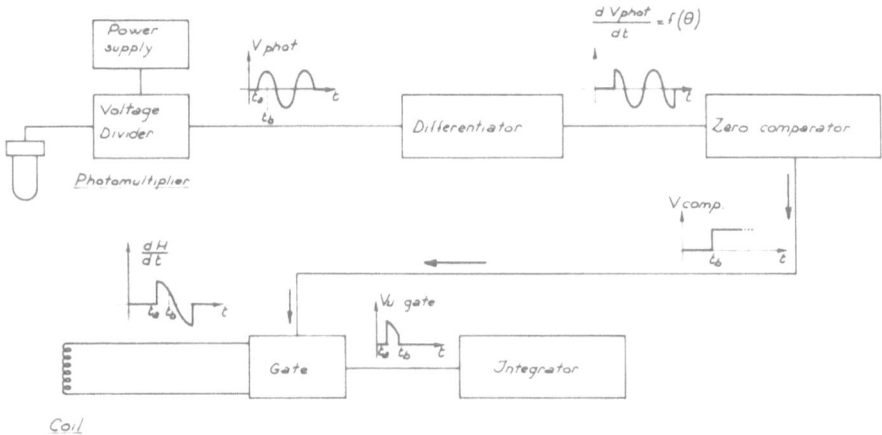

Fig. 8. Circuit of a digital system for detecting $H = H(\frac{1}{2}\pi - \theta_r)$ for Verdet's constant.

In the data reduction one may also try to use the value of $V_{\max}$ obtained when the polarizer and the analyzer are parallel. This value can be found from:

$$\theta_0 + \theta_2 = \text{arc cos } [V_u(t)/V_{u,\max}]^{1/2} = \theta_0 + V \int H \, dP \qquad (3)$$

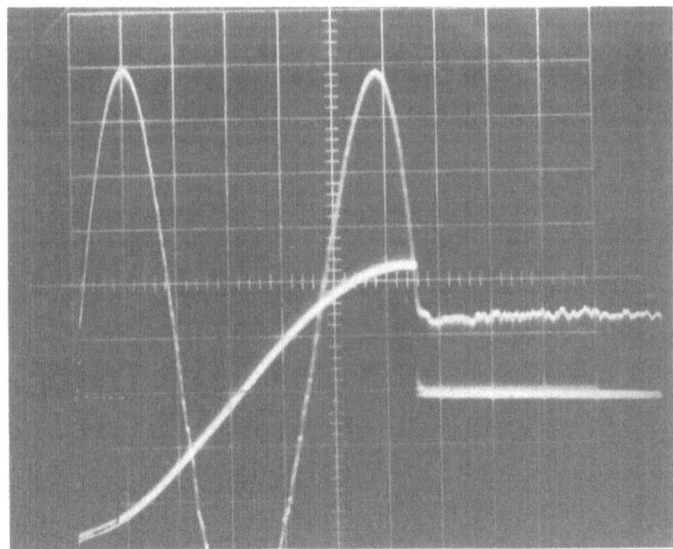

Fig. 9. Oscillograph traces of $V(t)$ and dH/dt.

(ii) To reveal the magnetic field and $V(t)$ on oscillograph changing from a photo to the other the value of θ_0. At the instant where $V(t)$ is minimum (i.e. ~ 0) it results that $\theta_r = (\pi/2) - \theta_0$, and this is determined independently from the fact that K is or is not constant (Fig. 8).

(iii) An integrator can be used to measure magnetic field. The integrator may be digital and the gate may be driven by the derivative of $V_u(t)$. When it changes sign the gate closes, and thus one obtains the value $H = H(\tfrac{1}{2}\pi - \theta_0)$. At Frascati we are working in this direction.

The first method is obviously less precise than the second. The third should be even more accurate.

Different problems arise when we work on materials that present a relaxation time longer than or of the same order of magnitude as the discharge time of the bank of condensers. In this condition it is better to work without a crowbar, as I will demonstrate.

We know that the following relation between the relaxation time of the magnetic field H and the magnetization M of the specimen is valid:

$$\frac{dM_1}{dt} = \frac{M_0(H_0) - M_1}{\tau(H_0)} \tag{4}$$

where M_1 is the magnetization of the specimen, $M_0(H_0)$ is the equilibrium

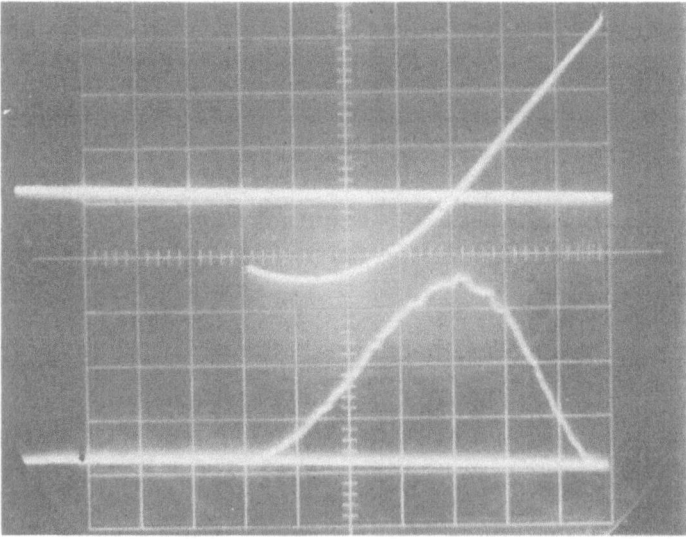

Fig. 10. Oscillographic traces of $V(t)$ and dH/dt with $\theta_0 = 20°$.

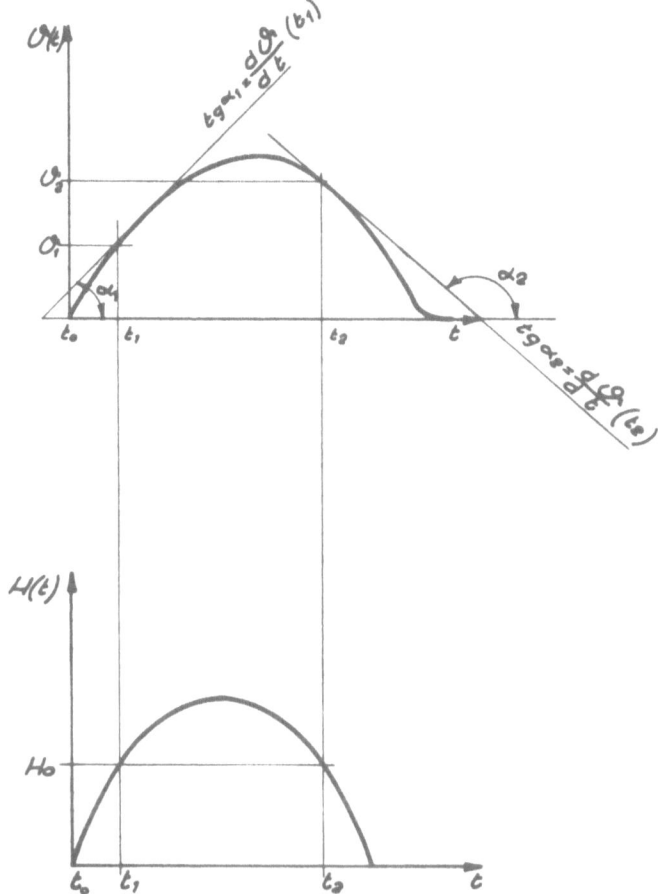

Fig. 11. Scheme for analyzing the output signal from a photomultiplier so as to obtain $\tau(H)$.

magnetization of the specimen at the magnetic field H_0, and $\tau(H_0)$ is the relaxation time.

Consider two points 1 and 2 on the curve $V(t)$ corresponding to the same value of H_0 (i.e., M_0) (see Fig. 11). We may write

$$\frac{dM_1}{dt} = \frac{M_0(H_0) - M_1}{\tau(H_0)} \tag{5}$$

$$\frac{dM_2}{dt} = \frac{M_0(H_0) - M_2}{\tau(H_0)} \tag{6}$$

Taking the difference, we obtain

$$\frac{dM_1}{dt} - \frac{dM_2}{dt} = \frac{M_2 - M_1}{\tau(H_0)} \tag{7}$$

and in the limit of proportionality between M and θ we may write

$$\frac{d\theta_1}{dt} - \frac{d\theta_2}{dt} = \frac{\theta_2 - \theta_1}{\tau(H_0)} \tag{8}$$

If there are many rotations (this may happen only for very long specimens) we obtain

$$\frac{d\theta_1}{dt} = \frac{\pi}{\Delta t_1}, \qquad \frac{d\theta_2}{dt} = \frac{\pi}{\Delta t_2}$$

where Δt_1 and Δt_2 are the time intervals between through the minimum of voltage $V(t)$ and θ_1 and θ_2 are obtained by interpolation. If the rotation is $<2\pi$ it is necessary to get the terms in Eq. (8) by many registrations in which one changes θ_0 ($H_{\max} =$ const). It is thus possible to get terms proportional to $d\theta_1/dt$, $d\theta_2/dt$, θ_1, and θ_2, and so to obtain $\tau(H)$.

It is possible to think of dividing the light in different paths and of analyzing each one with different θ_{0i} (i indicates the path, as shown in Fig. 12) and to obtain sufficient data from the different curves registered simultaneously to obtain the value of $\tau(H)$; this may be a difficult job.

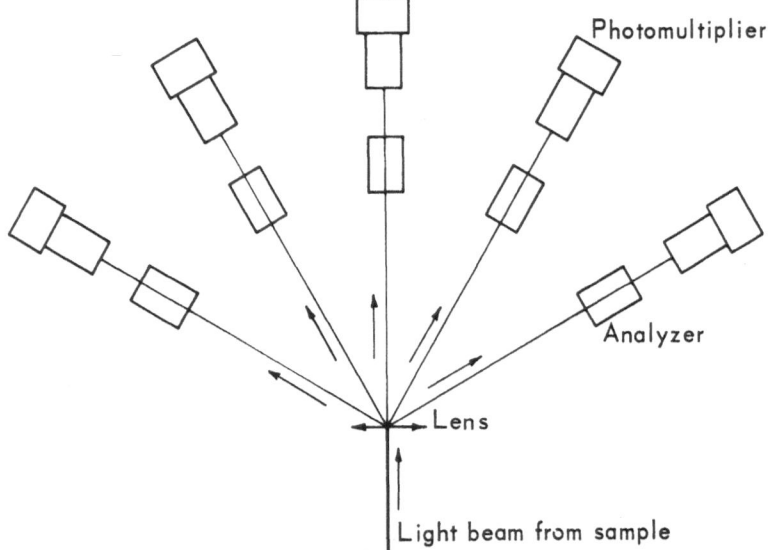

Fig. 12. Scheme of experimental apparatus for obtaining parallel analyses of rotation with different angles.

4. The causes of errors in this kind of measurement are many. Some of the more important error sources are:

(a) Depolarization of light due to imperfections and impurities in the specimens and other glass windows. This is not a cause of strong errors if the intensity of the source remains constant. By this assumption the only effect here is to increase the voltage $V_u(t)$ by a constant, and if the experiment works by changing θ_0 this should not introduce errors.

In addition, the bubbles of gas in the liquid helium may cause random depolarization at temperature beyond the λ-point.

(b) Fluctuation in the intensity of the light source due, for example, to the discharge of the condenser bank and noise induced in the photomultiplier and in the wires to the oscillograph.

TABLE I

Faraday Rotation Measurements*

Temperature (°K)	B (Wb/m²)	θ (deg)
77	1.1	7.9
	2	13.1
	3.1	19.8
	4.2	25.3
	5.2	37.1
4.2	10 ± 1	0.55 ± 0.08
	20 ± 1	1.10 ± 0.08
	30 ± 1	1.60 ± 0.09
	40 ± 1	2.12 ± 0.10
	50 ± 1	2.71 ± 0.11
	60 ± 1	3.70 ± 0.11
	70 ± 1	3.69 ± 0.20
1.2	10	0.084
	20	0.139
	40	0.308
	70	0.610
	250	2.105

* Sample: Neodymium glass (length 22.25 mm) plus Pyrex (length 8 mm). Light: He–Ne laser.

In Table I we give some results we obtained. In some cases noise was ± 0.1 V, in other cases ± 0.3 V, compared to 50 V of maximum signal (polarizer and analyzer parallel). It should seem possible in principle to decrease this noise by using filters at frequencies 20–30 times the frequency of discharge of the condenser bank. However, this would decrease the precision in the determination of the point at which the value of $V(t)$ passes through a minimum because the filter itself will produce a delay in the signal of the same order of magnitude as the noise shut down. The noise is often comparable to a rotation of $2°$; in the better photographs it corresponds to only $1°$ ($\theta_0 = 0$).

To reduce the noise of the photomultiplier we also cool it and its voltage divider by liquid nitrogen.

(c) The mechanical vibration of the Dewar because of the pulsed magnet. These vibrations are not a problem, because the frequency is much lower than that of the pulsed field. A photograph of the output with $\theta_0 = 20°$ and with the time scale amplified is shown in Fig. 10.

(d) Errors in the measurement of the magnetic field. When we obtain the magnetic field by graphical integration of the derivative versus time from the oscillographic record most of the error is due to the scale x, y of the oscilloscope and the precision in the determination of the instant t at which the rotation is equal to $\frac{1}{2}\pi - \theta_0$.

(e) Residual rotation in the glass windows. This rotation has been measured at different fields and temperatures, and we have corrected the measurements on the specimen by the corresponding amount.

(f) Nonuniformity of the magnetic field in the magnet used. The magnetic field was uniform in the specimen to better than 1%; it is not an important cause of errors.

5. We think that with a larger condenser bank it should be possible to increase the field to 200 kG with the same period of discharge, and to 300 kG with a shorter period of 20-30 μsec. We have already tested pulsed magnets with fields of this order of magnitude.

Using fields of 30 kG preliminary measurements of Verdet's constant have been made on a neodymium-doped glass 2 cm in length. The results obtained are shown in Table II. We have not yet made measurements of spin-lattice relaxation time on the ethyl sulfate hydrate of neodymium because we found many difficulties in preparing specimens and working and polishing their surfaces; only one of the thirty which we prepared seems good enough to be used for our measurements, i.e., it does not depolarize light. We have chosen that "difficult" material (if put in a vacuum

TABLE II

Verdet's Constant as a Function of Temperature in Neodymium Glass

V $\left(\dfrac{\text{deg}}{\text{cm-Wb/m}^2}\right)$	T (°K)
55.1	1.2
10.8	4.2
3.56	77

at room temperature it becomes opaque to light because the HO crystal ions go to the surface, creating a white layer; thus it is necessary before doing vacuum work to freeze the specimen, avoiding the creation of an opaque layer on the crystal by the residual humidity of air) because it was possible to develop a theory of the phenomenon and to test its validity. Our results [see ([6])] change Orbach's results [see ([7])] by an order of magnitude and seem to agree better with the few experimental data available on $\tau(H, T)$ at low fields 2 kG. We also calculated the "bottleneck effect," both geometrical and spectral, and the order of magnitude of our approximate calculations indicates that this does not affect our measurements at temperatures greater than 1 °K.

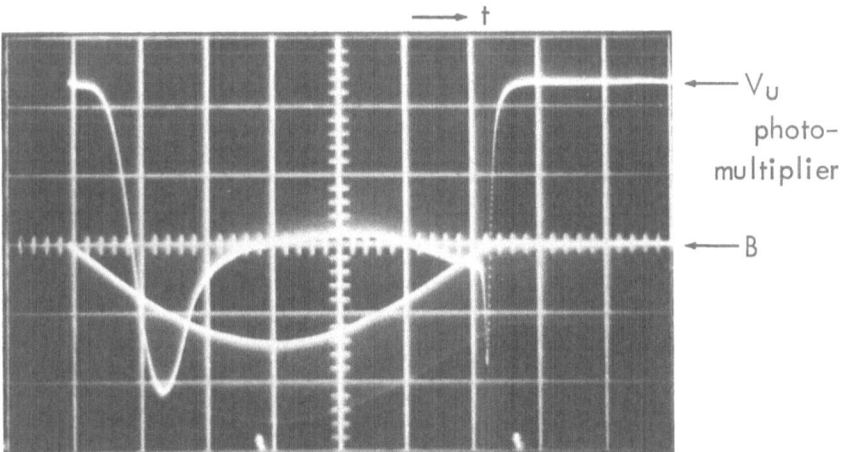

Fig. 13. Experimental result for neodymium ethyl sulfate hydrate. $T_H = 1.0$°K, the angle between the polarizer and the analyzer is 90°, $B_{\max} \approx 3.9$ Wb/m².

ACKNOWLEDGMENTS

The experimental work reported here has been supported by Consiglio Nazionale delle Ricerche.

REFERENCES

1. C. Faloci, M. Mancini, G. Sacerdoti, G. Toraldo di Francia, and F. Uccelli, INFN/ FM - 65/1 (4 January 1965).
2. R. Bruzzese, F. Frazzoli, M. Mancini, G. Sacerdoti, G. Toraldo di Francia, and F. Uccelli, INFN/FM - 65/3 (December 1965).
3. R. Bruzzese, F. Frazzoli, M. Mancini, G. Sacerdoti, G. Toraldo di Francia, and F. Uccelli, INFN/FM - 66/2 (December 1966).
4. C. Faloci, Graduation Thesis, 1964.
5. F. Frazzoli, Graduation Thesis, 1965.
6. F. Frazzoli and M. Mancini, "A Contribution to the Theory of Spin-Lattice Relaxation Time in Rare-Earth Salts," *Nuovo Cimento, Ser. X* **53B**: 471–484 (1968).
7. A. A. Manenkov and R. Orbach (eds.), *Spin-Lattice Relaxation in Ionic Solids*, Harper and Row, New York, 1966.

NOTE ADDED IN PROOF

Following the Crete meeting at which the paper on which this chapter is based was delivered, we experimented on neodymium ethyl sulfate hydrate. A characteristic photograph of the photomultiplier signal and magnetic field versus time is shown in Fig. 13. It is evident that there is a relaxation time and a saturation effect. We are now analyzing data, and our result will be published.

Chapter 13

ELECTRONS IN A MAGNETIC FIELD

W. Zawadzki

*National Magnet Laboratory**
Massachusetts Institute of Technology
Cambridge, Massachusetts
and
Institute of Electron Technology
Polish Academy of Sciences
Warsaw, Poland

In this chapter we shall describe the behavior of free nonrelativistic electrons in the presence of a uniform magnetic field. First we shall give the classical description of the motion and then present the quantum mechanical treatment. We hope that this presentation will be of some use to those interested in this subject, treatment of which is presently scattered piecemeal through many books and original publications.

CLASSICAL DESCRIPTION

We shall first consider the free electrons in a magnetic field according to classical mechanics. Instead of starting with the Lorentz equation of motion and then constructing the Hamilton function, we shall simply postulate the Hamiltonian and then show the well-known consequences that follow.

For a free nonrelativistic electron in an eletromagnetic field the Hamiltonian is

$$\mathcal{H} = \frac{1}{2m} \sum_k \left(p_k + \frac{e}{c} A_k \right)^2 - e\varphi \tag{1}$$

where the summation is over the three Cartesian coordinates x, y, z. The quantity

$$\mathbf{P} = \mathbf{p} + (e/c)\mathbf{A}$$

* Supported by the U. S. Air Force Office of Scientific Research.

is the canonical (also called generalized or kinetic) momentum, m is the free-electron mass, c the velocity of light, and e the absolute value of the electron charge. Magnetic and electric fields $\mathbf{H}$ and $\mathbf{E}$ are introduced by means of vector and scalar potentials $\mathbf{A}$ and φ through the relations

$$\mathbf{H} = \text{curl } \mathbf{A}, \qquad \mathbf{E} = -\text{grad } \varphi - \frac{1}{c} \frac{\partial \mathbf{A}}{\partial t} \tag{2}$$

Now, according to the Hamilton equations of motion

$$\dot{x}_i = \partial \mathcal{H}/\partial p_i, \qquad \dot{p}_i = -\partial \mathcal{H}/\partial x_i \qquad (i = 1, 2, 3) \tag{3}$$

This gives

$$\dot{x}_i = \frac{1}{m}\left(p_i + \frac{e}{c} A_i\right) \tag{4}$$

and

$$\dot{p}_i = -\frac{e}{mc} \sum_k \left(p_k + \frac{e}{c} A_k\right) \frac{\partial A_k}{\partial x_i} + e \frac{\partial \varphi}{\partial x_i} \tag{5}$$

From Eq. (4) we have

$$p_i = m\dot{x}_i - (e/c)A_i \tag{6}$$

Hence

$$\dot{p}_i = m\ddot{x}_i - \frac{e}{c} \dot{A}_i = m\ddot{x}_i - \frac{e}{c}\left(\sum_k \frac{\partial A_i}{\partial x_k} \dot{x}_k + \frac{\partial A_i}{\partial t}\right) \tag{7}$$

Combining Eqs. (5) and (7) and using Eq. (4), we get

$$m\ddot{x}_i = e\left(\frac{\partial \varphi}{\partial x_i} + \frac{1}{c} \frac{\partial A_i}{\partial t}\right) + \frac{e}{c} \sum_k \dot{x}_k\left(\frac{\partial A_i}{\partial x_k} - \frac{\partial A_k}{\partial x_i}\right)$$

which, according to relations (2), can be written as

$$m\ddot{x}_i = -eE_i - \frac{e}{c} (\mathbf{v} \times \mathbf{H})_i \tag{8}$$

Thus we have obtained the Lorentz equation of motion.

We notice that according to Eq. (4) the Hamiltonian (1) can be rewritten in the form

$$\mathcal{H} = \sum_k \tfrac{1}{2}mv_k^2 - e\varphi = T - e\varphi \tag{9}$$

where T is the kinetic energy. One can see that the total energy (represented here by the Hamiltonian) does not depend on the vector potential. This means that the magnetic field does not influence the energy of a particle. The energy can be changed only by the electric field.

We shall now find the motion of the free electron in a constant uniform magnetic field. This can be done using the equation of motion (8); we shall, however, employ the Hamilton formalism to emphasize the similiarity to the quantum mechanical description.

We now have to specify the vector potential $\mathbf{A}$. Since it is the magnetic field that is the observable physical quantity, all vector potentials which satisfy Eq. (2) should be physically equivalent (gauge invariance). There are two commonly used forms of the vector potential: asymmetric (or Landau) gauge and symmetric gauge. For a magnetic field directed along the z axis, $\mathbf{H} = [0, 0, H]$, these read

$$\mathbf{A} = [-Hy, 0, 0] \qquad \text{(Landau gauge)}$$
$$\mathbf{A} = [-\tfrac{1}{2}Hy, \tfrac{1}{2}Hx, 0] \qquad \text{(symmetric gauge)} \tag{10}$$

In the following considerations we shall be using the Landau gauge, but as has been mentioned, the choice of gauge is immaterial for all physically observable quantities.* Our Hamiltonian now becomes

$$\mathscr{H} = \frac{1}{2m}\left(p_x - \frac{e}{c}Hy\right)^2 + \frac{p_y^2}{2m} + \frac{p_z^2}{2m} \tag{11}$$

Hence, according to Eq. (3),

$$\dot{p}_x = 0, \qquad \dot{p}_y = -\omega_c\left(p_x - \frac{e}{c}Hy\right), \qquad \dot{p}_z = 0 \tag{12}$$

and

$$\dot{x} = \frac{1}{m}\left(p_x - \frac{e}{c}Hy\right), \qquad \dot{y} = \frac{p_y}{m}, \qquad \dot{z} = \frac{p_z}{m} \tag{13}$$

where $\omega_c = eH/mc$ is the cyclotron frequency. Hence $p_x = \text{const} = p_x^0$ and $p_z = \text{const} = p_z^0$ Combining Eqs. (12) and (13),

$$m\ddot{y} = -\omega_c\left(p_x^0 - \frac{e}{c}Hy\right) \tag{14}$$

Denoting $Y = y - (cp_x^0/eH)$, we get $\ddot{Y} = -\omega_c^2 Y$; hence

$$Y = r_c \sin(\omega_c t + \alpha) \tag{15}$$

---

* The approach based on the equation of motion (which deals only with the field H and does not require the specification of the vector potential) can be found in many textbooks. See, e.g., ([1]).

where r_c and α are constant. Thus

$$y = r_c \sin(\omega_c t + \alpha) + (c p_x^0/eH) \tag{16}$$

and from Eq. (13) we now get

$$\dot{x} = \frac{1}{m}\left(p_x^0 - \frac{e}{c}Hy\right) = -\omega_c r_c \sin(\omega_c t + \alpha) \tag{17}$$

Integrating over time

$$x = r_c \cos(\omega_c t + \alpha) + x_0 \tag{18}$$

so that x and y satisfy the relation

$$(x - x_0)^2 + (y - y_0)^2 = r_c^2 \tag{19}$$

Thus we have arrived at the result that the projection of the motion on the plane perpendicular to the magnetic field direction is a circle with radius r_c and center $\mathbf{r}_0(x_0, y_0)$. We want to determine r_c and x_0. From Eq. (16)

$$\dot{y} = r_c \omega_c \cos(\omega_c t + \alpha) \tag{20}$$

which, combined with Eq. (17), gives

$$r_c^2 = (\dot{x}^2 + \dot{y}^2)/\omega_c^2 = 2\varepsilon_\perp/\omega_c^2 m \tag{21}$$

where $\varepsilon_\perp$ is the energy of the transverse motion in xy plane. We conclude from the above relation that the velocity of the electron on the cyclotron orbit is constant in time. The cyclotron radius is determined by this velocity and the magnetic field intensity, $r_c = v_c/\omega_c$.

According to Eqs. (16) and (13)

$$y_0 = c p_x^0/eH = (\dot{x}/\omega_c) + y \tag{22}$$

and x_0 can now be determined from Eq. (19) to give

$$x_0 = -(\dot{y}/\omega_c) + x = -(c p_y/eH) + x \tag{23}$$

It follows from Eqs. (12) and (13) that along the z direction we have the free motion with the constant velocity $v_z^0 = p_z^0/m$. Thus the motion along the magnetic field is not affected by the field, this being of course due to the fact that as long as the velocity is parallel to the field there is no force acting on the electron.

There are three independent constants of motion: the energy of the transverse motion $\varepsilon_\perp$, the energy of the parallel motion ε_z, and the y component of the center y_0 (or p_x^0).

$$\varepsilon_\perp = \tfrac{1}{2}m(\dot{x}^2 + \dot{y}^2) = \tfrac{1}{2}mr_c^2\omega_c^2 = \text{const} \tag{24}$$

$$\varepsilon_z = \tfrac{1}{2}m\dot{z}^2 = p_z^{02}/2m = \text{const} \tag{25}$$

The energy of the motion does not depend on the position of the center. The angular momentum

$$L_z = (\mathbf{r} \times \mathbf{p})_z = \tfrac{1}{2}m\omega_c(r_c^2 - r_0^2) \tag{26}$$

also remains constant in time. However, since r_c^2 can be expressed by the energy $\varepsilon_\perp$, L_z is not an independent constant of the motion.

QUANTUM MECHANICAL DESCRIPTION

We shall now consider the problem of the free electron placed in a uniform magnetic field according to the nonrelativistic quantum mechanics. The Hamiltonian for our problem (the Pauli equation) is

$$\mathscr{H} = (1/2m) \sum_k \left(p_k + \frac{e}{c} A_k\right)^2 + \mu_B H \sigma_z \tag{27}$$

where σ_z is the Pauli spin matrix, $\mu_B = e\hbar/2mc$ the Bohr magneton, and we have chosen as before the magnetic field along the z direction. The momentum operator is now given by $p_k = (\hbar/i)\, \partial/\partial x_k$. We have introduced the spin into the Hamiltonian (27), as it is strictly a quantum mechanical concept. Since σ_z acts only on the spin variables, the Hamiltonian (27) can be separated into coordinate and spin parts by looking for a solution in the form of a product. In the following considerations we shall be concerned only with the coordinate part.

The eigenenergies of our problem can be found without specifying the gauge of the vector potential. We observe that for $\mathbf{H}$ in the z direction $A_z = 0$ [see Eq. (10)]. Hence we can separate the z coordinate in our equation by looking for a solution in the form

$$\Psi(x, y, z) = \exp(ik_z z)f(x, y)$$

The eigenvalue problem now becomes

$$(1/2m)(P_x^2 + P_y^2)f(x, y) = \varepsilon' f(x, y) \tag{28}$$

where as before $P_i = p_i + (e/c)A_i$, and $\varepsilon' = \varepsilon - (\hbar^2 k_z^2/2m)$. Using Eq. (2) the commutation relation for the canonical momenta P_x and P_y can be obtained as

$$(1/2m)[P_x, P_y] = -i(\hbar\omega_c/2) \tag{29}$$

Relations (28) and (29) are now in the form analogous to the familiar one-dimensional harmonic oscillator, and both eigenvalues and eigenfunctions may be obtained by factorization (also called the canonical quantization) [see, e.g., (²)]. Thus we get $\varepsilon' = \hbar\omega_c(n + \frac{1}{2})$ and the total energy is

$$\varepsilon = \hbar\omega_c(n + \tfrac{1}{2}) + (\hbar^2 k_z^2/2m) \qquad n = 0, 1, 2\ldots \tag{30}$$

The last term is the energy of the free motion along the z direction with the constant momentum $p_z = \hbar k_z$. The first term is the energy of the traverse cyclotron motion, which is now quantized, with adjacent energy levels separated by $\hbar\omega_c$. These are called Landau levels (³). The operators P_+ and P_- given by

$$P_\pm = P_x \pm iP_y$$

are raising (creation) and lowering (annihilation) operators for the eigenfunctions of our problem.*

If the spin part of the Hamiltonian (27) is also taken into account, we get

$$\varepsilon_{\uparrow\downarrow} = \hbar\omega_c(n + \tfrac{1}{2}) + (\hbar^2 k_z^2/2m) \pm \tfrac{1}{2}\hbar\omega_c \tag{31}$$

the plus sign corresponding to spin value $m_s = +\frac{1}{2}$ and the minus sign to $m_s = -\frac{1}{2}$. It follows then that for the free electrons in a magnetic field the spin splitting is equal to the Landau splitting of the levels. We can also say that the spectroscopic spin splitting factor $g = (\varepsilon_\uparrow - \varepsilon_\downarrow)/\mu_B H = 2$ for the free electrons.

In order to obtain the eigenfunctions explicitly, and make connection with the classical solutions, we shall now solve the problem using the Landau gauge in Eq. (10). Hence we again deal with the Hamiltonian (11), and the Schrödinger equation is

$$\left[\frac{1}{2m}\left(p_x - \frac{e}{c}Hy\right)^2 + \frac{p_y^2}{2m} + \frac{p_z^2}{2m}\right]\psi(x, y, z) = \varepsilon\psi(x, y, z) \tag{32}$$

---

* A very elegant and general treatment of the motion in a magnetic field based on the canonical momenta formalism can be found in (⁴).

It is easy to see that, looking for the solution in the form

$$\psi(x, y, z) = \frac{1}{4\pi^2} \exp(ik_x x + ik_z z)\varphi(y) \tag{33}$$

we can separate the variables in the Hamiltonian. In other words $p_x = \hbar k_x$ and $p_z = \hbar k_z$ are good quantum numbers (p_x and p_z commute with the Hamiltonian). The equation for $\varphi(y)$ becomes

$$\left(-\frac{\hbar^2}{2m}\frac{\partial^2}{\partial y^2} - \hbar\omega_c k_x y + \frac{m\omega_c^2}{2}y^2\right)\varphi = \left(\varepsilon - \frac{\hbar^2 k_x^2}{2m} - \frac{\hbar^2 k_z^2}{2m}\right)\varphi \tag{34}$$

By changing the variable

$$y = y' + k_x L^2, \qquad L = (\hbar c/eH)^{1/2} \tag{35}$$

we can transform the equation into

$$\left[-\frac{\hbar^2}{2m}\frac{\partial^2}{\partial y'^2} + \frac{m\omega_c^2}{2}y'^2\right]\varphi = \left(\varepsilon - \frac{\hbar^2 k_z^2}{2m}\right)\varphi \tag{36}$$

which is identical with the problem of the one-dimensional harmonic oscillator. For the eigenvalues we immediately get again Eq. (30), and the eigenfunctions are

$$\varphi = \Phi_n\left(\frac{y - y_0}{L}\right), \qquad n = 0, 1, 2\ldots \tag{37}$$

From Eq. (35) $y_0 = k_x L^2$, and the Φ_n are the harmonic oscillator functions

$$\Phi_n(x) = C_n \exp(-x^2/2)H_n(x) \tag{38}$$

where the H_n are the Hermite polynomials and the C_n are the normalization factors:

$$C_n = (2^n n! \sqrt{\pi} L)^{-1/2} \tag{39}$$

L defined in Eq. (35) is called the magnetic radius and its length is characteristic for the quantum mechanical treatment of our problem.

With this normalization our wave functions, defined in Eq. (33), satisfy the orthonormality relations

$$\int_{-\infty}^{+\infty} \psi_{n, k_x, k_z}(x, y, z)\psi_{n', k_x', k_z'}(x, y, z)d^3r = \delta_{nn'}\delta(k_x - k_x')\delta(k_z - k_z') \tag{40}$$

There are three independent constants of motion

$$\varepsilon_\perp = \hbar\omega_c\left(n + \frac{1}{2}\right), \qquad \varepsilon_z = \frac{\hbar^2 k_z^2}{2m} = \frac{p_z^{02}}{2m}, \qquad y_0 = k_x L^2 = \frac{p_x^0 c}{eH} \tag{41}$$

where y_0 is the y coordinate of the center of oscillations. This is in complete analogy to the classical description of the problem (except for the quantization). We notice that the quantity $x_0 = (-cp_y/eH) + x$ is also a constant of the motion, (its operator commutes with the Hamiltonian), and it represents the x coordinate of the center, just as in the classical description. Now, however, the two operators do not commute with each other,

$$[x_0, y_0] = iL^2 \tag{42}$$

i.e., there is an uncertainty in the measurement of the center, and L^2 represents the minimum area in the xy plane in which the center may be located. The wave function (37) indicates, that we deal with the harmonic oscillations along the y direction with the frequency ω_c around the center y_0. We also have harmonic motion along the x direction around x_0, and this is not in contradiction with the fact that $p_x^0 = const$, since $p_x = m\dot{x}$ $-(eA/c)$, and not $p_x = m\dot{x}$!. The wave function (37) seems to indicate, that we have the oscillatory motion only in the y direction, whereas the classical solution clearly shows the oscillations in both x and y directions. This apparent contradiction is due to the fact that the quantum mechanical solution (37) describes the states with an undetermined value of x_0 [since y_0 is exactly determined by Eq. (41)].

If we introduce an operator of the cyclotron radius r_c according to Eq. (19),

$$r_c^2 = (x - x_0)^2 + (y - y_0)^2 = (m\omega_c)^{-2}\left[\left(p_x - \frac{e}{c}Hy\right)^2 + p_y^2\right] \tag{43}$$

where we have used definitions (22) and (23). Thus in the energy representation r_c is known exactly:

$$r_c^2 = \frac{2\hbar\omega_c(n + \frac{1}{2})}{m\omega_c^2} \qquad r_c = (2n + 1)^{1/2}L \tag{44}$$

and the uncertainty in locating points on the orbit is solely due to the uncertainty in locating the center of the orbit. The relation (44) indicates that the permitted orbits have areas

$$S = \pi(2n + 1)L^2 \tag{45}$$

We also notice that for high quantum numbers n, when the energy quantum $\hbar\omega_c$ is small compared to the total energy (i.e., in the classical limit), we have $\hbar\omega_c(n + \frac{1}{2}) \approx \frac{1}{2}mv_c^2$. Hence from Eq. (44) $r_c = v_c/\omega_c$—the classical result!

We shall now determine the degeneracy of levels and the density of states for an electron in a magnetic field. First we observe that just as in the classical description the energy in Eq. (30) does not depend on the quantum number k_x. In other words, it does not depend on the position of the center y_0. This means that all the wave functions given in Eq. (33) with different k_x belong to the same eigenenergy. Since in free space there is no restriction on k_x, we conclude that each energy level has an infinite degeneracy.

Now we consider an electron confined in a rectangular box of dimensions being L_1, L_2, and L_3. If we apply periodic boundary conditions to the function (33) in x and z directions, then the allowed values of k_x and k_z are

$$k_x = (2\pi/L_1)l_1, \qquad k_z = (2\pi/L_3)l_3 \tag{46}$$

where l_1 and l_3 are integers.

Now we assume that the center of the orbit y_0 is confined between $y = 0$ and $y = L_2$ (this leads to modifications of the functions with centers located within the distance L from the boundaries, but the relative number of such functions is very small as long as $L \ll L_2$). According to Eq. (41) this means that $0 \leq k_x \leq L_2/L^2$. Since within the unit length of k_x there are $L_1/2\pi$ allowed k_x values, the total number of allowed k_x values is $N = L_1 L_2/2\pi L^2$, and this is the degeneracy of each Landau level (for a given value of k_z). When the magnetic field is strong, so that the interval $\hbar\omega_c$ between the levels for a given value of k_z is large, there is a strong bunching of the allowed values around the levels given by $\varepsilon = \hbar\omega_c(n + \frac{1}{2})$.

We now want to calculate the density of states per unit energy associated with each quantized level.* We write

$$\varepsilon_z = \varepsilon - \hbar\omega_c(n + \tfrac{1}{2}) = \hbar^2 k_z^2/2m \tag{47}$$

The number of allowed values of k_z such that $|k_z| < k_{z0}$ is $L_3 k_{z0}/\pi$. The number of allowed values of ε_z such that $\varepsilon_z < \varepsilon_{z0}$ is therefore also equal to $L_3 k_{z0}/\pi$. Expressing this by ε_z, we obtain $L_3(2m\varepsilon_{z0})^{1/2}/\pi\hbar$. The number of allowed levels in the interval $d\varepsilon_z$, $\varrho(\varepsilon_z)\,d\varepsilon_z$, is therefore given by

$$\varrho(\varepsilon_z)d\varepsilon_z = \frac{L_3(2m)^{1/2}}{2\pi\hbar\sqrt{\varepsilon_z}}\,d\varepsilon_z \tag{48}$$

---

* Here we follow rather closely the derivation given in (^{5}). While logically simpler derivations exist, they involve more advanced concepts [see, e.g., (^{6})].

In order to get the total density of states, we have to sum over all the levels for which $\varepsilon_z \geq 0$, remembering that each has degeneracy $N = L_1 L_2 / 2\pi L^2$. Using the definition (47), this finally gives

$$\varrho(\varepsilon) = V\left(\frac{2m}{\hbar}\right)^{1/2} \frac{1}{4\pi^2 L^2} \sum_{n=0}^{n_{max}} \left[\varepsilon - \hbar\omega_c\left(n + \frac{1}{2}\right)\right]^{-1/2} \tag{49}$$

where V is the volume of the box. As follows from the derivation, the summation extends over the values of n for which the square root is still positive. This means that the nth Landau level contributes to the density of states only for energies $\varepsilon \geq \hbar\omega_c(n + \frac{1}{2})$. Thus, due to the fact that the magnetic field quantization breaks the three-dimensional $\varepsilon(k)$ relation into a series of one-dimensional $\varepsilon(k_z)$ relations given by Eq. (30), the nonsingular density of states for the free electrons, $\varrho(\varepsilon) \propto \varepsilon^{1/2}$, now becomes singular whenever $\varepsilon = \hbar\omega_c(n + \frac{1}{2})$. This effect is readily observed in a number of phenomena in solids. If the spin splitting is not included, the density of states should be multiplied by the factor 2.

Finally it should be mentioned that the problem of a free electron in a magnetic field can also be solved with the symmetric gauge of the vector potential $\mathbf{A}$ given in Eq. (10). It is then convenient to use the cylindrical coordinate system, and the solutions are given in terms of the associated Laguerre polynomials [7]*.

REFERENCES

1. L. D. Landau and E. M. Lifshitz, *Classical Theory of Fields*, Addison-Wesley Inc., Reading, Massachusetts, 1959.
2. P. A. M. Dirac, *Principles of Quantum Mechanics*, Oxford Clarendon Press, 1958.
3. L. D. Landau, *Z. Physik* **64**: 629 (1930).
4. M. H. Johnson and B. A. Lippmann, *Phys. Rev.* **76**: 828 (1949).
5. R. A. Smith, *Wave Mechanics of Crystalline Solids*, Chapman and Hall, London, 1963, p. 382.
6. R. Kubo et al., in: *Solid State Physics*, Vol. 17 (F. Seitz and D. Turnbull, eds.), Academic Press, New York, 1965, p. 291.
7. R. B. Dingle, *Proc. Roy. Soc.* **A211**: 500 (1952).
8. J. C. Chapman, Ph. D. Thesis, Massachusetts Institute of Technology, 1966.

* A detailed discussion of these solutions can be found in [8].

Chapter 14

BLOCH ELECTRONS IN CROSSED ELECTRIC AND MAGNETIC FIELDS

W. Zawadzki

*National Magnet Laboratory**
Massachusetts Institute of Technology
Cambridge, Massachusetts
and
Institute of Electron Technology
Polish Academy of Sciences
Warsaw, Poland

The purpose of this chapter is to describe recent theoretical and experimental studies on the properties of electrons in solids in the presence of crossed electric and magnetic fields. Semiconducting materials are best suited for such studies for the following reasons. First, the relatively high resistivity of semiconductors makes it possible to apply electric fields of the order of 10^3 V/cm in pure materials and 10^4–10^5 V/cm in reverse-biased p–n junctions. Second, the small effective masses of electrons makes it possible for one to observe the magnetic field effects clearly, since the separation between Landau levels is given in terms of $\hbar\omega_c = \hbar eH/mc$. For a magnetic field of $H = 100$ kG and the free-electron mass this is equal to ∼0.001 eV, so that in semiconductors like InSb ($m \approx 0.013\, m_0$) energy separations of the order of 0.1 eV are achievable. Finally, due to low free-carrier densities a thin slab of a semiconducting material can transmit a light beam fairly well, which greatly facilitates the optical studies. Also, as we shall see later, narrow forbidden energy gaps, with resulting strong interaction between the conduction and valence bands, introduce a variety of interesting physical effects.

* Supported by the U. S. Air Force Office of Scientific Research.

ELECTRON MOTION IN CROSSED FIELDS

We begin by using the one-band effective-mass approximation (EMA) for a simple nondegenerate energy band. In other words, we treat the Bloch electrons in an energy band as free electrons, with the free-electron mass m_0 replaced by the effective mass m, which reflects the interaction of other energy bands with the band in question. In this approximation the cross-field problem bears many similiarities to the magnetic field case, which has been treated in detail in the previous chapter in this volume.

The Schrödinger equation for the envelope function of the conduction electron in crossed fields is then

$$\left[\frac{1}{2m}\left(\mathbf{p} + \frac{e}{c}\mathbf{A}\right)^2 + e E r\right] f(\mathbf{r}) = \varepsilon f(\mathbf{r}) \tag{1}$$

The magnetic field is introduced by means of the vector potential $\mathbf{A}$. For the magnetic field in the z direction we choose $\mathbf{A} = [-Hy, 0, 0]$, so that $\mathbf{H} = \operatorname{curl} \mathbf{A} = [0, 0, H]$. We take the transverse electric field in the y direction, $\mathbf{E} = [0, E, 0]$.

The full solution of the problem is given by

$$\Psi_l(\mathbf{r}) = f(\mathbf{r}) u_{l0}(\mathbf{r}) \tag{2}$$

where $u_{l0}(\mathbf{r})$ is the periodic part of the Bloch function for the lth band taken at $k = 0$, i.e., at the bottom of the band.

With the above choice of the vector and scalar potentials the coordinates in Eq. (1) can be separated, just as in the case of magnetic field alone. We look for the solution in the form

$$f(x, y, z) = (1/4\pi^2) \exp(ik_x x + ik_z z)\varphi(y) \tag{3}$$

It is well known that the solutions of Eq. (1) without the electric field term are given by the harmonic oscillator functions (see previous chapter). It can be seen that the electric field, introducing the term linear in y, does not change the character of the solutions. It simply shifts the potential well. Thus the solutions of our problem are

$$\varphi_1(y) = C_{n_1}\Phi_{n_1}\left(\frac{y - k_{x1}L^2 + (eEL^2/\hbar\omega_{c1})}{L}\right) \tag{4}$$

where the Φ_n are the harmonic oscillator functions. The eigenenergies are

$$\mathscr{E}_1 = \hbar\omega_{c1}\left(n_1 + \frac{1}{2}\right) + \frac{\hbar^2 k_{z1}^2}{2m} - eE k_{x1}L^2 - \frac{m_1 c^2}{2}\frac{E^2}{H^2} \tag{5}$$

where $L = (\hbar e/eH)^{1/2}$ is the radius of the first Landau orbit and the sub-sript 1 denotes the conduction band. Here C_n is the normalization factor. For $E = 0$ Eqs. (4) and (5) reduce to the eigenfunctions and the eigenvalues for the electron in a magnetic field alone.

In order to interpret the results obtained we consider the classical motion of an electron in crossed fields. The nonrelativistic equation of motion

$$m\ddot{\mathbf{r}} = -e\mathbf{Er} - (e/c)(\mathbf{v} \times \mathbf{H}) \tag{6}$$

can easily be solved [see, e.g., (¹), p. 54]. The resulting motion for the initial condition $\dot{\mathbf{r}} = \mathbf{r} = 0$ is a superposition of the oscillatory cyclotron motion in xy plane with the cyclotron frequency $\omega_c = eH/mc$ and the motion with constant velocity $v_{tr} = cE/H$ *transverse* to both fields (Fig. 1). The kinetic energy of this motion $\mathscr{E}_{\text{kin}} = \frac{1}{2}mv_{tr}^2$ is developed due to shift of the center of the cyclotron orbit in the direction of the electric field. The term $k_x L^2$ represents the center of the cyclotron motion in the absence of the electric field. Thus $eEL^2/\hbar\omega_c$ in Eq. (4) corresponds to the shift of the center due to electric field. The term $eEk_x L^2$ in Eq. (5) is the energy of the magnetic orbit in the electric field, and the last term is the energy due to the shift of the orbit in electric field.

Two points should be emphasized here. First, the motion described is

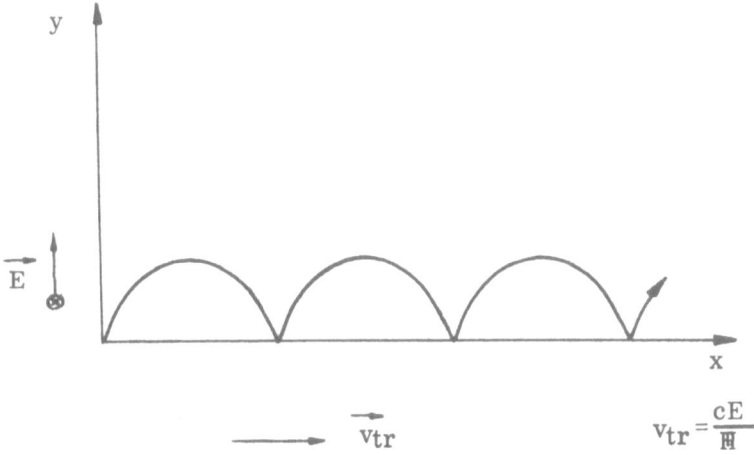

Fig. 1. Motion of a free nonrelativistic electron in crossed electric and magnetic fields. (The magnetic field is perpendicular to the plane of the paper.) The motion is a superposition of the cyclotron motion with the angular frequency $\omega_c = eH/m_0c$ and the motion with constant velocity $\vartheta_{tr} = cE/H$ transverse to the direction of both fields.

essentially of the magnetic-field type: the cyclotron motion is still the dominant feature and there is no average acceleration along the electric field. Second, if the transverse velocity becomes comparable to the light velocity (for free electrons), the presented description is no longer valid, since the relativistic equation of motion should be used. This suggests that the above description holds only for small values of the E/H ratio. This is confirmed by Eq. (5), which leads into obvious trouble when this ratio becomes too large. It should be borne in mind, however, that the scale of values is completely different for the free electrons and electrons in a solid. We shall return to this point later.

INTERBAND MAGNETOOPTICAL EFFECTS IN CROSSED FIELDS

It can be seen from Eq. (3) that the electric field lowers magnetic Landau levels in the conduction band (we do not consider the linear E term, since it is of no importance for the transitions in which the momentum is conserved). Aronov ([2]) observed that the hole magnetic levels are *raised* by the electric field. The eigenfunctions and eigenenergies for holes can be obtained from Eqs. (4) and (5) by changing the sign and value of the effective mass. Thus for the holes in a simple band we obtain

$$f_2(\mathbf{r}) = \frac{1}{4\pi^2} \exp(ik_{x2}x + ik_{z2}z)\Phi_{n2}\left[\frac{y - k_x{}^2L^2 - (eEL^2/\hbar\omega_{c2})}{L}\right] \quad (7)$$

and

$$\mathscr{E}_2 = -\mathscr{E}_g - \hbar\omega_{c2}\left(n_2 + \frac{1}{2}\right) - \frac{\hbar^2 k_{z2}^2}{2m} - eEk_{x2}L^2 + \frac{m_2c^2}{2}\frac{E^2}{H^2} \quad (8)$$

where the zero energy is chosen at the bottom of the conduction band.

Hence the shift of Landau levels in crossed electric field can be observed most readily in interband transitions between the valence and the conduction band. If the electric field is strong enough, these transitions should be observable even for photon energies below that of the direct gap of a material. There is, however, an additional effect due to the electric field, which we shall now consider. For this purpose we have to examine more closely the probability of optical transitions between two states.

It is well known [see, e.g. ([3])] that the perturbation due to the light wave has, in the electric dipole approximation, the form

$$\mathscr{H}' = (eE_0/2\omega m_0)\mathbf{\epsilon P} \quad (9)$$

where E_0 is the amplitude of the electric vector of radiation, ϵ its polarization vector, ω its frequency, and $\mathbf{P}$ is the kinetic momentum operator in the presence of a magnetic field: $\mathbf{P} = \mathbf{p} + (e/c)\mathbf{A}$. The transition probability is proportional to the matrix element of the perturbation $\mathcal{H}'$ between initial and final states. For our perturbation using the wave functions given by Eq. (2) we get the matrix element in the form

$$\langle \psi_f \mid \mathcal{H}' \mid \psi_i \rangle = \alpha \langle u_f \mid u_i \rangle \langle f_f \mid \epsilon \mathbf{P} \mid f_i \rangle + \alpha \langle u_f \mid \epsilon \mathbf{p} \mid u_i \rangle \langle f_f \mid f_i \rangle \qquad (10)$$

where we have separated the integrals containing the periodic parts u, as they are rapidly varying functions (they are periodic with the lattice periodicity), as compared to the slowly varying envelope functions (they extend over many lattice sites). Here $\alpha = e\mathcal{E}_0/2\omega m_0$. It is easy to see that the first term represents the intraband part of the matrix element, since in general u_l is orthogonal to $u_{l'}$, so that unless $u_f \equiv u_i$ the first term vanishes. Here we are interested in the second term, which is responsible for the interband transitions. We notice that this term involves the scalar product of envelope functions for the initial and final states of the electron. Thus we have to evaluate the quantity

$$\frac{1}{(4\pi^2)^2} C_{n_1} C_{n_2} \int_{-\infty}^{+\infty} \exp i(k_{x2} - k_{x1})x \, dx \int_{-\infty}^{+\infty} \exp i(k_{z2} - k_{z1})z \, dz$$

$$\times \int_{-\infty}^{+\infty} \Phi_{n_2}\left(\frac{y - y_2}{L}\right) \Phi_{n1}\left(\frac{y - y_1}{L}\right) dy$$

where

$$y_1 = k_{x1}L^2 - (eEL^2/\hbar\omega_{c1}) \qquad \text{and} \qquad y_2 = k_{x2}L^2 + (eEL^2/\hbar\omega_{c2})$$

The first two integrals yield the selection rules $k_{x2} = k_{x1}$ and $k_{z2} = k_{z1}$, i.e., we deal with the direct transitions. In the absence of the electric field $y_1 = y_2$, and taking into account the orthonormality of the harmonic oscillator functions, we get the well-known selection rule for the interband magnetooptical transitions $n_1 = n_2$. However, in the presence of an electric field the arguments of the functions are different, and there is no selection rule for the quantum number n. We shall not give the final result of the integration here; this can be found in the original paper (^{2}). We just state that for small electric fields the formerly allowed transitions ($\Delta n = 0$) should decrease in intensity, whereas formerly forbidden transitions ($\Delta n = \pm 1, \pm 2, \ldots$) should now be possible. There should be a peak in the optical absorption whenever the photon energy is equal to the energy of an interband transition:

$$\hbar\omega = \mathscr{E}_1 - \mathscr{E}_2 = \mathscr{E}_g + \hbar\omega_{c1}(n_1 + \tfrac{1}{2}) + \hbar\omega_{c2}(n_2 + \tfrac{1}{2})$$
$$- \tfrac{1}{2}(m_1 + m_2)c^2(E^2/H^2) \tag{11}$$

Here we have used Eqs. (5) and (8), taking into account the fact that the k_z terms determine the shape of the absorption peaks rather than the energy at which they occur. The change of selection rules due to the presence of the transverse electric field has been observed experimentally on pure germanium samples at low temperatures [4]. However, electric fields which can be applied to the pure material (up to 10^3 V/cm) are too small to shift the energy levels appreciably according to Eq. (11), especially at the high magnetic fields (100 kG) necessary for good resolution of Landau levels.

Experiments with higher electric fields have been performed on a germanium p–n junction which, if biased in a reverse direction, made it possible to achieve fields up to 10^5 V/cm. It turned out to be more convenient to measure the dispersion of light rather than the absorption. We cannot go into the theory of dispersion here [5]. It will be sufficient to say that whenever the photon energy is equal to that of an interband transition between Landau levels there should be a peak in the Voigt phase shift, its sign de-

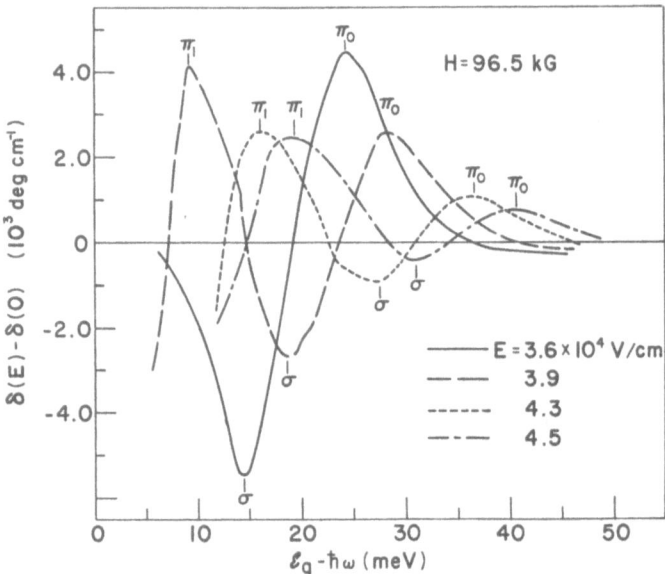

Fig. 2. Voigt effect in crossed fields for photon energies below the direct gap of germanium for $H = 96.5$ kG and various values of the electric field strength. The transitions shift to lower energies and their intensities decrease as E increases.

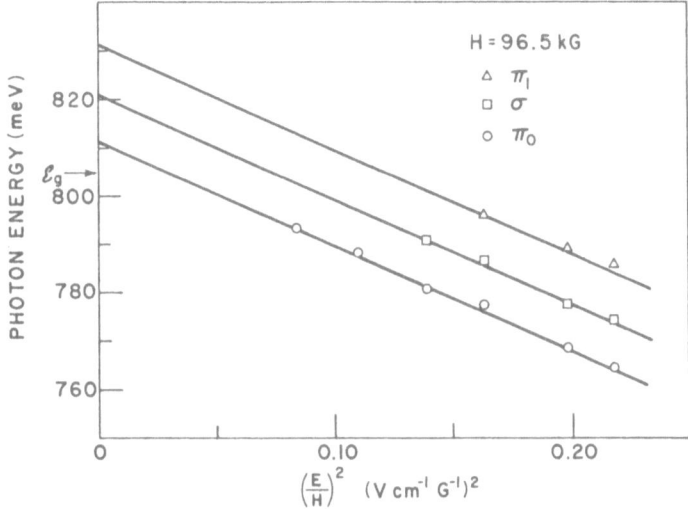

Fig. 3. Photon energies of Voigt-effect peaks in crossed fields for $H = 96.5$ kG as a function of $(E/H)^2$. All three peaks undergo equally large shifts proportional to $(E/H)^2$.

pending on a spin selection rule for the transition in question. Thus, following the shift of peaks, we can follow the change in the position of Landau levels. Figure 2 shows the Voigt effect measured for photon energies below the direct gap of germanium (^{6}). We deal with electric fields strong enough to pull the interband transitions *into* the energy gap.

As can be seen, for increasing electric field the transitions shift to the lower energies. According to Eq. (11), the shift should be proportional to E^2/H^2. Figure 3 shows this to be in agreement with experiment.

It turns out that the transitions π_0 and σ are the ones allowed at $E = 0$ ($\Delta n = 0$), whereas π_1 is a transition made possible by the electric field ($\Delta n = -1$). In view of this the fact that the intensity of the π_1 transition also decreases with increasing electric field seems to be in contradiction with the theoretical predictions. The detailed examination shows, however, that we deal here with such high electric fields that in this range the intensity of the formerly forbidden transitions should also decrease. In another experiment both E and H have been varied in such a way that E/H was kept constant (see Fig. 4).

Thus the last term in Eq. (11) is constant, and the energies vary only due to the change of the two magnetic terms, and are proportional to the magnetic field. This again is in agreement with the experiment, as shown in Fig. 5.

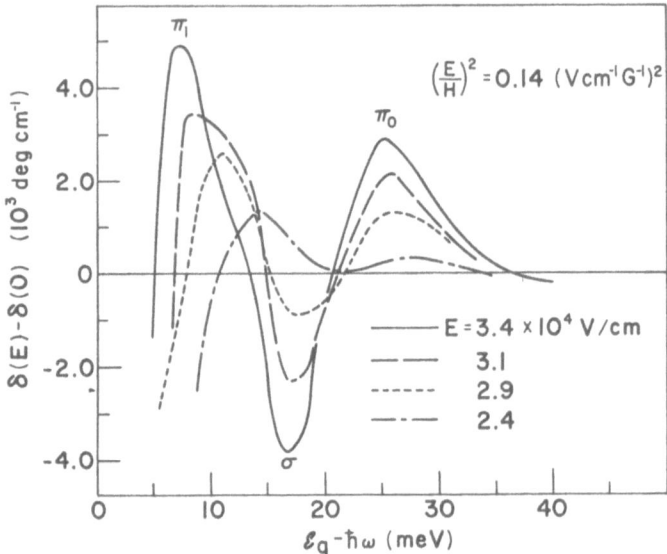

Fig. 4. Voigt effect in crossed fields for photon energies below the direct gap for a constant value of $(E/H)^2$ and various values of E and H. With increasing magnetic field the peaks shift to higher photon energies.

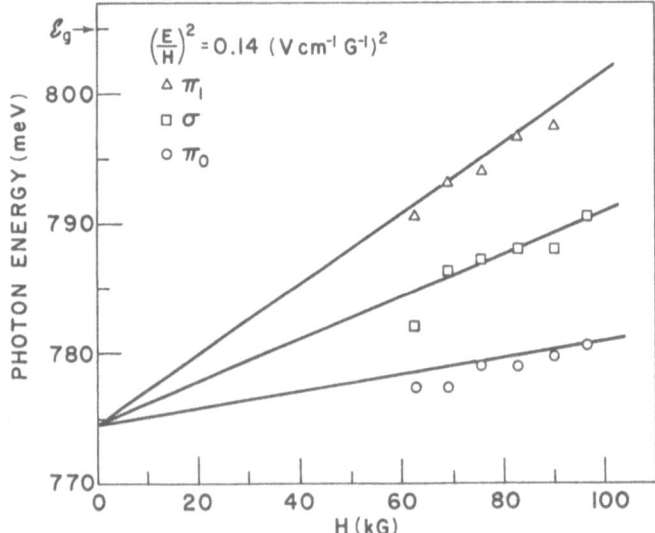

Fig. 5. Photon energies of Voigt-effect peaks as a function of H in crossed fields for $(E/H)^2 = 0.14$ (V-cm^{-1}/G)^2. All three peaks undergo a shift proportional to the magnetic field.

There are some interesting effects due to the fact that the valence band of germanium, from which the transitions occur, is not simple but is a degenerate band. Thus the outline of the theory presented above is, strictly speaking, applicable only to the conduction band (with spin included). The details of this problem and the exact comparison with the experiment is presented in the original publication ([6]).

THE TWO-BAND MODEL

As mentioned above, the one-band effective-mass approximation (EMA) leads to trouble when E/H ratio becomes too large. For example, it would lead to the conclusion that the Landau levels in conduction and valence bands could cross each other as E/H is increased. This obviously does not make much sense physically. It is well known that the one-band effective mass approximation is valid only as long as the considered energies counted from the bottom of a band are small compared to the gaps separating energy bands. Without external fields this criterion is: $\hbar^2 k^2/2m \ll \mathscr{E}_g$, and it restricts the values of the wave vector k for which the band is still parabolic (i.e., $\mathscr{E} \propto k^2$). In the case of crossed fields this means that we have to have independently [see Eq. (5)]

$$\hbar\omega_c\left(n + \frac{1}{2}\right) \ll \mathscr{E}_g, \qquad \frac{mc^2}{2}\,\frac{E^2}{H^2} \ll \mathscr{E}_g, \qquad \frac{\hbar^2 k_z^2}{2m} \ll \mathscr{E}_g \qquad (12)$$

[the restriction $eEk_xL^2 \ll \mathscr{E}_g$ can be shown to be equivalent to the second restriction in Eq. (12)]. The intuitive reasoning presented above has been confirmed by a rigorous examination of conditions restricting the validity of the one-band EMA ([7]).

If the restrictions in Eq. (12) are not satisfied, one has to go beyond the one-band EMA. The simplest adequate description is based on a so-called two-band model, which we shall outline presently. This model was first developed by Kane ([8]) for InSb in the absence of external fields, and then applied by Bowers and Yafet ([9]) to the magnetic field case. It turned out to be very useful in the case of crossed fields ([10]).

We start with the full Hamiltonian for our problem (excluding spin)

$$\mathscr{H} = \frac{1}{2m_0}\left(\mathbf{p} + \frac{e}{c}\,\mathbf{A}\right)^2 + e\mathbf{E} \cdot \mathbf{r} + V(\mathbf{r})$$

where $V(\mathbf{r})$ is the periodic potential of the lattice. We calculate this Hamiltonian in the Kohn–Luttinger representation,

$$\chi_{nk} = \exp(i\mathbf{k} \cdot \mathbf{r})u_{n0}(\mathbf{r})$$

where u_{n0} are the periodic parts of the Bloch functions for the nth band taken at $k = 0$. They satisfy the eigenvalue problem

$$\left[\frac{1}{2m_0}\mathbf{p}^2 + V(\mathbf{r})\right]u_{n0} = \mathscr{E}_n u_{n0}$$

where $\mathscr{E}_n$ is the energy at the bottom of the nth band (at $k = 0$).

After returning to the coordinate representation* we get the set of equations for the envelope functions in the form

$$\sum_{n' \neq n} [(\mathbf{P}^2/2m_0) + e\mathbf{E} \cdot \mathbf{r} + \mathscr{E}_n - \mathscr{E})\delta_{n'n} + \pi_{n'n} \cdot \mathbf{P}]f_{n'}(\mathbf{r}) = 0 \quad (13)$$

Here $\pi_{n'n} = (1/m_0)\langle u_{n'} \mid \mathbf{p} \mid u_n \rangle$ are determined by the interband matrix element of momentum. The summation is over all energy bands. The normal way of arriving at the one-band EMA ([11]) is to remove the nondiagonal part of the Hamiltonian to the second order in the nondiagonal elements $\pi_{n'n} \cdot \mathbf{P}$ using a suitable canonical transformation. The final result is the set of separated equations (each for one envelope function f_n) with the free mass replaced by an effective one

$$\frac{1}{m_n} = \frac{1}{m_0} + 2 \sum_{n' \neq n} \frac{\mid \pi_{n'n} \mid^2}{\mathscr{E}_{n'} - \mathscr{E}_n} \quad (14)$$

This would be the approximation with which we started [Eq. (1)]. However, this procedure is possible only if the restrictions (12) are satisfied. Since we want to go beyond this, we shall apply another approximation —namely, we neglect all the energy bands but two. Strictly speaking, we take into account one s-like conduction band and three degenerate (at $k = 0$) p-like valence bands, separated by the gap $\mathscr{E}_g$,

$$u_1 = \mid S \rangle, \qquad u_2 = -\frac{1}{\sqrt{2}} \mid X - iY \rangle,$$

$$u_3 = \mid Z \rangle, \qquad u_4 = \frac{1}{\sqrt{2}} \mid X + iY \rangle \quad (15)$$

where S, X, Y, and Z denote the symmetry properties of the atomic-like orbitals.

---

* The details of this calculation can be found in ([11]). Also see ([7]).

In our spinless model there is no spin-orbit interaction. Thus we are left with the set of four differential equations. To facilitate the computations, we neglect the $P^2/2m_0$ term in Eq. (13). This is equivalent to the omission of the free-mass term in Eq. (14), which normally is just a small correction.

We first solve our model in the absence of both fields $E = 0$ and $H = 0$, i.e., $\mathbf{P} = \mathbf{p}$. After a simple calculation we arrive at the following relation between the energy and momentum

$$\mathscr{E} = -\frac{\mathscr{E}_g}{2} \pm \left[\left(\frac{\mathscr{E}_g}{2} \right)^2 + \mathscr{E}_g \frac{\hbar^2 k^2}{2m} \right]^{1/2} \tag{16}$$

where $1/m = 2 \mid \pi_{12} \mid^2 / \mathscr{E}_g$, in agreement with Eq. (14) for two interacting bands. Equation (16) represents a simplified Kane's formula (the plus and minus signs correspond to the conduction and valence bands, respectively). We not that for $\hbar^2 k^2/2m \ll \mathscr{E}_g$ the square root can be expanded to give $\mathscr{E} = \hbar^2 k^2/2m$, in agreement with the one-band EMA.

INTRABAND MAGNETOOPTICAL EFFECTS IN CROSSED FIELDS

As follows from Eq. (8), in the one-band EMA all Landau levels in the conduction band are lowered at the same rate due to the electric field. This means that the intervals between the levels in the same band do not change. Calculation shows that the probabilities of the cyclotron resonance transitions, i.e., the transitions between two adjacent Landau levels, also are not changed by the electric field. Thus, to this approximation, the transverse electric field does not have any effect on the intraband transitions.

However, this is not true as soon as magnetic field is high enough that the first of criteria (12) is not satisfied. This is the case we are going to consider now. Namely, we assume that the magnetic field is high but that the electric field is small and may be treated as a perturbation. Thus we first have to solve the two-band model in the presence of the magnetic field alone to get the functions and energies that will serve as a basis for the perturbation treatment.

Again, the set of four differential equations obtained from Eq. (13) can be solved by substitution if the free-electron term is neglected. The complete solutions are now given in terms of linear combinations of the envelopes, in the form of harmonic oscillator functions, and the periodic parts

$$\Psi_n = \frac{1}{4\pi^2} \exp(ik_x x + ik_z z)(C_1 \phi_n u_1 + C_2 \phi_{n+1} u_2 + C_3 \phi_n u_3 + C_4 \phi_{n-1} u_4) \quad (17)$$

where $\phi_n = \phi_n[(y - y_0)/L]$ denote the normalized harmonic oscillator functions, and $y_0 = k_x L^2$. The eigenenergies corresponding to these solutions are

$$\mathscr{E}_n = -\frac{\mathscr{E}_g}{2} \pm \left[\left(\frac{\mathscr{E}_g}{2}\right)^2 + \mathscr{E}_g D_n\right]^{1/2} \quad (18)$$

where

$$D_n = \hbar\omega_c (n + \tfrac{1}{2}) + (\hbar^2 k_z^2/2m) \quad (19)$$

and the effective mass is defined as before. It is easy to see the close analogy between Eqs. (16) and (19), with the free energy replaced by the Landau energy. In terms of the above energies the coefficients C for the conduction band solutions are given as

$$C_1 = -\frac{\mathscr{E}_g + \mathscr{E}_n}{[\mathscr{E}_g(\mathscr{E}_g + \mathscr{E}_n + 2D_n)]^{1/2}}, \qquad C_3 = i\left[\frac{\hbar^2 k_z^2/2m}{\mathscr{E}_g + \mathscr{E}_n + 2D_n}\right]^{1/2}$$

$$C_2 = -\frac{i}{\sqrt{2}}\left[\frac{\hbar\omega_c(n + 1)}{\mathscr{E}_g + \mathscr{E}_n + 2D_n}\right]^{1/2}, \qquad C_4 = \frac{i}{\sqrt{2}}\left[\frac{\hbar\omega_c n}{\mathscr{E}_g + \mathscr{E}_n + 2D_n}\right]^{1/2}$$

where for the energy $\mathscr{E}_n$ in Eq. (18) the plus sign is to be taken. One can see that for magnetic energies small compared to the energy gap we have $\mathscr{E}_n \to D_n$, $|C_1| \to 1$, and $|C_2| = |C_3| = |C_4| \to 0$, so that we are again in the limit of one-band EMA [cf. Eq. (2)]. However, for strong magnetic fields or high quantum numbers n the mixing of the p-like functions in the solution for the conduction band is appreciable.

It can be seen from Eq. (18) that now the Landau levels are no longer uniformly spaced. It can, however, be shown with the use of the functions of Eq. (17) and the procedure similar to that outlined in Eq. (10) that for intraband transitions still only the transitions with the selection rule $\Delta n = \pm 1$ are allowed. In other words, even in the two-band model only the fundamental cyclotron resonance transitions between two adjacent Landau levels are possible.

The solutions given by Eq. (17) and energies Eq. (18) can now be used for the perturbation theory with the electric field term used as the perturbing potential

$$\mathscr{H}_{\text{pert}} = eEy \quad (20)$$

There are very few nonvanishing matrix elements of our perturbation.

Using the notation $\phi_n[(y - y_0)/L] = |n\rangle$, we get

$$\langle n \mid eEy \mid n \rangle = eEk_x L^2,$$

$$\langle n \mid eEy \mid n - 1 \rangle = \langle n - 1 \mid eEy \mid n \rangle = eEL(n/2)^{1/2} \tag{21}$$

all other matrix elements vanish. It is now very easy to calculate the perturbed functions and energy levels. It turns out that for the first correction to the energy we get

$$\mathscr{E}_n^{(1)} = eEk_x L^2 \tag{22}$$

i.e., the same as for the simple one-band model (it does not depend on n). The second corrections to the two lowest levels are obtained as, approximately,

$$\mathscr{E}_0^{(2)} = -\tfrac{1}{2}(eEL)^2/(\mathscr{E}_1 - \mathscr{E}_0) \tag{23}$$

$$\mathscr{E}_1^{(2)} = -\frac{(eEL)^2}{2}\left[\frac{2}{\mathscr{E}_2 - \mathscr{E}_1} - \frac{1}{\mathscr{E}_1 - \mathscr{E}_0}\right] \tag{24}$$

We see that in the limit of small magnetic energies, when $\mathscr{E}_n = D_n$, we get $\mathscr{E}_1^{(2)} = \mathscr{E}_0^{(2)} = -(eEL)^2/2\hbar\omega_c$, which is exactly equal to the last term in Eq. (5). Therefore in this limit the second-order perturbation theory gives results identical with those obtained from the exact solutions. In our case, however, as follows from Eq. (18), $\mathscr{E}_1 - \mathscr{E}_0 > \mathscr{E}_2 - \mathscr{E}_1$, so that for the increasing electric field the first Landau level is lowered *faster* than the zeroth one, and the higher levels are lowered still faster. The situation for the parabolic band (one-band EMA) and for the nonparabolic two-band model is shown in Fig. 6. Thus we can see that if the first restriction of Eq. (12) is not fulfilled, even small electric fields affect different Landau levels differently, and this effect should be observable in the intraband cyclotron resonance transitions. Moreover, it should be possible to observe this change in transition probability. The solutions of Eq. (17) perturbed by the potential (20) can be used for the calculation of intraband (cyclotron resonance) transitions following the procedure outlined in Eq. (10). The numerical results obtained this way for InSb ($\mathscr{E}_g = 0.22$ eV, $m = 0.013 \, m_0$) for a magnetic field of 70 kG are shown in Fig. 7.

Two features can be observed. First, the transition probability for formerly allowed transitions ($\Delta n = +1$) is decreased by the electric field. Second, new (formerly forbidden) transitions with $\Delta n = +2$ are now possible. In general, all higher harmonics of cyclotron resonance ($\Delta n = 2, 3, \ldots$) are possible, but for small electric fields $\Delta n = 2$ are the most important

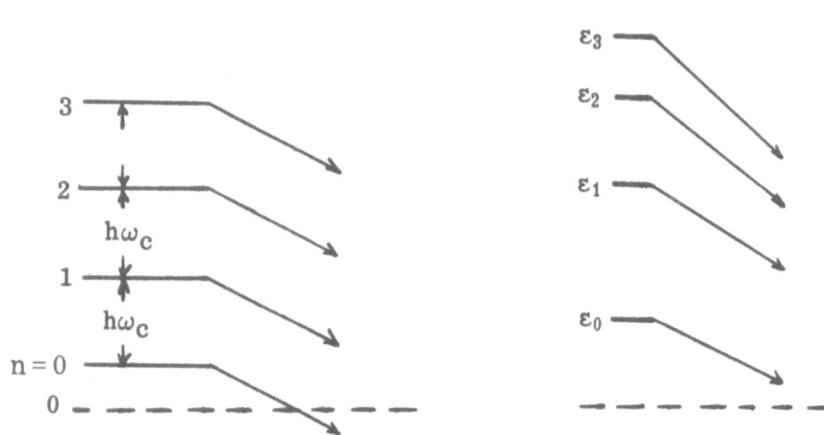

Fig. 6. Landau levels and their shift in a transverse electric field for a simple energy band and in the two-band model (schematically). The levels in a simple band are equally spaced, and they all undergo the same shift in the electric field. According to the two-band model the spacing is not uniform, and the higher levels are lowered with the electric field faster than the lower ones.

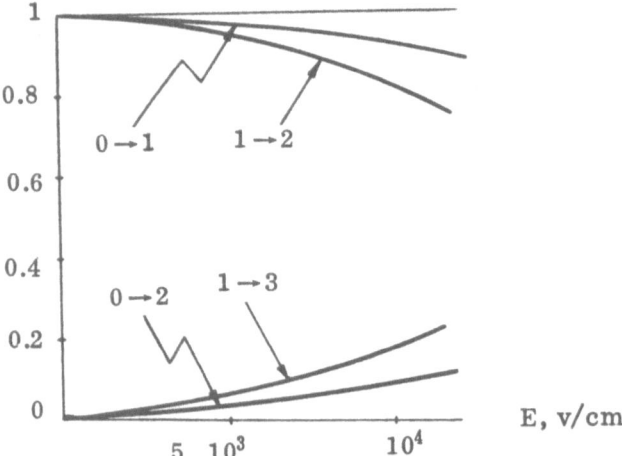

Fig. 7. Probabilities of two lowest fundamental cyclotron resonance transitions ($\Delta n = 1$) and two transitions forbidden for $E = 0$ ($\Delta n = 2$) as functions of the electric field strength according to the two-band model (schematically). (Both fundamental transitions are normalized to 1 at $E = 0$.) The transitions with $\Delta n = 1$ decrease in intensity, whereas the formerly forbidden transitions become possible.

ones. The above effects are in close analogy with the interband magneto-optical effects in crossed fields predicted by Aronow and discussed above.

The appearance of harmonics in cyclotron resonance can also be explained classically. According to the two-band model, we can regard the electron as having a mass depending on velocity. This is due to the fact that the relation (16) is strictly analogous to the relativistic dependence, with the following correspondences ([10])

$$2m_0 c^2 \to \mathscr{E}_g, \qquad m_0 \to m \tag{25}$$

We now go back to Fig. 1, which shows the classical motion in crossed fields. As long as the mass does not depend on velocity we have strictly harmonic motion in the y direction with the cyclotron frequency ω_c. This corresponds to the absorption or radiation of only this frequency ω_c. If, however, the mass does depend on velocity, the electron is heavier at the apex than at the turning point on the x axis. This obviously introduces unharmonicity to the motion, resulting in the possibility of absorbing and emitting harmonics of the fundamental frequency.*

MAGNETIC- AND ELECTRIC-TYPE MOTION

We now consider the case of electric fields high enough that the second of the restrictions in Eq. (12) is also not satisfied. First, we shall try to gain an intuitive understanding of the problem by using classical concepts. Namely, we consider the classical motion of the free electron in crossed fields according to relativistic mechanics, since we know from Eq. (16) that our electron should behave in a "relativistic" way, though in a different scale of values ([10]). The solutions of classical equations for the free electrons are well known [see, e.g. ([1]), p. 59]. It turns out that as long as the electric and the magnetic field strengths are such that

$$cE/H < c \tag{26}$$

the motion is essentially of the magnetic-field type, i.e., there are oscillations and in average there is no acceleration in the electric field direction. If, however, the applied electric field is so strong (or the magnetic field so weak), that the reverse inequality holds,

$$cE/H > c$$

---

* I am greatly indebted to Professor A. B. Pippard for raising this point in private discussion.

the motion is essentially of the electric-field type, i.e., there are no oscillations and in average there is an acceleration in the electric field direction.

Thus the relativistic equation of motion gives the physically reasonable result that for high magnetic and low electric fields the motion is of the magnetic type, and in the reverse situation the character is also reversed.

In order to apply this result to our electron in a semiconductor, we have to determine the maximum velocity in the two-band model, corresponding to the light velocity for the free electrons. We have, according to Eq. (25),

$$c = (2m_0c^2/2m_0)^{1/2} \rightarrow (\mathscr{E}_g/2m)^{1/2} = v_{max} \qquad (27)$$

Now the criterion (26) becomes

$$\frac{cE}{H} < v_{max} = \left(\frac{\mathscr{E}_g}{2m}\right)^{1/2}, \qquad \text{or} \qquad \omega_c^2 > \frac{2e^2E^2}{m\mathscr{E}_g} \qquad (28)$$

We note that $v_{max} \approx 10^8$ cm/sec (it does not vary strongly from one material to the other). Thus it is much smaller than the maximum velocity for a free electron in vacuum.

The same result, i.e., the transition from one character of the motion to the other, can be obtained by a rigorous treatment using our two-band model. We start again from the set (13), with both fields present, and restrict the band scheme to one conduction and three degenerate valence bands, just as in the magnetic-field case. If the free-electron term is neglected, the set of four equations can again be solved, and we obtain the differential equation for the envelope function in the form [10]

$$\left\{-\frac{\hbar^2}{2m}\frac{\partial^2}{\partial y^2} - \alpha y + \frac{m}{2}\left[\left(\frac{eH}{mc}\right)^2 - \frac{2e^2E^2}{m\mathscr{E}_g}\right]y^2\right.$$
$$\left. + \frac{3}{8}\frac{\hbar^2}{m}\left[\frac{eE}{eEy - \mathscr{E}_g - \mathscr{E}}\right]^2\right\}\varphi(y) = \lambda\varphi(y) \qquad (29)$$

where $\alpha = \hbar\omega_c k_x - eE(\mathscr{E}_g + 2\mathscr{E})/\mathscr{E}_g$ and $\lambda = \mathscr{E}(\mathscr{E} + \mathscr{E}_g)/\mathscr{E}_g - (\hbar^2k_x^2/2m) - (\hbar^2k_z^2/2m)$. The solution in terms of $\varphi(y)$ is

$$f(x, y, z) = \exp(ik_x x + ik_z z) \mid eEy - \mathscr{E}_g - \mathscr{E}\mid^{1/2}\varphi(y)$$

The last term on the left-hand side of Eq. (29) is normally very small and can be neglected for the purposes of our discussion. The first three terms constitute the Schrödinger eigenvalue problem, the second and third playing the role of potential energy. It is the third term, however, that determines the character of the solutions, since the term linear in y simply shifts the potential well, just as in the one-band EMA Hamiltonian. In general Eq. (29) looks very much like the one-band EMA, with one important exception: in Eq. (1) the width of the potential well (the coefficient in front of y^2) was $m\omega_c^2/2$, i.e., was determined only by the magnetic field. Hence the quantization was in terms of $\hbar\omega_c$. Now, however, the width of the potential well is also determined by the electric field. If the electric field is sufficiently small that $(eH/mc)^2 \gg 2e^2E^2/m\mathscr{E}_g$, we obtain the the one-band EMA Hamiltonian. But this inequality is exactly equivalent to the second restriction in Eq. (12) for the validity of the EMA! If the electric field term is not negligible, but not predominant either, we find that the cyclotron frequency depends on the electric field $\omega_{\text{eff}}^2 = \omega_c^2 - (2e^2E^2/m\mathscr{E}_g)$. Finally, if the electric field is so strong that the electric field term predominates, the whole coefficient in front of y^2 changes sign and we no longer have the potential well; in other words, there is no quantization, and the solutions are given in terms of the Weber functions. The critical point between the quantized (magnetic-type) and the nonquantized (electric-type) motion is given by

$$\omega_c^2 = 2e^2E^2/m\mathscr{E}_g \tag{30}$$

in exact agreement with the classical result (28).

The solutions of Eq. (29) for electric fields higher than the critical one given by Eq. (30) were found and used to calculate the properties of interband transitions both with and without photon assistance. They have the exponential tunneling behavior characteristic of the electric-field type solutions, and not the oscillatory behavior demonstrated above and characteristic of the magnetic-field type solutions. This is in agreement with the experimental results obtained in the high electric field range ([14]).

REFERENCES

1. L. D. Landau and E. M. Lifshitz, *Classical Theory of Fields*, Addison-Wesley Inc., Reading, Massachusetts, 1959.
2. A. G. Aronov, *Soviet Phys.–Solid State (English Transl.)* **5**: 402 (1963). The value of the matrix element calculated in this paper contains an error [see ([4])].
3. L. M. Roth, B. Lax, and S. Zwerdling, *Phys. Rev.* **114**: 90 (1959).
4. Q. H. F. Vrehen and B. Lax, *Phys. Rev. Letters* **12**: 471 (1964); Q. H. F. Vrehen, *Phys. Rev.* **145**: 675 (1966).
5. W. Zawadzki, Q. H. F. Vrehen, and B. Lax, *Phys. Rev.* **148**: 849 (1966).
6. Q. H. F. Vrehen, W. Zawadzki, and M. Reine, *Phys. Rev.* **158**: 702 (1967).
7. J. Zak and W. Zawadzki, *Phys. Rev.* **145**: 536 (1966).
8. E. O. Kane, *J. Phys. Chem. Solids* **1**: 249 (1957).
9. R. Bowers and Y. Yafet, *Phys. Rev.* **115**: 1165 (1959).

10. W. Zawadzki and B. Lax, *Phys. Rev. Letters* **16**: 1001 (1966).
11. J. M. Luttinger and W. Kohn, *Phys. Rev.* **97**: 869 (1955).
12. H. Weiler, W. Zawadzki, and B. Lax, *Phys. Rev.*, **163**: 733 (1967); W. Zawadzki, in: *Proceedings of the Advanced Summer Institute on Tunneling Phenomena in Solids, Risø, Denmark*, 1967, Plenum Press, New York, 1968.
13. A. G. Aronow and G. E. Pikus, *Soviet Phys.–JETP* (*English Transl.*) **24**: 188, 339 (1967); also, in: "Proceedings of the International Conference on the Physics of Semiconductors, Kyoto, 1966," *J. Phys. Soc. Japan Suppl.* **21**: 608 (1966).
14. M. Reine, Q. H. F. Vrehen, and B. Lax, *Phys. Rev.*, **163**: 726 (1967).

Chapter 15

GROUP-THEORETICAL APPROACH TO THE PROBLEM OF A BLOCH ELECTRON IN A MAGNETIC FIELD

J. Zak

Department of Physics, Technion
Israel Institute of Technology
Haifa, Israel

INTRODUCTION

Translational symmetry is a fundamental property of solids and is connected to their ordered structure. It is because of this property that solids are considered a special class of materials, and many phenomena have their origin in translational symmetry. The best known result that follows from symmetry is the Bloch theorem, which enables one to describe the motion of an electron in a crystal by means of wave-like Bloch functions.

In this chapter we will be interested in the information one can obtain about the dynamics of an electron in both a periodic potential and a constant magnetic field (Bloch electron in a magnetic field) by using translational symmetry. Although usual translations are no longer symmetry operations for the Hamiltonian of the problem, it is nevertheless possible to define magnetic translations which differ from the usual translations by a gauge factor and which do commute with the Hamiltonian. We will show here how these magnetic operators can be used for a qualitative description of the energy spectrum and the wave functions of a Bloch electron in a magnetic field.

A FREE ELECTRON IN A MAGNETIC FIELD

Let us first review different ways for describing the motion of an electron in a magnetic field. The Hamiltonian of the problem is

$$\mathcal{H} = \frac{[\mathbf{p} + (e/c)\mathbf{A}]^2}{2m} \tag{1}$$

329

where $\mathbf{p}$ is the momentum, e is the charge of the electron ($e > 0$), c is the velocity of light, $\mathbf{A}$ is the vector potential, for which we choose the gauge $\mathbf{A} = \frac{1}{2}[\mathbf{H} \times \mathbf{r}]$ ($\mathbf{H}$ is the magnetic field and $\mathbf{r}$ is the radius vector of the electron), and m is the mass of the electron. One can easily check that the operators

$$\mathbf{p} - (e/c)\mathbf{A} \tag{2}$$

commute with the Hamiltonian (1). If one chooses the magnetic field $\mathbf{H}$ in the z direction, then the x and y components of the operators (2) are simply connected with the center of the classical orbit $\mathbf{r}_0$ ([1])

$$\left[\frac{1}{m}\left(\mathbf{p} - \frac{e}{c}\mathbf{A}\right)\right]_{x,y} = -[\boldsymbol{\omega} \times \mathbf{r}_0]_{x,y} \tag{3}$$

where $\boldsymbol{\omega} = (e/mc)\mathbf{H}$. The z component of the operator (2) is just p_z. Since the operators (2) commute with the Hamiltonian (1), they are constants of motion and they can be used for specifying the quantum states. Classically we know that the x and y components of the center of the electron orbit in a magnetic field do not vary during the motion and that the velocity along the magnetic field $v_z = p_z/m$ is constant. Quantum mechanically not all three components of the operators (2) can be used for specifying the states, because the x and y components do not commute:

$$\left[p_x - \frac{e}{c}A_x, p_y - \frac{e}{c}A_y\right] = i\frac{\hbar^2}{L^2} \tag{4}$$

where $L = (\hbar c/eH)^{1/2}$ is the cyclotron radius. One usually specifies states by means of the eigenvalues of p_z and, say $p_x - (e/c)A_x$, i.e., one requires the eigenfunctions of the Hamiltonian (1) to also be eigenfunctions of p_z and $p_x - (e/c)A_x$. By requiring this one gets the well-known result for the wave functions ([1])

$$\psi_{nk_xk_z}(x, y, z) = A[\exp(-i(xy/2L^2) + ik_xx + ik_zz)]\phi_n(y - y_0) \tag{5}$$

where A is a normalization constant, $y_0 = k_xL^2$ is the y component of the center of the orbit, and n is the quantum number of the energy. Such a specification of the wave function is the conventional one. It turns out that for treating the problem of a Bloch electron in a magnetic field there is another way of presenting the function which is very often much more convenient. For doing this we shall use the following operators:

$$\tau(\mathbf{R}) = \exp\left\{\frac{i}{\hbar}\left(\mathbf{p} - \frac{e}{c}\mathbf{A}\right) \cdot \mathbf{R}\right\} \tag{6}$$

where $\mathbf{R}$ is a constant vector. Without the vector potential in the exponent the operators (6) would be usual translations; with the vector potential they are called magnetic translations. It is clear that the operators (6) commute with the Hamiltonian (1). The question now is, how can these operators be used for specifying the eigenstates of the Hamiltonian (1)? One can easily check that

$$\tau(\mathbf{R}_1)\tau(\mathbf{R}_2) = \tau(\mathbf{R}_2)\tau(\mathbf{R}_1) \exp\left[-2\pi i \frac{\mathbf{H} \cdot (\mathbf{R}_1 \times \mathbf{R}_2)}{(hc/e)}\right] \tag{7}$$

This means that in general the magnetic translation operators for different $\mathbf{R}$ do not commute. However, if we choose $\mathbf{R}_1$ and $\mathbf{R}_2$ (both perpendicular to $\mathbf{H}$) in such a way that

$$\frac{\mathbf{H} \cdot (\mathbf{R}_1 \times \mathbf{R}_2)}{(hc/e)} = 1 \tag{8}$$

and $\mathbf{R}_3 \parallel \mathbf{H}$ (otherwise arbitrary), then all the operators $T(\mathbf{R})$ defined on vectors

$$\mathbf{R} = n_1\mathbf{R}_1 + n_2\mathbf{R}_2 + \mathbf{R}_3 \tag{9}$$

with any integers n_1 and n_2 will commute. We therefore have an infinite set of magnetic translation operators for integer n_1 and n_2 and for any $\mathbf{R} \parallel \mathbf{H}$ that commute with the Hamiltonian (1) and also among themselves. Such a set of operators can be used for specifying the quantum states. We can require that the eigenstates of the Hamiltonian (1) satisfy the relations (in analogy with the Bloch theorem):

$$\tau(\mathbf{R})\psi_k = \exp(i\mathbf{k} \cdot \mathbf{R})\psi_k \tag{10}$$

where $\mathbf{k}$ is a wave vector and it assumes the following values:

$$\mathbf{k} = \alpha_1\mathbf{K}_1 + \alpha_2\mathbf{K}_2 + \mathbf{k}_3 \tag{11}$$

Here $0 \le \alpha_1, \alpha_2 < 1$, $\mathbf{k}_3$ is an arbitrary vector in the direction of the magnetic field, and $\mathbf{K}_1$ and $\mathbf{K}_2$ are the unit cell vectors of the reciprocal lattice to the lattice built on the vectors $\mathbf{R}_1$ and $\mathbf{R}_2$:

$$\mathbf{K}_i \cdot \mathbf{R}_j = 2\pi \, \delta_{ij}, \qquad i, j = 1, 2 \tag{12}$$

The requirement (10) does not fully specify the wave functions; the quantum number of the energy operator will be needed for doing so, as was also pointed out in (5). For comparison of (10) with (5) let us consider the motion only in the plane perpendicular to $\mathbf{H}$, because in the direction of

H there is really no difference in the specification of the states. The two-dimensional motion in (5) is specified k_x (apart from the energy index) on an infinite axis, while in (10) we have a two-dimensional vector $\mathbf{k}$ which varies in the area of the first Brillouin zone built on the vectors $\mathbf{K}_1$ and $\mathbf{K}_2$. It is well known that the energy does not depend on k_x in the specification (5), and it clearly cannot depend on the $\mathbf{k}$ that specifies the functions (10). To prove this statement, let us note that the operators $T(\mathbf{R})$ built on the vectors (9) are not all the operators that commute with the Hamiltonian (1) and that there are operators $T(\mathbf{R}')$ with $\mathbf{R}' \neq \mathbf{R}$ which commute with the Hamiltonian but do not commute with $T(\mathbf{R})$. It is known that this usually leads to degeneracy.

Indeed, assume that ψ_k corresponds to some energy E, then since $T(\mathbf{R}')$ commutes with the Hamiltonian, the function $T(\mathbf{R}')\psi_k$ will belong to the same eigenvalue E, and so also will a function $\tau(\delta\mathbf{R})\psi_k$, where $\delta\mathbf{R}$ is an infinitesimal vector. But as can be shown from the commutation relation (7):

$$\tau(\delta\mathbf{R})\psi_k = \text{const } \psi_{k+2\pi\frac{\mathbf{H}\times\delta\mathbf{R}}{(hc/e)}} \tag{10'}$$

Since $\delta\mathbf{R}$ assumes any arbitrary value, ψ_k with any $\mathbf{k}$ will belong to the same energy, e.g., the energy is $\mathbf{k}$-independent.

As was already mentioned, the function ψ_k has in addition to $\mathbf{k}$ another index which specifies the energy. Regardless of what this other quantum number is let us find the density of states that are connected with all the values of the $\mathbf{k}$ vector in the plane perpendicular to the magnetic field. For the elementary area $d^2\mathbf{k}$ the density of states is $d^2\mathbf{k}/(2\pi)^2$ and the whole density of states will be obtained by replacing $d^2\mathbf{k}$ by the area of the first Brillouin zone, which is $\mathbf{K}_1 \times \mathbf{K}_2$. One therefore finds for the density of states $|\mathbf{K}_1 \times \mathbf{K}_2|/(2\pi)^2$. According to the definition (12) of $\mathbf{K}_i$, this is equal to $1/|\mathbf{R}_1 \times \mathbf{R}_2|$ and from Eq. (8) we finally get for the density of states the usual result eH/hc. This way of obtaining the density of states cannot be used in the conventional description of the problem because, although we deal with a two-dimensional motion, there is only one component of the $\mathbf{k}$ vector, say k_x, in the wave function.

Let us now turn to the functions (10) and get an explicit expression for them. To do this we can use the general projection operator techniques (^{2}). Starting with any function ψ one can construct a function*

$$\psi_k = \sum_{\mathbf{R}} (\exp i\mathbf{k} \cdot \mathbf{R})\tau(-\mathbf{R})\psi \tag{13}$$

---

* For simplicity we omit a phase factor which can appear because the group of operators $T(\mathbf{R})$ is Abelian but not cyclic [see (^{2})].

which satisfies the relation (10). The summation vector in (13) is defined in (9). It is clear that in general the function ψ_k will not be a solution for the Hamiltonian (1). However, if ψ is a solution of the Hamiltonian (1), so will ψ_k. For example, we achieve this by choosing ψ equal the function (5). We then get

$$\psi_{nk} = \text{const} \sum_{n_2} \exp(ik_y n_2 R_2)\tau(-n_2 R_2)\psi_{nk_x k_z} \tag{14}$$

where $\mathbf{k}$ is a three-dimensional vector. There was no need to symmetrize along the x and z directions because the function $\psi_{nk_x k_z}$ was already properly symmetrized. As another example, take a two-dimensional motion and let ψ in (13) be a classical orbit-like solution of the Hamiltonian (1). The ψ_k will represent a network of orbits which is closely related to Pippard's model for describing magnetic breakdown ([3]).

To this point we have treated the free electron in a magnetic field, and what we did was just a different representation for the problem. However, things become much more interesting and the representation we introduced much more useful when one deals with the problem of a Bloch electron in a magnetic field.

A BLOCH ELECTRON IN A MAGNETIC FIELD

When in addition to the magnetic field a periodic potential $V(\mathbf{r})$ is present the Hamiltonian of the problem is

$$\frac{[\mathbf{p} + (e/c)\mathbf{A}]^2}{2m} + V(\mathbf{r}) \tag{15}$$

The magnetic translation operators that commute with this Hamiltonian are

$$\tau(\mathbf{R}_n) = \exp\left\{\frac{i}{\hbar}\left(\mathbf{p} - \frac{e}{c}\mathbf{A}\right) \cdot \mathbf{R}_n\right\} \tag{16}$$

where

$$\mathbf{R}_n = n_1\mathbf{a}_1 + n_2\mathbf{a}_2 + n_3\mathbf{a}_3 \tag{17}$$

with integer n_1, n_2, and n_3 are Bravais lattice vectors. We see therefore that by introducing a periodic potential the translational symmetry of the problem became lower because the operators (16) form a subset of the operators (6). In order to make the influence of the periodic potential on the energy spectrum and the wave functions clear, let us restrict ourselves again to a two-dimensional motion in a plane perpendicular to the magnetic field. For a free electron we chose two arbitrary vectors $\mathbf{R}_1$ and $\mathbf{R}_2$ that satisfied the condition (8) and by means of them we have constructed a commuting set of operators $T(\mathbf{R})$. When a periodic potential is present this is no longer possible, because the available translations $\mathbf{a}_1$ and $\mathbf{a}_2$ are fixed. What one can do, however, it to choose special magnetic fields that satisfy the condition

$$\frac{\mathbf{H} \cdot \mathbf{a}_1 \times \mathbf{a}_2}{(hc/e)} = \frac{1}{N} \tag{18}$$

where N is an integer. Condition (18) is called the rationality condition on the magnetic field [2]. To get the connection with the free-electron case we assume that

$$\mathbf{R}_1 = l_1 \mathbf{a}_1, \qquad \mathbf{R}_2 = l_2 \mathbf{a}_2, \qquad N = l_1 l_2 \tag{19}$$

By this assumption conditions (8) and (18) are identical. We can now compare the reciprocal lattice $\mathbf{K}_1$, $\mathbf{K}_2$ built on $\mathbf{R}_1$, $\mathbf{R}_2$ with the reciprocal lattice $\mathbf{b}_1$, $\mathbf{b}_2$ built on $\mathbf{a}_1$, $\mathbf{a}_2$:

$$\mathbf{b}_1 = l_1 \mathbf{K}_1, \qquad \mathbf{b}_2 = l_2 \mathbf{K}_2 \tag{20}$$

For the free-electron case the energy was completely $\mathbf{k}$-independent, i.e., when $\mathbf{k}$ varies over all the $\mathbf{K}_1$, $\mathbf{K}_2$ Brillouin zone the energy does not change. When the lattice is present this degeneracy is partially removed because in relation (10') we cannot use any arbitrary vector $\delta\mathbf{R}$ but have to restrict ourselves by $\mathbf{a}_1$ and $\mathbf{a}_2$. From (10'), (12), (18), (19), and (20) we have

$$\tau(\mathbf{a}_1)\psi_k = \text{const } \psi_{k+\mathbf{K}_2/l_1}$$
$$\tau(\mathbf{a}_2)\psi_n = \text{const } \psi_{k-\mathbf{K}_1/l_2} \tag{21}$$

Relations (21) show us that the area in the Brillouin zone $(\mathbf{K}_1, \mathbf{K}_2)$ where changes of $\mathbf{k}$ can lead to changes of the energy is given by the vectors $\mathbf{K}_1/l_2$ and $\mathbf{K}_2/l_1$. This area is N times smaller than the area built on $\mathbf{K}_1$ and $\mathbf{K}_2$. The energy is therefore N-fold degenerate, instead of the infinite degeneracy in the free-electron case. From symmetry arguments it thus follows that the periodic potential of the lattice broadens the Landau levels (the energy is $\mathbf{k}$-dependent) and the degeneracy becomes finite.

The above picture can be illustrated by the following example. Assume that the periodic potential in the Hamiltonian (15) can be treated as a perturbation. In order to find the influence of this perturbation on the Landau levels one has in general to solve a secular equation where the matrix elements are given by the periodic potential between different functions (5) for a free electron in a magnetic field. Having to solve the secular equation or the diagonalization process is avoided by using the symmetry-adapted functions (14). Since these functions are eigenfunctions of the magnetic translation operators the matrix elements

$$(\psi_k, V\varphi_{k'}) \qquad (22)$$

will vanish for $\mathbf{k} \neq \mathbf{k}'$. The influence of the periodic potential on the Landau levels will therefore be given by the diagonal matrix element (22):

$$(\psi_k, V\psi_k) \qquad (23)$$

This matrix element leads to an unequal spacing of Landau levels and to their broadening ([2]), the latter being a consequence of symmetry lowering by the periodic potential. The energy as a function of $\mathbf{k}$ will satisfy the periodicity condition

$$\varepsilon\left(\mathbf{k} + \frac{\mathbf{b}_1}{N}\right) = \varepsilon\left(\mathbf{k} + \frac{\mathbf{b}_2}{N}\right) = \varepsilon(\mathbf{k}) \qquad (24)$$

where $\mathbf{b}_1$ and $\mathbf{b}_2$ are the reciprocal lattice vectors (20). The Brillouin zone where the energy varies as a function of $\mathbf{k}$ is thus $N \times N$ smaller than the usual Brillouin zone of the periodic lattice. The lattice in real space that corresponds to such a small Brillouin zone is given by $N\mathbf{a}_1$, $N\mathbf{a}_2$. This is exactly the lattice on which Pippard's network ([3]) is built.

Although the results of the energy spectrum for a Bloch electron in a magnetic field were obtained here for a particular example, one expects that in a general case the qualitative picture will be similar.

In conclusion let us prove what we can call the Bloch theorem for a Bloch electron in a magnetic field. Because of the translational symmetry of the problem we could require that the eigenfunctions of the Hamiltonian (15) satisfy relation (10). For a free electron in a magnetic field the properly symmetrized eigenfunctions could be chosen in the form (14). When a periodic potential is present functions (14) will still be eigenfunctions of the magnetic translations but will no longer be eigenfunctions of the Hamiltonian. However, in analogy with the usual Bloch theorem, where the functions are given by a product of a free-electron function and a periodic part, we can construct the following functions:

$$\psi_k = \psi_k^F w \qquad (25)$$

where ψ_k^F is the symmetrized function for a free electron in a magnetic field and w is periodic in $\mathbf{R}_1$ and $\mathbf{R}_2$. The functions (25) are a complete analog of the usual Bloch functions, and they satisfy relation (10). The influence of the potential of the crystal is given through the w functions ([4]) in a similar way as in the usual Bloch case: it is contained in the periodic part of the function.

We see therefore that from point of view of symmetry the problem of a Bloch electron in a magnetic field has very much in common with the Bloch theorem for an electron in a crystal. The reason for this is that in both cases the symmetry is given by a commutative translation group.

REFERENCES

1. M. H. Johnson and B. A. Lippman, *Phys. Rev.* **76**: 828 (1949).
2. J. Zak, *Phys. Rev.* **136**: A776 (1964).
3. A. B. Pippard, *Proc. Roy. Soc. (London)* **A270**: I (1962): *Phil. Trans. Roy. Soc. London, Ser. A* **256**: 317 (1964).
4. J. Zak, *Phys. Rev.* **139**: A1159 (1965).

Chapter 16

TIGHT-BINDING APPROXIMATION FOR A BLOCH ELECTRON IN A MAGNETIC FIELD

A. Rabinovitch

Department of Physics, Technion
Israel Institute of Technology
Haifa, Israel

In this chapter electron states in a periodic potential and an applied magnetic field are calculated using group-theoretical considerations ([1]) and the so-called "tight-binding approximation". It turns out that under some specific simplifying assumptions the resulting equation conforms with that obtained [e.g., ([2])] by using effective Hamiltonian methods. We also show some results of the calculation.

Schrödinger's equation for an electron in a crystal with a magnetic field is

$$\mathcal{H}\Psi = \left\{ \frac{[\mathbf{p} + (e/c)\mathbf{A}]^2}{2m} + V(\mathbf{r}) \right\}\Psi = E\Psi$$

where $\mathbf{p} = (\hbar/i)\nabla$; $V(\mathbf{r})$ is the lattice periodic potential, which has the property $V(\mathbf{r} + \mathbf{R}_n) = V(\mathbf{r})$, with $\mathbf{R}_n = n_1\mathbf{a}_1 + n_2\mathbf{a}_2 + n_3\mathbf{a}_3$ a lattice vector; and $\mathbf{A}$ is the vector potential, which we take to be $\mathbf{A} = \frac{1}{2}\mathbf{H} \times \mathbf{r}$, where $\mathbf{H} = H\mathbf{1}_z$ is a constant magnetic field in the z direction. We take $V(\mathbf{r}) = \sum_{\mathbf{R}_n} V_a(\mathbf{r} - \mathbf{R}_n)$, where V_a can be, for instance, a modified Hartree-Fock atomic potential.

Let $\mathcal{H}_0 = (p^2/2m) + V_a(\mathbf{r})$, $V_a(\mathbf{r})$ being the atomic potential due to the atom in the same cell as the electron, and let

$$\mathcal{H}' = \frac{e}{mc}\mathbf{A} \cdot \mathbf{p} + \sum_{\mathbf{R}_n \neq 0} V_a(\mathbf{r} - \mathbf{R}_n) + \frac{e^2}{2mc^2}A^2$$

$$\mathcal{H} = \mathcal{H}_0 + \mathcal{H}'$$

As a set of initial functions we take the atomic (or rather the Wannier-type) functions, with the property:

$$\mathscr{H}_0 u(\mathbf{r}) = \varepsilon_a u(\mathbf{r})$$

We try to find the correct linear combinations for the problem with the magnetic field.

The operators which commute with the Hamiltonian $\mathscr{H}$ are ([3])

$$\tau(\mathbf{R}_n) = \exp\left\{\frac{i}{\hbar}\left(\mathbf{p} - \frac{e}{c}\mathbf{A}\right) \cdot \mathbf{R}_n\right\}$$

These operators form a group called the magnetic translation group (MTG) ([3]).

If we take the "rational" case, namely,

$$\frac{\mathbf{H} \cdot \mathbf{a}_1 \times \mathbf{a}_2}{hc/e} = \frac{1}{J}$$

where J is an integer, then some of the operators commute among themselves, namely,

$$\tau(\mathbf{R}_J), \qquad \text{where} \quad \mathbf{R}_J = n_1\mathbf{a}_1 + l_2 J\mathbf{a}_2 + n_3\mathbf{a}_3$$

and l_2 is an integer. The other operators, which commute with the Hamiltonian but do not commute with $\tau(\mathbf{R}_J)$, are

$$\tau(n\mathbf{a}_2), \qquad \text{where} \quad n = 0, 1, \dots (J-1)$$

In this case it is relatively easy to build symmetry adapted functions $\varphi_{i,\alpha}^{(\mathbf{k})}$ belonging to the ith row of the kth representation ([3]). There are J rows in the representation, and i takes the values $0, 1, \dots (J-1)$. The index α labels the different functions belonging to the same row and the same representation. There are J such functions, and α also takes the values $0, 1, \dots (J-1)$. We want to find linear combinations of the φ's, $\Psi'^{(\mathbf{k})} = \Sigma_n C_n\varphi_n^{(\mathbf{k})}$, so that H will be diagonal in them. The reason for choosing combinations of functions with the same $\mathbf{k}$ will soon be obvious.

We should mention here that $\mathbf{k}$ varies in the small Brillouin zone*

$$\mathbf{k} = m_1(\mathbf{K}_1/J) + m_2(\mathbf{K}_2/J) + m_3\mathbf{K}_3$$

---

* See Eq. (10a) in ([1]) and the subsequent discussion.

where m_1, m_2, and m_3 go from 0 to 1, and the **K** are the usual Brillouin zone vectors.

We now want the Ψ's to obey Schrödinger's equation

$$\mathcal{H}\Psi = \mathcal{E}\Psi$$

and, as is well known, we get a secular equation:

$$\text{Det}\{\mathcal{H}_{nn'} - \mathcal{E}I_{nn'}\} = 0$$

where $\mathcal{H}_{nn'} = \langle\varphi_n|\mathcal{H}|\varphi_{n'}\rangle$ and $I_{nn'} = \langle\varphi_n|\varphi_{n'}\rangle$.

Now, from group-theoretical considerations we know that

$$\langle\varphi_{i,n}^{(k)}|\mathcal{H}|\varphi_{i',n'}^{(k')}\rangle = \delta(\mathbf{k}-\mathbf{k}')\delta_{ii'}\langle\varphi_{i,n}^{(k)}|\mathcal{H}|\varphi_{i,n'}^{(k)}\rangle$$

Thus by choosing these correct linear combinations of functions a part of the task has been completed. We now have to diagonalize only the n, n' part. We get a $J \times J$ determinant, which we equate to zero.

We make the following assumptions:

1. We use only one-band atomic functions in order to build the φ's (this is a common assumption in the tight-binding approximation).
2. We take into account only "nearest-neighbor interactions."
3. We neglect the term $(e/mc)\mathbf{A} \cdot \mathbf{p}$, which in our gauge is $\frac{1}{2}i\hbar\omega \times [y\,\partial/\partial x - x\,\partial/\partial y]$, between neighbors in comparison with the term containing $V(\mathbf{r}+\mathbf{a})$ between the same neighbors. (Note that the Wannier functions are concentrated around atomic sites, and also that even for fields as high as 10^5 G $\hbar\omega$ is of the order of 0.001 eV.)

We finally get, for othorombic symmetry, the following determinantal equation:

$$0 = \begin{vmatrix} \lambda+2\nu\cos\alpha_1 & \exp\{i\alpha_2\} & 0 & \cdots & 0 & \exp\{-i\alpha_2\} \\ \exp\{-i\alpha_2\} & \lambda+2\nu\cos\left(\alpha_1+\dfrac{2\pi}{J}\right) & \exp\{i\alpha_2\} & 0 & \cdots & 0 \\ 0 & \exp\{-i\alpha_2\} & & & & \\ \vdots & 0 & & & & \\ 0 & & \vdots & & & \\ \exp\{i\alpha_2\} & 0 & & & & \end{vmatrix}$$

where $\alpha_1 = \mathbf{k} \cdot \mathbf{a}_1$; $\alpha_2 = \mathbf{k} \cdot \mathbf{a}_2$; ν is a measure of the asymmetry of the lattice (ν is 1 for cubic crystals); and $\lambda = 4(1+\nu)(\Delta - \mathcal{E})/\Delta E$, with Δ

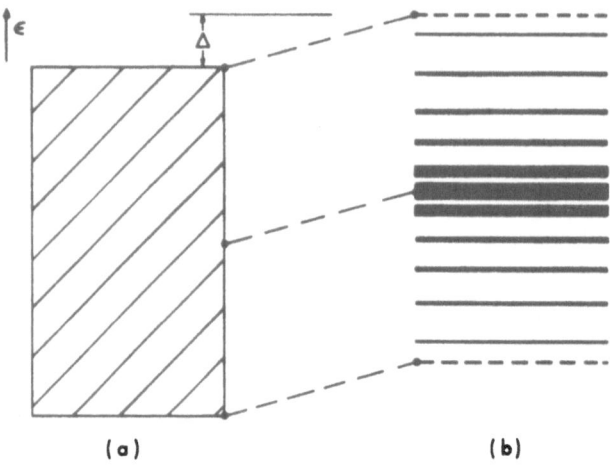

Fig. 1. Two-dimensional lattice. (a) Without magnetic field. One band: $E = A \cos \ (\mathbf{k}' \cdot \mathbf{a}_1 + \mathbf{k}' \cdot \mathbf{a}_2)$. (b) With magnetic field. J narrow bands, each level J-times degenerate.

the shift in energy of the whole band from that without a magnetic field and ΔE the width in energy of that band.

This result was obtained by Langbein and Gerlach ([5]), who used effective Hamiltonian methods with the same restrictions on the magnetic field.

The solution of this equation should give J values for λ, J energy levels (Landau levels), each of which is broadened depending on $\mathbf{k}$ via α_1 and α_2. Each level (or narrow band) is still J-times degenerate, due to the existence of the J operators $\tau(n\mathbf{a}_2)$.

We get the picture shown in Fig. 1 for a two-dimensional lattice. Let us count the number of states. We know that in two dimensions the number of states is proportional to the area in $\mathbf{k}$ space. Now, without the magnetic field $\mathbf{k}'$ changes in the entire Brillouin zone. With a magnetic field $\mathbf{k}$ changes only in a part which is $1/J^2$ of the Brillouin zone, so we have $(1/J^2) \times$ number of states without $\mathbf{H}$ in each narrow band. However, we have J narrow bands, and each level is J-times degenerate, so the number of the states does not, as it should not, change with the application of a magnetic field.

We can derive many useful properties about the energy levels from this determinantal equation, some of which are given in ([5]). On the other hand, it turns out that we can transform this equation into a finite difference equation derived before by Harper ([4]) and others using the so-called Peierl's theorem. This equation is as follows:

$$y_{n+1} + y_{n-1} + \left[2v \cos\left(\frac{2\pi n}{J} + \alpha_2\right) - \lambda\right]y_n = 0 \tag{1}$$

with the boundary condition

$$y_{n+J} = \exp\{iJ\alpha_1\}y_n, \tag{2}$$

where in our case the y's are partial determinants.

We must find those λ's for which there is a solution of (1) obeying (2).

Until now every solution of this equation was obtained by using WKB methods, whose accuracy on delicate points is not entirely clear.

Now, for large J we can approximate (1) and (2) by a differential equation, which turns out to be of a Mathieu type. This equation has been thoroughly investigated by mathematicians. The differential equation is

$$\frac{d^2u}{dz^2} + \left\{\frac{J^2}{\pi^2}(2 - \lambda) + \frac{2J^2}{\pi^2}v \cos 2z\right\}u = 0$$

To connect the above with the literature we let

$$s = 4v\left(\frac{J}{\pi}\right)^2 \qquad K = \frac{\lambda}{1 + v}, \qquad d = \frac{J^2}{\pi^2}[2 - K(1 + v)]$$

and the equation becomes

$$\frac{d^2u}{dz^2} + \left(d + \frac{1}{2}s \cos 2z\right)u = 0 \tag{1'}$$

which is the form used in the tables of the NBS ([6]), with the difference that the term containing $\cos 2z$ appears with a minus sign instead of a plus sign. This difference only means that here the minima of this term occur for $z = 0, \pm\pi, \pm 2\pi, \ldots$, instead of at the points $z = \pm\pi/2, \pm 3\pi/2, \ldots$, as in ([6]).

Now, we are looking for solutions of (1') that are "periodic of the second kind" ([7]), namely, that have the property

$$u(z + \pi) = \exp\{iJ\alpha_1\}u(z) \tag{2'}$$

These solutions u_r with the corresponding energies K_r turn out to be the bounded solutions of (1'). From the difference equation we know ([4]) that the values of K must lie between ± 2 and are symmetric with respect to $K = 0$; hence the values for d for cubic crystals ($v = 1$) lie between $-2J^2/\pi^2$ and $+2J^2/\pi^2$, i.e., between $-s/2$ and $+s/2$.

Plotting K as a function of J from the data for d versus s ([6]) we get Fig. 2. For a given value of J we get a vertical line which cuts the shaded regions and gives the allowed energy values.

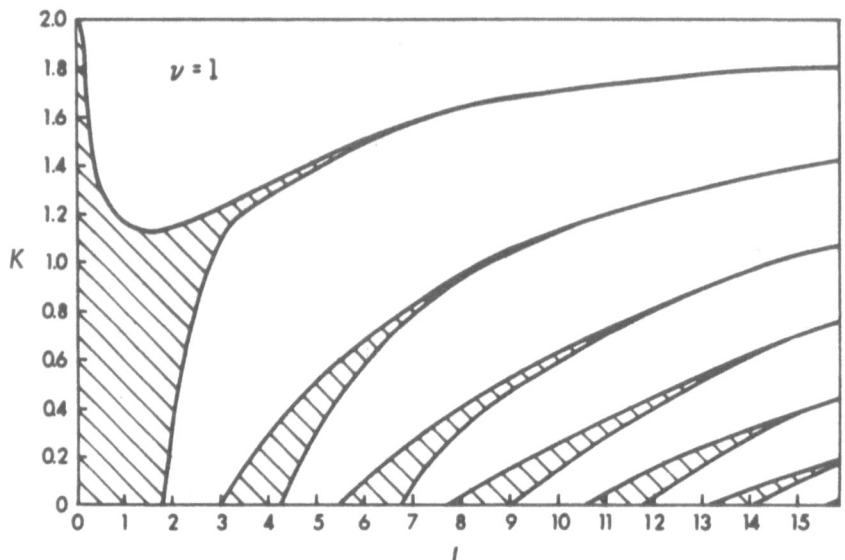

Fig. 2. Plot of K as a function of J.

For analytic results we consult Blanch ([8]). According to Chapter 20.2.30 of this work, we get for large values of J and for positive K (for negative K the picture is symmetric, as we already know)

$$K_r \approx 2 - \frac{2\pi}{J} \frac{\sqrt{\nu}}{1+\nu} (2r+1) + \frac{\pi^2}{J^2} \frac{1}{1+\nu} \frac{(2r+1)^2+1}{8} + \ldots \qquad (3)$$

and the width of these levels:

$$\Delta K_r \approx \frac{\nu^{1/2(r-1/2)}}{1+\nu} \frac{2^{(4r+11/2)}}{\pi^r} J^{r-1/2} \frac{\exp\{-4\sqrt{\nu}\,(J/\pi)\}}{r!} \qquad (4)$$

These results hold for $r < J$. For $r \approx J$ the accuracy of (3) and (4) is poor.

For cubic crystals ($\nu = 1$) we get:

$$K_r \approx 2 - \frac{2\pi}{J}\left(r + \frac{1}{2}\right) + \frac{\pi^2}{J^2} \frac{(2r+1)^2+1}{16} + \ldots \qquad (3')$$

$$\Delta K_r \approx \frac{2^{4r+9/2}}{\pi^r} J^{r-1/2} \frac{\exp\{-4(J/\pi)\}}{r!} \qquad (4')$$

We see that for small values of r, namely, for the uppermost and lowermost energy levels, we can take for K_r

$$K_r \approx 2 - (2\pi/J)(r + \tfrac{1}{2}), \qquad (r \text{ small, cubic crystals})$$

which is the effective-mass result. The difference in energy between two adjacent levels is $(\pi/2J)(\Delta E)$. According to the effective-mass theory this should be equal to $\hbar\omega = \hbar eH/m^*c$. Thus we get (taking into account the "rational" condition):*

$$m^* = 4\hbar^2/(\Delta E)a^2 \tag{5}$$

where a is the length of one side of the unit cell.

Thus we see that the main features of the solution of a Bloch electron in a magnetic field are as follows: We get narrow bands which are spaced symmetrically with respect to the center. Broadening is increased as one approaches the center. The outer narrow bands are very narrow, and the whole spectrum is slightly shifted and slightly squeezed. For the outer levels we get the spacing (3), (3'), the width (4), (4'), and also the "effective mass" (5).

REFERENCES

1. J. Zak, *Phys. Rev.* **136**: A776 (1964).
2. G. E. Zil'berman, *Soviet Phys.–JETP* (*English Transl.*) **5**: 208 (1957).
3. J. Zak, *Phys. Rev.* **134**: A1602, 1607 (1964).
4. P. G. Harper, *Proc. Phys. Soc.* **A68**: 874 (1955).
5. E. Gerlach and D. Langbein, *Phys. Rev.* **145**: 449 (1966).
6. Tables Relating to Mathieu Functions, U. S. National Bureau of Standards, Columbia University Press, New York, 1951.
7. T. Fort, *Finite Differences*, Clarendon Press, Oxford, 1948, p. 178.
8. G. Blanch, Mathieu Functions, in: *Handbook of Mathematical Functions*, National Bureau of Standards, Applied Math. Series Vol. 55, 1964, p. 721.

---

* For $a = 3$ Å we get $m^* = 2.5\, m_e/(\Delta E)$, where m_e is the electronic mass and ΔE is measured in eV.

Chapter 17

THE ZEEMAN EFFECT IN CRYSTALS

W. A. Runciman

Solid State Division
Atomic Energy Research Establishment
Harwell, Didcot, Berks

INTRODUCTION

Absorption and fluorescence lines of ions in crystals have linewidths appreciably greater than those of free ions. Even with crystals cooled to liquid helium temperature, 4.2 °K, the zero-phonon lines have half-widths in the range 1–10 cm^{-1} due, at least partly, to inhomogeneous strain broadening in the crystalline environment. Thus the optical spectroscopy of crystals is naturally suited to the use of intense magnetic fields of 100 kG and upward. Hyperfine effects will be neglected, since they are not normally detected in the optical spectroscopy of crystals, and the following discussion will be limited to first-order Zeeman effects in cubic crystals. First-order effects are found when the energy difference between the electronic levels is large compared with the Zeeman splitting. In the first account of the Zeeman effect in crystals Bethe ([2]) restricted himself to tetragonal symmetry. Here attention will be focused on cubic crystals and the discussion will be divided into sections depending on whether there is orientational as well as electronic degeneracy and on whether there is an even or odd number of electrons in the center. Orientational degeneracy occurs when the defect centers being studied have less than cubic symmetry, but are randomly distributed among equivalent directions in the crystal, for example, the cube axes. This presentation follows a contribution to the Zeeman Centennial Conference held in Amsterdam in September 1965. However, it was published only in abstract form ([17]), and was found to be similar to the more extensive Russian papers which will be referred to in the appropriate sections. The approach adopted is group theoretical, but no detailed theory is attempted, the emphasis being on the variety of cases which can occur.

For more detailed application of group theory to this problem reference may be made to the papers by Koster and Statz ([11]) and Statz and Koster ([18]). Only electric dipole transitions will be considered, as the theory is easily extended to magnetic dipole transitions by interchanging the role of the electric and magnetic vectors.

THEORETICAL BACKGROUND

As distinct from many of the other topics covered in this volume, optical absorption is here considered as occurring at point defects in the crystals. These point defects may be transition-metal ions with an incomplete d or f shells of electrons or centers produced by radiation damage. No translational symmetry is involved, since this is removed by the presence of the defect. It may be useful to introduce some concepts from the theory of finite groups and from elementary quantum mechanics.

As an example of a point group consider the group O. This contains the 24 rotational elements of an octahedron or cube. These are: (1) E, the identity operation, (2) $8C_3$, eight threefold rotations, (3) $6C_4$, six fourfold rotations, (4) $6C_2$, six twofold rotations, and (5) $3C_4^2$, three rotations through $180°$ about fourfold rotation axes. In the group O_h there is also an inversion operation, and this is expressed by the equation $O_h = O \times i$. This product means that in addition to the operations (1)–(5) there are those involving both rotation and inversion, making a total of 48 group operations in all. A simple cube has symmetry O_h and we leave as an exercise the construction of an object having symmetry O but not O_h. There are five irreducible representations of point group O, and examples will be quoted of wave functions having the requisite transformation properties; i.e., the components transform among themselves under operations of the group:

A_1	1	Singly degenerate
A_2	xyz	Singly degenerate
E	$\dfrac{1}{\sqrt{2}}(x^2 - y^2), \dfrac{1}{\sqrt{6}}(2z^2 - x^2 - y^2)$	Doubly degenerate
T_1	x, y, z	Triply degenerate
T_2	yz, zx, xy	Triply degenerate

It might be better to write the wave functions in the form x/r rather than x, but the purely radial part is suppressed in order to emphasize the angular properties of the wave functions. Here A_1 has full cubic symmetry and the spherically symmetric function 1 fulfils this condition. As a check on the number of irreducible representations it can be noted that the sum of squares of the orders of the representations equals the number of elements ($1^2 + 1^2 + 2^2 + 3^2 + 3^2 = 24$). The wave functions can be constructed out of free-atom wave functions of orbital angular momentum L. Those for T_1 are $L = 1$ wave functions, those for E and T_2 together comprise the five $L = 2$ wave functions, and the A_2 wave function is one of the set of $L = 3$ wave functions. The A_1 wave function corresponds to $L = 0$. It is sometimes thought that crystal fields quench the orbital angular momentum, but it is seen that there is no effect on the $L = 1$ wave functions. The $L = 2$ wave functions split into two sets, the T_2 wave functions, effectively possessing unit angular momentum, and the E wave functions, which have none. Thus the quenching may be complete, partial, or nonexistent depending on the wave functions involved. The five $L = 2$ wave functions furnish a good example of a reducible representation of group O. Since those wave functions belonging to T_2 only transform among themselves and not into those belonging to the E representation, the five $L = 2$ wave functions are reducible into two sets transforming according to the irreducible E and T_2 representations.

For group O_h we use the same representation labels with a subscript g (gerade) for even parity states, which are unchanged by inversion, and u (ungerade) for odd parity states, which change sign on inversion.

In order to apply group theory to the Zeeman effect, it is necessary to know the symmetry properties of the magnetic field **H**. They will be compared with those of the electric field **E**, especially since the usual vector notation suggests they are similar quantities. The electric field **E** transforms as T_{1u}, since an electric field has x, y, z components. The magnetic field **H** transforms as T_{1g}, since it does not change sign under inversion. The magnetic field has a sense of rotation rather than of direction. This can best be seen by considering the field due to a small circular coil containing a current. Under inversion the current does not change direction, and hence the magnetic field due to it is also unchanged. Another way of considering the point is to note that if magnetic field is defined as a vector product using Ampère's law, then it will not change sign under inversion, since it is generally true that vector products are invariant under inversion. While **E** and **H** are sometimes referred to as polar and axial vectors, the terminology vector and pseudovector is preferable. Electric charge is then a scalar quantity and magnetic charge a pseudoscalar one because the force (a vector) is $q(\mathbf{H}$ or $\mathbf{E})$. Hence if magnetic monopoles exist they have different symmetry properties from electric charges. There is a more mathematical discussion of this point in a paper on optical rotatory power by Condon [(^{3}), pp. 445–6].

In order to find the wave functions of energy levels split by a magnetic field perturbation, we need to find solutions of the Schrödinger equation,

$$\mathcal{H}\psi = E\psi$$

where $\mathcal{H}$ is the Hamiltonian operator, E the energy, and ψ the wave function. The Zeeman term is $H_z(L_z + 2S_z)$ for a magnetic field along the z direction. For the sake of illustration we make the restriction to orbital angular momentum, and the wave functions must be eigenstates of L_z. The differential form of the orbital angular momentum operator can be used, i.e.,

$$L_z = -i\hbar\left(x\frac{\partial}{\partial y} - y\frac{\partial}{\partial x}\right)$$

By substitution in the Schrödinger equation it can be seen that $(1/\sqrt{2})$ $\times (x + iy)$, z, and $(1/\sqrt{2})$ $(x - iy)$ are eigenfunctions, whereas x and y are not solutions of the equation. The factors $1/\sqrt{2}$ are introduced for normalization of the wave functions.

Having obtained wave functions, it is now possible to calculate the intensities and polarizations of Zeeman transitions. Consider the simplest case of $A_{1g} \rightarrow T_{1u}$ transition, which is an allowed electric dipole transition.

The electric dipole operator, $e\mathbf{r}$, has components ex, ey, and ez. For a transverse Zeeman experiment the magnetic field is along the z direction and the light travels in the y direction. Using polarizers, the π spectrum, with the light $\mathbf{E}$ vector parallel to z, may be separated from the σ spectrum, which has the light $\mathbf{E}$ vector parallel to the x direction. The matrix elements are proportional to the integral

$$I = \int \psi^* e\mathbf{r}\phi \, d\mathbf{r}$$

In this example $\psi^* = 1$ and ϕ is one component of the T_1 wave function. The relative intensities are proportional to the modulus squared of the matrix elements, and the Zeeman pattern shown in Fig. 1 is easily verified. The intensities have been multiplied by a factor to yield the simplest integers. It is generally true that, apart from thermal effects for levels populated in accordance with a Boltzmann distribution, transverse Zeeman patterns are symmetrical to first order. This follows from the symmetrical splitting of the energy levels in a magnetic field, which is necessary on ac-

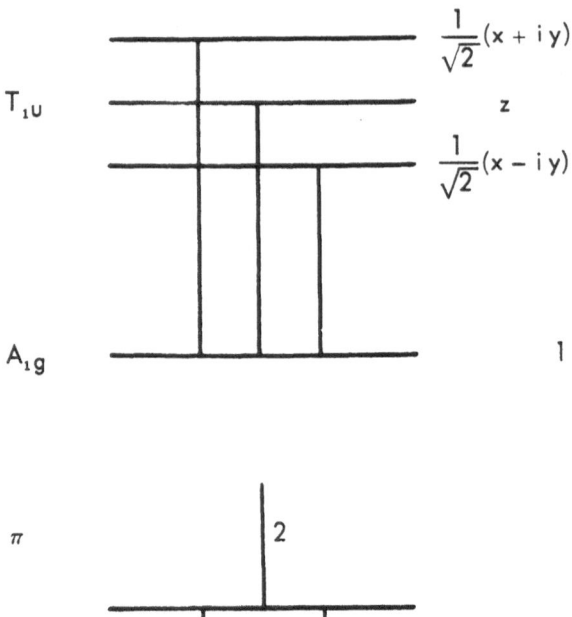

Fig. 1. Zeeman spectrum for an $A_{1g} \rightarrow T_{1u}$ transition.

count of time-reversal symmetry. Since the line with zero magnetic field is unpolarized, the total π intensity is equal to the total σ intensity, again apart from thermal factors.

The preceding discussion has been a simple outline of the normal Zeeman effect. The calculation has been made using specific wave functions. In fact, the intensities are correct for any $A_1 \rightarrow T_1$ transition, a result following from the Wigner–Eckart theorem. In this example if we calculate the product of the representations $A_1 \times T_1 \times T_1$ we can gain information on the uniqueness or otherwise of the $A_1 \rightarrow T_1$ transition. The central T_1 stands for the electric dipole operator. The product table for representations of the group O is shown in Table I. It can be derived from the character table for group O, and the reader is referred to standard texts on group theory for details ([7,8,20]) or to the tables given in Koster et al. ([12]). From Table I we see that $A_1 \times T_1 \times T_1 = A_1 + E + T_1 + T_2$. Since the product contains the identity representation A_1 once only, the intensities are unique. If the identity representation does not appear, the transition is forbidden, and if it appears more than once, the intensity ratios are no longer unique.

In the preceding paragraph parity labels have been dropped, as it is

TABLE I

Multiplication Table of Irreducible Representations of Group O

O	A_1	A_2	E	T_1	T_2
A_1	A_1	A_2	E	T_1	T_2
A_2	A_2	A_1	E	T_2	T_1
E	E	E	$A_1 + A_2 + E$	$T_1 + T_2$	$T_1 + T_2$
T_1	T_1	T_2	$T_1 + T_2$	$A_1 + E + T_1 + T_2$	$A_2 + E + T_1 + T_2$
T_2	T_2	T_1	$T_1 + T_2$	$A_2 + E + T_1 + T_2$	$A_1 + E + T_1 + T_2$

convenient to consider them separately. Incidentally, they illustrate the power of an approach based on symmetry. The equations

$$g \times u \times g = u \quad \text{and} \quad u \times u \times u = u$$

follow from the product rule for parity: the product of an odd number of odd parity representations is odd, and multiplication by even parity representations produces no change. They can be interpreted in two ways:

1. Regarding the central u as the electric dipole operator and the outer labels the parity labels of wave functions ψ and ϕ, then the equations show the Laporte or parity selection rule for free-atom transitions. Transitions between levels of the same parity are forbidden, since the product does not contain the identity representation g.

2. Regarding the central u as the representation of an electric field $\mathbf{E}$ and the outer labels as the parity labels of the wave functions ψ of a degenerate energy level, then the equations show that the energy level is not split by an electric field perturbation in first order.

This illustrates one of the advantages, and possible sources of confusion, of group theory, that the same type of argument applies to transitions and perturbations.

Finally, we make a few remarks on the magnitude of the Zeeman effect. For the case of the normal triplet pattern the energy levels in a magnetic field are, at E_0, $E_0 \pm g\beta H$, where g is the Landé g-factor, β is the Bohr magneton, and H is the magnetic field intensity. The size of the effect is such that $g\beta H$ is approximately 1 cm^{-1} for $g = 2$ and a field of 10 kG.

For Russell–Saunders coupling

$$g = 1 + \frac{J(J+1) - L(L+1) + S(S+1)}{2J(J+1)}$$

It is interesting to note that g can be zero, e.g. for $^5F_1$, ($J = 1$, $L = 3$, $S = 2$, and hence $g = 0$); or even negative, e.g., for $^7G_1$ ($J = 1$, $L = 4$, $S = 3$, and hence $g = -0.5$).

The following special cases are important:

1. $S = 0$, $J = L$: $g = 1$ when there is only orbital angular momentum.
2. $L = 0$, $J = S$: $g = 2$ when there is only spin angular momentum.
3. $S = L$: $g = 1.5$ for all possible values of J.

For further details of the Zeeman effect in free atoms Condon and Shortley ([4]) may be consulted; for crystals then Griffith ([7]) gives fuller details.

Case (2) illustrates the power of magnetic field experiments, as spin degeneracy is not removed by either uniaxial stress or electric field effects.

CUBIC CENTERS

For additional information on cubic centers reference should be made to the paper by Zakharchenya and Rusanov ([23]).

Cubic Centers with an Even Number of Electrons

The Zeeman operator, $L + 2S$, transforms according to the T_1 representation. Hence an energy level belonging to the doubly degenerate E representation does not split in a magnetic field, since the produce $E \times T_1 \times E$ does not contain the identity representation. Two classes of transition are analogous to those in a free ion, $A_1 \to T_1$ or $A_2 \to T_2$ being like $J = 0 \to J = 1$ and $T_1 \to T_1$ or $T_2 \to T_2$ being like $J = 1 \to J' = 1$. With light propagation perpendicular to the magnetic field direction the patterns are symmetrical, and the magnitude of the Zeeman effect is independent of the crystal orientation. In those cases such as $A_1 \to T_1$ and $T_1 \to T_1$ which correspond to free-ion transitions there can be no anisotropy, but for cases such as $E \to T_1$ or T_2 and for $T_1 \to T_2$ the intensities of the different polarized components can be functions of the orientation of the crystal with respect to the magnetic field. The longitudinal Zeeman effect can give rise to some interesting features in crystals. For some transitions there are equal right- and left-circularly polarized components, giving linearly polarized light, but when a degenerate E state is involved there is no longer a definite phase relationship between these components, and the light can be regarded as a mixture of the two circular polarizations. In other cases the two components are unequal, and there is a resultant elliptic polarization if there is a definite phase relationship, and a mixture otherwise. The longitudinal Zeeman effect also distinguishes between electric and magnetic dipole transitions. The longitudinal spectrum is similar to the σ transverse spectrum, with the electric vector perpendicular to the applied magnetic field, for electric dipole radiation and to the π transverse spectrum, with the electric vector parallel to the applied magnetic field, for magnetic dipole radiation. Divalent samarium in calcium fluoride provides examples of $A_1 \rightarrow T_1$ electric dipole transitions in both absorption and fluorescence ([13,21]). Zeeman experiments were crucial for deciding that the upper states belonged to the f^5d rather than the f^6 configuration. In this system Crozier ([5]) has detected an $A_{1g} \rightarrow A_{1u}$ transition which is only observed in the presence of a magnetic field and has an intensity proportional to the square of the magnetic field.

Cubic Centers with an Odd Number of Electrons

So far the discussion has concentrated on orbital angular momentum. When electron spin is introduced into the theory half-integral units of angular momentum appear and necessitate the introduction of double-groups. Thus for group O there are three cubic irreducible representations for an odd number of electrons: Γ_6 and Γ_7, which are two-dimensional, and Γ_8, which is four-dimensional. Examples of Γ_6 and Γ_8 are $J = \frac{1}{2}$ and $J = \frac{3}{2}$, respectively. However, since the product $\Gamma_8 \times T_1 \times \Gamma_8$ contains the identity representation twice, the Γ_8 representation does not necessarily behave like $J = \frac{3}{2}$. For instance, a ^2E state belongs to Γ_8 and yet splits into two doubly degenerate components in a magnetic field. Both these types of Γ_8 are observed in the $^4A_2 \rightarrow {}^2E$ transition for Cr^{3+} in an octahedral site in MgO studied by Sugano et al. ([19]). In general, a Γ_8 may behave as any linear combination of these two cases having four Zeeman components unequally spaced. Then the magnitude of the Zeeman effect depends on the orientation of the crystal with respect to the magnetic field. Hellwege ([9]) stated that the Γ_8 representation always splits into three components for O symmetry, but corrected this point in an erratum. The intensity ratios are not unique for $\Gamma_8 \rightarrow \Gamma_8$ transitions, since $\Gamma_8 \times T_1 \times \Gamma_8$ contains A_1 twice, so that any theory in these cases requires a detailed knowledge of the wave functions. In this case the T_1 occurring in the triple product is the representation of the electric dipole operator.

TABLE II

Intensities of Zeeman Components for $\Gamma_6 \to \Gamma_8$ and $\Gamma_7 \to \Gamma_8$ Transitions with the Magnetic Field along [111]

O_h		Γ_8			
	S_6	Γ_4	Γ_5	Γ_6	Γ_6
Γ_6	Γ_4	π 4	σ_+ 1	σ_- 0	σ_- 3
	Γ_5	σ_- 1	π 4	σ_+ 3	σ_+ 0
Γ_7	Γ_4	π 12	σ_+ 3	σ_- 8	σ_- 1
	Γ_5	σ_- 3	π 12	σ_+ 1	σ_+ 8

The intensity ratios are unique for the $\Gamma_6 \to \Gamma_8$ case, and therefore must be isotropic since the special example of $J = \frac{1}{2} \to J' = \frac{3}{2}$ has isotropic properties. However, the Zeeman splitting will in general be dependent on the crystal orientation through the anisotropy of the Γ_8 splitting. Although the intensity ratios for $\Gamma_7 \to \Gamma_8$ are similar to those of $\Gamma_6 \to \Gamma_8$, for a [100] magnetic field direction this is not generally true, and the intensity ratios have to be calculated for each orientation. The results of calculations for the magnetic field oriented along the [111] axis are shown in Table II; these results were not obtained by Zakharchenya and Rusanov ([23]). The isotropic nature of the intensities for $\Gamma_6 \to \Gamma_8$ is confirmed. The only remaining theoretical calculations needed to round off this work are those for a magnetic field directed along the [110] axis. The numbers of components for different classes of Zeeman spectra are summarized in Table III.

The Zeeman effect of the resonance line at 4130 Å of divalent europium in calcium fluoride has been studied both in absorption and emission ([14,24]). Here the transition is from a ground state $^8S_{7/2}$ of the f^7 configuration to a Γ_8 belonging to f^6d. The crystal field splitting of the $^8S_{7/2}$ is small compared with the Zeeman splitting for magnetic fields yielding resolved Zeeman spectra, so that special considerations apply. Due to the large Zeeman splitting of the ground state for $S = \frac{7}{2}$ with $g = 2$, there is thermal

TABLE III

Twenty-One Classes of Zeeman Spectra*

Type of center	Magnetic field direction			
	[100]	[111]	[110]	[hkl]
Electronic degeneracy				
1. $A_1 \to T_1$, $A_2 \to T_2$	3		Isotropic	
2. $E \to T_1$, $E \to T_2$	3	3	3	3
3. $T_1 \to T_2$	6	9	9	9
4. $T_1 \to T_1$, $T_2 \to T_2$	6		Isotropic	
5. $\Gamma_6 \to \Gamma_6$, $\Gamma_7 \to \Gamma_7$	4		Isotropic	
6. $\Gamma_6 \to \Gamma_8$	6	6	6	6
7. $\Gamma_7 \to \Gamma_8$	6	8	8	8
8. $\Gamma_8 \to \Gamma_8$	16†	16†	16†	16†
Electronic + orientational degeneracy				
9. Tetragonal $\langle 100 \rangle$, $A \to E$, $\pm\frac{3}{2} \to \pm\frac{3}{2}$	3	2	3	6
10. Trigonal $\langle 111 \rangle$, $A \to E$, $\pm\frac{3}{2} \to \pm\frac{3}{2}$	2	4	3	8
11. Tetragonal $\langle 100 \rangle$, $E \to E$	3	2	3(2[001])‡	6
12. Trigonal $\langle 111 \rangle$, $E \to E$	4	8	5	8
13. Tetragonal $\langle 100 \rangle$ $\pm\frac{1}{2} \to \pm\frac{3}{2}$	4	2	4	6
14. Trigonal $\langle 111 \rangle$ $\pm\frac{1}{2} \to \pm\frac{3}{2}$	2	4	4	8
15. Tetragonal $\langle 100 \rangle$ $\pm\frac{1}{2} \to \pm\frac{1}{2}$	8	4	8	12
16. Trigonal $\langle 111 \rangle$ $+\frac{1}{2} \to \pm\frac{1}{2}$	4	8	8	16
17. Orthorhombic I $\langle 110 \rangle$	8	8	12	24
18. Orthorhombic II $\langle 100 \rangle$ C_2	12	4	12	24
19. Monoclinic I $\langle hhl \rangle$	8	12	16	48
20. Monoclinic II $\langle hko \rangle$	12	8	16	48
21. Triclinic $\langle hkl \rangle$	12	16	24	96

* The number of allowed components in unpolarized transverse spectra with the magnetic field along the direction indicated. The wave functions are labeled according to the appropriate subgroup for classes 9–12. Crystal quantum numbers are indicated for classes 9, 10, and 13–16.
† Only 12 components for a transition of type $J = \frac{3}{2} \to J' = \frac{3}{2}$.
‡ Only 2 components visible for viewing along the [001] direction.

depopulation of some states at liquid helium temperature, 4.2 °K. However, all the lines can be observed using both fluorescence and absorption results. A rather simpler case is the emission line at 1.116 μ due to divalent thulium in calcium fluoride studied by Zakharchenya *et al.* ([22]). This is a simple example of a magnetic dipole transition $(^2F_{5/2})\Gamma_7 \to (^2F_{7/2})\Gamma_7$.

NONCUBIC CENTERS IN CUBIC CRYSTALS

Further details of the account in this section can be found in the paper by Arkhangel'skaya and Feofilov ([1]).

Noncubic Centers with an Even Number of Electrons

There are not many cases to be considered here since only centers with a three- or fourfold symmetry axis have two-dimensional representations. The application of uniaxial stress on a crystal removes pure orientational degeneracy, but the effect is zero to first order for an applied magnetic field. For centers with a fourfold symmetry axis there is no doublet splitting when the magnetic field is perpendicular to the symmetry axis, i.e., $g_\perp = 0$. This follows since the magnetic field transforms as $A_2 + E$ for D_4 symmetry, and using tables appropriate for D_4 symmetry the product $E \times E \times E$ does not contain the identity representation A_1. This result holds for any group containing fourfold symmetry. A similar result applies for D_3 symmetry, where again $g_\perp = 0$. A proof can be constructed using results in ([7]), [Theorem 5, p. 218, and the tables on p. 405]. This is a consequence of time-reversal symmetry. Applying similar arguments the result can also be proved for C_3 symmetry. Using a crystal quantum number approach Hellwege ([9]) erroneously derived a first-order Zeeman effect in this case. It is possible to calculate the relative intensities and splittings of the Zeeman pattern for different crystal orientations with respect to the magnetic field for the tetragonal $A \to E$ and $E \to E$ transitions and for the trigonal $A \to E$ transition (Fig. 2). However, these are not uniquely defined for the trigonal $E \to E$ transition. Trivalent rare earth ions in a calcium fluoride crystal produce trigonal centers when an oxygen ion replaces a nearest-neighbor fluorine ion as charge compensation and produce tetragonal centers when charge compensation is provided by an interstitial fluorine ion.

Noncubic Centers with an Odd Number of Electrons

For centers with an odd number of electrons and with less than cubic symmetry, which in this context includes tetrahedral symmetry, the only degeneracy normally present is the twofold Kramers degeneracy which must occur for all states. This degeneracy is not removed by uniaxial stress or by an electric field, so that the Zeeman effect is especially useful in deciding whether to assign absorption and fluorescence lines to centers with an odd number of electrons. The types of orientational degeneracy

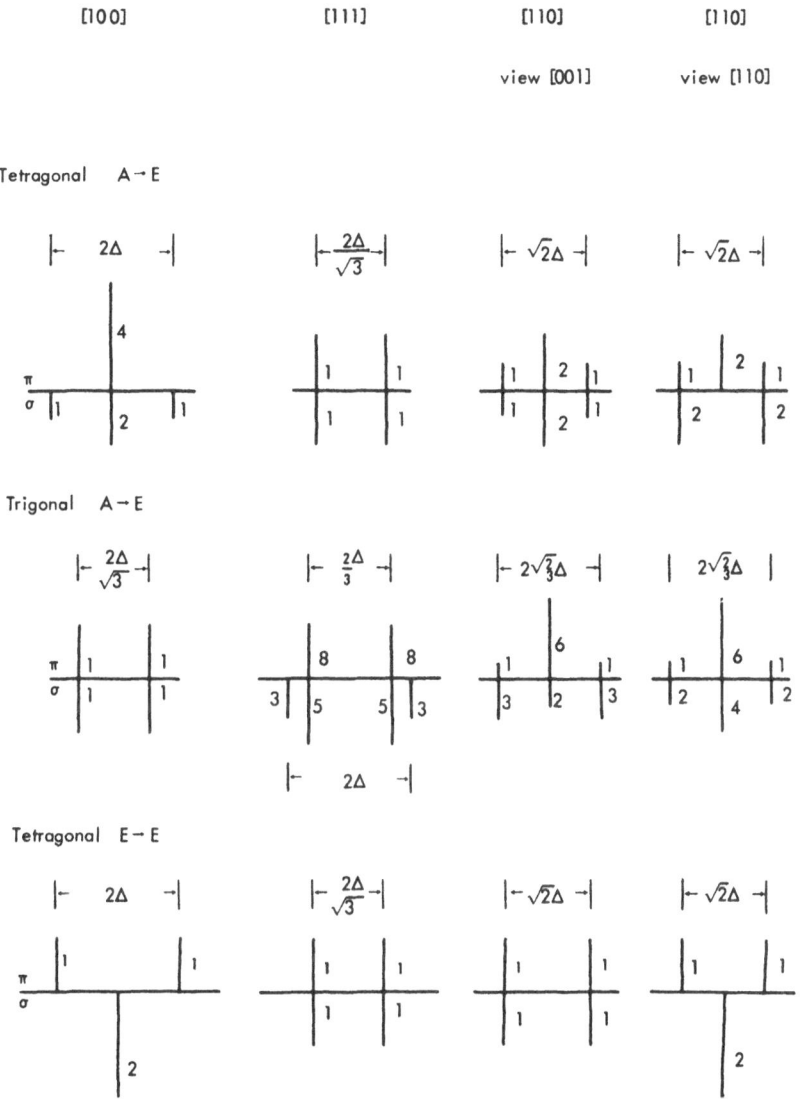

Fig. 2. Theoretical Zeeman spectra for trigonal and tetragonal centers in cubic crystals with the magnetic field along [100], [111] and [110].

which can be differentiated are the same as those which can be detected by the application of uniaxial stress ([10,16]). The difference is that the orientational Zeeman effect occurs only for an odd number of electrons, and in general there will be four Zeeman components for each orientation.

For some doublets the splitting is zero when the magnetic field is perpendicular to the axis of the center, but this is not so for doublets with the crystal quantum number $\mu = \pm \frac{1}{2}$. Crystal quantum numbers have been extensively discussed by Hellwege ([9]). For groups C_n the crystal quantum number is the magnetic quantum number J_z subject to modulus n. Both in a review paper ([6]) and in the original account ([1]) the assumption is made that $g_\perp = 0$ in all cases, and in consequence some splittings are regarded as anomalous. There are considerable simplications in the spectra for tetragonal and trigonal centers except when both the doublets involved in the transition have the crystal quantum number $\mu = \pm \frac{1}{2}$. In some cases Zeeman patterns are observed similar to those for an even number of electrons, which were just described above. Table III lists the numbers of components expected for the various classes of Zeeman spectra.

Some color centers fall into the category of noncubic centers with an odd number of electrons. It might be thought that the R center, consisting of an equilateral group of three F centers, should show a Zeeman effect. No such effect has been observed though many attempts to find it must have been made. The explanation seems to be that for small spin-orbit coupling a selection rule $\Delta M_s = 0$ operates, so that if both the upper and lower doublet states have a $\dot{g} = 2$ splitting, no splitting is observed, since transitions occur only between corresponding levels whose separation is unchanged. For a doublet observed in additively colored calcium fluoride ([15]) the Zeeman splitting obtained was comparable with the separation between the lines, making deductions very difficult.

DISCUSSION

From the preceding discussion it is clear that Zeeman spectra can give a wealth of information about isolated ions in crystals. The emphasis has been on determining the degeneracy and orientation of the centers, and the references given should be consulted for further information on relative spacings and intensities. In many cases the calculation of the g-factor governing the absolute splitting can be calculated using Russell–Saunders or intermediate coupling. For spin-only cases $g = 2$, but for weak spin-orbit coupling, as in color centers, this may lead to a negative Zeeman effect. No attempt has been made to cover exciton spectra, in which a case of quadrupole radiation has been found, or the spectra of concentrated magnetic salts. There is likely to be interest in these areas for many years.

Finally, a few words about experimental details. The low-temperature cryostats and the grating spectrographs normally used are now standard

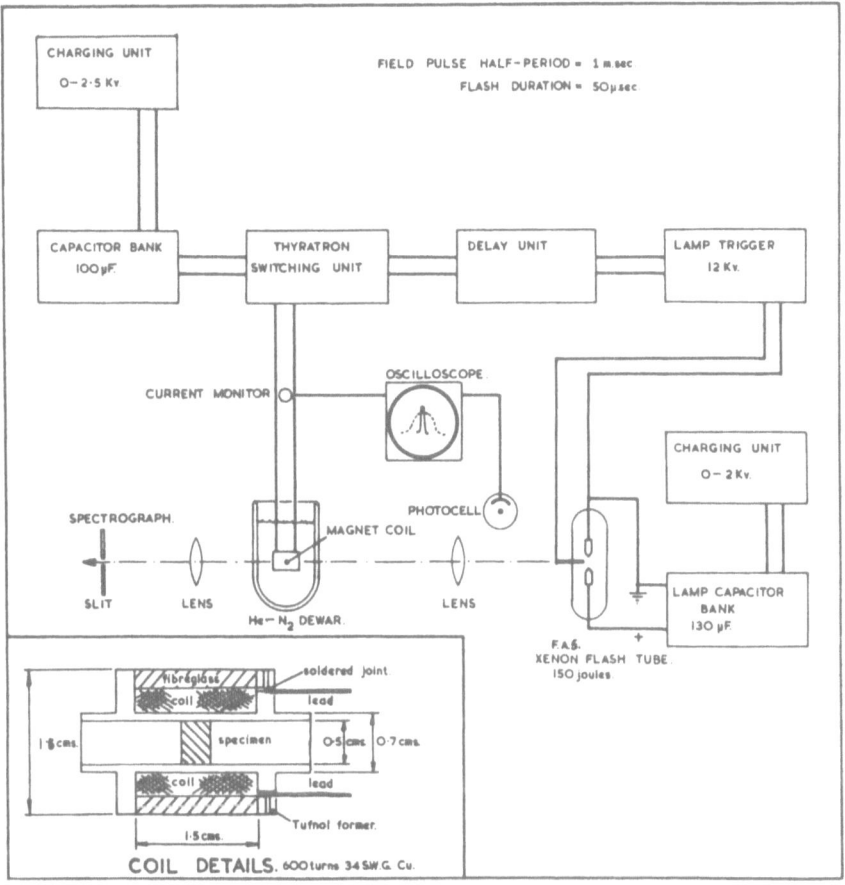

Fig. 3. Diagram of pulsed field apparatus.

equipment. The magnetic field required to obtain resolved spectra may exceed the 30 kG normally obtained with iron core magnets. Some absorption spectra can be obtained by a pulse technique using a copper coil at liquid helium temperature. Fields of 200 kG have been obtained with the apparatus shown in outline in Fig. 3. The xenon lamp is synchronized to flash when the field is near maximum, and repetitive flashes may be used to obtain long exposures. For fluorescence spectra and for absorption spectra requiring very long exposures steady fields are required, and these can be obtained with superconducting solenoids up to about 125 kG. Higher steady fields require larger water-cooled magnets at installations which have power supplies of many megawatts.

ACKNOWLEDGMENTS

I wish to thank Dr. A. E. Hughes for collaboration in pulsed field experiments and Dr. D. F. Johnston and Mr. S. Marlow of the Theoretical Physics Division for unpublished calculations on which Table II is based, and which when published will include tables appropriate to the case of the magnetic field oriented along the [110] direction. I have also benefitted greatly from discussions with Drs. Hughes and Johnston.

REFERENCES

1. V. A. Arkhangel'skaya and P. P. Feofilov, *Opt. i Spektroskopiya* **4**: 602 (1958) [translated by Associated Technical Services, Inc.].
2. H. Bethe, *Z. Phys.* **60**: 218 (1930); AERE-Trans. 1041 (1965).
3. E. U. Condon, *Rev. Mod. Phys.* **9**: 432 (1937).
4. E. U. Condon and S. H. Shortley, *The Theory of Atomic Spectra*, Cambridge University Press, 1951.
5. M. H. Crozier, *Phys. Rev. Letters* **13**: 394 (1964).
6. P. P. Feofilov and A. A. Kaplyanskii, *Soviet Phys.–Usp.* (*English Transl.*) **5**: 79 (1962).
7. J. S. Griffith, *The Theory of Transition Metal Ions*, Cambridge University Press, 1961.
8. V. Heine, *Group Theory in Quantum Mechanics*, Pergamon Press, New York and London, 1960.
9. K. H. Hellwege, *Z. Phys.* **127**: 513 (1950); AERE-Trans. 1045 (1965); Erratum: *Z. Phys.* **128**: 172 (1950).
10. A. A. Kaplyanskii, *Opt. Spectry.* (*USSR*) (*English Transl.*) **16**: 329 (1964).
11. G. F. Koster, H. Statz, *Phys. Rev.* **113**: 445 (1959).
12. G. F. Koster, J. O. Dimmock, R. G. Wheeler, and H. Statz, *Properties of the Thirty-Two Point Groups*, MIT Press, 1963.
13. W. A. Runciman and C. V. Stager, *Bull. Am. Phys. Soc.* **7**: 85 (1962); *J. Chem. Phys.* **37**: 196 (1962).
14. W. A. Runciman and C. V. Stager, *J. Chem. Phys.* **38**: 279 (1963).
15. W. A. Runciman, C. V. Stager and M. H. Crozier, *Phys. Rev. Letters* **11**: 204 (1963).
16. W. A. Runciman, *Proc. Phys. Soc.* **86**: 629 (1965).
17. W. A. Runciman, *Physica* **33**: 287 (1967).
18. H. Statz and G. F. Koster, *Phys. Rev.* **115**: 1568 (1959).
19. S. Sugano, A. L. Schawlow, and F. Varsanyi, *Phys. Rev.* **120**: 2045 (1960).
20. M. Tinkham, *Group Theory and Quantum Mechanics*, McGraw-Hill Book Co., New York, 1964.
21. B. P. Zakharchenya, V. P. Makarov, and A. Ya. Ryskin, *Opt. Spectry.* (*USSR*) (*English Transl.*) **17**: 116 (1964).
22. B. P. Zakharchenya, V. P. Makarov, A. V. Varfolomeyev, and A. Ya. Ryskin, *Opt. Spectry.* (*USSR*) (*English Transl.*) **16**: 248 (1964).
23. B. P. Zakharchenya and I. B. Rusanov, *Opt. Spectry.* (*USSR*) (*English Transl.*) **19**: 207 (1965).
24. B. P. Zakharchenya, I. B. Rusanov and A. Ya. Ryskin, *Opt. Spectry.* (*USSR*) (*English Transl.*) **18**: 563 (1965).

Chapter 18

MAGNETIC BREAKDOWN

A. B. Pippard

Cavendish Laboratory
Cambridge

LECTURES COMPILED BY K. A. McEWEN

The behavior of the conduction electrons of a hypothetical two-dimensional metal perpendicular to a magnetic field may be usefully analyzed with semi-classical techniques. When no electric field is present the dynamics of an individual electron are prescribed by the equation of motion

$$\hbar \dot{\mathbf{k}} = e\mathbf{v} \times \mathbf{H} \tag{1}$$

where

$$\mathbf{v} = \hbar^{-1} \operatorname{grad}_k E \tag{2}$$

Conventionally, the z direction is defined by the uniform magnetic field $\mathbf{H}$. Thus

$$dk_z = \dot{k}_z \, dt = 0 \tag{3}$$

$$dE = \operatorname{grad}_k E \cdot d\mathbf{k} = \hbar \mathbf{v} \cdot \dot{\mathbf{k}} \, dt = 0 \tag{4}$$

Hence an electron moves in $\mathbf{k}$ space around trajectories of constant energy in a plane normal to $\mathbf{H}$. By writing $\mathbf{v}$ as $\dot{\mathbf{r}}$ Eq. (1) may be integrated in the form:

$$k_x = (-eH/\hbar)(y - y_0) \tag{5}$$

$$k_y = (eH/\hbar)(x - x_0) \tag{6}$$

Therefore in real space the electron describes a similar orbit rotated through $\pi/2$ and multiplied by the factor $s^{-1} = \hbar/eH$ (Onsager's theorem). The scheme for the quantization of closed orbits results from the application of the Bohr–Sommerfeld phase integral rule:

$$\oint \mathbf{p} \, d\mathbf{q} = (n + \phi)h \tag{7}$$

Transforming to **k** space, we find that the areas of permitted orbits are given by

$$A_k = (n + \phi)(2\pi e H/\hbar) = (n + \phi)2\pi s \tag{8}$$

i.e., the quantized orbit in real space encloses $n + \phi$ flux quanta h/e.

Let us imagine that the Fermi surface of the metal is circular and intersects only one pair of Brillouin zone boundaries:

$$k_x = \pm \frac{\pi}{a} = \pm\tfrac{1}{2}\mathbf{g} \tag{9}$$

For a magnetic field along z two types of electron orbit exist in the presence of a nonvanishing lattice potential V_g. Both the open and closed orbits shown are formed via Bragg reflection at each Brillouin zone boundary (Fig. 1).

Magnetic breakdown occurs when an electron passes through the region of Bragg reflection so quickly that only partial reflection takes place. As breakdown becomes complete the normal path is circular, as if the electrons were strictly free ($V_g = 0$).

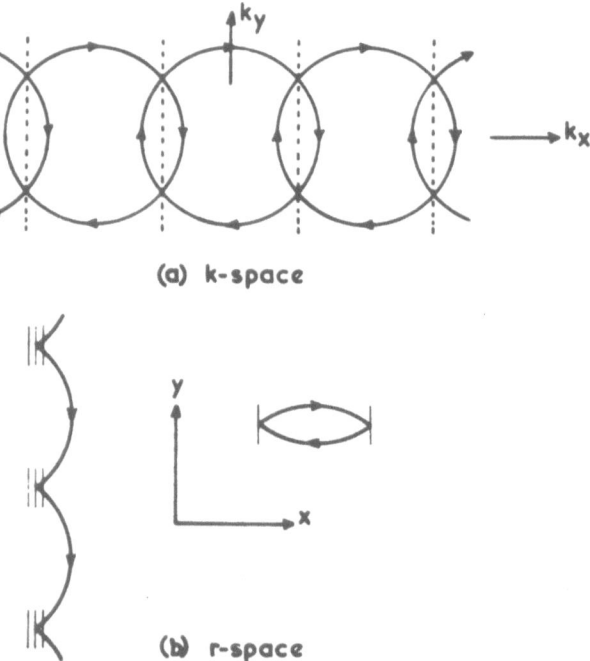

(a) k-space

(b) r-space

Fig. 1. Electron orbits in (a) **k** space and (b) **r** space.

Magnetic breakdown can be visualized in these semiclassical terms since the wave function of an orbiting electron has a representation in the form of a circular track. To look for these race-track solutions of the Schrödinger equation for a free electron we take the vector potential in the circular gauge:

$$\mathbf{A} = \tfrac{1}{2}(\mathbf{H} \times \mathbf{r}) = \tfrac{1}{2}H(-y, x, 0) \tag{10}$$

With (r, ϕ) polar coordinates in the (x, y) plane and azimuthal quantum number l solutions of the form

$$\psi = e^{il\phi}r^{l}e^{-1/4sr^{2}} \tag{11}$$

exist, with corresponding energy

$$E_l = (l + \tfrac{1}{2})\hbar\omega_c \tag{12}$$

Such a solution closely corresponds to the classical orbit. It has a maximum of $|\psi|^2$ at a radius $(2l/s)^{1/2}$ in place of the classical orbit radius $[(2l+1)/s]^{1/2}$ for an electron with energy E_l. The radial width of the orbit is $\sim s^{-1/2}$. Thus at a field of 100 kG the width is $\sim 10^{-6}$ cm, a small fraction of the orbit radius $\sim 10^{-4}$ cm. Since the relative width varies at $l^{-1/2}$, the two dimensions become comparable as the quantum limit is approached.

Chambers ([5]) has shown that adding a periodic potential to the appropriate Schrödinger equation does not alter the width of the race-track solutions. Bragg reflection switches an electron to another orbit, but there arc only a limited number of lattice planes satisfying the Bragg condition, especially when H is large and the orbit small.

An alternative approach to magnetic breakdown may be obtained by another representation of the wave function using linear combinations of Cartesian functions. We write the vector potential in the Landau gauge:

$$\mathbf{A} = (0, Hx, 0) \tag{13}$$

Solutions of the Schrödinger equation exist in the form

$$\psi(x, y) = e^{ik_v y}X(x - x_0) \tag{14}$$

$X(x - x_0)$ satisfying the harmonic oscillator equation in a potential well $\tfrac{1}{2}(e^2H^2/m)(x - x_0)^2$:

$$\frac{\hbar^2}{2m} X'' + \left[E - \frac{\hbar^2}{2m} \left(\frac{eH}{h} x + k_y \right)^2 \right]X = 0 \tag{15}$$

$$x_0 = -k_y/s \tag{16}$$

The WKB solution of (15) yields

$$X(x) \sim e^{i\phi(x)}$$

with

$$\phi(x) = \pm \int^x \{(2m/\hbar^2)[E - (e^2H^2/2m)x^2]\}^{1/2}\,dx$$

$$= \pm \int^x (k_0^2 - s^2x^2)^{1/2}\,dx; \qquad k_0^2 = 2mE/\hbar^2 \tag{17}$$

Now if solutions like (14) (same E, similar k_y) are combined linearly to make a localized wave function, we find the terms add to produce a maximum of $|\psi|^2$ at just the classical orbit.

The classical range of the electron is seen from (15) to be

$$x_{\max}^2 = 2mE/e^2H^2 \tag{18}$$

The next step is to add the contribution of the lattice potential and evaluate the effect of Bragg reflection within $x < x_{\max}$. The modification of (15) to include a component V_g,

$$\frac{\hbar^2}{2m} X'' + \left[E - V_g \cos gx - \frac{\hbar^2}{2m}(sx + k_y)^2 \right] X = 0 \tag{19}$$

leads to a solution (for $V_g \ll E$) in which ψ decays in forbidden bands around the positions of Bragg reflection (Fig. 2).

The central and outer sections represent the lens and open orbits of Fig. 1, respectively. Breakdown is permissible if the wave functions at x_1 and x_2 can couple. Thus from our WKB solution the tunneling probability is of the form:

$$P = A \exp\left[-2 \int_{x_1}^{x_2} \mu(x)\,dx \right]$$

with

$$\mu(x) = \left[s^2x^2 + \frac{2m}{\hbar^2}(V_g \cos gx - E) \right]^{1/2} \tag{20}$$

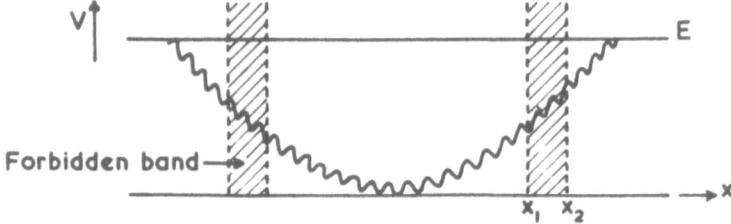

Fig. 2. Parabolic free-electron potential plus lattice potential.

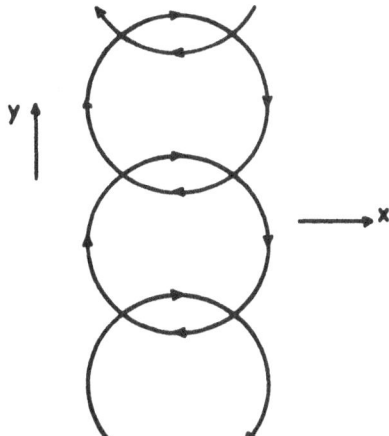

Fig. 3. Network representation of coupled orbits.

Both μ and the width $(x_1 - x_2) = w$ are proportional to the energy gap $\mathscr{E}$; w is also inversely proportional to H. Hence

$$P = A \exp(-\gamma\mathscr{E}^2/H) \tag{21}$$

More detailed analysis shows that actually $A = 1$ and

$$P = \exp(-H_0/H) \tag{22}$$

with

$$H_0 = \frac{n\mathscr{E}^2}{4e\hbar \mid v_x v_y \mid} \tag{23}$$

here v_x and v_y are the velocity components of a free electron at Bragg reflection. Substituting values $\mathscr{E} \approx 10^{-2}$ eV and $v \approx 10^8$ cm sec^{-1} yields $H_0 \approx 1$ kG.

Since the wave function for an electron in a magnetic field may be represented as a wave running on a closely defined track in **r** space, we may consider just the form of the track and the phase of the wave at any point on it. When an electron reaches a position such that it may follow one of two paths (see Fig. 1), a coupling of tracks produces a network whose degree of coupling is determined by the probability of breakthrough (Fig. 3).

We now calculate the amplitude and phase relations for waves on different segments of the network meeting at a junction. A wave of unit amplitude and zero phase arrives at the junction shown in Fig. 4a. Waves leave the junction with amplitudes and phases represented by p and q.

Fig. 4. The network junction.

A similar junction is shown in Fig. 4b. Conservation of particles for the two junctions considered together requires $1 = |p + q|^2$, i.e.,

$$pp^* + qq^* = P + Q = 1 \qquad (24)$$

and

$$pq^* + qp^* = 0 \qquad (25)$$

Hence p and q are in phase quadrature: we take p real as if the wave followed the free-electron orbit. The exact phase of q depends on $\mathscr{E}$ and the lattice position. Shifting the lattice alters the phase of Bragg reflection, which is determined by the phase of the appropriate lattice potential component at the point of reflection. If some lattice planes within the orbit are displaced, the rematching of the nodes of the standing wave patterns set up by reflection needs an energy level shift for the orbit.

We have chosen a wave function to represent a free electron in a circular orbit by solving the Schrödinger equation with the vector potential in the form $\mathbf{A} = \frac{1}{2}(\mathbf{H} \times \mathbf{r})$. Only if the gauge center corresponds to the orbit center does the phase of ψ vary uniformly around the orbit. Consequently, we must consider the rules for determining the phase change along a segment of the network which can, in the presence of breakdown, be part of several closed orbits. In particular, a gauge transformation from $\mathbf{A} = \frac{1}{2}(\mathbf{H} \times \mathbf{r})$, centered at the origin, to $\mathbf{A} = \frac{1}{2}\mathbf{H} \times (\mathbf{r} - \mathbf{R})$, centered at $\mathbf{R}$, gives a phase change:

$$\Delta\phi = \tfrac{1}{2}\mathbf{s} \cdot (\mathbf{r} \times \mathbf{R}) \qquad (26)$$

This is represented by s times the area of the triangle of Fig. 5. Using these ideas we calculate the net phase change of the closed orbits of Fig. 1b to be just s times the lens area. We remember that for an observable de Haas–van Alphen effect similar orbits must yield the same energy level schemes. Now the effect of an orbit displacement $\delta\mathbf{r}$ with respect to the perfect lattice is to produce a phase change $2\mathbf{g} \cdot \delta\mathbf{r}$ at a Bragg reflection involving $\mathbf{g}$.

However, the vector joining orbit centers (**R**) coupled by **g** is given by Onsager's theorem:

$$\mathbf{g} = \mathbf{s} \times \mathbf{R} \tag{27}$$

But for a closed orbit $\Sigma\,\mathbf{R}$ and hence $\Sigma\,\mathbf{g}$ sum to zero. Thus there is no net phase shift for the complete orbit. The situation of a dislocation inside the orbit leads to different values of $\delta\mathbf{r}$ and, in general $\Sigma\,\mathbf{g}\cdot\delta\mathbf{r}$ remains nonzero. The consequent shift in the level scheme breaks the degeneracy of similar orbits. An average dislocation density of a few dislocations per orbit is sufficient to produce phase shifts of several 2π between orbits and drastically reduce the amplitude of the de Haas–van Alphen effect. For example, the free-electron orbit with wave number $\sim 10^8$ cm^{-1} in a field of 10 kG has area $\sim 10^{-6}$ cm^2, so a dislocation density of 10^6 cm^{-2} is likely to cause damage. Any amplitude reduction should thus be more noticeable for orbits involving Bragg reflection. There is as yet no conclusive evidence to confirm this point.

Let us continue the network theory of orbits coupled by partial reflection. We draw in **k** space the repeated Brillouin zone scheme of our two-dimensional metal, keeping only those zone boundaries which allow breakdown. The form of the coupled orbits between which the electrons may switch is found by rotating the lattice through $\pi/2$ and dividing by s. The switching vector joining orbit centers is given by (27). Now the reciprocal lattice vectors are obtained from the appropriate ionic (I) lattice basis vectors **a** using

$$\mathbf{g} = 2\pi\,\frac{\mathbf{s}\times\mathbf{a}}{s\Sigma} \tag{28}$$

where Σ is the area of I-lattice unit cell. Thus

$$\mathbf{R} = J\mathbf{a}; \qquad J = 2\pi/s\Sigma = 2\pi\hbar/eH\Sigma \tag{29}$$

Hence the centers of orbits coupled by the lattice potential form a lattice

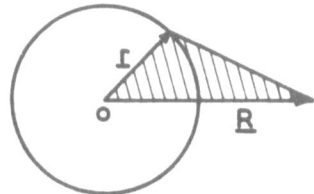

Fig. 5. Geometrical interpretation of the phase change in a gauge transformation.

(the O-lattice) similar to the I-lattice but scaled up by J. We note that $H\Sigma$
$\div 2\pi\hbar/e$ flux quanta pass through each I-lattice cell: the number of quanta
per unit cell of the O lattice is simply

$$(eH\Sigma/2\pi\hbar)J^2 = J \qquad (30)$$

Second, we may readily evaluate the number of levels per Brillouin zone as J.
Since the number of states in a Brillouin zone is $1/\Sigma$ and the Landau level
degeneracy is $s/2\pi$ per unit area of metal, the number of levels is

$$(2\pi/s)(1/\Sigma) = J \qquad (31)$$

Certain fields make J integral and the orbit centers become similarly placed
on the I lattice: in these cases junctions that look equivalent are equivalent.
For fields leading to J in the form M/N the O-lattice cell becomes JN
times the I-lattice cell. We see the important property of integral values of
J—periodic solutions of the network problem exist. Bloch's theorem applies
and phase changes may be characterized by the wave vector $\mathbf{K}$, i.e., the
phase difference between equivalent points on two orbits is $\mathbf{K} \cdot \mathbf{R}$, where $\mathbf{R}$
is the vector joining orbit centers. Now for a given choice of orbit centers
$\mathbf{K}$ leads to the energy of the state, since the phase relations at all junctions
can only be satisfied for particular values of the energy E.

In the case of the hexagonal network the I lattice is developed from basis
vectors $\mathbf{a}_1$, $\mathbf{a}_2$, and $\mathbf{a}_3$ oriented at 120° apart, and correspondingly the
O lattice (Fig. 6) has basis vectors $J\mathbf{a}_1$, $J\mathbf{a}_2$, and $J\mathbf{a}_3$. The figure represents
the wave function of coupled circular race tracks. In looking for periodic

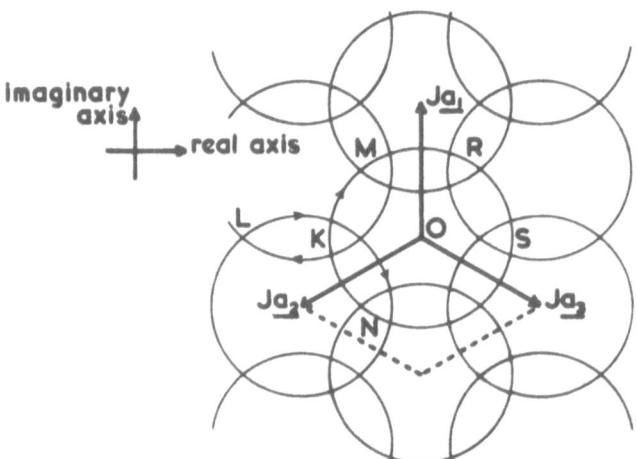

Fig. 6. O-lattice for the hexagonal network, showing unit cell.

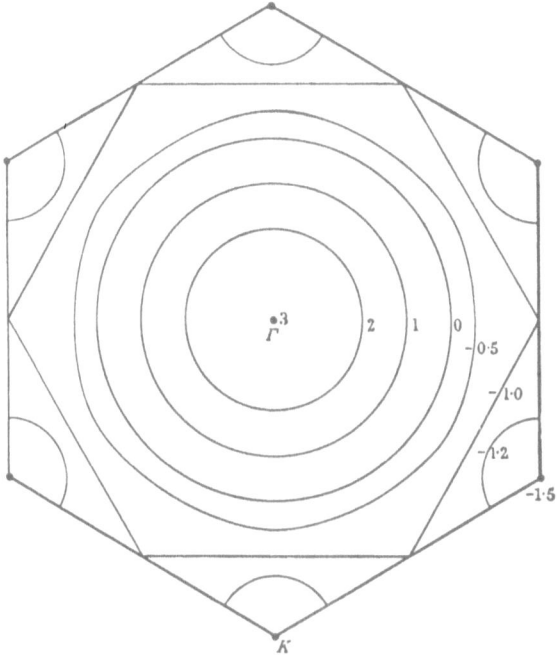

Fig. 7. Contours of constant C on the $\mathbf{K}$ plane.

solutions we assign complex amplitudes to each of the 12 orbital segments making the junctions consistent with p and q, the transmission and reflection ratios. Neighboring orbits have the amplitude assignments phase shifted by ω_1, ω_2, and ω_3, as appropriate ($\omega_i = J\mathbf{a}_i \cdot \mathbf{K}$). The six junctions in an O-lattice unit cell lead to 12 relations for consistency. Eliminating the 12 amplitudes from these equations yields

$$\cos \omega_1 + \cos \omega_2 + \cos \omega_3 = C \tag{32}$$

where C is a complicated function of p, q, and the phase angles corresponding to areas on the network diagram. It is physically restricted to the range $3 \geq C \geq -\frac{3}{2}$; contours of constant C in $\mathbf{K}$ space are contours of constant energy (Fig. 7). We see that Eq. (32), containing the band-structure of the network, resembles the expression for a tight-binding calculation for a hexagonal array of atoms with only nearest-neighbor interactions.

The Brillouin zone for the variation of $\mathbf{K}$ is the reciprocal of the O-lattice cell with area $1/J^2$ times that of the I-lattice Brillouin zone. Since we have noted that J independent networks exist and the O-lattice Brillouin zone holds $1/J^2$ states per ion of the I lattice, a filled O-lattice zone holds

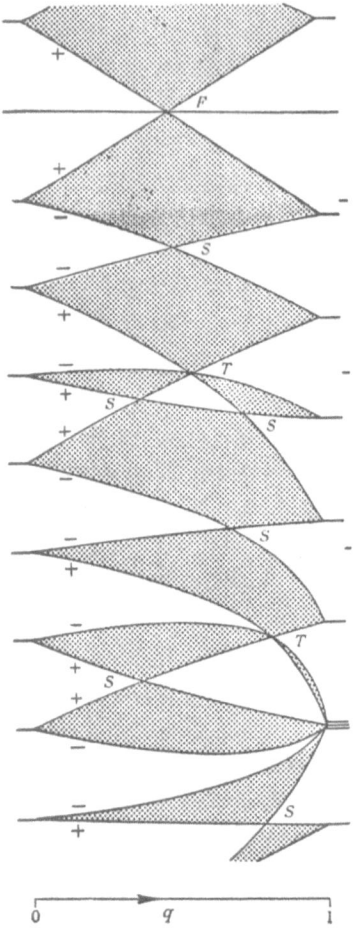

Fig. 8. Typical energy band structure for the
hexagonal network.

$1/J$ states per ion, consistent with the Landau level degeneracy of free
electrons.

The energy band structure of the network for a particular ratio of
orbital segments exhibits sharp free electron levels for $q = 0$ (high fields)
and two sets of levels for $q = 1$ (low fields) spaced apart in inverse propor-
tion to the hexagonal hole and triangular electron orbit areas [Fig. 8 of
([2])]. As the magnetic field is changed slowly the free-electron levels shift at
one speed while the levels on the right drift past at different speeds. The
complexity of the intermediate case makes clear the difficulties of any
perturbation theory approach to the band structure calculation.

The extended nature of the network wave functions allows us to form wave packets from a range of **K**. These current-carrying network functions propagate in straight-line trajectories at the group velocity $\hbar^{-1} \operatorname{grad}_k E$ as if the band structure $E(\mathbf{K})$ applied to particles in zero magnetic field. Every electron state can transport charge and all directions of transport are possible. Since in addition the propagation velocity is a considerable fraction of the free-electron velocity, these quasiparticles can, in principle, have a profound effect on the conductivity. But the presence of a few dislocations per orbit is sufficient to disturb the phase coherence of the network and scatter the quasiparticles. We shall investigate the transport properties of the two-dimensional network in more detail later.

The situation for nonintegral J may be summarized as follows. If the fractional part of J is M'/N, the dimensions of the O-lattice unit cell become NJ times those of the I-lattice cell. Each band of Fig. 8 splits into N subbands, of which M' go one way and $N - M'$ go the other, as shown schematically in Fig. 9.

The subbands reunite for each integral value of J. This splitting of broad bands into sharper subbands is analogous to the formation of Landau levels for electrons when a magnetic field is switched on. Although the real problem is complex, the energy separation of the subbands is masked under practical conditions by the broadening effects of temperature, collisions, and field inhomogeneity. We may develop the quasiparticle approach for the case of nonintegral J. Suppose

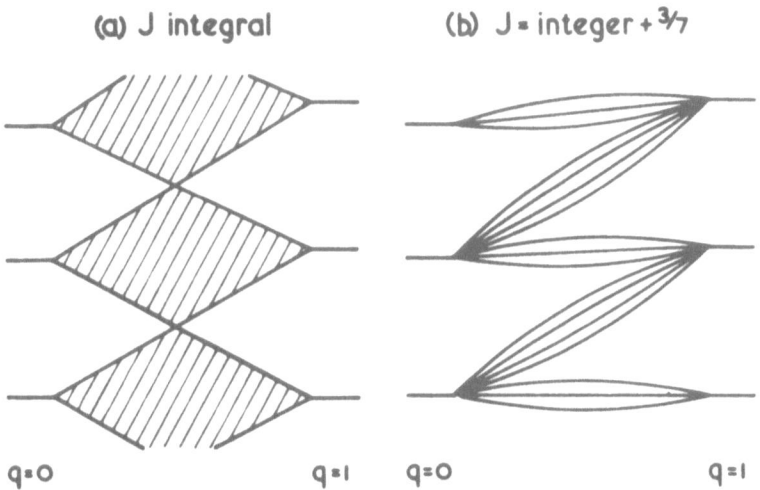

Fig. 9. The splitting of energy bands for nonintegral J.

$$J = J_0 + (1/J_1) \tag{33}$$

with J_0 and J_1 integral. Now we have $H = 2\pi\hbar/e\Sigma J$, so

$$\delta H = \frac{2\pi\hbar}{e\Sigma} \frac{1}{J_0{}^2 J_1}$$

i.e.,

$$\frac{2\pi\hbar}{e} = J_1 \, \delta H \, J_0{}^2 \Sigma \tag{34}$$

Thus an incremental field δH leads to as many subbands (J_1) as cells of the O-lattice required to hold one flux quantum. Azbel' ([8]) has pointed out how a hierarchy of field strengths may be generated by a continuous fraction of the form:

$$J = J_0 + \cfrac{1}{J_1 + \cfrac{1}{J_2 + \ldots}} \tag{35}$$

We may interpret J_0 as defining quasiparticles which are formed by breakdown and bent by J_1 into hyperorbits as if acted on by the appropriate incremental field. These in turn can break down to produce a coupled network of hyperorbits whose own quasiparticles are bent into hyperhyperorbits through J_2, and so on. Thus linear transport is still possible, though with a reduced velocity due to the narrowness of the subbands.

EXPERIMENTAL CONSEQUENCES OF MAGNETIC BREAKDOWN

The de Haas-van Alphen Effect

The de Haas–van Alphen effect may be evaluated from the free energy $F(\mathbf{H})$ once the density of states $N(E, \mathbf{H})$ of the electron system of coupled orbits has been calculated.

$$F(\mathbf{H}) = NE_F - 2 \int dE f(E) \int^E dE' \, N(E', \mathbf{H}) \tag{36}$$

$$\mathbf{M}(\mathbf{H}) = -\partial F/\partial H$$

where $f(E)$ is the Fermi–Dirac distribution function and $\mathbf{M}(\mathbf{H})$ the magnetization vector. We determine the oscillatory part of the density of states by studying the property of a pulse on a network returning to its origin as a series of pulses. For the example of free-electron orbits we analyze such a wave packet pulse into eigenfunctions of the network; since they are evenly spaced, they all add up in phase after one or more cyclotron periods (we have applied a classical interpretation of pulses traveling around orbits). Hence a δ-function develops as a regular series of pulses at the origin which continues in the absence of scattering. The same idea holds for more complex orbits, which also generate evenly spaced levels, and we may extend this to a network, remembering that there are many paths returning to any chosen origin.

Falicov and Stachowiak ([6]) considered the relation between the pulse pattern and the density of states by examining the evolution of a Green's function $G(\mathbf{r}, \mathbf{r}_0, t)$. At $t = 0$ let

$$\Psi(\mathbf{r}, 0) = \delta(\mathbf{r} - \mathbf{r}_0) = \sum_m \psi_m{}^*(\mathbf{r}_0)\psi_m(\mathbf{r}) \tag{37}$$

[$\psi_m(\mathbf{r})$ are the complete set of network eigenfunctions].

Then at time t the time-dependent wave function takes the form $G(\mathbf{r}, \mathbf{r}_0, t)$ and each $\psi_m(\mathbf{r})$ evolves with phase factor given from the appropriate eigenvalue E_m. Thus

$$\Psi(\mathbf{r}, t) = \sum_m \psi_m{}^*(\mathbf{r}_0)\psi_m(\mathbf{r}) \exp\left(-\frac{iE_m t}{\hbar}\right) = G(\mathbf{r}, \mathbf{r}_0, t) \tag{38}$$

and hence

$$\int G(\mathbf{r}_0, \mathbf{r}_0, t)d^3\mathbf{r}_0 = \sum_m \exp\left(-\frac{iE_m t}{\hbar}\right) = \Omega \int N(E) \exp\left(-\frac{iEt}{\hbar}\right) dE \tag{39}$$

where $N(E)$ is the density of states per unit volume of material and Ω is the volume. For a system exhibiting full translational symmetry

$$\int G(\mathbf{r}_0, \mathbf{r}_0, t)d^3\mathbf{r}_0 = \Omega G(0, 0, t) \tag{40}$$

and (40) becomes

$$G(0, 0, t) = \int N(E) \exp(-iEt/\hbar) \, dE \tag{41}$$

and the Fourier transform:

$$N(E) = (1/2\pi\hbar) \int G(0, 0, t) \exp(iEt/\hbar) \, dt \tag{42}$$

A succession of pulses $\delta[t - (2\pi n/\omega_c]$, alternating in sign due to the phase correction $\phi = \frac{1}{2}$ in (8), gives Fourier components of equal amplitude, so that the density of states adds up to a series of sharp levels spaced regularly $\hbar\omega_c$ apart. Collisions decrease the strength of the returning pulses, and if this decrease in intensity can be accounted for by a relaxation time τ, the Fourier transform produces a Lorentzian broadening of the previously sharp levels.

We apply the above analysis to calculate the de Haas–van Alphen effect for the network of coupled orbits. The pulses due to any one orbit have amplitudes multiplied by appropriate factors $p^n q^m$ per circuit, and so for most orbits harmonics are faint unless either p or q is near zero. However, different orbits with the same phase length contribute with the same periodicity but with the proviso that since p and q are complex, similar orbits may tend to reinforce or cancel others, leading to field-dependent interference effects. The precise value of J has not been considered in this method. Each orbit area is composed of orbital sectors plus unit cells. Now, when J is integral the phase contribution from an O-lattice unit cell $(sJ^2\Sigma)$ is just a multiple of 2π. At these values the Fourier components have a higher phase linking than otherwise.

Every collision must be regarded as destructive for the de Haas–van Alphen effect. We require only a very small fractional error in the phase change $2\pi l$ between pulse circuit returns to cause randomization (if l is large). So all small-angle collisions contribute to the relaxation time and the associated line broadening will be similar to that calculated using the thermal conductivity (not electrical conductivity) relaxation time. Even small-angle elastic phonon collisions which do not affect thermal conductivity will destroy the phase coherence of the returning pulses.

The influence of magnetic breakdown on the de Haas–van Alphen effect has been illustrated by the amplitude measurements on two hexagonal metals, magnesium and zinc. The low-frequency oscillation in Zn ([7]) due to the triangular orbit is very prominent. Three reflections mean the fundamental de Haas–van Alphen oscillation is governed by the factor [from (22)]

$$q^3 = (1 - e^{-H_0/H})^{3/2} \tag{43}$$

One may expect H_0 to vary along the length of the orbit perpendicular to the plane of the network, but the oscillatory effects should be dominated by the value of H_0 at the extremal cross section. If we compare the field variation of the de Haas–van Alphen amplitude with the theory for perfect orbits, we find the amplitude falling away noticeably above $H \approx 4 \, \text{kG}$. This can be interpreted with $H_0 \approx 2 \, \text{kG}$ but has not been a very sensitive method of measuring H_0.

Transport Effects

We have seen how a very small density of dislocations disrupts the phase coherence of the network model and negates the quasiparticle approach. We therefore return to the semiclassical picture of electrons constrained to move along segments of the network. At each junction the paths taken are obtained from the appropriate switching probabilities. The effective path method serves in the calculation of the conductivity as follows. An electric-field ($\mathscr{E}$) impulse shifts the electron distribution in **k** space at a rate $e\mathscr{E}/\hbar$. This displacement generates electrons or holes at the Fermi surface, as appropriate. After creation these particles move through the metal and are scattered. The conductivity is determined by the average distance they finally travel ($\bar{L}$). A uniform current density may be synthesized from

$$\mathbf{j}_{\text{uniform}} = \int_0^\infty \mathbf{j}_{\text{impulse}}(t)\, dt$$

$$= \Sigma \int_0^\infty e\mathbf{v}(t)\, dt$$

$$= \Sigma\, e\bar{L} \tag{44}$$

here $\bar{L}$ is the effective path length. Then the conductivity tensor is given by

$$\sigma_{ij} = (e^2/4\pi^3\hbar) \oint L_i\, dS_j \tag{45}$$

We proceed to calculate the effective path on the hexagonal network neglecting the effect of collisions. We assume the junctions randomize the electron motion and that all electrons generated on a given segment have the same probability of undergoing a particular random walk on the network. An electron entering the triangular knot may leave by the three arms indicated in Fig. 10a with probabilities A, B, and C, respectively.

To evaluate A, B, and C in terms of the junction transmission and reflection probabilities P and Q we let the particle flux along the arm leaving T (Fig. 10b) be Z and make the knot self-consistent. Hence at T, $Z = P + Q^3Z$, i.e., $Z = P/(1 - Q^3)$. Thus

$$A = 1 - B - C, \qquad B = P^2/(1 - Q^3), \qquad C = QB \tag{46}$$

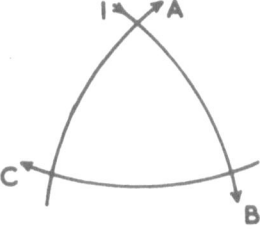

Fig. 10a. Triangular knot of the hexagonal network.

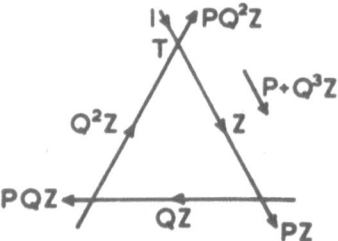

Fig. 10b. Schematic knot with particle flux along arms.

Figure 6 illustrates the progress of an electron approaching a knot; we use a complex notation with the electric field direction defining the real axis. The terminal point of an electron traveling along arc LK must be consistent with the terminal point of electrons leaving the knot at K along arcs KM, KN, and KL. If this point is called X with respect to L, i.e., $X - r$ with respect to K (r being the orbit radius), we have

$$X - r = AXe^{i\pi/3} + BXe^{-i\pi/3} + CXe^{i\pi} \tag{47}$$

and from (46) and (47)

$$X = \frac{re^{i\pi/3}}{[1 + \sqrt{3}\,iB(Q + e^{i\pi/3})]} \tag{48}$$

Referring again to Fig. 6, we consider the semicircle $KMRS$. Since the number of particles generated on each arc is proportional to its component in the electric field direction, we readily obtain the mean terminal point for the semicircle as the weighted mean of the individual terminal points. Arcs KM, MR, and RS have weights $\frac{1}{4}$, $\frac{1}{2}$, and $\frac{1}{4}$ respectively. The mean terminal point, measured from the center of the circle, is then

$$(Xe^{i\pi/3} - r)(\tfrac{1}{4} + \tfrac{1}{2}e^{-i\pi/3} + \tfrac{1}{4}e^{-2i\pi/3}) = \tfrac{3}{4}(X - re^{-i\pi/3}) = \bar{L}_T \tag{49}$$

To find the effective path length $\bar{L}$ for the particles we must subtract the mean starting point ($\bar{L}_S$) from $\bar{L}_T$. For the electrons generated on the semicircle $\bar{L}_S$ is given by the weighted centroid of the arc KS:

$$\bar{L}_S = i\pi r/4 \tag{50}$$

Then

$$\bar{L} = \bar{L}_T - \bar{L}_S = \frac{r}{4}\left[\frac{3e^{i\pi/3}}{1 + \sqrt{3}\,iB(Q + e^{i\pi/3})} - 3e^{-i\pi/3} - i\pi\right] \tag{51}$$

For low fields (when no breakdown occurs) $Q = 1$, $P = 0$, $B = 0$, and we have

$$\bar{L} = i(3\sqrt{3} - \pi)r/4 \tag{52}$$

For complete breakdown in high fields $P = 1$, $Q = 0$, $B = 1$, and

$$\bar{L} = -i\pi r/4 \tag{53}$$

Thus in the limits when the electron has no choice of path at the junctions and in the absence of other scattering mechanisms $\bar{L}$ (and therefore the conduction current) remains perpendicular to the electric field, producing nondissipative conduction (Fig. 11). Between these extreme cases the existence of a component of current parallel to $\mathscr{E}$ leads to dissipation. This may be viewed as a consequence of the entropy production due to the choice of exit paths from each junction.

Electrons and holes are compensated for low fields in the hexagonal metals Zn and Mg. Breakdown destroys this by converting hole orbits into electron orbits. Figure 12 demonstrates the quadratic rise of ϱ_{11} at low fields and the saturation at high fields of ϱ_{11} and the Hall coefficient $\sigma_{12}H$. The position of the mid-point of the saturation value of $\sigma_{12}H$ yields a sensitive measure of H_0 [see (23)]. The value $H_0 = 2.7$ kG agrees well with the peak in ϱ_{11}.

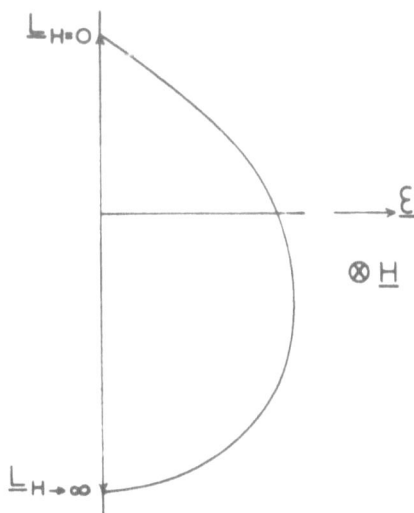

Fig. 11. Variation of effective path with magnetic field. From Pippard ([3]).

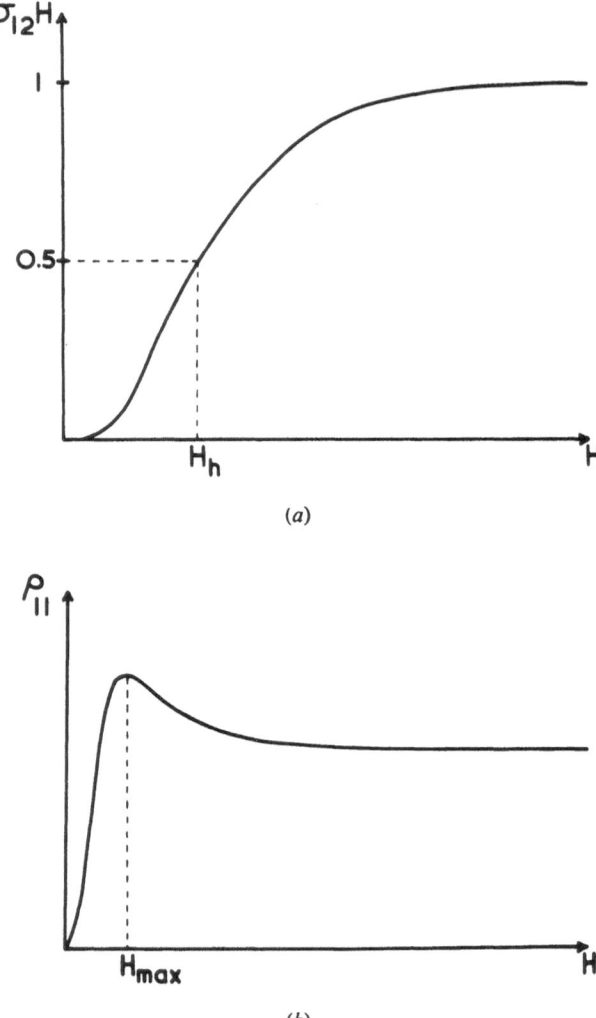

Fig. 12. Form of ϱ_{11} and $\sigma_{12}H$ for Mg and Zn.

Experimental measurements with crystals of moderate dislocation density follow Fig. 12, and at lower temperatures oscillations based on these curves become evident. These oscillations result from a partial phase coherence of the network. We regard the triangular knots as coherent units (Fig. 13), but not the longer communicating arms. Instead of the analysis leading to (46) we must match amplitudes rather than intensities.

If $\theta = \phi/3$ is the phase length of each arc of the knot, the equation for

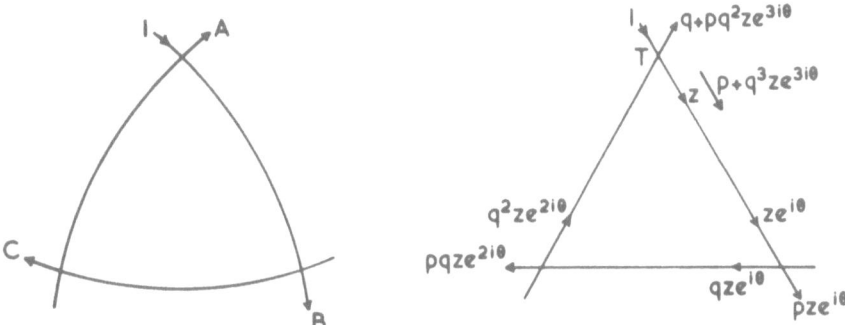

Fig. 13. The coherent knot of the hexagonal network.

consistency at T yields:

$$z = p + q^3 z e^{i\phi} \tag{54}$$

Thus $z = p/(1 - q^3 e^{i\phi})$. We have $B = |pze^{i\phi}|^2$, $pp^* = P$, etc., giving

$$B = P^2/(1 + Q^3 - 2Q^{3/2}\cos\phi); \quad C = QB; \quad A = 1 - B - C \tag{55}$$

Now, for small breakthrough probability $Q \approx 1$, and the electron can only enter the junction if the phase length around the knot is near a multiple of 2π. Thus the effective path oscillates with H at the de Haas–van Alphen frequency of the triangular orbit. Calculations of ϱ_{11} and $\sigma_{12}H$ using (55) agree reasonably well with experimental measurements.

This theory assumes no phase coherence around any orbit larger than the triangle. If we had a perfect sample we should have to abandon this semiclassical approach and treat the network conduction problem with the quasiparticle picture. Such a theory becomes less tractable when a few dislocations are introduced to scatter the quasiparticles. Stark has found greatly enhanced magnetoresistance oscillations in a Mg sample of low dislocation density. However, the oscillations did not survive thermal cycling of the specimen.

REFERENCES

1. A. B. Pippard, *Proc. Roy. Soc. (London) Ser. A* **270**: 1 (1962).
2. A. B. Pippard, *Phil. Trans. Roy. Soc. (London) Ser. A* **256**: 317 (1964).
3. A. B. Pippard, *Proc. Roy. Soc. (London) Ser. A* **287**: 165 (1965).
4. L. M. Falicov, A. B. Pippard, and P. R. Sievert, *Phys. Rev.* **151**: 498 (1966).
5. R. G. Chambers, *Proc. Phys. Soc.* **88**: 701 (1966); **89**: 695 (1966).
6. L. M. Falicov and H. Stachowiak, *Phys. Rev.* **147**: 505 (1966).
7. J. S. Dhillon and D. Shoenberg, *Phil. Trans. Roy. Soc. (London) Ser. A* **248**: 1 (1955).
8. M. Ya. Azbel', *Soviet Phys.–JETP (English Transl.)* **19**: 634 (1964).

Chapter 19

ULTRASONIC PROPAGATION
IN HIGH-FIELD SUPERCONDUCTORS

Y. Shapira

*National Magnet Laboratory**
Massachusetts Institute of Technology
Cambridge, Massachusetts

In the last few years a new class of superconducting materials whose super-conducting-to-normal transition at $T = 0 \,°\text{K}$ occurs at $\sim 10^5$ G have been discovered ([1]). To date only a few experimental studies of ultrasonic propagation in such high-field superconductors (hereafter HFS) have been reported. The results of these studies led to the formulation of a phenomenological theory which appears to be successful in describing quantitatively, and without the use of adjustable parameters, the observed attenuation and velocity changes ([9]). A microscopic theory for sound propagation in HFS is not available at the present time.

SOME PROPERTIES
OF HIGH-FIELD SUPERCONDUCTORS

HFS are type-II superconductors, as distinguished from superconductors like lead, tin, or indium, which are type-I superconductors. A type-II superconductor excludes magnetic flux from its interior as long as the external magnetic field H is less than the lower critical field H_{c1}. At $H_{c1} < H < H_{c2}$ there is a partial penetration of the magnetic flux into the interior of the superconductor. At fields above the upper critical field H_{c2} a type-II superconductor is in the normal state.

In the following discussion we shall be primarily concerned with the field interval $H_{c1} < H < H_{c2}$. In this field interval a type-II superconductor

* Supported by the U. S. Air Force Office of Scientific Research.

is said to be in the "mixed state" because the superconductor contains filaments of normal material, directed along **H**, embedded in a supercon-ducting material [see, for example, (⁴)]. Each normal filament (often called normal core, vortex core, or flux line) is associated with a unit of magnetic flux $\varphi_0 = hc/2e$.

When a DC electric current flows through a type-II superconductor which is in the mixed state a finite resistance is observed. This resistance arises from the motion of the flux lines and is called "flow resistance." The flow resistivity, ϱ_f, was studied by Kim et al. (²), who found that if the electric current is normal to **H** and if $H_{c1} \ll H < H_{c2}$, then

$$\varrho_f = \varrho_n H/H_{c2}(0) \tag{1}$$

where ϱ_n is the normal-state resistivity and $H_{c2}(0)$ is the upper critical field at $T = 0\,°K$.

In actual type-II superconductors there are defects which inhibit the motion of the flux lines. The flux lines are then said to be "pinned" to the lattice. In order to observe the flow resistance, it is necessary to use suffi-ciently high current densities so that the driving force on a flux line will overcome the pinning force, thereby enabling the flux line to move and generate an electric field. Thus with a DC electric current an electrical resis-tance is observed only when the current density J exceeds the so-called critical current density J_c. With AC electric currents the situation is different.

The AC electrical resistivity of a type-II superconductor at $H_{c1} \ll H < H_{c2}$ was considered by Gittleman and Rosenblum (³). From their treat-ment it follows that the resistivity ϱ for a small-amplitude AC current flowing normal to **H** is a complex quantity $\varrho = \varrho_1 + i\varrho_2$, where

$$\varrho_1 = \varrho_f/(1 + r^2) \tag{2}$$

and

$$\varrho_2 = r\varrho_f/(1 + r^2) \tag{3}$$

Here ϱ_f is given by Eq. (1) and $r = \omega_0/\omega$, where ω_0 is a frequency which characterizes the strength of the pinning forces. The frequency ω_0 can be estimated from the critical current density J_c (³,⁹).

HFS VERSUS TYPE-I SUPERCONDUCTORS

The primary causes of ultrasonic attenuation (and velocity changes) in HFS which are subjected to intense magnetic fields are completely different from those responsible for the attenuation observed in pure type-I superconductors. This difference is due to two facts: (1) The normal-state electrical resistivity of a typical HFS is several orders of magnitude larger than the normal-state resistivity of pure type-I superconductors, and (2) the magnetic fields employed in studying HFS are one or two orders of magnitude higher than those employed in studying low-field superconductors.

To obtain a better understanding of the difference between the ultrasonic attenuation in HFS and in type-I superconductors, we first consider the situation in the latter class of materials. In studies of the attenuation in type-I superconductors one uses the expression

$$\frac{\alpha_s}{\alpha_n} = \frac{2}{1 + \exp(\varepsilon/kT)} \tag{4}$$

where α_s and α_n are the attenuation coefficients in the superconducting and normal states, respectively, and 2ε is the temperature-dependent energy gap.

It should be pointed out that the attenuation coefficient α_n which appears in Eq. (4) is, strictly, the attenuation coefficient that would have been present at zero magnetic field had the material been normal. In other words, α_n does not contain any magnetic-field effects on the normal attenuation. A similar remark also holds for α_s. The fact that α_n and α_s refer to the normal and superconducting attenuation coefficients *at zero field* is often of little consequence in the case of type-I superconductors. The reason for this is that the magnetic fields which are used in studies of these materials are quite low ($H \lesssim 1$ kG), and aside from destroying the superconductivity have little effect on the attenuation. Thus it is a common procedure to apply a magnetic field to destroy the superconductivity and then measure the attenuation coefficient and assume that this attenuation coefficient is very nearly equal to α_n. The situation in the case of HFS is radically different.

For metals with an electron mean free path which is short compared to the ultrasonic wavelength, α_n is inversely proportional to the electrical resistivity. In general, the normal-state resistivity of HFS is so high that α_n is very small at frequencies in the 10-MHz range. Therefore at these frequencies one can ignore α_n and α_s. On the other hand, the high magnetic fields employed in studying HFS cause large changes in the attenuation. These changes are generally much larger at 10^4–10^5 G than α_n.

The preceding arguments indicate that in studying the ultrasonic attenuation in HFS subjected to intense magnetic fields one may ignore α_n (or the difference between α_n and α_s) and focus attention on the effects of the magnetic field on the attenuation. This approach is expected to be valid at frequencies in the 10-MHz range but may not be valid at much higher frequencies.

MAGNETIC-FIELD EFFECTS ON ULTRASONIC PROPAGATION IN THE NORMAL STATE

To understand the effects of a magnetic field on ultrasonic propagation in the mixed state, it is first necessary to examine the situation in the normal state. A theory which describes the influence of a magnetic field on the ultrasonic attenuation and velocity in metals *with short electron mean free path* was formulated by Alpher and Rubin ([5]). The basic assumptions of this theory are:

1. The equation of motion of a lattice in the presence of a magnetic field is

$$\frac{d^2\xi}{dt^2} = \frac{1}{d}\left[\mathbf{F}_e + \frac{1}{c}\,(\mathbf{J} \times \mathbf{B})\right] \tag{5}$$

where ξ is the displacement of the lattice, d is the density of the metal, $\mathbf{J}$ is the current density associated with the sound wave, $\mathbf{B}$ is the magnetic induction, and $\mathbf{F}_e$ is the elastic force at zero field. $\mathbf{F}_e$ can be expressed in terms of the Young's modulus Y, and the Poisson ratio ν:

$$\mathbf{F}_e = \frac{Y}{2(1 + \nu)}\left[\nabla^2\xi + \frac{1}{1 - 2\nu}\,\nabla(\nabla \cdot \xi)\right] \tag{6}$$

2. The current density $\mathbf{J}$ is given by

$$\mathbf{J} = \sigma\left[\mathbf{E} + \left(\frac{1}{c}\right)\left(\frac{d\xi}{dt} \times \mathbf{B}\right)\right] \tag{7}$$

where σ is the electrical conductivity and $\mathbf{E}$ is the electric field associated with the sound wave.

Using these two assumptions together with Maxwell's equations one can derive expressions for the changes in the ultrasonic attenuation and velocity caused by the magnetic field. It turns out that for a shear wave these changes are largest when the direction of sound propagation is along $\mathbf{H}$, whereas for a longitudinal wave the changes are largest when the sound propagates in a direction normal to $\mathbf{H}$. In these configurations the change in the amplitude-attenuation coefficient for a sound wave (shear or longitudinal) due to the magnetic field is given by

$$\alpha = \left(\frac{\omega H^2\mu}{8\pi dV^3}\right)\left(\frac{\beta}{1 + \beta^2}\right) \quad (\text{cm}^{-1}) \tag{8}$$

$$\beta = c^2\omega\varrho/4\pi\mu V^2 \equiv \varrho/\varrho_0 \tag{9}$$

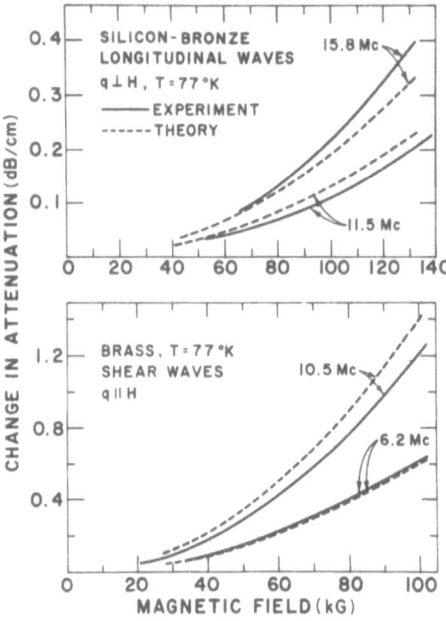

Fig. 1. Change of the ultrasonic attenuation
in a magnetic field. The dashed curves are calculated from Eqs. (8) and (9).

where ω is the sound frequency, μ is the permeability, V is the sound velocity, ϱ is the electrical resistivity and ϱ_0 is given by

$$\varrho_0 = 4\pi\mu V^2/c^2\omega \tag{10}$$

The corresponding change in the sound velocity is

$$\Delta V/V = \mu H^2/8\pi dV^2(1 + \beta^2) \tag{11}$$

Equation (8) was verified experimentally by Shapira and Neuringer [7] and
Eq. (11) was verified experimentally by Alers and Fleury [6]. Figure 1 shows
some results of attenuation measurements in brass and in silicon–bronze.

MAGNETIC-FIELD EFFECTS ON ULTRASONIC PROPAGATION IN THE MIXED STATE

A phenomenological theory for ultrasonic propagation in HFS at
$H_{c1} \ll H < H_{c2}$ was developed by Shapira and Neuringer [9]. The basic
assumption of this theory is that the magnetic-field effects on the propaga-

tion of shear waves with $q \parallel H$ and of longitudinal waves with $q \perp H$ can be calculated by modifying the Alpher–Rubin theory in the following way. Wherever the electrical resistivity appears in the Alpher–Rubin theory one should use the AC resistivity in the mixed state [Eqs. (2) and (3)] instead of the ordinary DC resistivity. This assumption is reasonable because the sound wave gives rise to AC currents rather than to DC currents.

When a metal is in the normal state there is practically no difference between the DC resistivity and the AC resistivity at frequencies of the order of 10 MHz. Therefore in dealing with metals which are in the normal state the question of which type of resistivity should be used in the Alpher–Rubin theory is of little practical importance. On the other hand, in HFS which are in the mixed state the distinction between the two types of resistivities is essential.

Using the above assumption together with Eqs. (2) and (3), one

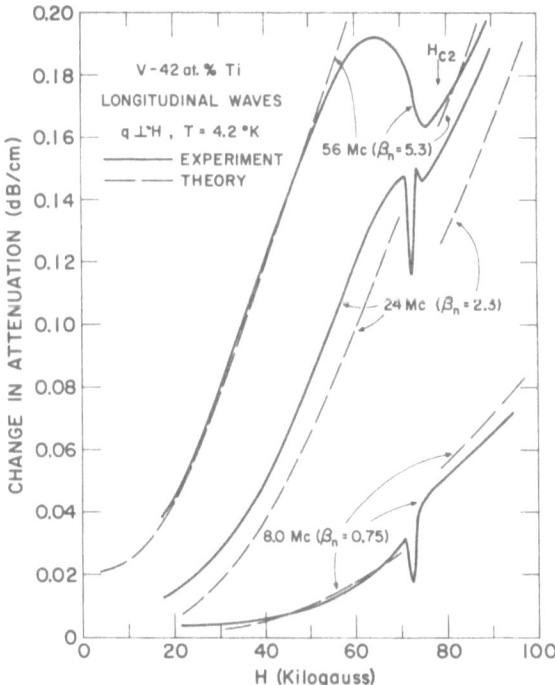

Fig. 2. Magnetic-field variation of the attenuation of longitudinal waves in V–42 at.% Ti at 4.2°K. The dashed curves for $H < H_{c2}$ and $H > H_{c2}$ are calculated from Eqs. (15) and (8), respectively. The curves for the 56-MHz wave are shifted upward by 0.02 dB/cm.

obtains the following expressions

$$\alpha = \left(\frac{\omega H^2 \mu}{8\pi d V^3}\right) \frac{\beta_f(1 + r^2)}{[(1 + r^2 + r\beta_f)^2 + \beta_f^2]} \quad (\text{cm}^{-1}) \tag{12}$$

and

$$\frac{\Delta V}{V} = \left(\frac{\mu H^2}{8\pi d V^2}\right)\left[\frac{(1 + r^2 + r\beta_f)(1 + r^2)}{(1 + r^2 + r\beta_f)^2 + \beta_f^2}\right] \tag{13}$$

where

$$\beta_f = \varrho_f/\varrho_0 \tag{14}$$

It should be noted that at $H \gg H_{c1}$ the permeability μ is approximately equal to unity.

There are several interesting cases in which Eq. (12) takes a simple form. When the pinning forces are weak, i.e., when $r = \omega_0/\omega \ll 1$, Eq. (12) can be written as

$$\alpha = \left(\frac{\omega H^2 \mu}{8\pi d V^3}\right)\left(\frac{\beta_f}{1 + \beta_f^2}\right), \quad r \ll 1 \tag{15}$$

Equation (15) can be interpreted physically as follows: When $\omega \gg \omega_0$ the expression for the attenuation in the mixed state has the same form as the expression for the attenuation in the normal state [Eq. (8)] except that the ordinary DC resistivity is replaced by the flow resistivity. Figure 2 shows the results of attenuation measurements in a V–42 at % Ti superconducting alloy for which $r \ll 1$ at $\omega/2\pi \geq 8$ MHz. It is apparent that there is good agreement between theory and experiment.

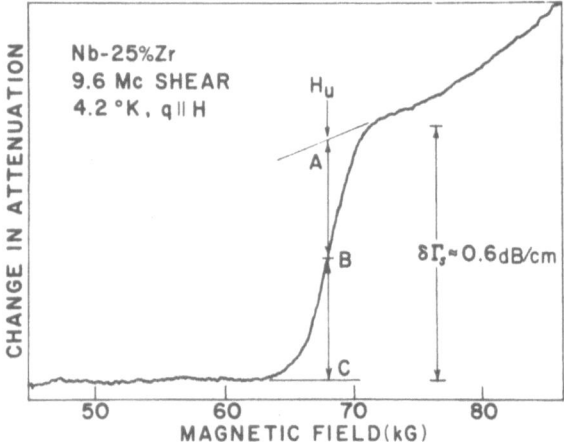

Fig. 3. Recorder tracing of the ultrasonic attenuation of shear waves in unannealed Nb–25% Zr.

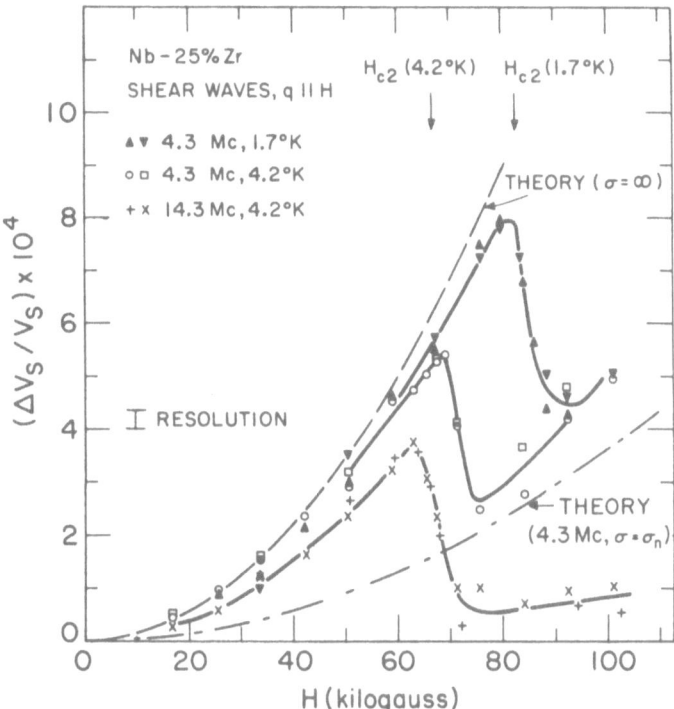

Fig. 4. Change in the ultrasonic velocity of shear waves in unannealed Nb–25% Zr. The theoretical curve with $\sigma = \infty$ is calculated from Eq. (16). The theoretical curve with $\sigma = \sigma_n$ is calculated for a 4.3-MHz wave using Eq. (11) and letting $\varrho = \varrho_n$.

It can be shown that the attenuation in the mixed state, at a given H and ω, is a monotonically decreasing function of r. Thus an increase in the strength of the pinning forces causes a decrease in the attenuation. In the limit $r \to \infty$ (or $\omega_0 \to \infty$) the attenuation in the mixed state does not depend on H, i.e., $\alpha = 0$. In this case when the metal becomes normal (at H_{c2}) the attenuation increases abruptly. This type of behavior was observed with ~10-MHz waves propagating in an unannealed specimen of Nb–25% Zr and in sintered V_3Ga. For both of these alloys $r \gg 1$ at ~10 MHz. Figure 3 shows the results in Nb–25% Zr.

When the pinning forces are sufficiently strong that $r \gg \beta_f$ Eq. (13) becomes

$$\Delta V/V = (\mu H^2/8\pi d V^2), \qquad (r \gg \beta_f) \qquad (16)$$

Expression (16) is identical to the expression for the velocity change in a metal with infinite conductivity which is in the normal state. Measurements of the velocity of 4.3-MHz shear waves in an unannealed Nb–25% Zr superconducting alloy, for which $r \gg \beta_f$, gave results which were in good agreement with Eq. (16). Figure 4 shows these results as well as the results for a 14.3-MHz wave. In the latter case the inequality $r \gg \beta_f$ was not satisfied and, consequently, Eq. (16) did not apply. The results for the 14.3-MHz wave are accounted for by Eq. (13).

REFERENCES

1. T. G. Berlincourt and R. R. Hake, *Phys. Rev.* **131**: 140 (1963).
2. Y. B. Kim, C. F. Hempstead, and A. R. Strnad, *Phys. Rev.* **139**: A1163 (1965).
3. J. I. Gittleman and B. Rosenblum, *Phys. Rev. Letters* **16**: 734 (1966).
4. P. G. de Gennes, *Superconductivity of Metals and Alloys*, W. A. Benjamin, New York, 1966.
5. R. A. Alpher and R. J. Rubin, *J. Acoust. Soc. Am.* **26**: 452 (1954).
6. G. A. Alers and P. A. Fleury, *Phys. Rev.* **129**: 2425 (1963).
7. Y. Shapira and L. J. Neuringer, *Phys. Letters* **20**: 148 (1966).
8. L. J. Neuringer and Y. Shapira, *Phys. Rev.* **148**: 231 (1966).
9. Y. Shapira and L. J. Neuringer, *Phys. Rev.* **154**: 375 (1967).

Chapter 20

GIANT QUANTUM OSCILLATIONS IN ULTRASONIC ABSORPTION

Y. Shapira

*National Magnet Laboratory**
Massachusetts Institute of Technology
Cambridge, Massachusetts

THE PHYSICAL ORIGIN OF THE GIANT QUANTUM OSCILLATIONS

The physical origin of the "giant quantum oscillations" (hereafter GQO) can be understood by combining two arguments, a kinematic one and a statistical one [1]. We start with the kinematic argument.

Consider the interaction of a sound wave with the conduction electrons in a very pure metal ($\tau \to \infty$) in the presence of a magnetic field $\mathbf{H}$ directed along the z axis. The sound wave can be regarded as a beam of coherent phonons each with energy $\hbar\omega$ and momentum $\hbar\mathbf{q}$. The sound absorption process is viewed as a collision between a phonon and an electron in which the phonon is absorbed and the electron undergoes a transition from some initial state to some final state. Let $E(n, k_z)$ be the energy of the initial state of the electron, where n is the Landau level quantum number and k_z is the crystal momentum along $\mathbf{H}$. Similarly, let $E(n', k_z')$ be the energy of the final state. Applying the momentum and energy conservation theorems, one obtains

$$k_z' = k_z + q_z \tag{1}$$

and

$$E(n', k_z') - E(n, k_z) = \hbar\omega \tag{2}$$

At high magnetic fields the phonon does not have sufficient energy to cause transitions between different Landau levels either by cyclotron resonance or

* Supported by the U. S. Air Force Office of Scientific Research.

by Doppler-shifted cyclotron resonance. Thus at high fields $n' = n$. Equations (1) and (2) then give

$$E(n, k_z + q_z) - E(n, k_z) = \hbar\omega = \hbar Vq, \qquad (3)$$

where V is the sound velocity. In general, the energy difference $E(n, k_z+q_z)$ $- E(n, k_z)$ depends on k_z so that Eq. (3) restricts the absorption of phonons to a particular value of k_z. We shall designate this particular value of k_z by k_0.

In most, but not all, experimental situations the ultrasonic frequency is sufficiently low that Eq. (3) can be approximated by

$$\frac{\cos\theta}{\hbar}\left[\frac{\partial E(n, k_z)}{\partial k_z}\right]_{k_z=k_0} \approx V \qquad (4)$$

where θ is the angle between $\mathbf{H}$ and $\mathbf{q}$. Equation (4) can be interpreted physically as follows. The group velocity $\mathbf{V}_g$ of an electron in the presence of a magnetic field is directed along $\mathbf{H}$ and is given by

$$V_g = \frac{1}{\hbar}\frac{\partial E(n, k_z)}{\partial k_z} \qquad (5)$$

Equation (4) therefore states that the only electrons which can absorb sound waves are those whose velocity along $\mathbf{q}$, given by $V_g \cos\theta$, is equal to the sound velocity. For this reason the condition $k_z = k_0$ is sometimes referred to as the "surf-riding condition."

We now turn from the kinematic argument to the statistical argument. In the process of absorbing a phonon an electron undergoes a transition from some initial state to some final state. Prior to this transition the initial state is occupied by an electron, whereas the final state is unoccupied. The probability that a given state with energy E is occupied is given by the Fermi-Dirac distribution function

$$f(E) = \frac{1}{\exp[(E - \zeta)/kT] + 1} \qquad (6)$$

where ζ is the Fermi energy. At low temperatures $f(E) = 1$ at energies well below the Fermi energy ζ, and $f(E) = 0$ at energies well above ζ. Near $E = \zeta$ the function $f(E)$ changes from unity to zero over an energy interval of order kT. Therefore the requirement that the initial state of the electron is occupied implies that the energy of this state is not higher than ζ by more than $\sim kT$. Similarly, the energy of the final state should not be lower than ζ by an amount exceeding $\sim kT$. For ultrasonic frequencies in the RF range and at temperatures which are not much below $1\,°K$ the energies of the initial and final states are separated by $\hbar\omega \ll kT$. It then follows from the above considerations that the energy of the initial (and final) states of the electron involved in absorbing a phonon must be within $\sim kT$ of the Fermi energy ζ.

The GQO in the ultrasonic absorption are due to the existence of the selection rule $k_z = k_0$ and the additional requirement that the energy of the electron involved in the absorption of a phonon must be within $\sim kT$ of the energy ζ. Consider Fig. 1. This figure shows several Landau levels at high fields ($\hbar\omega_c \gg kT$). The solid circles in the figure correspond to states with $k_z = k_0$, and the hatched area represents the region $|E - \zeta| < kT$. If the magnetic field is such that one of the solid circles is inside the hatched area, then sound absorption can take place. However, if H has a value such that none of the solid circles is inside the hatched area, then the electrons cannot absorb the sound wave. Thus in certain intervals of H the ultrasonic absorption is high, but in other intervals of H the absorption is very small. If H is increased gradually, one observes a series of absorption peaks separated by minima in the absorption. This phenomenon is called GQO. If $\hbar\omega_c \gg kT$, then the widths of the absorption peaks are small compared to the widths of the minima in the absorption, and the GQO have a spike-like character. In this respect the GQO are different from the de Haas–van Alphen (dHvA) oscillations, which generally have a sinusoidal shape.

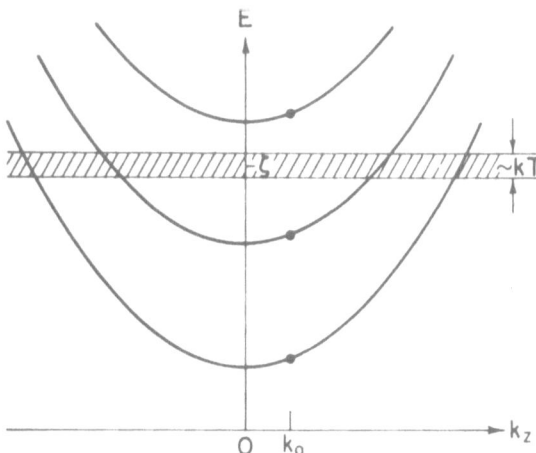

Fig. 1. Landau levels in a free-electron-type metal. The solid circles correspond to states with $k_z = k_0$. The hatched area represents energies within $\sim kT$ of the Fermi energy ζ.

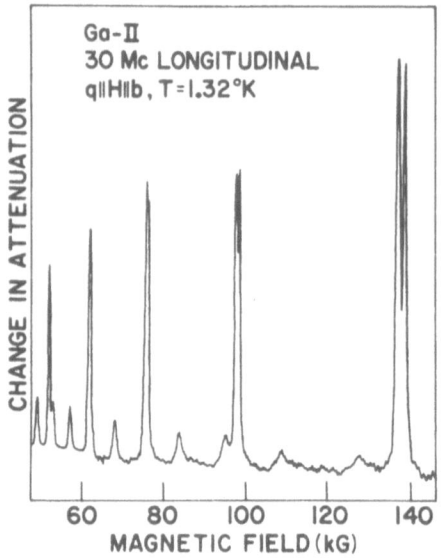

Fig. 2. Recorder tracing of the attenuation of
30-MHz longitudinal waves in gallium. The
peaks at the highest field are split due to spin.

This difference arises from the existence of the selection rule $k_z = k_0$ in the
case of GQO and from the absence of a similar selection rule for dHvA
oscillations. An example of GQO is shown in Fig. 2.

One can show that the GQO are periodic in H^{-1} and that the period
P is given by

$$P = 2\pi e\hbar/cA_0'$$ (7)

where A_0' is the cross-sectional area of the Fermi surface normal to **H** at
$k_z = k_0$ [see, for example, ([5])]. Equation (7) is similar to the Onsager
expression for the dHvA period. The latter expression contains the extremal
area, A_0, normal to **H** instead of A_0'. In most experimental cases A_0' is
nearly equal to A_0 unless θ is close to 90°.

THE LINE SHAPE OF THE GQO

The line shape of the GQO was calculated by several authors following
two different approaches. In the first approach one assumes a certain form
for the electron–phonon interaction Hamiltonian and then calculates the
rate of phonon absorption using the "golden rule." This method was first

used by Gurevich *et al.* ([1]), who predicted the GQO. The second approach was used by Quinn and Rodriguez ([3]), who calculated the magnetoconductivity tensor in the presence of a quantizing magnetic field and then related the absorption coefficient to this tensor. A brief outline of the first approach is presented here.

We consider the absorption of a sound wave in a pure metal ($\tau \to \infty$) in the presence of a magnetic field. It is assumed that the electron is spinless and that $\hbar\omega_c \gg kT$. Under these conditions the "giant" absorption peaks are completely separated from each other. Also, each absorption peak is due to electrons which are in a single Landau level.

The rate, R_+, at which phonons are absorbed by electrons which are in a single Landau level is given by

$$R_+ \propto HM^2 f_1(1 - f_2) \tag{8}$$

where M is a matrix element of the electron–phonon coupling and f_1 and f_2 are the values of the Fermi function for the energies $E(n, k_0)$ and $E(n, k_0) + \hbar\omega$, respectively. The factor H in Eq. (8) arises from the fact that there are many electron states with energy $E(n, k_0)$, the degeneracy being proportional to H. The factor $f_1(1 - f_2)$ in Eq. (8) is the product of the probability that the initial electron state is occupied and the probability that the final state is unoccupied.

In addition to absorbing phonons the electrons also emit phonons. The rate, R_-, at which phonons are emitted by electrons which are in a single Landau level is given by

$$R_- \propto HM^2 f_2(1 - f_1) \tag{9}$$

The ultrasonic attenuation coefficient, α_n, due to the electrons which are in the nth Landau level is proportional to $R_+ - R_-$. Therefore

$$\alpha_n \propto HM^2(f_1 - f_2) \tag{10}$$

Assuming that $\hbar\omega \ll kT$, which is the case in most experimental situations, one can approximate $(f_1 - f_2)$ by $-\hbar\omega[\partial f(E)/\partial E]$. The magnetic-field variation of α_n is therefore given by

$$\alpha_n \propto -HM^2[\partial f(E)/\partial E] \tag{11}$$

where the derivative $\partial f/\partial E$ is evaluated at $E = E(n, k_0)$.

At low temperatures the function $(\partial f/\partial E)$ is practically zero except at $E \approx \zeta$. At $E = \zeta$ this function has a sharp dip. Hence α_n has a peak when

$E(n, k_0) = \zeta$, a result which was obtained earlier from more elementary considerations.

If the absorption peak at $H = H_n$ due to the electrons which are in the nth Landau level is completely separated from other absorption peaks, then the line shape of the peak is given by Eq. (11). While the matrix element M which appears in this equation may depend on H, its field variation at $H \approx H_n$ is small compared to that of the factor $(\partial f/\partial E)$. Therefore for any *single* absorption peak it is permissible to regard M as a constant. The line shape of the absorption peak is therefore given by

$$\alpha_n = -\text{const } H[\partial f(E)/\partial E]_{E=E(n,k_0)} \tag{12}$$

Equation (12) can also be written as

$$\alpha_n = \text{const } (H/T) \cosh^{-2}\left[\frac{E(n, k_0) - \zeta}{2kT}\right] \tag{13}$$

It can be shown that the energy $E(n, k_0)$ at $H \approx H_n$ is given by [5]

$$\varepsilon = E - \zeta = e\hbar(H - H_n)/m^*cPH_n \tag{14}$$

where m^* is the cyclotron effective mass at the Fermi energy. Equation (14) is rigorous for a parabolic band. For a nonparabolic band this equation is still valid provided the mass m^* does not vary appreciably when the energy E of the electron, at $E \approx \zeta$, changes by several kT. Using Eqs. (13) and (14), one obtains

$$\alpha_n = \text{const } (H/T) \cosh^{-2}\left[\frac{\beta^*(H - H_n)}{2kTPH_n}\right] \tag{15}$$

where $\beta^* = e\hbar/m^*c$. The full width at half-height, $(\delta H)_n$, of the peak at $H \approx H_n$ can be obtained from Eq. (15). It is

$$(\delta H)_n = 3.53kTPH_n/\beta^* \tag{16}$$

In the preceding calculations the effects of electron collisions on the line shape of the GQO were neglected. This is justified only when the following inequalities are satisfied:

$$ql \gg (\zeta/\hbar\omega_c)^{1/2} \tag{17}$$

and

$$\omega\tau \gg (m^*V^2)^{1/2}/[(m^*V^2 + 2kT)^{1/2} - (m^*V^2)^{1/2}] \tag{18}$$

where l is the electron mean free path and τ is the electron relaxation time.

In practice, these inequalities are often not satisfied, so that the above treatment of the line shape of the GQO is not valid. However, in some experimental studies the inequalities (17) and (18) were apparently satisfied and the line shape of the GQO was in agreement with the above theory.

INFORMATION WHICH CAN BE OBTAINED FROM GQO

The Period. The most readily measured property of the GQO in the ultrasonic absorption is the period P. In most experimental situations this period is virtually identical to the de Haas–van Alphen period, which is, in turn, simply related to an extremal cross-sectional area of the Fermi surface [see Eq. (7) and subsequent discussion].

The g-Factor. Due to the spin of the electron each absorption peak is split, at low temperatures and high magnetic fields, into two subpeaks associated with the two spin directions. Let H_1 and H_2 be the fields at the centers of the spin-up and spin-down subpeaks, respectively. Then it can be shown that the g-factor of the electrons is related to $H_2 - H_1$ by the expression

$$g = 2 \frac{m_0}{m^*} \left[\frac{H_2 - H_1}{PH_1H_2} + l \right] \tag{19}$$

where m_0 is the free-electron mass and l is an integer or zero. The case $l = 0$ corresponds to a situation in which the two subpeaks are due to electrons which are in the same Landau level. The more general case $l = r$ corresponds to a situation in which the peak at H_1 is due to electrons in the nth Landau level with spin up, and the peak at H_2 is due to electrons in the $(n + r)$th Landau level with spin down. In general, it is not possible to determine the sign of $H_2 - H_1$ or the integer l from experiments on GQO, and therefore it is not possible to determine the g-factor uniquely from such experiments.

The Cyclotron Mass. When the "giant" absorption peaks are not significantly broadened by electron collisions, i.e., when the inequalities (17) and (18) are satisfied, it is possible to determine the cyclotron mass m^* at the Fermi energy from the line shape of the GQO. This can be done either by using the width of the absorption peaks and Eq. (16) to determine $\beta^* = e\hbar/m^*c$, or by fitting the line shape to Eq. (15) and obtaining the value of β^* which gives the best fit. Because of the spin splitting of the absorption peaks these two procedures can be used only when the two subpeaks in a given absorption doublet are completely separated from each other. Equations (15) and (16) then apply to either subpeak. When the two subpeaks in a given doublet are not completely separated from each other it is still possible to determine m^* from the line shape of the absorption peaks ([5]).

Matrix Elements of the Electron–Phonon Coupling. The height of the "giant" absorption peaks depends on, among other factors, a matrix element of the electron–phonon coupling. As a consequence one can determine this matrix element from the height of the absorption peaks. The procedure for doing this is outlined in a paper by Shapira and Lax ([5]) and will not be repeated here. It should be remarked, however, that this procedure is valid only if the GQO are not significantly broadened by electron collisions.

The Electron Distribution Function. Perhaps the most interesting property of the electrons which can be determined from GQO in the ultrasonic absorption is the electron distribution function, $F(E)$. In many discussions of the electronic properties of metals, including the one above, it is assumed that $F(E)$ is equal to the Fermi–Dirac function $f(E)$. This assumption leads to many experimentally verified conclusions, such as the linear dependence of the electronic specific heat on the absolute temperature T. However, it is usually not possible to invert the procedure and to derive the detailed variation of $F(E)$ with the energy E from experimental data on a given property of a metal. The difficulty in obtaining $F(E)$ from experimental data arises from the fact that the expression for a measurable parameter, Q, of a metal usually involves an integral over momentum space whose integrand contains a product of $F(E)$, or $\partial F/\partial E$, and other functions. In the presence of a magnetic field **H** such an integral changes to a sum over Landau levels and an integral over k_z. The relation between Q and $F(E)$ is, however, still a complicated one. Thus it is difficult to test the assumption $F(E) = f(E)$ directly. In this respect, the line shape of the GQO is an exception.

It was shown earlier that at high magnetic fields a sound wave can interact only with electrons which possess a particular value of k_z. Because of this unique property the expression for the ultrasonic attenuation coefficient α takes the simple form of a single sum over Landau levels. Each term in this sum gives rise to an absorption peak at a different magnetic field. The line shape of each of these absorption peaks is given by [see Eq. (12)]

$$\alpha_n = -\text{const } H[\partial F(E)/\partial E]_{E=E(n,k_0)} \tag{20}$$

Using Eqs. (14) and (20) and assuming that $F(E) = 1$ for $\varepsilon/kT \ll -1$ and that $F(E) = 0$ for $\varepsilon/kT \gg 1$, one obtains

$$F(\varepsilon) = 1 - \left(\int_{h_1}^{H(\varepsilon)} \alpha H^{-1}\, dH \Big/ \int_{h_1}^{h_2} \alpha H^{-1}\, dH \right) \tag{21}$$

where $H(\varepsilon)$ is related to ε by Eq. (14), and $H_{n+1} \ll h_1 \ll H_n \ll h_2 \ll H_{n-1}$. Note that the integral from h_1 to h_2 starts at a field well below the nth absorption peak but well above the $(n+1)$st peak, and ends at a field well above the nth absorption peak but well below the $(n-1)$st peak. It has been assumed that the absorption peaks are well separated from each other.

In order to determine $F(E)$ *experimentally*, by the use of Eq. (21), two conditions must be met: (1) The collision broadening of the "giant" absorption peaks must be negligibly small, and (2) the spin splitting of the absorption peaks must be completely resolved, since Eq. (20) applies to the line shape of either subpeak in an absorption doublet but not to the line shape of the entire doublet.

Figure 3 shows the function $F(E)$ as derived from the line shape of the

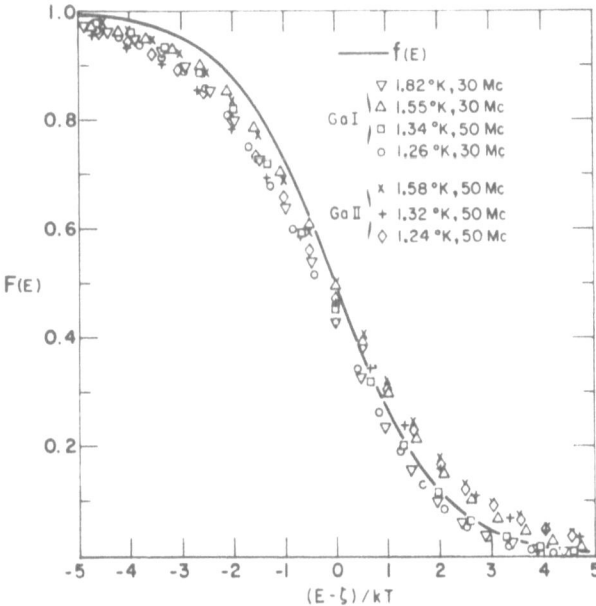

Fig. 3. The electron-distribution function in gallium as derived from the line shape of GQO. The solid line is the Fermi–Dirac function.

GQO in gallium. The solid line in this figure represents the Fermi–Dirac function $f(E)$. As can be seen, there is a good overall agreement between experiment and theory. The slight discrepancy at $E < \zeta$ is probably due to residual effects of the electron collisions on the line shape of the GQO.

REFERENCES

1. V. L. Gurevich, V. G. Skobov, and Yu. A. Firsov, *Soviet Phys.–JETP* (*English Transl.*) **13**: 552 (1961).
2. S. V. Gantsevich and V. L. Gurevich, *Soviet Phys.–JETP* (*English Transl.*) **18**: 403 (1964).
3. J. J. Quinn and S. Rodriguez, *Phys. Rev.* **128**: 2487 (1962).
4. A. P. Korolyuk, *Soviet Phys.—JETP* (*English Transl.*) **24**: 461 (1967).
5. Y. Shapira and B. Lax, *Phys. Rev.* **138**: A1191 (1965).
6. Y. Shapira and L. J. Neuringer, *Phys. Rev. Letters* **18**: 1133 (1967).
7. Y. Sawada, E. Burstein and L. Testardi, *J. Phys. Soc. Japan Suppl.* **21**: 760 (1966).
8. S. Mase, Y. Fujimori, and H. Mori, *J. Phys. Soc. Japan* **21**: 1744 (1966).

Chapter 21

HELICON PROPAGATION IN METALS: QUANTUM OSCILLATIONS IN TIN*

W. L. McLean

Rutgers University
New Brunswick, New Jersey

The propagation of electromagnetic waves through a medium containing free carriers in a magnetic field is discussed in other chapters of this volume. The emphasis there, however, is on the case of high-frequency waves in semiconductors, whereas here we shall be concerned with low-frequency waves in metallic single crystals. The special considerations that arise in this case will be reviewed and illustrated by reference to experiments in tin[1], a metal which exhibits most of the special effects. In addition, the origin of and some deductions from the quantum oscillations observed in the propagation in tin [1,2] will be discussed. An excellent series of papers on many aspects of electromagnetic waves in solids can be found in [3].

THE BASIC VIEWPOINT FOR METALS

The main difference in the basic approach in discussing the properties of metals as opposed to semiconductors arises from the failure of the effective-mass approximation [see, e.g. [4] and the references therein]. Even when the magnitude of the pseudopotential [[5] and the references therein] through which the electrons in a metal move is much less than the Fermi energy, the dynamical behavior of the electrons can be quite different from free carriers having constant effective masses. This can be seen by noting that the orbits in **k** space of the electron wave-packets in a magnetic field lie on constant energy surfaces and also in planes perpendicular to the magnetic field. The spherical constant-energy surfaces can be drastically changed topologically by only a comparatively weak pseudopotential, as

* Work supported by the National Science Foundation.

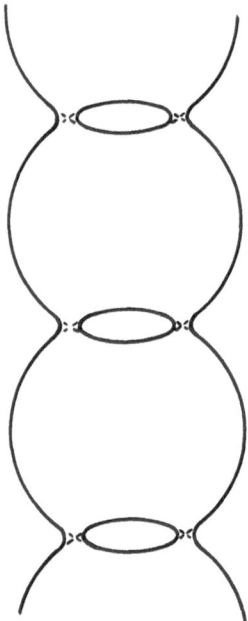

Fig. 1. Illustration of how even a weak lattice
potential can profoundly change the Fermi
surface from a sphere.

can be seen in Fig. 1 (see Chapter 18 by Pippard). The energy in **k** space is
a periodic function of **k** with the periodicity of the reciprocal lattice. For
free electrons there are therefore many spherical surfaces of constant energy
—not just one centered on the origin. The effect of a weak pseudopotential
is to distort the free-electron spherical constant-energy surfaces most strongly where they intersect with similar spheres of the same energy centered on
adjacent reciprocal lattice points, so that the distorted spheres join to form
surfaces which bear little resemblance to spheres. Of the many surfaces of
constant energy that exist it is those at the Fermi level that prove to be
important, since they divide the states that are occupied from those that
are unoccupied at the temperatures that are relevant here, namely, close
to absolute zero. It is only states near the Fermi level that can have their
populations changed by the low-energy, low-momentum-transfer perturbations that are most useful in investigating the dynamical behavior of the
electrons. Since we shall need to refer later to the Fermi surface of tin, we
show in Fig. 2 the surface that was suggested by Gold and Priestley ([6])—
partly by observing how the intersecting spheres would rejoin after being
slightly distorted by a weak lattice potential and partly by deducing various

extremal cross-sectional areas of the Fermi surface from their de Haas–van Alphen effect experiments. A slightly different version has been obtained by Weisz ([7]) by combining a pseudopotential calculation with the caliper dimensions of various sheets of the Fermi surface deduced by Gantmakher ([8]) from the size effect resonances in the radio-frequency skin effect in thin plates. Only one cell in reciprocal space has been shown in Fig. 2. The topological connectivity can only be appreciated by imagining the whole of **k** space packed with such cells. The importance of the change from spherical topology can be seen particularly in the case of the zone 4*a* hole surface,

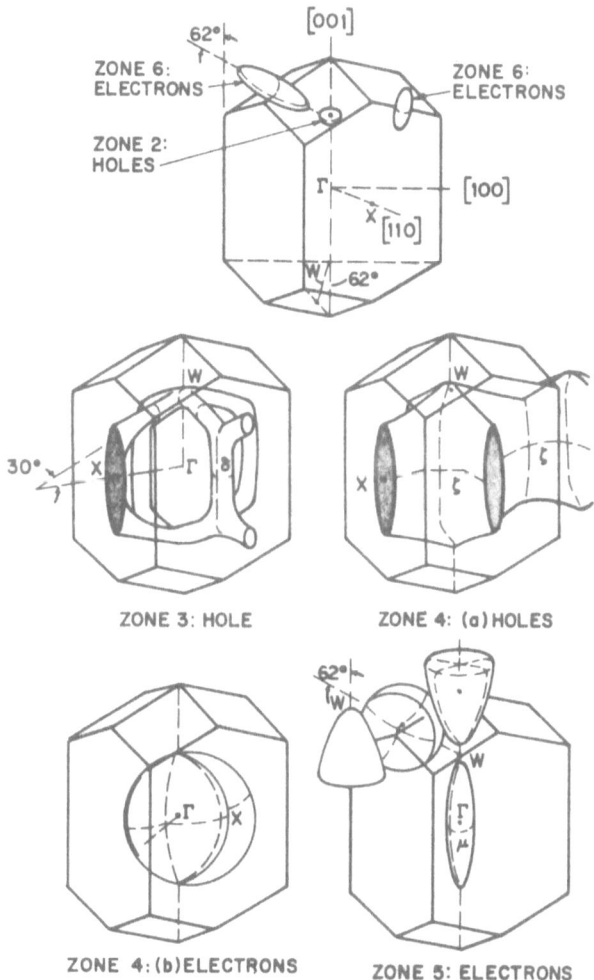

Fig. 2. The Fermi surface of tin according to Gold and Priestley ([8]).

on which the electron orbits defined by the intersection of planes perpendicular to the magnetic field with the surface can in some cases be "open" —passing from one cell in reciprocal space to the next without ever closing on themselves, as orbits on closed sheets like spheres or ellipsoids would do. The most profitable way of interpreting experimental results and of comparing band structure calculations with experiments in metals is by referring to the topology of the Fermi surface.

A further point of departure here from the general discussion in other chapters of waves propagating through a collection of free carriers is that even if the scattering of the carriers is so strong that $\omega\tau \ll 1$ and $ql \ll 1$, where ω and q are the angular frequency and wave vector of the waves, respectively, and τ and l are the relaxation time and the mean free path of the carriers, respectively, it is still possible to learn something about the electron orbits on the Fermi surface providing the cyclotron frequency ω_c is sufficiently large that $|\omega_c\tau|$ is of the order of unity or greater. In principle, such information could be obtained from DC experiments—such as magnetoresistance and Hall effect measurements—but in practice the study of electromagnetic waves sometimes has advantages. For instance, it is well known that attaching leads to samples can cause enough strain to seriously change the properties of the system being studied [1]. Experiments with helicon waves are free from this disadvantage. Also, in systems of low magnetoresistance, changes in the small attenuation of the helicon waves are more readily measurable than changes in the low magnetoresistance by DC methods.

EXPERIMENTAL METHODS

At this stage we digress from the discussion of general principles to briefly describe some of the experimental arrangements. The most commonly used and the simplest method of studying the propagation of waves through metals is the crossed-coil arrangement, which makes the best use of the fact that in many cases of interest the waves are circularly polarized, as is discussed later. A sample in the form of a flat plate is placed inside a flat coil with its axis parallel to the large faces of the plate. Both are placed inside a second similar coil with its axis perpendicular to the axis of the first, thus forming a crossed-coil transformer. A constant-amplitude alternating current is passed through one of the coils (the primary) producing an oscillating linearly polarized field at the surface of the sample, while a strong steady magnetic field is applied perpendicular to the large faces. The alternating field excites helicon waves, the two circular components of which propagate differently through the plate (as in the Faraday effect), leading to an oscillating flux parallel to the axis of the other coil (the secondary). Standing wave patterns are thus set up inside the plate, with resonances occurring approximately when an odd number of half-wavelengths fits in between the opposite faces. The induced e.m.f. in the secondary is measured with a vacuum-tube voltmeter or some similar detector or by a phase-sensitive detector (lock-in amplifier) with which the components of the secondary e.m.f. in-phase and out-of-phase with the primary current can be measured separately. Figure 3 shows the typical variation with frequency of the amplitude of the secondary e.m.f. obtained in tin with the crossed-coil arrangement. It is often more convenient to keep the frequency fixed and to sweep the magnetic field, in which case the higher harmonics occur at the lower fields and may not be observable if $| \omega_c \tau | \ll 1$. One disadvantage of the crossed-coil method is that the wave equation for the electromagnetic fields is difficult to solve exactly, and consequently it is not possible to calculate exactly the relation between the measured mutual inductance of the two coils and the properties of the sample. The approximate solution of Chambers and

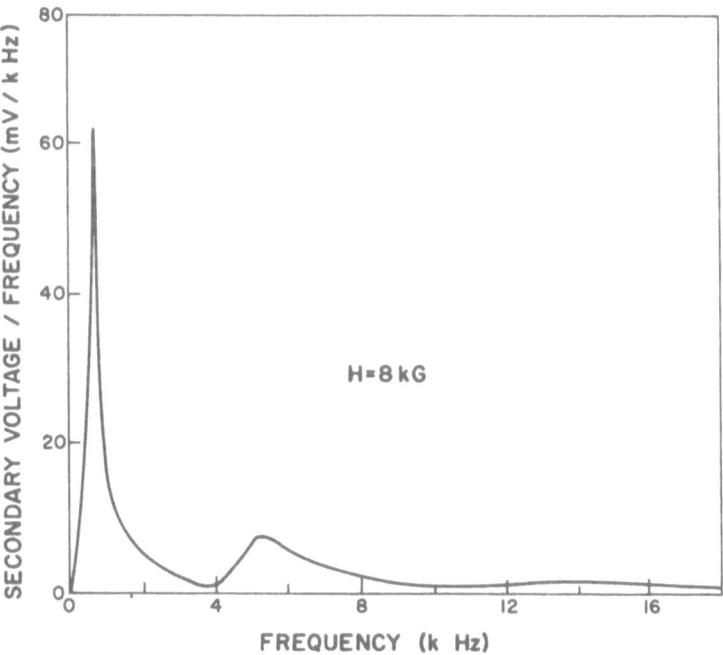

Fig. 3. Standing-wave helicon resonances in tin. The secondary voltage in the crossed-coil arrangement plotted against frequency. From Hays and McLean [1].

Jones ([9]) predicts fairly well but not exactly the line shape of the fundamental resonance (one half wavelength $\approx$ the thickness of the plate) found experimentally. Although in earlier work ([10]) it was merely assumed that a slab of the metal would act in a similar way to a microwave resonator of the same shape and have similar resonant modes, it has not been possible to show this generally from solutions of the wave equation. Solutions for some ideal geometries have been used by Harding and Thonemann ([11]) and others have been worked out by Legéndy ([12]), but it is still not easy to make the best use of these in practice.

Another method that has proved extremely satisfactory, at frequencies sufficiently high that the sample is many wavelengths thick, is the transmission method ([13]). A coil on one side of the plate acts as a source of waves which partly pass through and partly around (in spite of careful shielding) the sample plate. A detector is placed on the other side of the plate, where the interference of the waves that have traveled by two different paths can be observed. If the wavelength is small in comparison with the lateral dimensions of the plate, a good approximation is achieved to an infinite plate of finite thickness with plane wavefronts parallel to its faces. Another method that is suitable at high frequencies is the measurement of the surface impedance by means of a radio frequency bridge ([14]). The main disadvantage in principle of such a method is that generally the surface impedance is not simply related to the parameters of the medium, but at least qualitative, if not accurate quantitative, results can be reliably obtained. The use of high frequencies may not always be possible owing to additional damping effects—such as the Doppler-shifted cyclotron resonance [see, e.g. ([15])], and a type of geometrical resonance analogous to one type of resonance that can occur in ultrasonic propagation through a metal with a nonspherical Fermi surface in a magnetic field ([16]).

The study of helicon waves and Alfvén waves in semiconductors and semimetals requires the use of much higher frequencies than in metals owing to the smaller effective density of carriers; microwave techniques can then be used to advantage. We shall be concerned here only with metals and shall therefore return now to a discussion of the general principles of the propagation of waves in such a case.

EFFECTS OF THE TOPOLOGY OF THE FERMI SURFACE ON THE WAVE PROPAGATION

The response of metals to the fields of electromagnetic waves is more naturally discussed in terms of an effective resistivity tensor ρ rather than the effective dielectric constant tensor ε used in other chapters. The two, in Gaussian units which we use throughout, are simply related for a wave of angular frequency ω and time-dependent phase factor $\exp(-i\omega t)$ by $\omega \rho = 4\pi i \, \varepsilon^{-1}$.

Since it is of interest to study such things as standing wave resonances, it will be important to know how the dispersion formula for the waves depends on the elements of ρ. For free carriers it is shown in other chapters that the dispersion formula for a plane wave of wave-vector $\mathbf{q}$ inclined at an angle θ to the magnetic field and of angular frequency ω much less than the cyclotron frequency of the carriers ω_c is

$$- q^2 \cos \theta = \pm \omega \omega_p^2 / \omega_c c^2$$

where ω_p is the plasma frequency and c the velocity of light in vacuum. The two alternative solutions to the dispersion equation correspond to two different wave polarizations. It is convenient here, when the wave velocity is negligible in comparison with the velocity of light in vacuum, to describe the polarization of the waves by the behavior of their current vector, which is substantially perpendicular to the propagation direction. It is easy to show that the dispersion formula with the upper sign applies to a right circularly polarized wave ($J_x = -iJ_y$), while the lower sign applies to a left circular wave ($J_x = +iJ_y$), where we have chosen the z axis along the propagation direction and x, y, and z to be mutually orthogonal. In the free-electron gas ω_c is negative, so that if $0 \leq \theta \leq \pi/2$, only the right circular component would propagate, since the wave vector for the other is pure imaginary and it would consequently be attenuated with a decay factor $e^{-|q|z}$. The name "helicon" comes, of course, from the fact that the current vector of the propagating component traces out a helix as the wave travels forward.

In a medium with resistivity tensor ρ it follows from Maxwell's equations that the dispersion formula is

$$- 4\pi\omega/iq^2c^2 = \tfrac{1}{2}[(\varrho_{xx} + \varrho_{yy}) \pm \{(\varrho_{xx} - \varrho_{yy})^2 + 4\varrho_{xy}\varrho_{yx}\}^{1/2}]$$

where again the plane waves travel along the z axis (not necessarily parallel to the strong magnetic field) and it has been assumed that the wave velocity is much less than c so that the displacement current can be neglected. The polarizations corresponding to the two alternative signs in the dispersion formula are now in general no longer right and left circular but either (1) linear, when $(\varrho_{xx} - \varrho_{yy})^2 \geq \varrho_{xy}^2$, with $J_x^{(+)}/J_y^{(+)} = $ real number, $J_x^{(-)}/J_y^{(-)}$ $= $ another real number, or (2) elliptical, when $(\varrho_{xx} - \varrho_{yy})^2 < \varrho_{xy}^2$, with

$J_x^{(+)} = e^{i\alpha} J_y^{(+)}$ and $J_x^{(-)} = e^{-i\alpha} J_y^{(-)}$, where α is real. It is only when $\varrho_{xx} = \varrho_{yy}$ that $\alpha = -\pi/2$ and the two modes are circularly polarized as in the free-carrier case. This situation is only likely to occur when the wave vector and the strong magnetic field are parallel to a symmetry axis of the crystal.

When the fields of the wave are so rapidly varying with time or position that, respectively, the period or wavelength is comparable with the relaxation time or the mean free path (i.e., $\omega\tau \gg 1$ or $ql \gg 1$), the electric current density at a given instant or at a given position depends not only on the local value or the electric field but also on its value at neighboring times or positions, in which case the elements of the resistivity tensor depend on ω or on q. Such a "nonlocal" relation is important in many phenomena, such as: the two main kinds of cyclotron resonance effect (when $\omega \approx |\omega_c|$) [[17,18] and the references therein]; the Doppler-shifted cyclotron resonance effect (when $qv_F \approx |\omega_c|$, where v_F is the Fermi velocity) [15]; the giant quantum oscillations (when $qv_F \approx \omega$) [19]; and the skin effect in pure metals at low temperatures (when $ql \approx 1$) [20]. In all these cases it is usually possible to find more detailed information about the Fermi surface than can be obtained in the opposite limit when $\omega\tau \ll 1$ and $ql \ll 1$. For the present we shall be concerned with the low-frequency, long-wavelength limit, when the elements of $\rho(\omega, q)$ are independent of ω and q, and therefore we need only discuss $\rho(0, 0)$, the DC resistivity tensor, which of course is dependent on the magnetic field applied to the metal.

It is worthwhile at this point to consider a simple example, the free-electron gas with a scattering mechanism that has the same relaxation time for all electrons. Taking the set of orthogonal axes ξ, η, and ζ, with ζ parallel to the magnetic field, we have $\varrho_{\xi\xi} = \varrho_{\eta\eta} = \varrho_{\zeta\zeta} = \varrho_0$, say, $\varrho_{\eta\zeta} = \varrho_{\zeta\eta} = \varrho_{\zeta\xi} = \varrho_{\xi\zeta} = 0$, and $\varrho_{\xi\eta} = -\varrho_{\eta\xi} = -RH$, where $R = 1/Nec$ is the Hall coefficient and N the electron density. The elements of ρ based on the axes x, y, z can be found by the usual unitary transformation, and the dispersion formula then becomes

$$-\frac{4\pi\omega}{q^2 c^2} = \left(\frac{i\varrho_0}{\cos\theta} \pm RH\right) \cos\theta$$

We see that the attenuation is governed by the effective resistivity $\varrho_0/\cos\theta$, showing that no propagation can occur in a direction perpendicular to the field. Also, if $\varrho_0 = 0$, we obtain again the free-carrier dispersion formula.

The general procedure for obtaining $\rho(0, 0)$ for a real metal from its band structure has been dealt with in the review articles of Pippard [20], Chambers [21] and Fawcett [22], and we shall quote here only some of the results that are particularly relevant to our later discussion. The most straightforward way of formulating the problem is again to take a set of axes ξ, η, and ζ, with ζ parallel to the magnetic field direction, and first of all to calculate (by an analysis of the response in **k** space of the electrons to a constant electric field superimposed on the constant magnetic field) the elements of the conductivity tensor σ based on these axes. The elements ϱ_{xx}, etc., required for the dispersion formula are then obtained from the elements of σ^{-1} by a simple rotation of axes. The main effects of interest here in real metals can be understood from the dependence on field of the elements $\varrho_{\xi\xi}$, $\varrho_{\eta\eta}$, and $\varrho_{\xi\eta} = -\varrho_{\eta\xi}$, the only elements involved in the dispersion formula when **q** is parallel to the magnetic field. We shall only discuss qualitatively the case when **q** is not parallel to the field.

The dependence of these four elements on field is least complicated at very high fields, when the condition $|\omega_c\tau| \gg 1$ is satisfied for all electrons at the Fermi level. Here the topology of the Fermi surface is very important, and we shall have to consider it in some detail. It is first necessary to distinguish between what is meant by "electron" and "hole" sheets of the Fermi surface. A sheet is an electron sheet if all the electron orbits on it in the presence of the magnetic field enclose occupied states. If all the orbits enclose unoccupied states, it is a hole sheet. Multiply connected sheets of the Fermi surface, for some orientations of the magnetic field, may not fit into either of these two categories, since the electron orbits can enclose occupied states on some parts of such sheets and unoccupied states on other parts. Also, some of the orbits may be open rather than closed, in which case it is no longer possible to define the states enclosed by the orbit. For other orientations of the field all the orbits on a sheet might be closed and of the same character; then the sheet could be described as being either an electron or a hole sheet. The number of (filled) states in unit volume of the metal enclosed by electron sheets is called the density of electrons, while the number of (empty) states in unit volume of the metal enclosed by hole sheets is called the density of holes. It is only if every sheet of the Fermi surface is itself classifiable as either an electron or a hole sheet that the familiar simple formula $H\sigma_{\xi\eta} = (N_e - N_h)ec$ relates the difference between the densities of electrons and holes, $(N_e - N_h)$, to the element $\sigma_{\xi\eta}$ of the conductivity tensor at high fields. We recall that when $N_e \neq N_h$ (an "uncompensated" metal) the Hall coefficient is $\varrho_{\xi\eta}/H = 1/(N_e - N_h)ec$ but if $N_e = N_h$ ("compensated" metal) the limit as $H \to \infty$ of $\varrho_{\xi\eta}/H$ is zero. If there is at least one sheet of the Fermi surface on which for certain field orientations there are both electron and hole orbits, while for other orientations the sheet is classifiable as either an electron or a hole sheet and the metal as a whole happens to be compensated, then in the former orientations a careful evaluation of $\varrho_{\xi\eta}$ in the high-field limit shows that the metal will in general behave as though it were uncompensated. This is the so-called geometrical discompensation ([22]) that occurs in tin when the magnetic field is along directions of high symmetry—along the tetragonal axis [001], for example. There are other directions of the field for which tin can be unambiguously described as compensated. There is good agreement between the experimental value ([1]) of $(\varrho_{\xi\eta}/H)$ and an estimate from Weisz's ([7]) calculation in tin with the magnetic field along [001]. Further details about the dependence of the resistivity tensor on the topology of the Fermi surface in general can be found in ([22]).

In an uncompensated metal, and also in the discompensated case, the propagating mode of electromagnetic waves at high fields is the circularly polarized helicon wave, while in a compensated metal the linearly polarized Alfvén wave is the propagating mode if the metal is sufficiently pure. In tin the former occurs when the magnetic field is along the tetragonal axis of the crystal lattice, while the latter situation obtains at most other orientations when the field is inclined at more than about 37° to this axis.

Further complications arise when some of the electron orbits on the

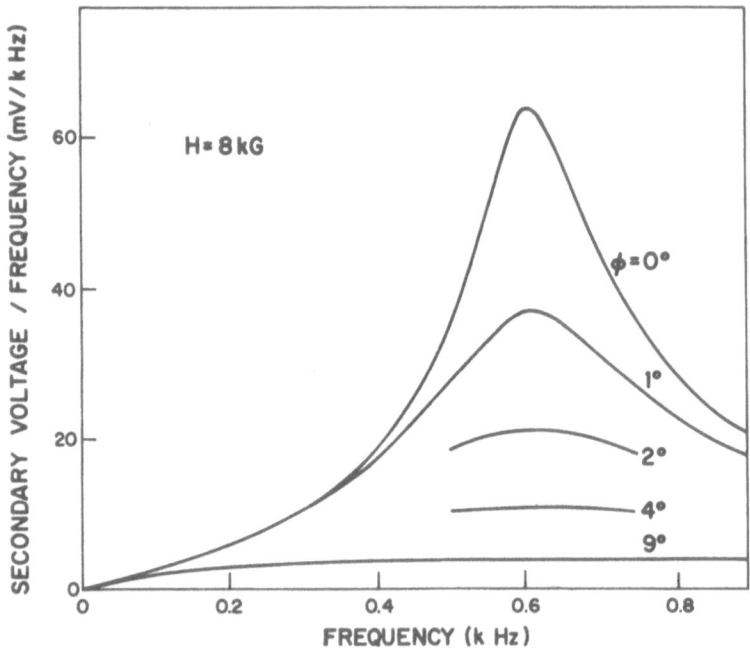

Fig. 4. The effect on the fundamental resonance of open orbits. ϕ is the angle between the magnetic field and [001]. From Hays and McLean ([1]).

Fermi surface are open. These, together with the cases just mentioned above, have been discussed by Buchsbaum and Wolff ([23]) and investigated experimentally by Grimes *et al.* ([24]). Briefly, open orbits prevent the saturation of the magnetoresistance which is found in an uncompensated metal—the magnetoresistance increases steadily (proportionally to H^2 in the high-field limit) instead of reaching a constant value as H is increased. The coefficient of H^2 is typically of such a magnitude that the nonsaturating magnetoresistance is much higher than the saturating value. This causes heavy damping of helicon waves in a pure metal ([1]) as is shown in Fig. 4, where the fundamental standing wave resonance is shown as a function of the angle ϕ between the magnetic field and the tetragonal axis of tin. When the field is parallel to the symmetry axis there are no open orbits and the magnetoresistance is relatively small. When the field is rotated away from the axis the number of open orbits steadily increases, causing a rapid increase in the damping. For a fixed value of ϕ there are a few special orientations of the magnetic field relative to the crystal axes, for which more than can be dealt with here should be said, and we refer the reader to the work of Alekseevskii *et al.* ([25]) for further discussion.

Apart from open-orbit damping, the criterion on purity for small damping of the helicon waves is much less stringent than for Alfvén waves. Heavily damped Alfvén waves can be elliptically polarized, so that it would be difficult in practice to distinguish between damped helicon and damped Alfvén waves merely from the nature of their polarizations.

MAGNETIC BREAKDOWN EFFECTS

We now turn to the effects of magnetic breakdown (see Chapter 18 by Pippard) on the resistivity tensor, and thus on the characteristics of the electromagnetic wave propagation through metals. The orbits in tin which are most affected by breakdown are those on the zone 3 and zone 4a hole surfaces (see Fig. 2) that pass near to the points X on the Brillouin zone faces, and we shall first consider such orbits for the magnetic field along the tetragonal axis [001]. The orbits of main interest are those lying in planes with $k_c \approx 0$, where k_c is the component of the electron wave-vector parallel to [001]. Figure 5 shows a section in **k** space (with $k_c = 0$) through the Fermi surface that occurs when no interaction of the electrons with the lattice is taken into account and the surface consists of free-electron constant-energy spheres with radii which are fixed by the electron density of tin. The effects of including the electrostatic interaction between the electrons and the periodic potential of the lattice are shown in Fig. 6. It should be

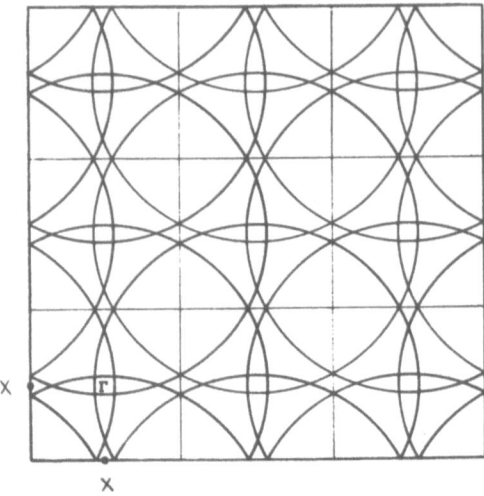

Fig. 5. Sections of free-electron Fermi surfaces for a metal having the same crystal structure and the same electron density as tin.

noted that the degeneracy at X is not removed owing to the "sticking together" ([26]) of the bands. The degeneracy is removed when the spin-orbit interaction is taken into account ([7]) and the resulting section of the Fermi surface is shown in Fig. 7. In weak magnetic fields the orbits of electrons on the Fermi surface in the plane $k_c = 0$ would be as shown in Fig. 7. As

Fig. 6. The effect of including the electrostatic lattice potential (schematic).

Fig. 7. The effect of including the spin-orbit interaction (schematic).

the magnitude of the field is increased breakdown becomes probable first near X since the band gap due to the spin-orbit interaction is much less than the gaps introduced by the electrostatic interaction. These are too large to allow any appreciable breakdown to occur at even the highest magnetic fields that can at present be reached. When breakdown does occur at X the orbits that had been independent at low fields link up to form a two-dimensional network similar to Fig. 6. As Pippard discusses in Chapter 18, the behavior of a network of orbits coupled by breakdown can be analyzed by treating the electron wave packets as being narrowly confined in real space to a network obtained by rotation of the k-space network through 90° about the field direction and by scaling its linear dimensions by the factor $\hbar c/eH$. Partial transmission along two paths occurs at each intersection with transmission coefficients that depend on H. Interesting oscillatory effects arise through interference of the electron waves around the small diamond-shaped parts of the network, while scattering mechanisms are usually sufficiently potent to prevent these from occurring around the larger parts. Figure 8 illustrates the splitting of an incident wave packet into two parts with transmission coefficients t_1 and t_2. The change of phase as a wave travels once around the diamond is equal to the flux passing through the orbit divided by $\hbar c/e$. When the flux changes by one unit hc/e the phase changes by 2π, and t_1 and t_2 consequently oscillate if the field is continuously varied. The metal thus oscillates between a tend-

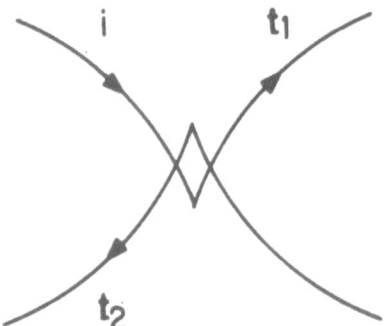

Fig. 8. The orbit responsible for phase coherence effects.

ency for open-orbit behavior of its magnetoresistance and closed-orbit
behavior. The effect of oscillations in the two-dimensional hexagonal
network of coupled orbits of magnesium and zinc on the elements of their
conductivity tensors has been calculated in detail by Falicov et al. ([27]),
who find satisfactory agreement with the pronounced oscillations observed
by Stark ([28]).

Oscillations in the elements of the resistivity tensor are also reflected
through the dispersion formula in the characteristics of the propagation of
electromagnetic waves through the metal. Figure 9 shows the effect of
varying the magnetic field on the amplitude of a standing wave set up in
a tin slab ([2]). At 4.2 °K, when the phase coherence effects are destroyed,
a similar curve is obtained but without the rapid oscillations. The fluctuations were periodic in $1/H$ with a period of $5.7 \times 10^{-7}\,G^{-1}$, from which
the area in k space of the diamond-shaped orbit can be found. This area
had already been found from pulsed-field de Haas–van Alphen effect
measurements by Gold and Priestley ([6]) (the orbit δ of Fig. 2) but greater
accuracy was possible in the helicon experiment ([1,2]) since the field, which
could be varied arbitrarily slowly, was measured by nuclear resonance
techniques. In order to check that there was no appreciable phase coherence
around the larger orbits (ζ orbit of Fig. 2) that link with the small diamond
orbits, the measurements were carried out up to 150 kG at the National
Magnet Laboratory, MIT. Such effects should produce a smaller-period
oscillation superimposed upon the main oscillations arising from the diamond orbits. In spite of indications that the field stability and instrumental
sensitivity were adequate, no additional oscillations were observed.

When the magnetic field is tilted a few degrees away from the [001]
axis of tin in a plane containing [001] and [110] the two-dimensional

network breaks up into a one-dimensional chain. This occurs because the band gap, which governs the probability of breakdown, varies rapidly along the [001] direction, so that any orbit that passes near the breakdown region but does not go through $k_c = 0$ is unlikely to undergo breakdown. This case has been analyzed by Young ([29]) using the general approach of Pippard but also including the possibility of a relaxation mechanism on the diamond orbit; excellent agreement is found with his experimental results. If the field is tilted away from [001] in a plane containing [001] but neither [100], [110], nor their symmetrical equivalents, the two-dimensional network breaks up completely, so that the only effect of breakdown is to join closed orbits into a single larger orbit. An orbit passing through $k_c = 0$ near X in one cell of reciprocal space will not do so in any of the neighboring cells, and consequently the band gap at its nearest approach to X will be so large that breakdown will be improbable. Thus the amplitude of the quantum oscillations observed in the helicon waves in tin is expected to diminish rapidly as the magnetic field is tilted away from [001] in a plane not containing any other axes of symmetry. This is illustrated in Fig. 10, which shows the amplitude of the quantum oscillations expressed as a fraction of the average helicon standing wave amplitude plotted as a func-

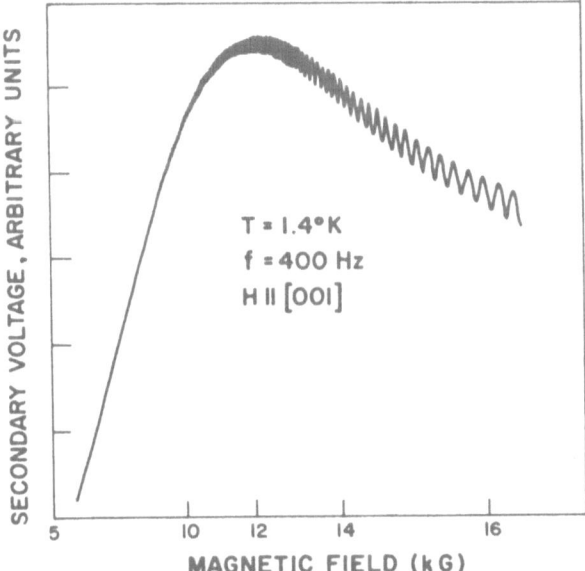

Fig. 9. The secondary voltage in the crossed-coil arrangement as a function of the applied field for tin at 1.2°K. From Hays and McLean ([2]).

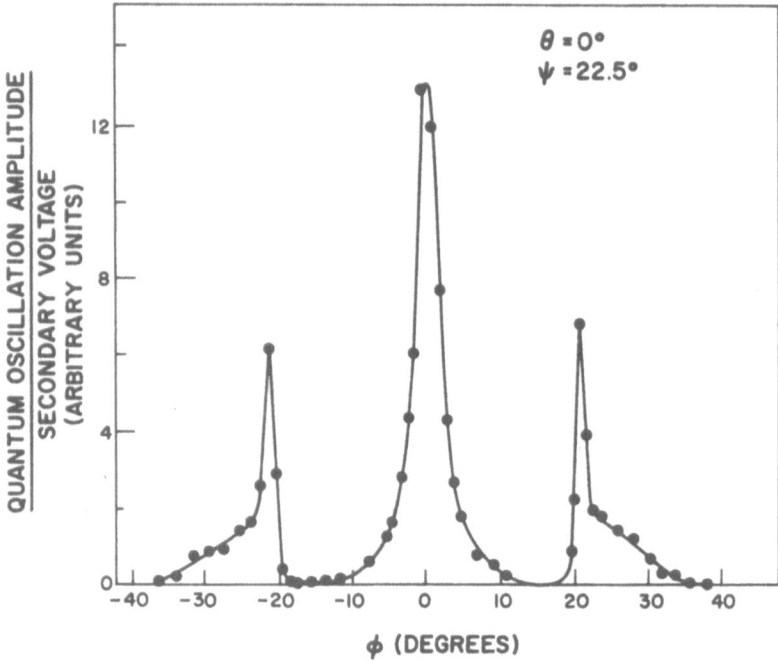

Fig. 10. The variation with the angle between the field and [001] of the quantum oscillation amplitude. From Hays and McLean ([1]).

tion of the angle ϕ between the field direction and [001]. The field was confined to a plane containing [001] and which passed midway between [100] and [110]. The effect of the breaking up of the two-dimensional network is further enhanced by the increase in the number of open orbits as $|\phi|$ increases from $0°$, thus decreasing the relative contribution of the network to the resistivity tensor. The sudden increase in the amplitude of the quantum oscillations at $\phi \approx 20°$ is associated with the open orbits first being able to pass through the breakdown region. When ϕ is less than $20°$ the open orbits run along the ridges near W on the zone $4a$ hole sheet and do not pass sufficiently close to X for breakdown to occur. Beyond $\phi \approx 37°$ no open orbits are present and if breakdown occurs, it again only results in the coupling of two closed orbits into a larger-sized orbit with no large coherent oscillatory effects on the elements of the conductivity tensor. The variation of the amplitude of the quantum oscillations with ϕ helps identify the oscillation of period $5.7 \times 10^{-7}\,G^{-1}$ with the δ orbit of the Fermi surface. The band structure calculation of Weisz ([7]) has given a period of $8 \times 10^{-7}\,G^{-1}$ for this orbit. Although such disagreement may seem large, it should be pointed out that the area of the δ orbit is only about 1% of the cross-sectional area of the Brillouin zone in the $k_c = 0$ plane, so that extremely careful calculation would be needed to provide better agreement with the experimental result, which is easily measurable with an accuracy of 1%. We have already noted ([1]) comparatively good agreement for a larger dimension of the Fermi surface related to the high-field Hall coefficient with the field parallel to [001]. A more thorough overall comparison with the theory would be better made by using the de Haas–van Alphen experiment. The helicon experiments are capable of supplementing such other experiments, as we have shown, and can provide information about magnetic breakdown effects if they occur. It would of course be desirable to also study the helicons at higher frequencies and purities, where more detail should be observable.

REFERENCES

1. D. A. Hays and W. L. McLean, *Phys. Rev.* **168**: 755 (1968).
2. D. A. Hays and W. L. McLean, *Phys. Letters* **17**: 215 (1965).
3. J. Bok (ed.), *Plasma Effects in Solids*, Academic Press, New York, 1965.
4. J. M. Ziman, *The Principles of the Theory of Solids*, Cambridge University Press, 1964.
5. B. J. Austin, V. Heine, and L. J. Sham, *Phys. Rev.* **127**: 276 (1962).
6. A. V. Gold and M. G. Priestley, *Phil. Mag.* **5**: 1089 (1960).
7. G. Weisz, *Phys. Rev.* **149**: 504 (1966).
8. V. F. Gantmakher, cited in ([7]).
9. R. G. Chambers and B. K. Jones, *Proc. Roy. Soc. (London) Ser. A* **270**: 417 (1962).
10. F. E. Rose, M. T. Taylor, and R. Bowers, *Phys. Rev.* **127**: 1122 (1962).
11. G. N. Harding and P. C. Thonemann, *Proc. Phys. Soc. (London)* **85**: 317 (1965).
12. C. R. Legéndy, *Phys. Rev.* **135**: A1713 (1964).
13. C. C. Grimes and S. J. Buchsbaum, *Phys. Rev. Letters* **12**: 357 (1964).
14. M. T. Taylor, *Phys. Rev.* **137**: A1145 (1965).
15. J. L. Stanford and E. A. Stern, *Phys. Rev.* **144**: A534 (1966).
16. A. B. Pippard, private communication.
17. M. Ya. Azbel' and E. A. Kaner, *J. Phys. Chem. Solids* **6**: 113 (1958).
18. J. K. Galt, W. A. Yager, F. R. Merritt, B. B. Cetlin, and A. D. Brailsford, *Phys. Rev.* **114**: 1396 (1959).
19. P. B. Miller, *Phys. Rev. Letters* **11**: 537 (1963); J. J. Quinn, *Phys. Letters* **7**: 235 (1963).
20. A. B. Pippard, *Reports on Progress in Physics XXIII*, 176 (1960).
21. R. G. Chambers, in: *The Fermi Surface* (W. A. Harrison and M. B. Webb, eds.), John Wiley and Sons, 1960.
22. E. Fawcett, *Advan. Phys.* **13**: 139 (1964).
23. S. J. Buchsbaum and P. A. Wolff, *Phys. Rev. Letters* **15**: 406 (1965).
24. C. C. Grimes, G. Adams, and P. H. Schmidt, *Phys. Rev. Letters* **15**: 409 (1965).

25. N. E. Alekseevskii, Yu. P. Gaidukov, I. M. Lifshitz and V. G. Peschanskii, *Soviet Phys.—JETP (English Transl.)* **12**: 837 (1961).
26. V. Heine, *Group Theory in Quantum Mechanics*, Pergamon Press, New York and London, 1960.
27. L. M. Falicov, A. B. Pippard, and P. R. Sievert, *Phys. Rev.* **151**: 498 (1966).
28. R. W. Stark, see [27].
29. R. C. Young, *Phys. Rev.* **152**: 659 (1966).

Chapter 22

THE SHUBNIKOV–DE HAAS EFFECT AND QUANTUM-LIMIT PHENOMENA IN SEMICONDUCTORS

G. Landwehr*

Physikalisch-Technische Bundesanstalt
Braunschweig, Germany

INTRODUCTION

In 1930 Shubnikov and de Haas [1] investigated the magnetoresistance of bismuth single crystals at liquid hydrogen temperatures. They observed an oscillatory behavior of the resistance as a function of the magnetic field which was not understood at that time. The foundation for the explanation was laid by Landau [2]. Peierls [3] suggested in 1931 that quantum effects might cause the phenomenon. However, not too much attention was paid to the effect, possibly because knowledge of the peculiar behavior was restricted to bismuth for quite some time. Not until 1956 was an oscillatory magnetoresistance, which is now frequently called the Shubnikov–de Haas effect, observed in other substances, in the semiconductor InSb [4] and in Zn [5]. Soon experiments on InAs [6,7] followed, and to the present the Shubnikov–de Haas effect has been found in more than a dozen semiconductors. At the end of the 1950's the first quantitative theories were offered [8-10] which showed much similarity with the theory of the de Haas–Van Alphen effect. Excellent review articles by Kahn and Frederikse [11] and by Adams and Keyes [12] have been published which cover the literature up to about 1961. Consequently, in this chapter emphasis will be on the newer results. In the first section, on oscillatory quantum effects, the theoretical situation will briefly be reviewed, followed by a presentation of recent experimental data.

---

* Present address: Physikalisches Institut der Universität Wuerzburg, Germany.

The final section will deal with quantum-limit phenomena, which arise when only the lowest Landau-level is occupied and the resistance changes with magnetic field at low temperature are no longer oscillatory, but monotonic. Several theoretical papers have dealt with this situation. Recently a review on this subject by Kubo *et al.* ([13]) appeared. The experimental data are sparse, however, in spite of the improved possibility for the generation of strong magnetic fields.

OSCILLATORY EFFECTS

The Period

Although the oscillatory magnetoresistance is a fairly complicated phenomenon if all details are taken into account, the origin of the periodicity of the oscillations in H^{-1} (H is the magnetic field strength) can easily be grasped starting from the properties of a free-electron gas in a magnetic field and considering the scattering processes of crystal electrons in a qualitative manner only. If electrons with an isotropic effective mass m^* are exposed to a magnetic field along the z direction, the energy eigenvalues are:

$$E = (n + \tfrac{1}{2})\hbar\omega + (\hbar^2 k_z^2/2m^*) \pm \tfrac{1}{2}g\mu H, \qquad n = 0, 1, 2, \ldots \qquad (1)$$

where ω is the cyclotron frequency eH/m^*c, k_z is the wave vector in the z direction, g is the spectroscopic splitting factor and μ is the Bohr magneton. The other symbols have their usual meaning. Due to the quantization of the electron orbits in **k** space in the plane perpendicular to the magnetic field the parabolic band is split into magnetic subbands which differ in energy by the amount $\hbar\omega$. The last term is due to the lifting of the spin degeneracy by the magnetic field. The uniform distribution of quantum states in **k** space in the field-free case is replaced by a series of interlocking Landau cylinders with cross sections A_n in the direction perpendicular to H:

$$A_n = (n + \tfrac{1}{2})(2\pi eH/\hbar c) \qquad (2)$$

The spin has been neglected for the moment. The discrete nature of the Landau cylinders is smeared out unless their energy separation $\hbar\omega$ is larger than kT (k is Boltzmann's constant). Another requirement for the observation of quantum effects is that complete orbits in **k** space be performed by the electrons before they are scattered, which can be formulated by the condition $\omega\tau > 1$, where τ is the relaxation time. The condensation of quantum states in a magnetic field has drastic consequences for the density of states. The total density of states ([11]) per unit energy and unit volume is of the form:

$$g(E) = \frac{1}{2} \left(\frac{1}{2\pi} \right)^2 \left(\frac{2m^*}{\hbar^2} \right)^{3/2} \sum_{n=0}^{n_{max}} \frac{\hbar\omega}{[E - (n + \tfrac{1}{2})\hbar\omega]^{1/2}} \tag{3}$$

The summation has to be done over all occupied quantum states. Whenever the energy coincides with that of a Landau level the density of states diverges. This is due to an oversimplification which neglects the finite width of the Landau levels. If collision broadening is taken into account, the divergence disappears. In order to demonstrate this effect, the density of states has been computed for various degrees of level broadening. A Lorentzian shape has been chosen for the energy levels. It is clearly seen in Fig. 1 that the sharp spikes in $g(E)$ have disappeared for $\omega\tau = 10$ and that the oscillations are almost completely damped out for $\omega\tau = 1$. Actually, the model chosen is too primitive, because the assumption of a uniform τ in the whole energy range is unrealistic. Nevertheless, the consequences of collision broadening for the region around the Fermi surface can be visualized. The oscillations in the density of states as a function of the magnetic field can strongly affect the scattering rates of electrons and produce oscillations in the transport properties. Suppose that in a solid the energy levels are filled up to the

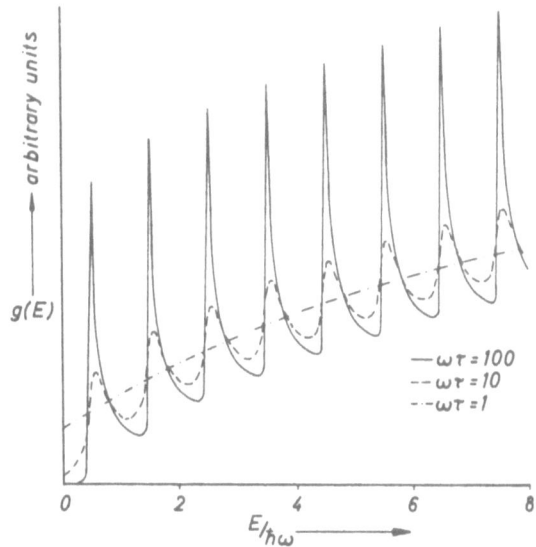

Fig. 1. Density of states as a function of energy (in units of $\hbar\omega$) for various degrees of level broadening. A Lorentzian shape has been assumed for the Landau levels.

Fermi level ζ with electrons. Provided that $\zeta > \hbar\omega$ and that the previously discussed conditions $\omega\tau > 1$ and $\hbar\omega > kT$ hold, maxima in the scattering (and consequently, maxima in the resistance) will occur in a longitudinal magnetic field whenever the Fermi level coincides with the Landau level. If the Fermi level stays constant (which it does only approximately, as will be discussed later), the oscillations in the resistivity are periodic in H^{-1}. The oscillations in the transverse magnetoresistance are also periodic in H^{-1}, as will be shown later. The period in both cases is

$$\Delta(1/H) = \hbar e/m^*\zeta c \tag{4}$$

If for ζ we substitute the usual low-temperature expression

$$\zeta = (h^2/2m^*)(3/8\pi)^{2/3}n^{2/3} \tag{5}$$

it is seen that the period no longer depends on the effective mass, but only on the carrier density n:

$$\Delta(1/H) = (2e/\hbar c)(3\pi^2 n)^{-2/3} \tag{6}$$

For ellipsoidal surfaces of constant energy the right-hand side is multiplied by a factor m_d^*/m_c^*, the ratio of density-of-states mass to cyclotron mass. By definition, the cyclotron mass is an average of the effective mass values which the electron takes along its orbit. Hence the anisotropy of the effective mass tensor can be explored by investigating the angular dependence of the Shubnikov–de Haas effect. This makes the effect an important tool for band structure work on semiconductors especially, because it applies to degenerate materials, for which other methods, like cyclotron resonance, sometimes fail.

By applying the semiclassical Bohr–Sommerfeld quantum condition to crystal electrons in a magnetic field

$$\oint (\hbar \mathbf{k}_\perp + (e/c)\mathbf{A}) \, d\mathbf{q} = (n + \alpha)h \tag{7}$$

where $\mathbf{k}_\perp$ is the wave vector perpendicular to the magnetic field, $\mathbf{A}$ is the vector potential, and α is a phase constant which is $\frac{1}{2}$ for free electrons.

Onsager[14] showed that Eq. (2) is valid for Fermi surfaces of rather general shape. In this case quantum oscillations in different parts of the Fermi surface will interfere destructively except in areas where the cross section is stationary, i.e., at extremal cross sections. Therefore the following relation holds:

$$\Delta\left(\frac{1}{H}\right) = \frac{2\pi e}{\hbar c} \frac{1}{A_{\text{extr}}} \tag{8}$$

where A_{extr} is the extremal area of the Fermi surface perpendicular to the magnetic field. Equation (8) is a useful tool for investigating the shape of Fermi surfaces.

Deviations in the Periodicity

So far it has been assumed that the Fermi level ζ is independent of the magnetic field. This is not strictly true. Kahn and Frederikse ([11]) have calculated the field dependence of ζ for isotropic energy surfaces, neglecting spin, on the basis of extreme degeneracy. Collision broadening was not taken into account. The Fermi energy is determined by the condition that the total number of electrons N must be constant,

$$N = 2 \int_{\frac{1}{2}\hbar\omega}^{\infty} g(E)f(E)\,dE \tag{9}$$

where $g(E)$ is the density of states according to Eq. (3) and $f(E)$ is the Fermi–Dirac distribution function. For $T = 0$ this leads to the following expression for the number of electrons n per unit volume:

$$n = \frac{\hbar\omega}{2\pi^2} \left(\frac{2m^*}{\hbar^2}\right)^{3/2} \sum_{n}^{n_{max}} \left[\zeta - \left(n + \frac{1}{2}\right)\hbar\omega\right]^{1/2} \tag{10}$$

From this the Fermi level as a function of H can be computed. The resulting ratios $\zeta/\hbar\omega$ have been plotted in Table I for the quantum numbers 1 to 10. The quantity $H^{-1}/\Delta(H^{-1})$ is abbreviated for the two cases $\zeta = $ const and $\zeta = f(H)$ as C_n and S_n, respectively. It is evident that for the larger quantum

TABLE I

Quantum number	$\zeta = $ const	$\zeta = f(H)$	
n	C_n	S_n	$S_{n+1} - S_n$
1	1.500	1.310	1.048
2	2.500	2.358	1.024
3	3.500	3.382	1.015
4	4.500	4.397	1.010
5	5.500	5.407	1.008
6	6.500	6.415	1.006
7	7.500	7.421	1.005
8	8.500	8.426	1.004
9	9.500	9.430	1.004
10	10.500	10.434	—

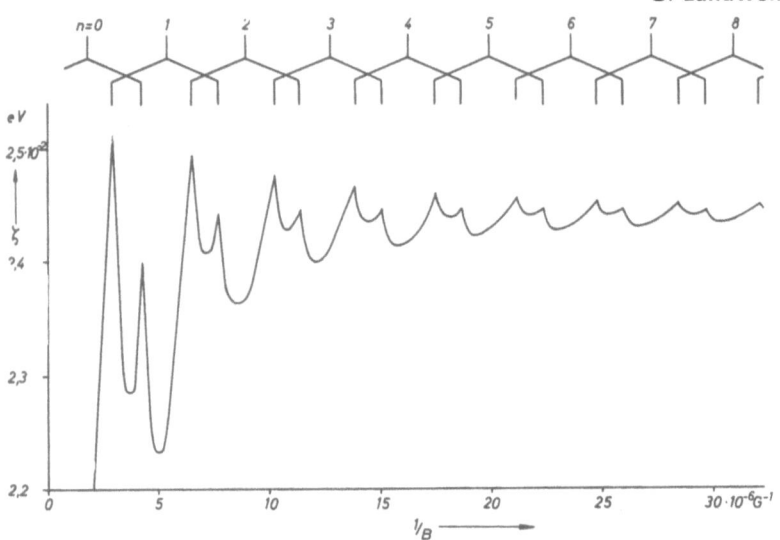

Fig. 2. Oscillations in the Fermi energy at $T = 0$ as a function of the reciprocal magnetic induction B^{-1} for a p-type Bi_2Te_3 single crystal. B is parallel to the trigonal axis, $m_c^* = 0.13\ m_0$, $g = 20$. The corresponding spin-split Landau ladder has been plotted at the top for quantum numbers $n = 0$ to 8. After Freudenhammer [52].

numbers the difference in the period is quite small. For low n the deviations should be observable. However, it should be remembered that collision broadening, which will tend to diminish the differences, was neglected.

The effect of spin splitting can easily be incorporated. It is only necessary to replace $(n + \frac{1}{2})$ in Eq. (10) by $(n + \frac{1}{2} \pm \frac{1}{2}M)$, where M stands for $m_c^*g/2m_0$, with m_c^* being the cyclotron mass and m_0 the free-electron mass. A calculation of the Fermi energy as a function of the reciprocal magnetic field for $T = 0$ has been made for a special Bi_2Te_3 crystal for which spin splitting of the Landau levels was actually observed [50]. The result for a g-factor of 20 is shown in Fig. 2. The sharp spikes are somewhat unrealistic because broadening of the Landau levels was not considered. But even in the present approximation the variation of the Fermi level is rather small for quantum numbers not too small.

For bismuth the variation of the Fermi level with magnetic field was computed [15] for different crystallographic axis with the result that for high fields appreciable deviations from the zero field value can occur.

If the variation of the Fermi level with magnetic field can be neglected, it is possible to evaluate the g-factor from the observed period of the oscillatory magnetoresistance and the positions $H_n{}^+$ and $H_n{}^-$ of the spin-split peaks which belong to the quantum number n:

$$M = \frac{m^*g}{2m_0} = \frac{(H^{-1})^+ - (H^{-1})^-}{\Delta(1/H)} \tag{11}$$

The field dependence of ζ can easily be taken into account at $T = 0$ if collision broadening of the Landau level is disregarded. Defining

$$S_n{}^+(M) = (H_n^{-1})^+/\Delta(1/H), \qquad S_n{}^-(M) = (H_n^{-1})^-/\Delta(1/H)$$

one obtains

$$
\begin{aligned}
S_n{}^+(M) &= 0.826\left[\sum_{k=0}^{n} (\sqrt{k} + \sqrt{k+M})\right]^{2/3} \\
S_n{}^-(M) &= 0.826\left[\sum_{k=0}^{n} (\sqrt{k} + \sqrt{k-M})\right]^{2/3}
\end{aligned}
\tag{12}
$$

Both factors have been plotted in Fig. 3 for $n = 0, 1$, and 2 as a function of M. The spin splitting is not symmetrical (except for $M = 1$) with respect

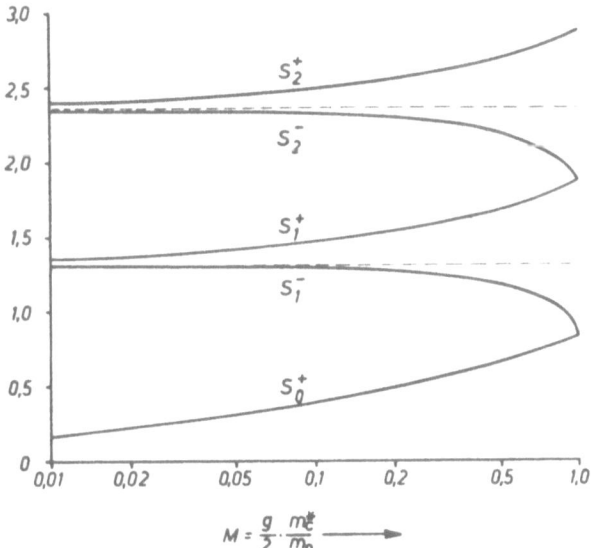

Fig. 3. Spin splitting of Landau levels at $T = 0$ as a function of the splitting parameter M for the quantum numbers $n = 0$, 1, and 2. The dotted lines indicate the quantum numbers for the unsplit Landau levels.

to the unsplit level. The 0^+ level is the last one which may be detected in an experiment of the Shubnikov–de Haas type. But even for low quantum numbers Eq. (11) is still a good approximation.

QUANTITATIVE THEORY

Transport properties in a magnetic field are usually calculated by solving an appropriate Boltzmann equation. It has been pointed out ([16]) that this method can no longer be employed in the quantum range for evaluating the conductivity in crossed electric and magnetic fields. Instead, rigorous quantum statistical methods (density matrix) have been applied to the problem by various workers [see ([10–13]) for a review of the literature]. All the recent theories agree in their results for the period and the temperature dependence of the amplitude of the oscillations. There is only some variation in the predictions concerning the dependence of the amplitude on magnetic field. The most elaborate theory, which is most easily compared with experiment, is due to Adams and Holstein (AH) ([10]). The essential results of this work will briefly be reviewed. One reason for doing this is that the AH theory has been subjected to a thorough test by Soule et al. ([17]), who compared their experimental results for graphite with the predictions of Adams and Holstein and found good agreement. Only the results of AH for the very last quantum oscillations have been questioned. Soule et al. modified the AH theory for very small quantum numbers by treating collision broadening of the Landau levels in a different way and came to somewhat different conclusions.

Finally, Argyres ([18]) theory of magnetoconductivity in longitudinal, strong magnetic fields will be outlined.

The Theory of Adams and Holstein

The theory applies to free electrons with isotropic effective mass in magnetic fields which are sufficiently strong that the condition $\omega\tau \gg 1$ holds. Spin splitting has been neglected. The results have been computed for lattice scattering of the electrons. As Adams and Holstein point out, they should also be valid for ionized impurity scattering. For the symmetric conductivity they find:

$$\sigma_{xx}(H) = \sigma_{\text{class}} + \Delta\sigma_1 + \Delta\sigma_2 \qquad (13)$$

where σ_{class} is the classical magnetoconductivity, which is proportional to H^{-2}, and $\Delta\sigma_1$ and $\Delta\sigma_2$ are quantum corrections due to the Shubnikov–de Haas effect. The correction $\Delta\sigma_1$ should be important for relatively large quantum numbers, when scattering between different Landau levels is possible. The correction $\Delta\sigma_2$ should dominate for the lowest quantum numbers, when transitions occur only in the Landau level closest to the Fermi level. For $T = 0$ and no collision broadening the following results were obtained:

$$\frac{\Delta\sigma_1}{\sigma_{\text{class}}} = \frac{5}{2}\left(\frac{\hbar\omega}{\zeta}\right)^{1/2}\left[\frac{1}{2}\left(\frac{\hbar\omega}{\zeta - E_r}\right)^{1/2} - \left(\frac{1}{2} + \frac{\zeta - E_r}{\hbar\omega}\right)^{1/2}\right] \quad (14a)$$

$$\frac{\Delta\sigma_2}{\sigma_{\text{class}}} = \frac{3}{2}\left(\frac{\hbar\omega}{\zeta}\right)\left[\frac{1}{2}\left(\frac{\hbar\omega}{\zeta - E_r}\right)^{1/2} - \left(\frac{1}{2} + \frac{\zeta - E_r}{\hbar\omega}\right)^{1/2}\right]^2 \quad (14b)$$

where ζ is the Fermi energy and E_r is the energy of the Landau level just below or equal to the Fermi level. The first term in the square bracket of Eq. (14a) has the same singular behavior as the density of states; $\Delta\sigma_2$ is even more singular, behaving as $1/[\zeta - (n + \frac{1}{2})\hbar\omega]$. Equations (14a) and (14b) indicate that there will be maxima in the symmetric conductivity whenever a Landau level coincides with the Fermi level. This means that for this condition, maxima in the resistivity ϱ_{xx}, the quantity that is actually measured experimentally, will be observed. The reason for this is that in strong magnetic fields the nondiagonal conductivity σ_{xy} is much larger than σ_{xx}, so that in the expression $\varrho_{xx} = \sigma_{xx}/(\sigma_{xx}^2 + \sigma_{xy}^2)$ the term σ_{xx}^2 in the denominator can be neglected. Because the Hall coefficient for simple bands is independent of the magnetic field for large $\omega\tau$, ϱ_{xx} will be proportional to σ_{xx}. Thus one expects that the oscillatory magnetoresistance in longitudinal and transverse magnetic fields will be in phase.

So far, collision broadening and the finite temperatures at which experiments are performed have been neglected. If they are taken into account, Adams and Holstein obtain the following results*:

$$\frac{\Delta\sigma_1}{\sigma_{\text{class}}} = \frac{5\sqrt{\pi}}{2}\left(\frac{\hbar\omega}{\zeta}\right)^{1/2}\sum_r \frac{(-1)^r(2\pi^2 rkT/\hbar\omega)}{(2\pi r)^{1/2}\sinh(2\pi^2 rkT/\hbar\omega)}$$
$$\times \cos\left(\frac{2\pi r\zeta}{\hbar\omega} - \frac{\pi}{4}\right)\exp\left(-\frac{2\pi}{\omega\tau_c}\right) \quad (15a)$$

$$\frac{\Delta\sigma_2}{\sigma_{\text{class}}} = \frac{3\pi}{8}\left(\frac{\hbar\omega}{\zeta}\right)\sum_r (-1)^r \frac{2\pi^2 rkT/\hbar\omega}{\sinh(2\pi^2 rkT/\hbar\omega)}$$
$$\times \cos\left(\frac{2\pi r\zeta}{\hbar\omega} - \frac{\pi}{2}\right)\exp\left(-\frac{2\pi}{\omega\tau_c}\right) \quad (15b)$$

* A numerical error which was pointed ont by Soule et al. [17] has been corrected here.

Each term in the harmonic series in (15a) and (15b) is multiplied by a factor $u/\sinh u$, with $u = 2\pi^2 rkT/\hbar\omega$. As in the theory of the de Haas–Van Alphen effect, collision broadening is taken into account by an exponential damping factor, which was first introduced by Dingle [19]. The collision time is in general different from the relaxation time which determines the mobility, because of the modification of the scattering whenever a Landau level approaches the Fermi level. The exponential term in (15a) and (15b) may also be written as $\exp -(2\pi^2 kT_D/\hbar\omega)$, with T_D, equal to $\hbar/\pi k\tau_c$, called the Dingle temperature. Because collision- and temperature-broadening produce about the same effects on the oscillations it is possible to characterize collision broadening by an appropriate temperature. There is a phase difference between the cosine factor in $\Delta\sigma_1$ and in $\Delta\sigma_2$ of $\pi/4$. This factor depends on the way collision broadening is introduced. Whereas Eq. (14a) can be developed in a Fourier series as it stands, this cannot be done with Eq. (14b) because of its logarithmic divergence. To circumvent this difficulty, Adams and Holstein [10] replaced in (14b) the expression $1/(\zeta - E_r)$ by $(\zeta - E_r)/[(\zeta - E_r)^2 + (1/\omega^2\tau^2)]$ and then Fourier transformed. Soule et al. [17] broadened the density-of-states function prior to working out the conductivity and obtained a different result. For details the reader is referred to the original paper. It is interesting to note that in their special case of graphite at the highest fields used ($\zeta/\hbar\omega \approx 2.5$) the $\Delta\sigma_2$ contribution was not important. The same conclusion can be drawn from recent work on n-type Bi_2Te_3 by Drath and Landwehr [54]. Even for the Landau level with the quantum number $n = 1$ no phase shift with respect to higher levels could be observed. Obviously, more work on this subject is desirable. For a satisfying comparison between theory and experiment it will be necessary to take into account the field dependence of the Fermi energy in a realistic manner, i.e., collision- and temperature-broadening must be properly included. At any rate, the conditions under which $\Delta\sigma_2$ should dominate—that, crudely speaking, the oscillatory component of ϱ is larger than the classical component—are not too easily obtained experimentally. In most practical cases $\Delta\sigma_1$ will be the important contribution and Eq. (15a) applies. Inspection of Eq. (15a) shows that it is possible to determine the effective mass of charge carriers by measuring the temperature dependence of the oscillations at constant field or the field dependence at constant temperature. The former method has been widely used with success. If the relaxation time is independent of temperature in the range investigated and if the harmonics can be neglected, the evaluation of m^* is especially simple. Complications arise from the presence of several kinds of charge carriers which contribute to the quan-

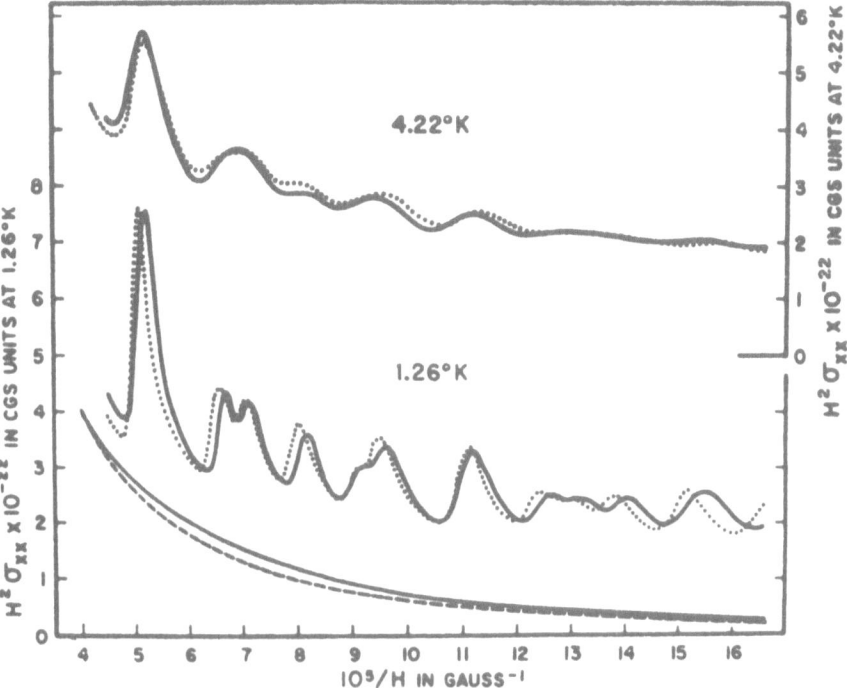

Fig. 4. Experimental and theoretical diagonal magnetoconductivities for graphite with the magnetic field H parallel to the hexagonal axis at 4.22 and 1.26°K. The points represent the experimental data. The oscillatory curves represent a linear combination of the Adams and Holstein theory and an empirical correction shown at the bottom of the diagram. The empirical correction was approximated by the function $64/(10^5/H)^2$, and is plotted as the dashed curve. From Soule et al. [17].

tum oscillations, or from more complicated bands, e.g., of the many-valley type. In this case it is necessary to fit experimental data on a trial and error basis using a computer. The matching of the field dependence data has to be done following the same principle. Simplifications are possible if the harmonics can be neglected. It should be mentioned that Soule et al. were able to fit the data on field and temperature dependence for graphite with the same parameters using the fundamental and the first harmonic, and that the agreement between theory and experiment was very good. This may be seen from Fig. 4. The AH theory also gave the correct ratio of $\Delta\sigma_1/\sigma_{class}$ if the average "low" field data were taken. This is significant, because other existing theories make different predictions for the relative amplitude of the oscillations. At "high" fields, however, a monotonic background magnetoresistance was observed which is not contained

in the AH theory. Several explanations have been offered for this effect ([20]).
Further work seems necessary to clarify the situation.

Kahn ([21]) has calculated the electrical conductivity in a strong magnetic
field for scattering by noninteracting impurity centers, treating the scattering
process more properly and not in the Born approximation, as Adams and
Holstein did. The divergence in $\Delta\sigma_2$ disappears without the need for introducing collision broadening. Whether the AH approach or Kahn's
approach applies will depend on the situation.

Oscillatory Magnetoresistance in Longitudinal Fields

The longitudinal magnetoresistance in fields which are sufficiently
strong that the quantization of the electron orbits becomes important can
be treated by common methods. This is possible because the electron motion
parallel to the electric field $\mathbf{E}$ is not inhibited in the longitudinal configuration
when $\mathbf{E}$ and $\mathbf{H}$ are parallel. A distribution function and a relaxation time
can be defined for each Landau level and a Boltzmann equation can be
formulated and solved for each level with quantum number n. The overall
current is the sum of the individual contributions. At low temperatures
one obtaines for the conductivity σ_z ([11])

$$\sigma_z = \frac{e^2}{m^*} \sum_n N_n \tau_n(\zeta) \tag{16}$$

where N_n is the number of electrons in the Landau level n and $\tau_n(\zeta)$ is the
relaxation time in the same level at the Fermi energy. An oscillatory behavior arises through the dependence of the density of states on the magnetic
field. Whenever a Landau level approaches the Fermi level the density of
states increases and the relaxation time decreases accordingly. The longitudinal relaxation time has been computed by Argyres ([18]) for lattice
scattering and by Kahn and Frederikse ([11]) for ionized impurity scattering.
For the conductivity in a magnetic field Argyres got the following result:

$$\frac{\Delta\sigma}{\sigma_0} = -\sqrt{\pi}\left(\frac{\hbar\omega}{\zeta}\right)^{1/2} \sum_{r=1}^{\infty} \frac{(-1)^r (2\pi^2 rkT/\hbar\omega)}{(2\pi r)^{1/2}\sinh(2\pi^2 rkT/\hbar\omega)} \cos\left(\frac{2\pi r\zeta}{\hbar\omega} - \frac{\pi}{4}\right) \tag{17}$$

The effective mass is assumed isotropic. Equation (17) has the same structure as the formula which describes the oscillations of the $\Delta\sigma_1$ type in transverse fields except that collision broadening was not taken into account.
In order to observe oscillations, it is necessary that the conditions $\omega\tau \gg 1$,
$\hbar\omega > kT$, and $\zeta > \hbar\omega$ are satisfied.

The calculation of Kahn and Frederikse has been carried out for InSb at $T = 0$. Sharp spikes appear in the resistance if a Landau level reaches the Fermi energy. Collision broadening and the inhomogeneity in the impurity distribution will round off the peaks.

EXPERIMENTAL METHODS

To meet the condition $\hbar\omega > kT$ for the observability of the Shubnikov–de Haas effect, it will always be necessary to employ low temperatures, preferably in the liquid helium and hydrogen range. The condition $\omega\tau > 1$ is roughly equivalent to $\mu B > 1$, where μ is the mobility. In semimetals with low effective masses magnetic fields of the order 10 kG or less are sufficient to get into high-field range at 4.2 °K. In semiconductors, however, one quite often needs larger fields. To obtain $\mu B > 10$ for a mobility of 10^4 cm^2/V sec one requires an induction of at least 10^{-3} V sec/cm^2 = 100 kG. The additional condition $\zeta > \hbar\omega$ can be very restrictive. To get a sufficient high Fermi level heavy doping is required, which reduces the low-temperature mobility quite drastically in many cases. For this reason germanium and silicon are not especially favorable for the observation of the Shubnikov–de Haas effect.

Ordinary DC methods using separate current and potential leads are usually employed in measurements of the oscillatory magnetoresistance. Modern methods to improve the signal to noise ratio of the data have been used in connection with conventional electromagnets. Lerner [22,23] and Ketterson and Eckstein [24] have made measurements of the Shubnikov–de Haas effect in bismuth using AC modulation of the magnetic field and synchronous detection. The measured resistance is usually plotted on an xy reorder versus the magnetic field or versus the reciprocal of H, which greatly simplifies the data handling. In order to get rid of the classical background magnetoresistance, a bucking technique [17,25] is advisable. In his measurements of the SdH effect in bismuth Brown [25] employed a second Bi crystal at N$_2$ temperature. Moreover, filtration and differentiation was used to enhance superimposed high frequency oscillations of small amplitude. Whereas pulsed magnetic fields with a duration of the order of a millisecond are not too well suited for the investigation of the magnetoresistance in metals at helium temperatures because of eddy current heating and the skin effect, this restriction is normally absent in semiconductors. Care has to be taken to compensate spurious induction voltages. In order to improve the sensitivity, the current through the sample can be pulsed. Mechanical vibrations during the pulse can be troublesome. Part of the loss in sensitivity which results from a relatively high noise level in pulsed field measurements is compensated by the larger signals which are obtained in fields of the order 100 kG or higher.

EXPERIMENTAL RESULTS

In this section recent results on the Shubnikov–de Haas effect in semiconductors will briefly be reviewed. Metals and semimetals, except bismuth and graphite, will be excluded.

Bismuth

Bismuth is still the material in which quantum oscillations have been most intensively studied. The experimental data may be explained on the basis of two sets of carriers. The electrons are located in three ellipsoids arranged symmetrically about the z axis and tilted in the bisectrix–trigonal plane. The hole Fermi surface is a prolate ellipsoid which lies on the trigonal axis. For both carriers spin splitting has been observed in experiments of the Shubnikov–de Haas type. Using a HF impedance bridge which was originally laid out to measure the capacitance of a bismuth–metal condensor, Smith et al. [15] measured resistance oscillations in Bi in fields up to 88 kG. The spin splitting of the hole levels was found to be very large (about twice the orbital splitting) in the trigonal direction and small in the binary-bisectrix plane. The anisotropy of the g-factor for the electrons was also measured. Ketterson and Eckstein [24] confirmed the two-carrier model for Bi (which had been questioned) by SdH measurements, and did see spin splitting of the hole Landau levels in smaller magnetic fields.

Graphite

The work of Soule et al. [17] has already been quoted. Analyzing the data on the anisotropy of the electron and hole periods, they came to the conclusion that the electron and hole surfaces are closed in all directions and that the results are incompatible with a Fermi surface which is open in the hexagonal direction. The surfaces were found to be highly prolate in the hexagonal direction, with anisotropy ratios of 12 for electrons and about 17 for holes. The electron surface turned out to be nearly ellipsoidal.

As mentioned above, the data were also used to check the theory of Adams and Holstein. It might be interesting to note that the Dingle temperatures obtained of $T_{De} = 0.81\ °K$ and $T_{Dh} = 0.56\ °K$ for electrons and holes, respectively, are considerably larger than the corresponding values of $T_\mu = 0.17\,°K$ and $0.07\,°K$ which can be calculated from the average mobilities. For InSb and InAs, which have spherical energy surfaces, a much better agreement between the broadening temperatures T_D and T_μ was found ([12]).

Indium Antimonide

Due to its simple band structure and its high electron mobility InSb is well suited for the comparison of experimental results on the oscillatory magnetoresistance with theoretical predictions. Another interesting feature of InSb is its large negative electronic g-factor. During the last few years spin splitting of the last few Landau levels has been observed by various workers. The 0^- maximum (the minus sign is a consequence of the negative g-factor in InSb) in the transverse magnetoresistance was first reported by Amirkhanov et al. ([26]), who employed pulsed fields up to 300 kG. The splitting of the $n = 0$ peak was observed by Bresler et al. ([27]) and by Amirkhanov and Bashirov ([28]). An example is given in Fig. 5, which shows the resistance as a function of a transverse magnetic field for an InSb single crystal with a carrier density of $6.5 \times 10^{16}/cm^3$ at $4.2\,°K$. It can be seen that the resistance at the position of the 0^- peak is several times larger than at zero magnetic field.

Whereas at large quantum numbers the longitudinal and transverse oscillatory magnetoresistances are in phase, as expected, differences occur close to the quantum limit. The most striking difference is the absence of a 0^- maximum in the longitudinal magnetoresistance. The absence of the 0^+ level in the transverse case has its origin in the fact that all carriers condense in the last Landau level when the 0^- level has passed the Fermi level and that they remain there unless freezing-out occurs. The additional absence of the 0^- peak in the longitudinal configuration is explained by Efros ([29]) with a low probability of electron scattering which is accompanied by spin flip. Another peculiarity is the existence of a rather strong 0-maximum in the Hall constant, the origin of which is not clear. The attempts to explain the position of the spin-split peaks theoretically in a quantitative way have not been too successful. Gurevich and Efros ([30]) have calculated expressions like (12) for the maxima with corrections for the actual shape of the Fermi distribution at finite temperatures. They predict too high magnetic fields for the positions of the peaks. This can be understood, because Landau-level broadening was disregarded. It is difficult to take this effect into account quantitatively for impurity scattering,

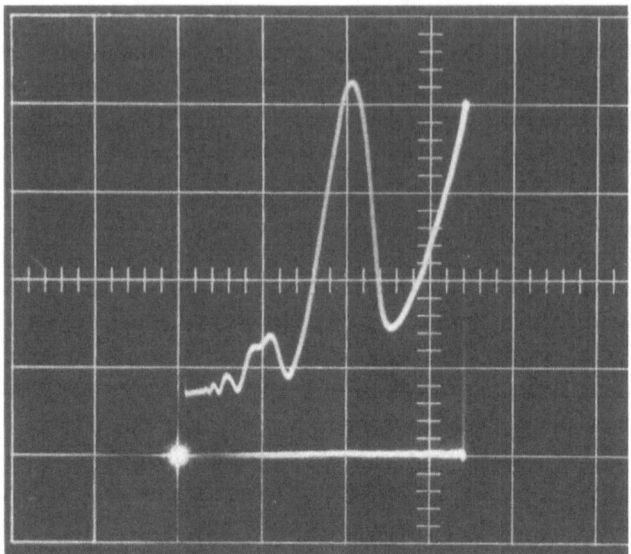

Fig. 5. Resistance of an *n*-type InSb single crystal as a function of a pulsed transverse magnetic field at 4.2°K. Horizontal scale: 35 kG/large division. The electron density is $6.5 \times 10^{16}/cm^3$. The current through the sample is switched on when the magnetic field is triggered and switched off when the magnetic field has reached its maximum. The flat trace for decreasing magnetic field indicates that spurious induction voltages are absent. The spin-split peak at about 35 kG has the quantum number 1, the sharp single peak belongs to the level 0^-. From L. M. Bliek, unpublished data.

which dominates in the heavily doped samples at low temperatures. For the 0^- maximum another difficulty arises through the oscillation in the Hall coefficient, for it cannot be excluded that the carrier density changes, which would shift the maximum. Because the position of the last quantum oscillations depend on the scattering processes, caution in the deduction of *g*-factors from the spin splitting of Shubnikov–de Haas data seems appropriate. For larger quantum numbers, if the field dependence of the Fermi level can be neglected, the derivation of *g*-factors should be reliable. Bresler *et al.* ([31]) have tried to circumvent the difficulties at low quantum numbers and have evaluated *g*-factors from spin-split peaks in the magnetothermal e.m.f. The existence of phonon-drag effects at helium temperatures leads to problems in the interpretation, however. Recently Antcliffe and Stradling ([31]) have detected a small warping of the Fermi surface of heavily doped InSb by means of the SdH effect.

Indium Arsenide

InAs is another semiconductor in which SdH oscillations are observed fairly easily. However, somewhat less effort was devoted to this substance than to InSb, probably because it is more difficult to produce homogeneous crystals. The first to see the 0^- maximum in the transverse magnetoresistance were Amirkhanov et al. [26]. Recently Bresler et al. [32] investigated quantum oscillations in the resistance and the thermal e.m.f. In principle, most of the problems which arise in InSb in the low-quantum-number region also exist for InAs, so that no more will be said about this substance here.

Gallium Antimonide

The Shubnikov–de Haas effect in gallium antimonide was first measured by Becker and Fan [33]. The doping of the samples investigated was so high that both the $\mathbf{k} = 0$ conduction band minimum and the somewhat elevated [111] band contained electrons. However, only the spherically symmetric band contributed to the quantum oscillations. Later Yep and Becker [34] extended the work to less heavily doped lithium-diffused crystals, in which the Fermi level was well below the [111] minima. In addition, samples were studied in which the Fermi energy was close to the bottom of the [111] valleys, so that principally in a magnetic field electron transfer could occur, as a consequence of different shifting rates of the two kinds of minima in magnetic fields. In addition to the Shubnikov–de Haas effect quantum oscillations in the Hall coefficient were observed which were about an order of magnitude smaller in their amplitude ratio than the resistance oscillations. The periods of both oscillations agreed. Supplemental evidence indicated that both types were caused by [000] electrons. It was shown that only a small fraction of the amplitude of the Hall effect oscillations could be attributed to a carrier transfer effect, which has been discussed by Adams and Keyes [12]. Thus the physical reason for the periodic Hall coefficient variations remains unknown. In the lithium-doped crystals the effective mass derived from the SdH data was smaller than the mass calculated for Te-doped, as-grown crystals. The Dingle temperature essentially agreed with the nonthermal damping temperature computed from the mobility.

Germanium and Silicon

From the low mobility of heavily doped germanium one expects that the SdH effect should just be observable for magnetic fields of the order of 100 kG. Fakidov and Zavadskii [35] found magnetoresistance oscillations at 20 °K in pulsed transverse fields up to 110 kG in an *n*-type sample which was intrinsic at room temperature. Hence the condition $\zeta > \hbar\omega$ was not met and the oscillations cannot be of the SdH type. Recently Bernard *et al.* ([36]) measured weak magnetoresistance oscillations in heavily doped crystals with selected orientations in DC magnetic fields up to 110 kG. An attempt to observe the SdH effect in silicon under the same conditions failed, as expected, because the condition $\omega\tau > 1$ could not be met.

Tellurium

The transverse and longitudinal magnetoresistance of heavily doped tellurium single crystlas with hole densities between $1 \times 10^{17}/\text{cm}^3$ and $6 \times 10^{18}/\text{cm}^3$ was investigated in pulsed magnetic fields up to 220 kG at helium temperatures by Braun and Landwehr ([37]). Pronounced SdH oscillations

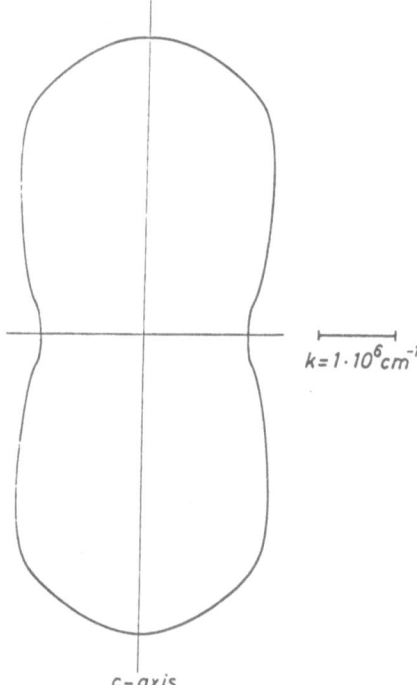

Fig. 6. Hole Fermi surface for a tellurium single crystal with a carrier density of $8 \times 10^{17}/\text{cm}^3$. The Fermi surface has rotational symmetry about the trigonal *c* axis. The "dumbbell"-like shape was derived from SdH data by Braun, Landwehr, and Neuringer ([64]).

were detected and used to explore the topology of the Fermi surface. Analysis of the anisotropy of the measured periods showed that the constant energy surfaces have rotational symmetry about the trigonal axis and that they are prolate with a ratio of the extremal cross sections $A(H \perp c)/A(H \parallel c)$ $= 2.2$. For samples with a carrier density of $1.5 \times 10^{18}/cm^3$ the Fermi surface could be examined in detail and turned out to be warped.

Recently, refined experiments have been performed ([64]) in DC magnetic fields up to 150 kG. They revealed two extremal cross sections in the direction $H \parallel c$. The SdH data for tellurium with a hole concentration of $8 \times 10^{17}/cm^3$ are compatible with a Fermi surface as shown in Fig. 6. Comparison with Hall data shows that the carriers are confined either in two surfaces which can accommodate holes with both spin directions or in two pairs of surfaces when carriers with different spin are separated. Shalyt *et al.* ([65]) recently measured the ShD effect in tellurium in the concentration range between 10^{16} and 10^{17} per cm³ and deduced four approximately ellipsoidal energy surfaces without spin degeneracy. Guthmann and Thullier ([66]) also measured the anisotropy of the SdH effect in tellurium and suggested that at high concentrations two ellipsoids of revolution grow together to form a "dumbbell"-shaped Fermi surface.

Probably all details of the Fermi surface have not yet been resolved. Improved cyclotron resonance measurements in tellurium at submillimeter wavelengths ([40]) strongly indicate deviations from ellipsoidal energy surfaces at small hole densities.

Mercury Telluride

Mecury telluride is another compound which is favorable for measuring an oscillatory magnetoresistance. Many experimental data support the view that HgTe in semimetallic and that conduction and valence bands are degenerate at $\mathbf{k} = 0$. As in HgSe, the electron mass is small and the mobility is high. The first observation of the SdH effect was reported by Furdyna ([42]). Stradling and Antcliffe ([43]) detected an oscillatory magnetoresistance in rather pure HgTe (electron concentration about $2 \times 10^{15}/cm^3$). Results of a systematic study of the SdH effect on samples with electron densities between $5 \times 10^{17}/cm^3$ and $3 \times 10^{18}/cm^3$ were published recently by Giriat ([44]). He found pronounced oscillations (see Fig. 7) and could detect no anisotropy in measuring the angular dependence of the effect. Hence the electron Fermi surface of HgTe should be spherical. The electron mass evaluated from the temperature dependence of the oscillations varied with concentration, and agreed well with results from other experiments. The nonparabolicity could be described by the Kane theory ([45]).

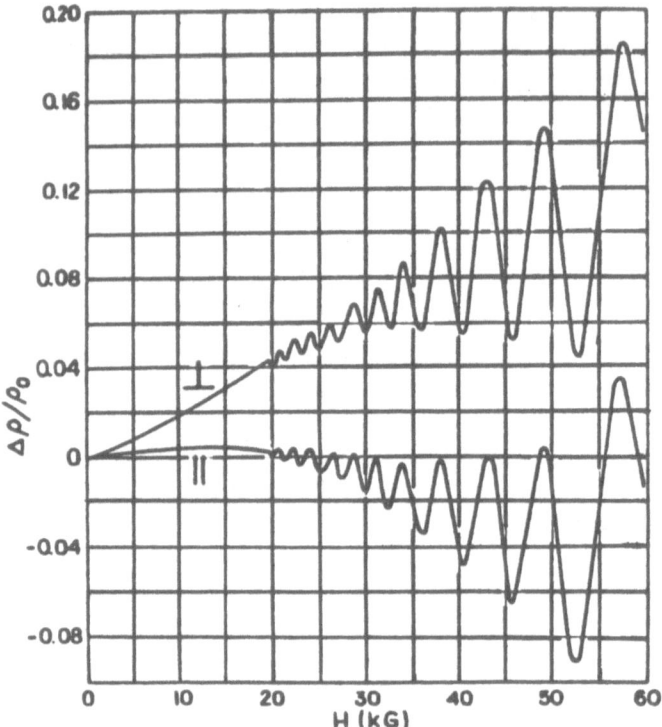

Fig. 7. SdH oscillations in mercury telluride in transverse ($\perp$) and longitudinal ($\parallel$) magnetic fields at 4.2°K. The electron density is 1.18 $\times 10^{18}$/cm³. Note that the oscillations are in phase. From Giriat ([44]).

Mercury Selenide

Mercury selenide is a material which is very suitable for measuring magnetotransport effects. It can be characterized as a semimetal. The room temperature electron mobility is about 18 000 cm²/V sec and for samples with a carrier density of $\sim 10^{17}$/cm³ electron mobilities of the order 10^5 cm²/V sec at 4.2 °K have been found. Whitsett ([46]) measured the SdH effect in single crystals with electron concentrations varying from 1.9 $\times 10^{17}$/cm³ to 4.5×10^{18}/cm³. Very well-developed magnetoresistance oscillations were detected, which allowed a detailed investigation to be made of the shape of the constant energy surface. Typical data are reproduced in Fig. 8. Analysis of the anisotropy of the period showed that the constant energy surfaces are almost spherical, with slight bulges in the [111] direction of **k** space. The conduction band was shown to be nonparabolic, the effective cyclotron mass varied from 0.033 to 0.068 m_0 in the concentra-

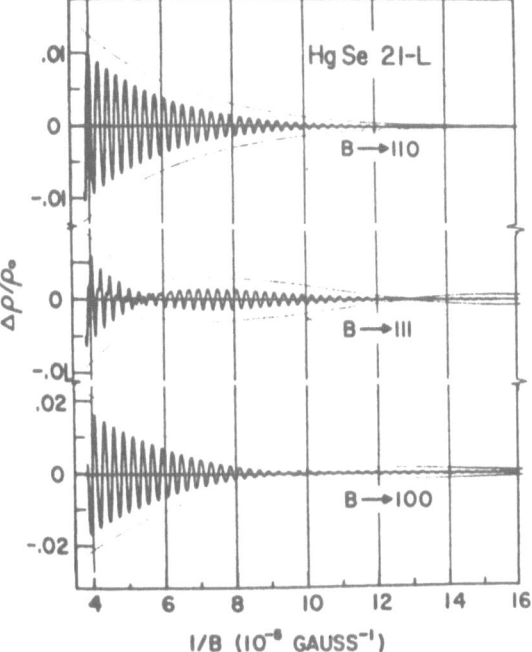

Fig. 8. SdH oscillations in a mercury selenide single
crystal with an electron density of $1.34 \times 10^{18}/cm^3$
for three orientations of the magnetic field B with
respect to the crystal axis as a function of B^{-1}. The
data were taken at 4.2°. The dotted curves indicate
the envelopes of the oscillations, when the temperature was reduced to 1.2°K. The beating effects can
be interpreted by bulges in the [111] directions of the
nearly spherical Fermi surface. From Whitsett ([46]).

tion range investigated, and was found to be in accord with the Kane ([45])
conduction band model for InSb. The collision time derived from the
nonthermal damping of the oscillations was at least two times smaller than
the collision time computed from the mobility.

Lead Telluride, Lead Selenide, and Lead Sulfide

For all these three materials the SdH effect has been observed by
Cuff and co-workers ([47,48]). The data served to elucidate the band structure
of these lead compounds. A unifying picture has emerged from this work.
The principal valence and conduction bands of PbTe, PbSe, and PbS can
be characterized by prolate ellipsoids of revolution centered at the L point

of the Brillouin zone. The anisotropy in the effective mass is large for PbTe and relatively small for the other compounds. The longitudinal mass decreases upon going from PbTe to PbS, and the transverse mass increases in this sequence. The bands are nonparabolic and the effective energy gaps derived agree with optical data. In all three semiconductors spin splitting has been observed for valence and conduction bands, amounting to nearly half of the Landau-level spacing.

Bismuth Telluride

The SdH-effect in p-type bismuth telluride was investigated by Auch and Landwehr ([49]) and by Landwehr and Drath ([50]). The validity of the six-valley model for the constant energy surfaces of the valence band originally proposed by Drabble ([51]) was confirmed. In samples of high homogeneity spin splitting was observed for the trigonal direction. A g-factor of 20 or 10 was derived, depending on whether the spin-up and spin-down levels belonging to adjacent quantum numbers have crossed or not. The experimental data did not allow a distinction to be made between these possibilities. Recently Freudenhammer ([52]) calculated the g-factor for Bi_2Te_3 using the method of Cohen and Blount ([53]). It was assumed that the effective mass (and the g-factor) was determined by the interaction

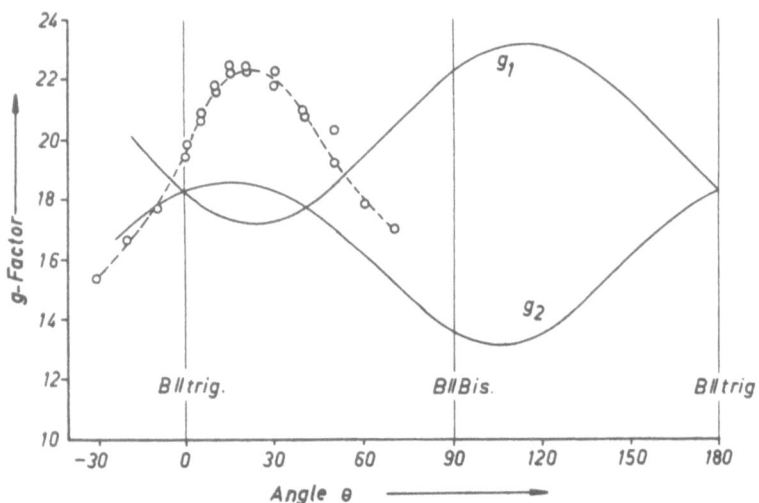

Fig. 9. Anisotropy of the g-factor for holes in Bi_2Te_3 when the magnetic field is rotated in the trigonal–bisectrix plane. The solid lines have been computed by Freudenhammer ([52]) and the circles represent experimental data ([54]).

of one valence and conduction band only. The predicted g-factor of 18.3 is remarkably close to the experimental value of $20 \pm 15\%$. At the same time the anisotropy of the g-factor in the trigonal–bisectrix plane was measured. Only one branch, belonging to the larger cyclotron mass, could be determined. The measured values and the g-factor calculated by Freudenhammer are shown in Fig. 9. There is qualitative, but not quantitative agreement between theory and experiment.

Magnetoresistance oscillations were also found in n-type Bi_2Te_3 by Drath and Landwehr [54]. The carriers causing the quantum effects are located in tilted ellipsoids in k space which are centered on mirror planes. However, the anisotropy differs appreciably from the values which had been deduced from low-field galvanomagnetic data. The origin of this discrepancy has to be sought in an additional conduction band which has its minimum slightly above the multivalley band and which obscures the low-field data. The additional band manifests itself, e.g., in an anomalous field dependence of the high-field Hall coefficient.

Tin Telluride

Complicated SdH oscillations were observed in the p-type semiconductor SnTe in fields up to 150 kG by Burke and co-workers [55,56]. The analysis of the data was difficult because in symmetry directions the oscillatory patterns were involved. In order to get insight into the structure of the data, they were Fourier-analyzed for their $1/H$ contents. It was found that the smallest cross section of the Fermi surface is in the [111] direction. However, not only two cross sections were observed in [111], which would be expected for an ellipsoidal many-valley model, but four. Taking into account the carrier density derived from Hall data, it was concluded that the Fermi surface of the SnTe samples investigated consists of four pieces which have maximum and minimum cross sections perpendicular to the [111] direction.

Cadmium Arsenide

Measurements of the SdH effect in $CdAs_2$ were recently reported by Rosenman [57]. From the anisotropy of the observed single period the shape of the constant energy surfaces of the conduction band was determined in the concentration range $n = 2$–$3 \times 10^{18}/cm^3$. The Fermi surface can be described by a single ellipsoid with rotational symmetry and rather small anisotropy.

From a computer fit of the temperature and field dependences of the data the effective-mass values and the Dingle temperature were derived. Good agreement was obtained between the anisotropy values for the period and the effective mass. The Dingle temperature was found to be about an order of magnitude larger than the characteristic broadening temperature computed from the mobility.

QUANTUM-LIMIT EFFECTS

If a semiconductor is subjected to a magnetic field sufficiently high that all conduction electrons are condensed in the Landau cylinder with the quantum number $n = 0$ (or $n = 0^-$ in the case of spin splitting), the resistance will no longer oscillate with increasing magnetic field. This situation has been called "quantum limit" by Argyres and Adams ([58]). The relevant energy E of the electrons kT or ζ for nondegenerate and degenerate material, respectively, is then smaller than the Landau-level spacing $\hbar\omega$. For the condition $\hbar\omega \gg E$ the term "extreme quantum limit" has been coined.

In the quantum limit the resistance will increase monotonically with the magnetic field. This is due to a highly anisotropic constant energy surface and a strongly field-dependent density of states. From qualitative considerations it can be seen that the detailed field dependence of the magnetoresistance in the quantum limit will depend on the scattering mechanism. The radius of the last Landau cylinder will increase with increasing magnetic field and its height will decrease. This means that the momentum of the electrons in the direction of **H** is more and more restricted, so that it can happen that an electron is scattered again and again by the same impurity atom. Hence this will have a strong influence on the resistance.

Not too much will be said about the theoretical situation. Recently an excellent review article by Kubo *et al.* ([13]) has been published which treats many facets of the problem in detail, especially those which involve collision broadening and inelastic scattering. The article by Adams and Keyes ([12]) is also concerned with quantum-limit effects. Moreover, in this work the influence of strong magnetic fields on shallow donor states has been considered. Not much progress has been made on the experimental side during the last few years. This may seem surprising, because strong magnetic fields have become more readily available recently. On the other hand, it must not be forgotten that the condition $\hbar\omega \gg E$ can be quite a handicap. Only for a few materials has the extreme quantum limit been achieved. So far InSb and Ge have been investigated most. Quite often it cannot be excluded with certainty that the results have been obscured by contact or freeze-out effects (or both), so that some care seems appropriate when theory and experiment are compared.

As in the oscillatory range, the resistance in longitudinal magnetic fields in the quantum limit can be treated by a modified Boltzmann technique. This has been done by Argyres and Adams [58] for lattice and ionized impurity scattering. Appel [59] investigated the longitudinal magnetoresistance in quantizing fields for nondegenerate semiconductors and phonon scattering of the electrons. In the quantum limit he finds $\varrho \sim HT^{1/2}$, in agreement with Argyres and Adams [58]. The same field dependence is also predicted by Kubo et al. [13]. Under nondegenerate conditions the strong field magnetoresistance has been studied for InSb [see, e.g. [12], p. 124] and for n-type Ge [60,61]. In both cases a linear increase of resistance was observed in longitudinal fields. This behavior is predicted by Adams and Holstein [10] for high-temperature acoustic and point-defect scattering. However, in InSb lattice scattering should not be the dominant mechanism. The published results have been discussed in detail by Adams and Keyes [12]. In n-type germanium a complication arises due to its many-valley structure. If the direction of a magnetic field is not symmetric with respect to the conduction-band valleys, different minima shift with different rates as a consequence of the difference in the cyclotron masses. Hence there will be changes in the population of the minima depending on the band parameters and the direction of the field. Because electrons will be transferred into high-mobility valleys, a decrease in resistance results at high fields. The effect has been quantitatively evaluated by Miller and Omar [62].

A treatment of the magnetoresistance in the quantum limit in transverse and longitudinal fields for isotropic energy surfaces is contained in the paper of Adams and Holstein [10]. They investigate the resistance as a function of the magnetic field for a variety of possible scattering mechanisms for degenerate and nondegenerate electrons in the extreme quantum limit. The results are compiled in tables which are also given in the article by Adams and Keyes [12], so that they will not be reproduced here. Various kinds of field dependence are predicted, from field-independence of ϱ for ionized impurity scattering in the nondegenerate range, to an $H^{5.5}$ law in transverse fields in the degenerate case for low-temperature acoustical scattering. Because of the large differences in behavior which should occur for different scattering mechanisms quantum limit effects could principally serve as a tool for studying scattering. However, the experimental difficulties in achieving this goal are considerable. Amirkhanov et al. [26] investigated InAs samples with electron concentrations between $4 \times 10^{16}/cm^3$ and $3 \times 10^{17}/cm^3$ at 20 °K in the quantum limit range using pulsed fields up to

300 kG. The field dependence of the longitudinal magnetoresistance varied from $H^{1.4}$ for the least doped sample to $H^{2.8}$ for the specimen with 3×10^{17} electrons/cm³. It was argued that for the most heavily doped sample the degeneracy prevailed at all fields, so that the results could be considered as a confirmation of the AH theory, which predicts an H^3 relation for ionized impurity scattering. However, since no Hall data were available for the samples, no firm conclusion can be drawn. Recently Beckman *et al.* [63] investigated quantum-limit phenomena in InSb in transverse static fields up to 150 kG. In addition to the resistance the Hall constant was measured. Data were taken between 1.3 and 4.2 °K. In their sample No. 116, with

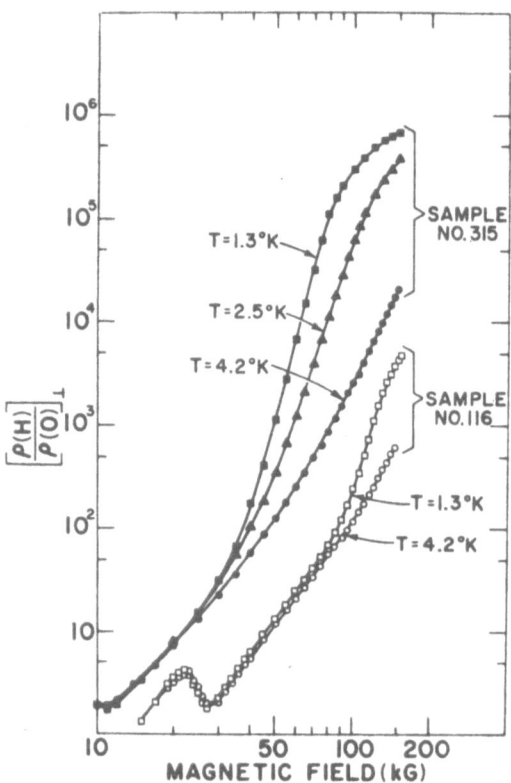

Fig. 10. Transverse magnetoresistance of *n*-type InSb at liquid helium temperatures in the quantum limit. The carrier concentrations in the exhaustion range are 2.6×10^{15}/cm³ and 1.1×10^{16}/cm³ for samples Nos. 315 and 116, respectively. From Beckman *et al.* [63].

$n = 1.1 \times 10^{16}/cm^3$, the resistance increased between 30 and 90 kG according to an $H^{3.3}$ law both at 1.3 and 4.2 °K. In this range the Hall constant was independent of the magnetic field. The data are reproduced in Fig. 10. Beyond 50 kG carrier freeze-out occurs, which causes the increase in the slope of the $\varrho-H$ curve above 50 kG. In the exhaustion range the magneto-resistance is independent of temperature, as predicted by Adams and Holstein. Also, the less heavily doped sample, No. 315, for which carrier freeze-out already occurs around 25 kG, an increase in ϱ close to an H^3 law was observed between 12 and 25 kG. The stronger increase in magnetoresistance beyond the upper field is especially evident for the 1.3 °K data. The initial rise is due to carrier freeze-out, and the leveling off at the highest magnetic fields is interpreted by the authors as the beginning of impurity conduction. In the freeze-out range the magnetoresistance at fixed magnetic fields does increase with T^{-1}, which proves the existence of bound states. The activation energy does increase proportional to $H^{1/3}$, in agreement with a theory of Keyes [cited in ([63])].

In conclusion, the experimental data in the quantum-limit range are not very numerous and much remains to be learned on this subject.

NOTE ADDED IN PROOF

Since the completion of the manuscript a number of new results have been obtained which are worth mentioning. The $H^{3.3}$ behavior in the quantum limit was confirmed for InSb by Bliek and Landwehr ([67]). In addition, it was found that not only the 0^- maximum was missing in the longitudinal magnetoresistance, but that no spin splitting at all was observable in the longitudinal configuration, provided that potential leads with sufficiently small diameter were employed. The same samples showed pronounced spin splitting in the transverse magnetoresistance and the Hall effect. The same result, i.e., strong spin-split peaks in the Hall voltage and transverse magnetoresistance, but no splitting in the longitudinal case, was observed for n-type HgSe ([68]). This behavior is not understood at present.

The HgSe data, together with data on the de Haas–Van Alphen effect, strongly support the suggestion by Roth et al. ([69]) that the beating SdH oscillations in the [111] and [100] directions as observed by Whitsett ([46]) are caused not by an anisotropic Fermi surface, but by inversion asymmetry splitting. This effect is attributed to the missing inversion symmetry with a relatively large spin–orbit coupling. The reader is referred to the original papers for details.

Guseva and Zyryanov ([70]) have carried out a calculation of the oscillatory components of the conductivity to a higher order and find oscillations in the Hall coefficient.

REFERENCES

1. L. Shubnikov and W. J. de Haas, *Nature* **126**: 500 (1930).
2. L. D. Landau, *Z. Phys.* **64**: 629 (1930).
3. R. Peierls, *Z. Phys.* **81**: 186 (1933).
4. H. P. R. Frederikse and W. R. Hosler, *Can. J. Phys.* **34**: 1377 (1956).

5. P. B. Alers, *Phys. Rev.* **101**: 41 (1956).
6. R. Sladek, *Phys. Rev.* **110**: 817 (1958).
7. H. P. R. Frederikse and W. R. Hosler, *Phys. Rev.* **110**: 880 (1958).
8. P. N. Argyres, *Phys. Rev.* **109**: 1115 (1958).
9. I. M. Lifshitz, *Phys. Chem. Solids* **4**: 11 (1958).
10. E. N. Adams and T. D. Holstein, *J. Phys. Chem. Solids* **10**: 254 (1959).
11. A. H. Kahn and H. P. R. Frederikse, in: *Solid State Physics, Vol. 9* (F. Seitz and D. Turnbull, eds.), Academic Press, New York and London, 1959, p. 257.
12. E. N. Adams and R. W. Keyes, *Progress in Semiconductors, Vol. 6*, (A. F. Gibson, ed.), Heywood & Co., London, 1962.
13. R. Kubo, S. J. Miyake, and N. Hashitsume, *Solid State Physics, Vol. 17* (F. Seitz and D. Turnbull, eds.), Academic Press, New York and London, 1965.
14. L. Onsager, *Phil. Mag.* **43**: 1006 (1952).
15. G. E. Smith, G. A. Baraff, and J. M. Rowell, *Phys. Rev.* **135**: A1118 (1964).
16. E. N. Adams, *Phys. Rev.* **85**: 41 (1952).
17. D. E. Soule, J. W. McClure, and L. B. Smith, *Phys. Rev.* **134**: A453 (1964).
18. P. N. Argyres, *J. Phys. Chem. Solids* **4**: 19 (1958).
19. R. B. Dingle, *Proc. Roy. Soc.* **A211**: 517 (1952).
20. J. W. McClure, *Phys. Rev.* **112**: 715 (1958).
21. A. H. Kahn, *Phys. Rev.* **119**: 1189 (1960).
22. L. S. Lerner, *Phys. Rev.* **127**: 1480 (1962).
23. L. S. Lerner, *Phys. Rev.* **130**: 605 (1963).
24. J. B. Ketterson and Y. Eckstein, *Phys. Rev.* **132**: 1885 (1963); *Phys. Rev.* **137**: A1777 (1965).
25. R. D. Brown, *IBM J. Res. Develop.* **10**: 462 (1966).
26. Kh. I. Amirkhanov, R. I. Bashirov and Yn. E. Zakiev, *Soviet Phys.—Solid State (English Transl.)* **5**: 340 (1963).
27. M. S. Bresler, R. V. Parfen'ev, and S. S. Shalyt, *Soviet Phys. – Solid State (English Transl.)* **7**: 1025 (1965); **8**: 1414 (1966).
28. Kh. I. Amirkhanov and R. I. Bashirov, *Soviet Phys.–JETP Letters (English Transl.)* **1**: 49 (1965).
29. A. L. Efros, *Soviet Phys.—Solid State (English Transl.)* **7**: 1206 (1965).
30. L. E. Gurevich and A. L. Efros, *Soviet Phys.—JETP (English Transl.)* **16**: 402 (1963).
31. G. A. Antcliffe and R. A. Stradling, *Phys. Letters* **20**: 119 (1966).
32. M. S. Bresler, N. A. Redko, and S. S. Shalyt, *Phys. Stat. Sol.* **15**: 745 (1966).
33. W. M. Becker and H. Y. Fan, in: *Proceedings of the International Conference on the Physics of Semiconductors, Paris, 1964*, Dunod, Paris, 1964, p. 663.
34. T. O. Yep and W. M. Becker, *Phys. Rev.* **144**: 741 (1966); *Phys. Rev.* **156**: 939 (1967).
35. I. G. Fakidov and E. A. Zavadskii, *Soviet Phys.—JETP (English Transl.)* **4**: 716 (1958).
36. W. Bernard, H. Roth, W. D. Straub, and J. E. Mulhern, *Phys. Rev.* **135**: A1386 (1964).
37. E. Braun and G. Landwehr, *Z. Naturforsch.* **21a**: 495 (1966); E. Braun and G. Landwehr, in: "Proceedings of the International Conference on the Physics of Semiconductors, Kyoto, 1966," *J. Phys. Soc. Japan Suppl.* **21**: 380 (1966); E. Braun, Dissertation, Technische Hochschule Braunschweig, February 1967.
38. J. H. Mendum and R. N. Dexter, *Bull. Am. Phys. Soc.* **9**: 632 (1964).

39. J. C. Picard and D. L. Carter, in: "Proceedings of the International Conference on the Physics of Semiconductors, Kyoto, 1966," *J. Phys. Soc. Japan Suppl.* **21**: 202 (1966).
40. K. J. Button and G. Landwehr, to be published.
41. C. Guthmann and J.-M. Thullier, *Compt. Rend.* **263**: 303 (1966).
42. J. Furdyna, *Phys. Rev. Letters* **16**: 646 (1966).
43. R. A. Stradling and A. Antcliffe, in: "Proceedings of the International Conference on the Physics of Semiconductors, Kyoto, 1966," *J. Phys. Soc. Japan Suppl.* **21**: 374 (1966).
44. W. Giriat, *Phys. Letters* **24A**: 515 (1967).
45. E. O. Kane, *J. Phys. Chem. Solids* **1**: 249 (1957).
46. C. S. Whitsett, *Phys. Rev.* **138**: A829 (1965).
47. K. F. Cuff, M. R. Ellett, and C. D. Kuglin, in: *Proceedings of the International Conference on the Physics of Semiconductors, Exeter, 1962*, p. 316.
48. K. F. Cuff, M. R. Ellett, C. D. Kuglin, and L. R. Williams, in: *Proceedings of the International Conference on the Physics of Semiconductors, Paris, 1964*, Dunod, Paris, 1964, p. 677.
49. K. Auch and G. Landwehr, *Z. Naturforsch.* **18a**: 424 (1963).
50. G. Landwehr and P. Drath, in: *Proceedings of the International Conference on the Physics of Semiconductors, Paris, 1964*, Dunod, Paris, 1964, p. 669; *Z. Angew. Phys.* **20**: 392 (1966).
51. J. R. Drabble, *Proc. Phys. Soc. (London)* **72**: 380 (1958).
52. A. Freudenhammer, Diplomarbeit T. H. Aachen, 1967.
53. M. H. Cohen and E. I. Blount, *Phil. Mag.* Vol. 115 (1960).
54. P. Drath and G. Landwehr, *Phys. Letters* **24A**: 504 (1967); *Z. Naturforsch.* **23a**: 1146 (1968); P. Drath, Dissertation T. H. Braunschweig, 1967.
55. J. R. Burke, R. S. Allgaier, B. B. Houston, J. Babiskin, and P. G. Siebenmann, *Phys. Rev. Letters* **14**: 360 (1965).
56. J. R. Burke, B. Houston, H. T. Savage, J. Babiskin, and P. G. Siebenmann, in: "Proceedings of the International Conference on the Physics of Semiconductors, Kyoto, 1966," *Phys. Soc. Japan Suppl.* **21**: 384 (1966).
57. I. Rosenmann, in: "Proceedings of the International Conference on the Physics of Semiconductors, Kyoto, 1966," *J. Phys. Soc. Japan Suppl.* **21**: 370 (1966).
58. P. N. Argyres and E. N. Adams, *Phys. Rev.* **104**: 900 (1956).
59. J. Appel, *Z. Naturforsch.* **11a**: 892 (1956).
60. W. F. Love and W. F. Wei, *Phys. Rev.* **123**: 67 (1961).
61. T. J. Diesel and W. F. Love, *Phys. Rev.* **124**: 666 (1961).
62. S. C. Miller and M. A. Omar, *Phys. Rev.* **123**: 74 (1961).
63. O. Beckman, E. Hanamura and L. J. Neuringer, *Phys. Rev. Letters* **18**: 773 (1967).
64. E. Braun, G. Landwehr, and L. J. Neuringer, to be published.
65. L. S. Dubinskaya, I. I. Farbstein, and S. S. Shalyt, *Zh. Eksperim. i Teor. Fiz.* **54**: 754 (1968).
66. C. Guthmann and J. M. Thullier, *Proceedings of the Symposium on the Physics of Selenium and Tellurium, Montreal*, 1962, in press.
67. L. M. Bliek and G. Landwehr, in: *Proceedings of the International Conference on the Physics of Semiconductors, Moscow*, 1968, in press.
68. L. M. Bliek and G. Landwehr, *Phys. Stat. Sol.* January, 1969.
69. L. M. Roth, S. H. Groves, and P. W. Wyatt, *Phys. Rev. Letters* **19**: 576 (1967).
70. G. I. Guseva and P. S. Zyryanov, *Phys. Stat. Sol.* **25**: 775 (1968).

Chapter 23

HALL EFFECT
AND TRANSVERSE VOLTAGES
IN TYPE-II SUPERCONDUCTORS
IN THE MIXED STATE*

S. J. Williamson†

Faculté des Sciences
Laboratoire Associé du C.N.R.S.
Orsay (Essone), France

and

J. Baixeras

L.C.I.E.
Fontenay-aux-Roses (Hauts de Seine), France

We have recently reported ([1]) the observation of "dip" effects in the Hall voltage and Hall angle of a type-II superconductor (Nb–1%Zr) which are analogous in some respects to the well-known dip effect in the longitudinal voltage exhibited by many superconductors in the mixed state ([2]). Such effects, which are sensitive to the applied magnetic field and current density, have been attributed to manifestations of the superconducting state in which vortex motion is influenced by surface or bulk defects of the specimens ([3,4]). Thus it should be possible to obtain information concerning vortex–defect and vortex–vortex interactions from the study of these effects. We shall describe experiments on ribbons of Nb–1%Zr which have been cold-rolled to a thickness of 19 μm which demonstrate (1) the effect of defects which reduce the Hall angle and may produce a dip effect in both the Hall voltage V_H and Hall angle θ_H as well as the longitudinal voltage V_L, and (2) the effect of aligned lattice or surface imperfections, which introduce transverse voltages which are not Hall voltages.

* Supported by D.G.R.S.T. (Comité "Electrotechnique Nouvelle").
† USA National Academy of Sciences–National Research Council Postdoctoral Fellow; present address: North American Aviation Science Center, Thousand Oaks, California.

Previously Reed *et al.* ([5]) found that the Hall angle for pure Nb in the mixed state *increases* monotonically as the magnetic field H is increased up to $H_{c2}(T)$ and is always *less* than the corresponding value obtained by extrapolation from the normal state $[H > H_{c2}(T)]$. In contrast, Niessen and co-workers ([6]) observed that for well-annealed Nb–50 at.% Ta θ_H is relatively insensitive to the applied current, unlike the results of Reed *et al.*, and *decreases* with increase in H, being always *greater* than the value extrapolated from the normal state. This latter behavior has defied explanation if one requires that the magnetic field in the core of the vortices should not be greater than $H_{c2}(0)$ ([7]). Recently Niessen *et al.* ([8]) reported that well-annealed, Ta-rich Nb–Ta alloys have a reduced Hall angle $\theta_H/\mu H_{c2}(0)$, where μ is the carrier mobility, which approaches a common value of about 2.3 as $H \rightarrow 0$, independent of T. However, their results for Nb-rich alloys show a generally lower value of $\theta_H/\mu H_{c2}(0)$ and are sensitive to the manner of sample preparation. It is not yet clear whether samples with a Nb concentration greater than 80% can have a Hall angle which exceeds the value extrapolated from the normal state.

These experiments on Nb and Nb–Ta alloys have focused upon the Hall effect in relatively well-annealed samples in an attempt to minimize vortex pinning and produce "ideal" vortex motion. In contrast, we have studied unannealed or slightly annealed, cold-rolled ribbons in order to emphasize these pinning effects. The present Nb–1%Zr samples, which were reduced 50 : 1 in the rolling process, exhibited fiber structure which was aligned along the rolling direction **R** similar to that studied in the Nb samples of Kramer and Rhodes ([9]). Typical fibers were from 1 μm to 5 μm in thickness. In addition, a slight undulation of the sample surface was evident, with ridges 1 μm apart running parallel to **R** and with a crest-to-trough height of less than 1 μm. Transmission electron microscopy of the subfiber structure in a plane parallel to the broad face of the sample revealed cellular networks with typical cell dimensions of 5000 Å and an irregularly spaced series of dislocation walls aligned parallel to **R**. Figure 1 shows this subfiber structure. No preferred orientation of the cell structure could be detected, unlike the case for the sample of Kramer and Rhodes ([9]), in which the cells were reported to be aligned at about 45° with respect to **R**.

Experiments were conducted with the 40 mm × 4 mm × 0.019 mm samples immersed in liquid helium at about 4.2 °K and with the magnetic field applied perpendicular to the broad face. Current densities from 10^4 A/cm^2 to 2×10^5 A/cm^2 were impressed upon different samples having various angles ϕ between the current I and the rolling direction **R**. The upper critical field of all samples was $H_{c2} = 9.2$ kOe. The voltage V_T

Fig. 1. Electron microscope examination by transmission through a plane parallel to the broad face of a sample. The dark narrow lines are formed by the intersection of dislocations and are parallel to the rolling direction **R**. Irregularly shaped light regions define the cellular structure, which is a substructure of larger fibers aligned along **R**. Typical fiber structure is illustrated in microphotographs given in [9] and [18].

from the transverse probes was then recorded as a function of field H for the four possible polarities of I and H. The crucial step is then to eliminate spurious effects and to determine which combination of voltages will give V_H. We define V_H by the classical formula $V_H = V_A + V_B$, where $V_A = V(+I, +H) - V(+I, -H)$ and $V_B = V(-I, -H) - V(-I, +H)$; the resulting V_H is an average of the amplitudes of the Hall voltages for the four combinations of I and H.

Since the transverse probes are generally not aligned precisely perpendicular to the current, the observed voltages are further complicated by the spurious introduction of a portion of the longitudinal voltage. Many factors such as bulk defects, dissimilar edges of the sample, and field inhomogeneities can influence vortex motion, thus introducing transverse voltages or affecting V_L. However, meaningful data analysis is made possible by the fact that the various causes of vortex motion will produce contributions to V_T which should have specific symmetries upon reversal of I or H. The following list includes the more important possible effects in an isothermal sample, showing the symmetry of the voltage upon reversal of I or H, respectively. We should remark that not all of these possible effects have been proven experimentally to exist.*

Longitudinal Voltages

1. Lorentz-type forces have $(-, +)$.

2. Lorentz-type forces with bulk or surface pinning which depend upon the direction of the Lorentz force; the dominant component, corresponding to (1), has $(-, +)$, and the component proportional to the difference in vortex response has $(+, -)$.

3. Transverse gradient in the perpendicular applied magnetic field has $(+, -)$.

4. Transverse gradient in the perpendicular field associated with the transport current has $(-, -)$.

Transverse Voltages

5. Hall effect has $(-, -)$.

6. Lorentz-type force with vortices guided by aligned defects has $(-, +)$ [11].

7. Longitudinal gradient in the perpendicular applied magnetic field has $(+, -)$.

While it is true that many of the above effects should be small, the Hall voltage in many conductors is also small, so these second-order effects can not be dismissed out of hand. It should be apparent that $V_H = V_A + V_B$ gives the true Hall voltage only if effect (4) is not appreciable.

The observed voltages can be interpreted in terms of vortex motion by use of the formula [12] $\mathbf{E} = -(\mathbf{v}/c) \times \mathbf{B}$ which relates the vortex velocity $\mathbf{v}$ to the magnetic induction $\mathbf{B}$ and the field $\mathbf{E}$ producing the observed potential. Two important assumptions upon which this formula is based are that all vortices are moving and that they have a common velocity $\mathbf{v}$. If some are trapped, as is likely in dirty materials at low fields, or if there

---

* For example, (3). However, the analogous effect, motion stimulated by temperature gradients, has been observed [see (10)].

is a distribution of velocities, then **v** should be replaced by the average $\langle n\mathbf{v}/n_0 \rangle$, where n is the density of vortices having a velocity **v** and n_0 is the total density of vortices. For the present samples, with a demagnetizing coefficient of approximately unity, the vortex density is $n_0 = B/\phi_0$, with $\phi_0 = hc/2e$. Henceforth "vortex motion" will signify the velocity average $\langle n\mathbf{v}/n_0 \rangle$.

The V_L voltage onset fields for $\phi = 0°$ and $90°$ reveal a marked observed anisotropy in the pinning effectiveness which depends upon the relative direction of the Lorentz force and **R**, the onset for $\phi = 90°$ being at $H/H_{c2} \approx 0.13$ and for $\phi = 0°$ at $H/H_{c2} \approx 0.5$. Vortices are preferentially restricted in the present samples from moving perpendicular to the rolling direction. Such a characteristic is typical of many cold-rolled superconductors [see, e.g. [13]]. However, the new feature of the present experiments is that V_H has the *same* respective onset field as V_L for $\phi = 90°$. This demonstrates that the vortices are able to move perpendicular to **R** even at low fields provided there is a substantial velocity component parallel to **R**; without the parallel component (the case $\phi = 0°$) no perpendicular motion is possible until much higher fields. One simple interpretation of these results is that the parallel motion assists vortices or flux bundles to thread their way around defects, whereas without the parallel motion vortices or flux bundles meet defects "head on" and are stopped. Another interpretation is that the vortex–defect interaction is dependent upon the speed of the vortices, being considerably reduced when the vortices are in motion.

The low-field Hall effect is illustrated in Fig. 2, which shows a finite (positive) Hall angle for fields considerably below $H/H_{c2} \approx 0.5$. For high current densities θ_H in slightly annealed samples (annealed at 525 °C for 1 hr in vacuo at 10^{-6} Torr) may be comparable at $H/H_{c2} \approx 0.3$ to θ_H predicted by linear extrapolation from the normal state. However, unannealed samples may have θ_H reduced by a factor of five or more, thus demonstrating that defects in this sample can limit the Hall effect. The low-field magnitude of θ_H for the slightly annealed specimens is in reasonable agreement with the theory of Bardeen and Stephen [14] if the field in the core of a vortex is comparable to the applied field, but it is smaller by a factor of five than the value predicted by Nozières and Vinen [15]. The disagreement is hardly surprising, since the latter theory, which includes a Magnus force acting to produce a Hall voltage, was developed for clean superconductors. We have never observed a value of θ_H which was greater than the value given by extrapolation from the normal state, and thus the present results are qualitatively contrary to the results of Niessen and co-workers [6,8] for Ta-rich Nb–Ta alloys. It should be remembered, however, that the present values

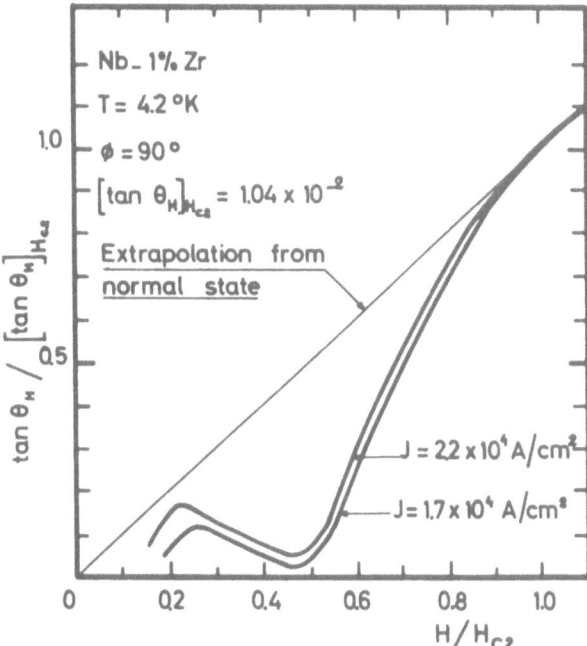

Fig. 2. Field dependence of the tangent of the Hall angle for a cold-rolled sample subsequently annealed for 1 hr at 525°C in vacuo at 10^{-6} Torr. The magnitude of the low-field maximum is sensitive to the state of anneal of the sample, although the simultaneous low-field maximum in the longitudinal voltage V_L is considerably less sensitive.

of θ_H represent a *lower limit* for the ideal θ_H in Nb–1%Zr, and are somewhat dependent upon the transport current, suggesting that they are defect-limited.

At higher fields V_H and V_L for $\phi = 90°$ both display a dip at $H/H_{c2} \approx 0.47$. The dip in V_H is relatively more pronounced and therefore is associated with a dip in θ_H, as is shown in Fig. 2. This has been interpreted [1] in terms of the Anderson–Kim explanation [16] for the dip effect in V_L in which the increased proximity of vortices at high fields leads to a constraining force on each vortex or between flux bundles which is in addition to the dynamic pinning forces produced by defects in the lattice or on the surface.* The enhanced effectiveness of pinning will reduce the vortex mo-

---

* It is perhaps remarkable that the present samples have a dip effect at a much lower value of H/H_{c2} than most samples; compare with [2].

tion and produce a dip in θ_H.* It is tempting to ascribe the cause of this dip to defects aligned along **R**, since V_L is relatively less affected at the dip than V_H; however, this is not proven by the present Hall experiments. It is also an open question whether a dip in V_L is a necessary or a sufficient condition for a dip in V_H. An alternate explanation for the dips in V_L and V_H, based upon the possibility that the surfaces of the sample which are parallel to H can support a higher current density than the bulk for $H/H_{c2} \approx 1$, does not seem to apply to the present specimen, for which the low-field side of the dip is at about $H/H_{c2} \approx 0.3$. Nor does the phenomenological explanation proposed by Swartz and Hart ([17]) seem to apply, as it is difficult to see why the surface critical current density should have a pronounced peak at such a low value of H/H_{c2}. However, the fact that they observed their peak (corresponding to a dip in V_L) disappear when their annealed Pb–5 at.%Tl sample was copper-plated can be explained in terms of a surface-pinning model for vortices instead of a superconducting surface sheath model, the former model being more reasonable for the present specimens. Copper plating may substantially reduce the order parameter within a distance ξ of the surface, thereby reducing the pinning effect of surface defects. Unfortunately, our attempts to plate the sample with a highly conducting metal to reduce the order parameter at the edges or on the broad face have proved unsuccessful, apparently due to the tenacious oxide layer at the surface.

With a sample for which ϕ has an intermediate value, $\phi = 30°$, quite a different field dependence for V_T is observed, as is illustrated in Fig. 3a. The Hall voltage is dominated at low fields by a voltage $V_T{}^e$ which is unchanged upon reversal of B, reverses upon reversal of I, and is similar to the observed V_L up to $H/H_{c2} \approx 0.5$. This voltage diminished to below the noise level for $H/H_{c2} \to 1$. We attribute this to guided motion of vortices by defects aligned along the rolling direction, as was first suggested by Staas et al. ([11]). Their experiments indicated that such defects could channel vortices in a direction which is not necessarily parallel to the Lorentz force and could thereby produce a transverse voltage. This voltage would have the same symmetry as V_L upon reversal of I or H, as observed.

If we define a "guiding efficiency" by the formula $\varepsilon = (E_T{}^e/E_L) \tan \phi$, where $E_T{}^e$ and E_L are respectively, the transverse and longitudinal electric fields resulting from the effects of guided motion, we find the remarkable field-dependence shown in Fig. 3b. The efficiency is essentially independent

---

* For the present specimens the dip corresponds to an intervortex spacing of 800 Å (for a triangular lattice). In comparison, an approximate calculation gives $\lambda = 670$ Å and $\xi = 370$ Å.

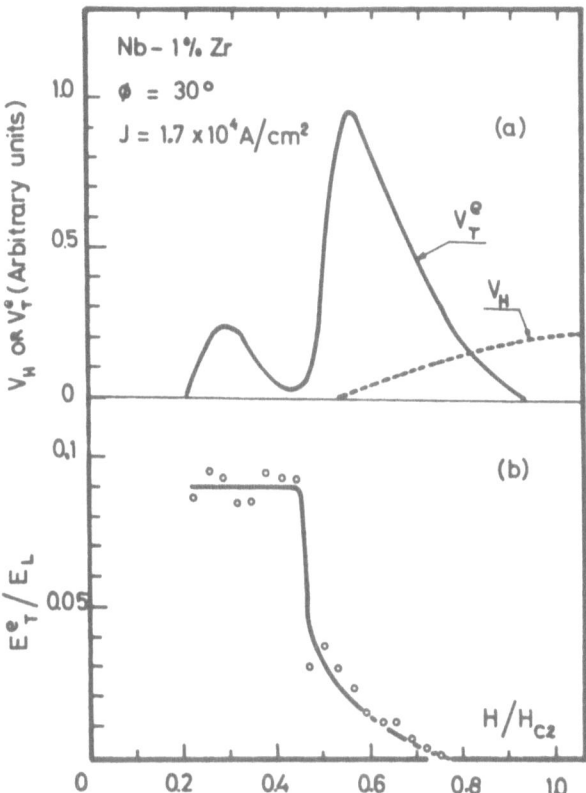

Fig. 3. (a) Transverse voltage V_T^e attributed to guided motion of vortices by defects aligned along the rolling direction compared with the Hall voltage V_H (on the same scale). The dip in V_T^e occurs simultaneously with a dip in V_L. (b) Ratio of transverse to longitudinal electric fields for the same sample.

of H for all fields below the minimum of the dip. For this unannealed specimen ε is small ($\varepsilon \approx 0.09$) compared with the value for ideal guided motion ($\varepsilon = 1$). But then at the minimum of the dip ε drops abruptly by a factor of two or more. Thus for $H/H_{c2} \gtrsim 0.4$ the defects have lost their ability to guide vortices, or the Lorentz force is sufficiently strong to push vortices out of the potential valleys. The occurence of the drop in ε and the onset of V_L for $\phi = 0°$ at approximately the same field suggests that the same type of defect is playing the dominant role in the two cases, although here again the fact that the onset of V_L is at a slightly higher field than the drop in ε indicates that the interaction between vortices and defects is stronger if the vortices are not moving.

A direct connection between the dip effect and aligned defects is suggested by the fact that the high-field side of the dip occurs when ε drops abruptly. On the other hand, the fact that at $H/H_{c2} \approx 0.3$ the dip begins with no essential change in ε suggests that the vortex–defect interaction does not substantially change during the lower-field portion of the dip, and thus by itself cannot cause the dip. An explanation consistent with these deductions is the one previously mentioned which is based upon the fact that vortices are closer together and interact more strongly at higher fields. The increased density may produce bottleneck effects for the movement of vortices or flux bundles between defects, so that the fraction n/n_0 decreases due to enhanced trapping at bottlenecks. At higher fields, where ε decreases, the vortices can move through as well as around defects. A higher defect concentration which opposes the motion of vortices perpendicular to the rolling direction thus accounts for the dip in θ_H for $\phi = 90°$. Although this explanation is consistent with the present results, it should be remembered that the present experiments do not define precisely what causes the low-field portion of the dip.

In conclusions, we wish to emphasize that these experiments have been performed on specimens having a high defect concentration, both in the bulk and at the surface, with a preferential alignment along the rolling direction. The field dependence of the Hall angle is qualitatively similar to that observed by Reed et al. [5] in pure Nb, except that the present Nb–1%Zr samples exhibit a pronounced dip at $H/H_{c2} \approx 0.4$ for $\phi = 90°$ and have a voltage-onset field which depends strongly upon ϕ. The field dependence of the Hall angle is qualitatively different from that in the well-annealed, Ta-rich alloys of Nb–Ta studied by Niessen and co-workers [6,8]. The efficiency of guided motion for 30° indicates that aligned defects play a role not only in the anisotropy of the voltage onset, but in the dip effect, although an additional factor not defined by the present experiments is responsible for initiating the dip at low fields.

ACKNOWLEDGMENTS

We wish to thank Mr. Berteaux for assistance with several of the experiments and also to express our appreciation to Compagnie Générale d'Electricité for providing the samples and performing the electron microscope studies. One of us (SJW) wishes to thank Professors J. Friedel and P. G. de Gennes and the Faculté des Sciences, Orsay, for their hospitality during his fellowship year.

REFERENCES

1. J. Baixeras and S. J. Williamson, *Solid State Commun.* **5**: 599 (1967).
2. T. G. Berlincourt, *Phys. Rev.* **114**: 969 (1959); M.A.R. Le Blanc and W. A. Little, in: *Proceedings of the 7th International Conference on Low Temperature Physics*, University of Toronto Press, Toronto, 1960, p. 362; T. G. Berlincourt and R. R. Hake, *Phys. Rev. Letters* **9**: 293 (1962); S. H. Autler, E. S. Rosenblum, and K. H. Gooen, *Phys. Rev. Letters* **9**: 489 (1962).
3. T. G. Berlincourt, R. R. Hake, and D. H. Leslie, *Phys. Rev. Letters* **6**: 671 (1961); C. S. Tedmon, Jr., R. M. Rose, and J. Wulff, *J. Appl. Phys.* **36**: 829 (1965).
4. G. J. van Gurp, *Philips Res. Rept.* **22**: 10 (1967).
5. W. A. Reed, E. Fawcett, and Y. B. Kim, *Phys. Rev. Letters* **14**: 790 (1965).
6. A. K. Niessen and F. A. Staas, *Phys. Letters* **15**: 26 (1965); F. A. Staas, A. K. Niessen, and W. F. Druyvesteyn, *Phys. Letters* **17**: 231 (1965).
7. A. G. van Vijfeijken and A. K. Niessen, *Philips Res. Rept.* **20**: 505 (1965).
8. A. K. Niessen, F. A. Staas, and C. H. Weijsenfeld, *Phys. Letters* **25A**: 33 (1967).
9. D. Kramer and C. G. Rhodes, *Trans. Met. Soc. AIME* **233**: 192 (1965).
10. F. A. Otter, Jr., and P. R. Solomon, *Phys. Rev. Letters* **16**: 681 (1966); J. Lowell, J. S. Munoz, and J. Sousa, *Phys. Letters* **24A**: 376 (1967).
11. F. A. Staas, A. K. Niessen, W. F. Druyvesteyn, and J. van Suchtelen, *Phys. Letters* **13**: 293 (1964).
12. B. D. Josephson, *Phys. Letters* **16**: 242 (1965).
13. R. R. Hake, D. H. Leslie, and C. G. Rhodes, in: *VIII International Conference on Low Temperature Physics*, Butterworth and Co., London, 1963, p. 342.
14. J. Bardeen and M. J. Stephen, *Phys. Rev.* **140**: A1197 (1965).
15. P. Nozières and W. F. Vinen, *Phil. Mag.* **14**: 667 (1966).
16. P. W. Anderson and Y. B. Kim, *Rev. Mod. Phys.* **36**: 39 (1964).
17. P. S. Swartz and H. R. Hart, Jr., *Phys. Rev.* **137**: A818 (1965).

Chapter 24

EXPERIMENTS ON THE CRITICAL BEHAVIOR OF TYPE-II SUPERCONDUCTORS IN HIGH MAGNETIC FIELDS

E. Saur

Institut für Angewandte Physik
Universität Giessen (West-Deutschland)

INTRODUCTION

For two different reasons type-II superconductors have attracted great interest during the last few years in experimental work as well as in theoretical considerations. At first there was mainly a practical reason, because all materials for superconducting coils are type-II superconductors. Thus work on the fabrication of these materials and the improvement of their properties was important and initiated many experiments. But then the fundamental scientific aspect took over. New theories appeared and many experimental results were found and checked with these theories. The main feature of a type-II superconductor is the existence of a mixed state of superconducting and normal material for each temperature T below the critical temperature T_c. The mixed state exists between the lower critical field $H_{c1(T)}$ and the upper critical field $H_{c2(T)}$. In the region between the two critical field values flux penetrates the type-II superconductor. For some groups of type-II superconductors the upper critical fields are very high, especially for compounds of the $\beta-W$ and NaCl-type structure. Experiments will be described and the results will be compared with recent theory for these materials.

The following books and articles may be recommended for an introduction into the field of type-II superconductors: de Gennes [1], Goodman [2], and Hake [3].

EXPERIMENTAL RESULTS ON THE CRITICAL BEHAVIOR OF TYPE-II SUPERCONDUCTORS

For the first time reliable measurements of critical field and quenching curves of some high-field type-II superconductors in transverse magnetic fields up to 230 kOe have been made, using Bitter-type solenoids of the MIT National Magnet Laboratory for the generation of the high fields. The critical field curves have been measured in a cryostat with a heating coil for intermediate temperatures [4] and the quenching curves at the temperature of boiling helium (4.2 °K). A proper mounting of the samples is very important for the measurements. The connections to the current leads have been made by ultrasonic soldering with indium [5].

Critical Field Curves

The results of measurements of the critical field curves for Nb_3Al, Nb_3Sn, V_3Si, V_3Ga, and NbN are given in Fig. 1. For the measurements on Nb_3Al, Nb_3Sn, and V_3Si samples with thin diffusion layers of these compounds have been used. The V_3Ga and NbN samples were both bulk

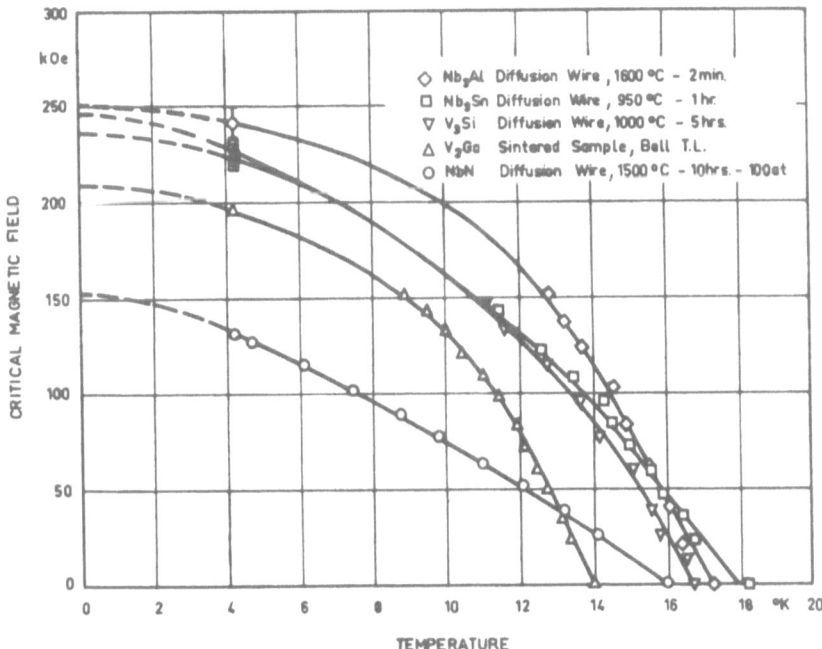

Fig. 1. Critical field curves of different high-field type-II superconductors.

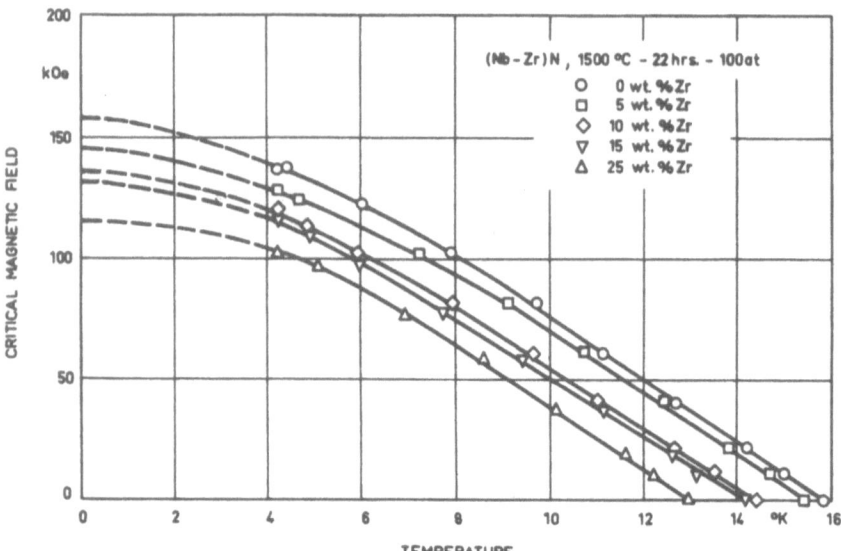

Fig. 2. Critical field curves of niobium nitride with different amounts of zirconium.

materials. The measuring points for 4.2 °K in the critical field curves of Nb$_3$Al, Nb$_3$Sn, V$_3$Si, and V$_3$Ga have been taken indirectly by the extrapolation of the quenching curves of these materials to very low currents. The results on V$_3$Ga and NbN have been published recently ([4,6,7]).

The values of the critical fields at zero temperature can be extrapolated from the experimental values measured at temperatures above 4.2 °K. These extrapolated values are: 250 kOe for Nb$_3$Al, 245 kOe for Nb$_3$Sn, 235 kOe for V$_3$Si, 208 kOe for V$_3$Ga, and 153 kOe for NbN. For the compounds of β-tungsten-type structure these results are in contradiction to earlier extrapolations of Kunzler et al. ([8]), especially concerning the order of these compounds with regard to the zero critical fields.

The results of critical field measurements on (Nb–Zr)N samples (Fig. 2) show a substantial decline of critical field values with increasing amounts of zirconium in the solid solution of the starting material.

Quenching Curves of Bulk Materials

Nb$_3$Sn

Quenching curves for different types of Nb$_3$Sn diffusion wires are represented in Fig. 3. The Nb$_3$Sn diffusion layers have been prepared with three different methods: cladding niobium wire with tin, dipping niobium wire into molten tin, with a post diffusion at high temperatures in both

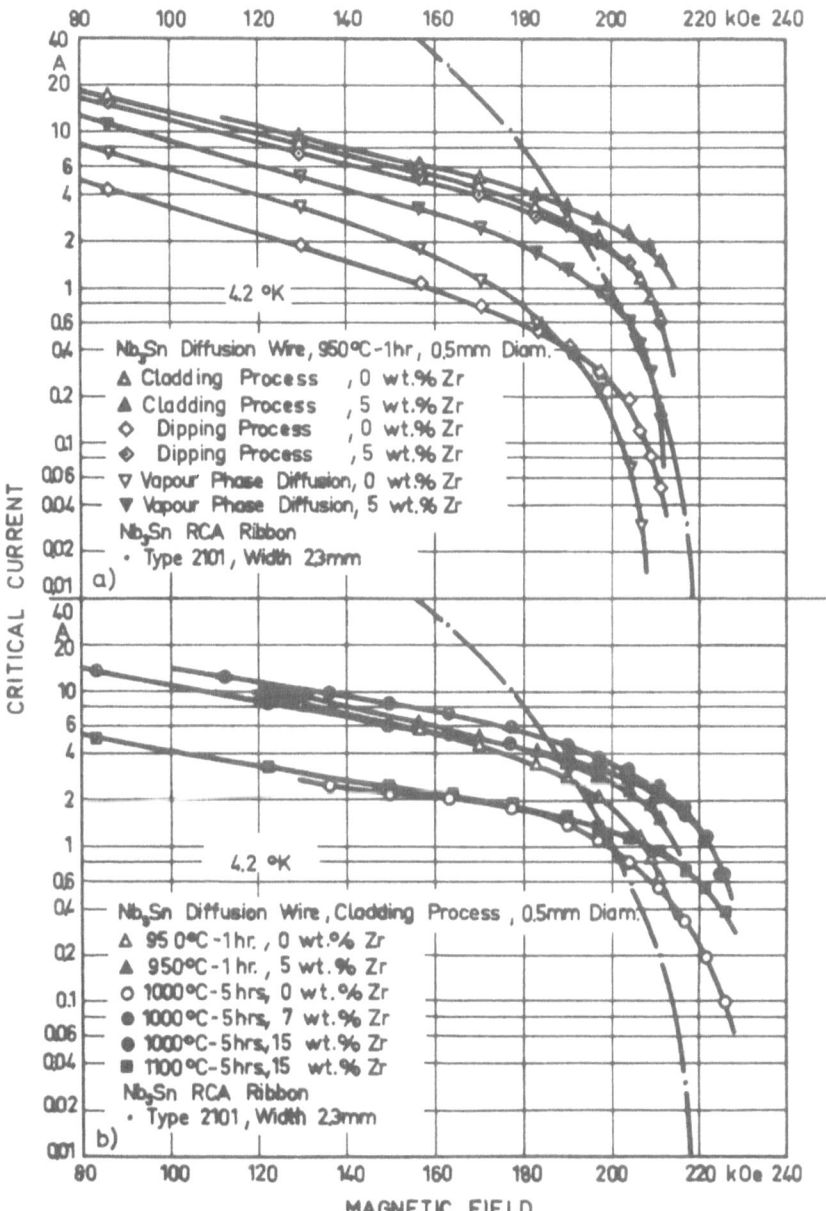

Fig. 3. Quenching curves of different Nb₃Sn samples in transverse magnetic fields.

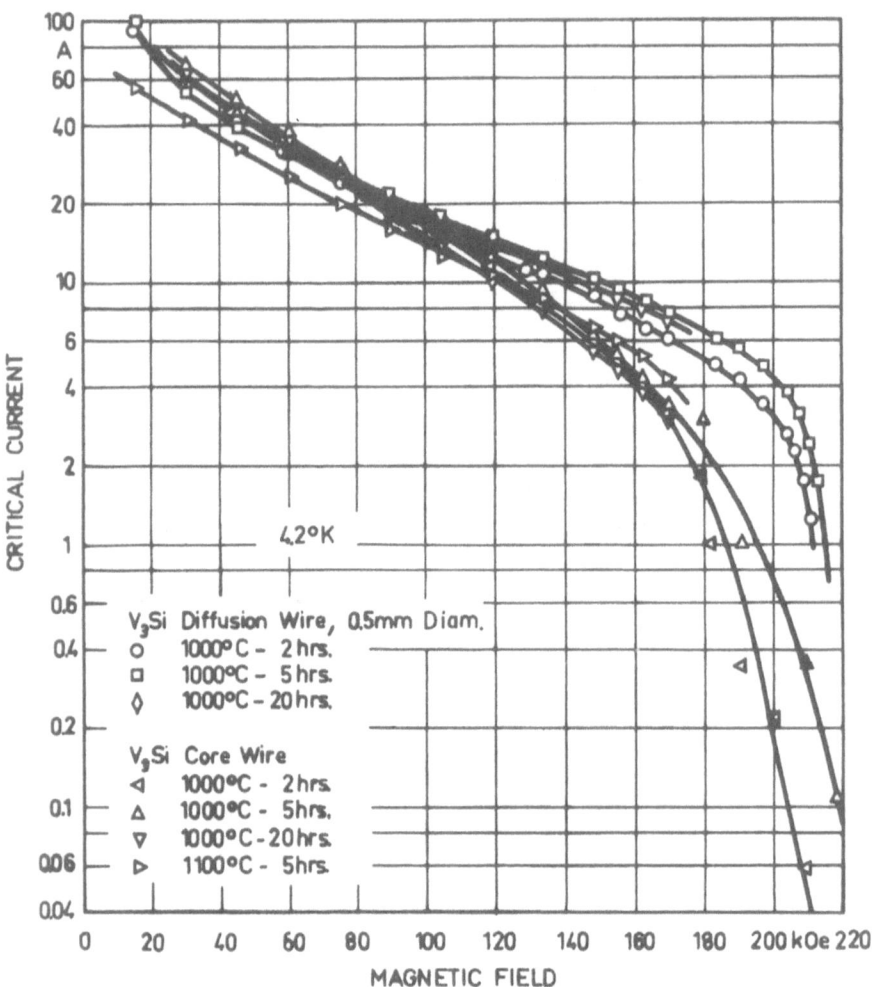

Fig. 4. Quenching curves of different V₃Si samples in transverse magnetic fields.

cases, and diffusion of tin into niobium from the vapor phase (⁹). Each
of these preparation methods results, with a small amount of zirconium
(up to 5 or 7 at.%), in diffusion layers with higher critical currents compared
to those of corresponding samples without zirconium. This is in accordance
with earlier observations of DeSorbo (¹⁰) in lower fields and on samples
with smaller amounts of zirconium, up to about 1 at.%. The quenching
curves for RCA ribbon show the behavior of this well-known commercial
material up to the highest fields.

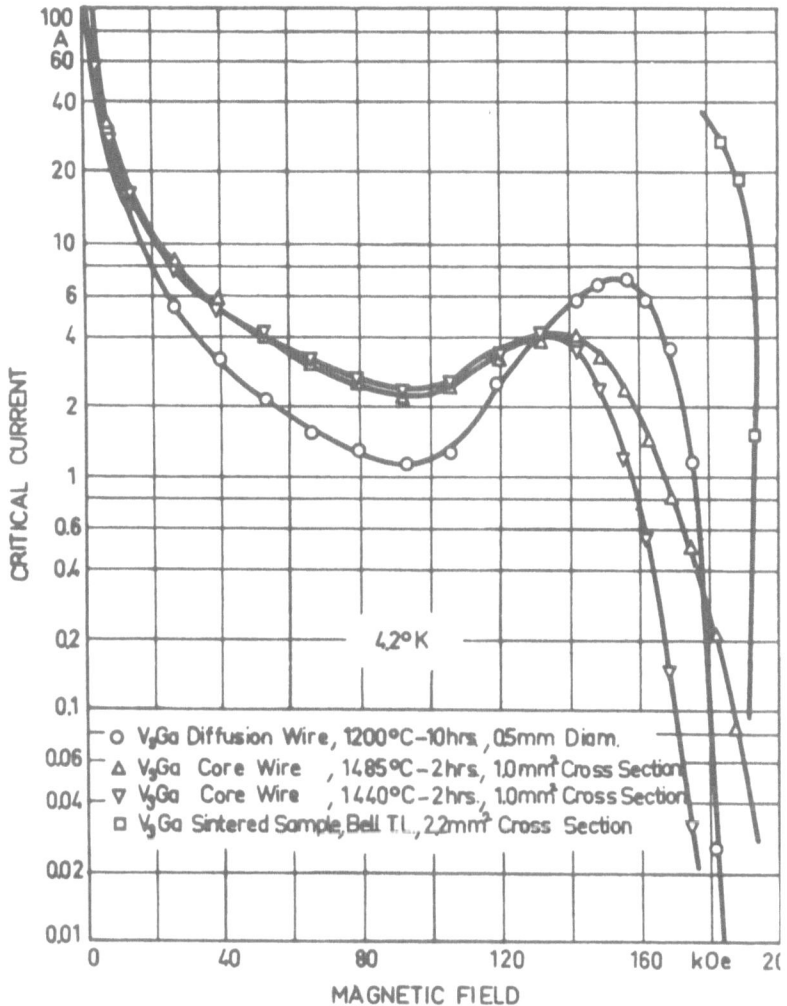

Fig. 5. Quenching curves of different V₃Ga samples in transverse magnetic fields.

V₃Si

The quenching curves of V_3Si have been investigated on two different types of samples: diffusion layers, prepared by vapor-phase diffusion of silicon into vanadium [11], and core wires of the Kunzler type [11]. The results are given in Fig. 4. In spite of the larger cross section of the active material in the core wire samples the critical currents are not higher than those of the corresponding diffusion layers, but are even lower.

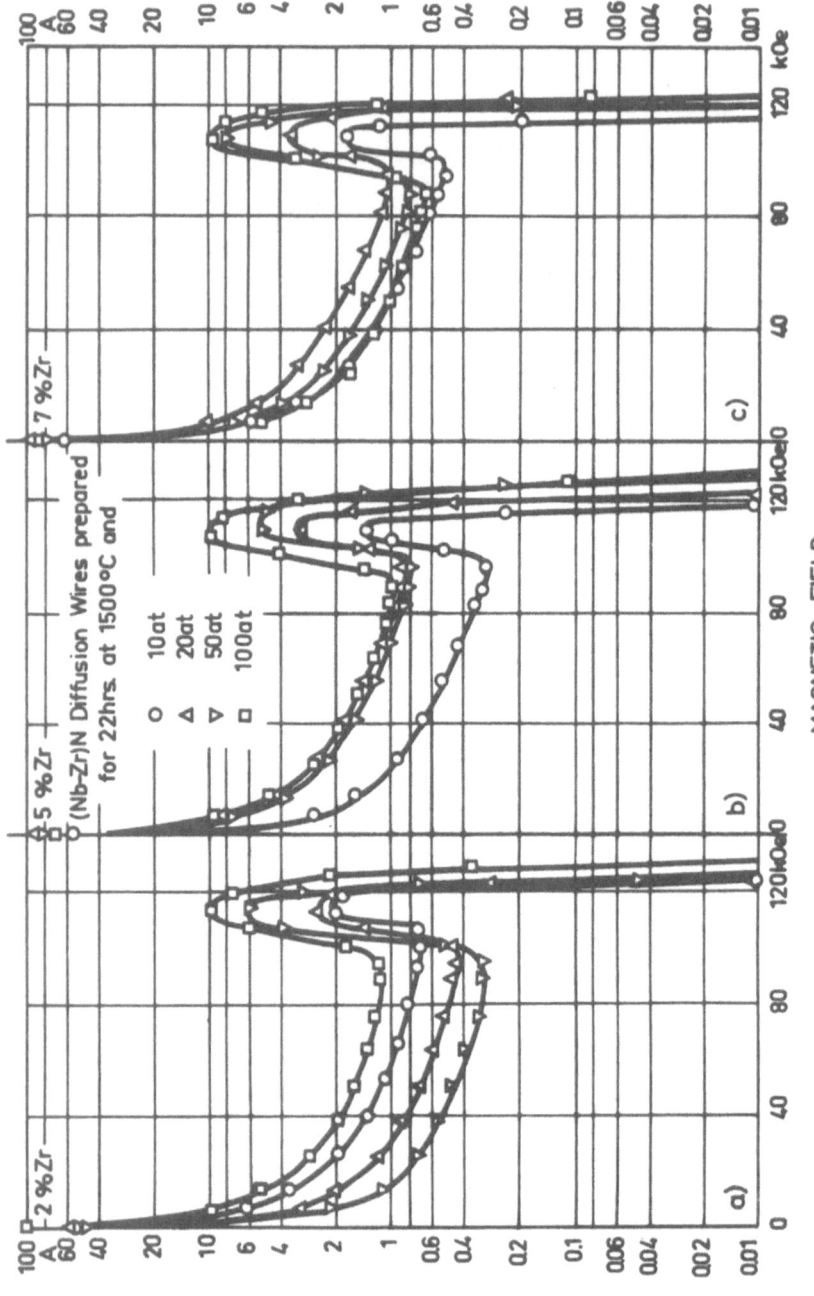

Fig. 6. Quenching curves of niobium nitride samples with different amounts of zirconium in transverse magnetic fields.

V_3Ga

Three different types of V_3Ga samples have been used: diffusion layers, core wires, and sintered samples [11]. The quenching curves of some samples (Fig. 5) have a pronounced peak effect in the high-field region with a steep slope in higher fields.

NbN

Diffusion wires of NbN are more or less homogeneous over the whole cross section. Results on the quenching behavior have recently been published [6,7]. Most of the quenching curves of NbN show very pronounced peak effects at about 115 kOe, with steep slopes on both sides. Quenching curves of NbN samples with small amounts of zirconium are shown in Fig. 6.

Quenching Curves of Materials with Artificial Structure

In earlier papers [12] we have described the preparation of "multiwires" with a special artificial inner structure based on Nb_3Sn by drawing down niobium tubes filled with some hundreds of tin-clad niobium wires before heating them for diffusion and by reaction of tin with niobium to form many very thin layers of Nb_3Sn within the tube. We found the highest critical currents in the high-field region for Nb_3Sn multiwire 522, which has 522 single wires drawn down to 0.7 mm in diameter. The quenching curves are given in Fig. 7a for Nb_3Sn multiwire 195 and in Fig. 7b for Nb_3Sn multiwire 522, compared again with the curve for RCA ribbon as a standard. The slope of the quenching curves of Nb_3Sn multiwire 522 for different conditions of heat treatment is very flat in the high-field region. The critical fields of Nb_3Sn multiwire 522 are much higher than those of the different samples of bulk Nb_3Sn. One possible explanation for this effect would be the creation of many internal Nb_3Sn layers with a very small thickness on the surface of the different unchanged niobium wires, so that surface superconductivity [1] with a much higher critical field may occur over the whole cross section. The creation of such an artificial inner structure is probably the only way to raise the critical field beyond the limit which is set for the bulk high-field type-II superconductors.

COMPARISON OF EXPERIMENTAL RESULTS WITH RECENT THEORIES OF TYPE-II SUPERCONDUCTORS

For an analysis of the critical field curves the results on critical fields are shown in Fig. 8a as a function of reduced temperature $t = T/T_c$; thus

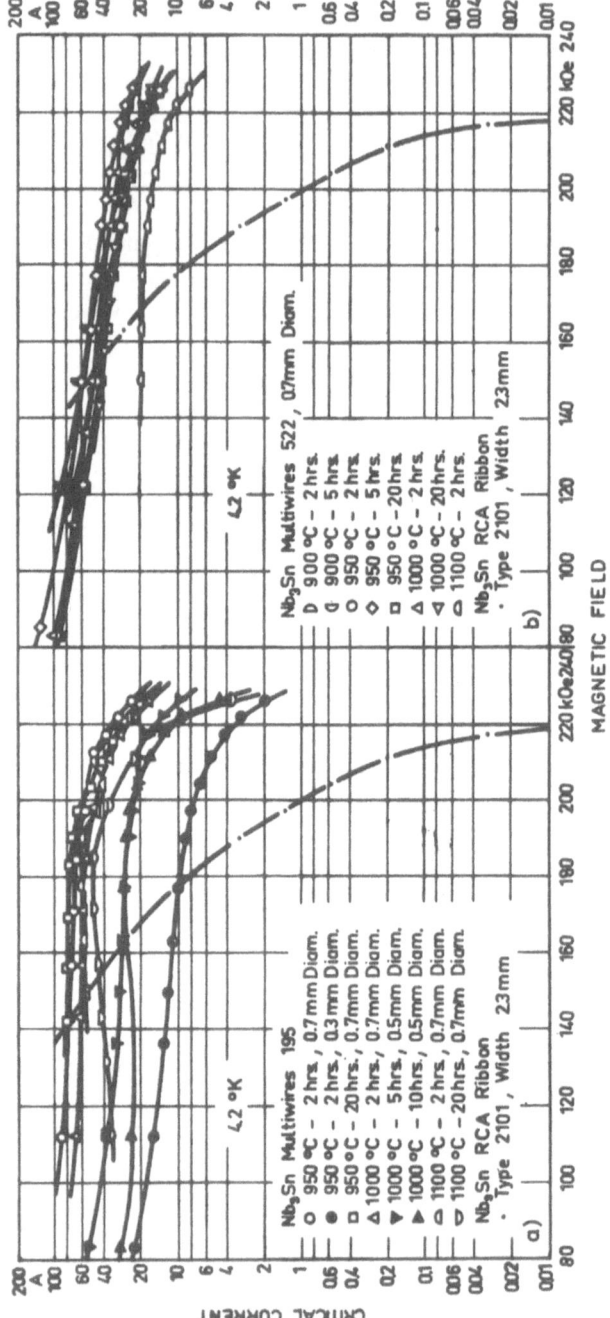

Fig. 7. Quenching curves of different Nb₃Sn multiwire samples in transverse magnetic fields.

these critical field curves start for all materials at $t = 1.0$ with different slopes. In Fig. 8b the critical field values are plotted as a function of the square of this reduced temperature. This representation shows that none of the type-II superconductors investigated has a parabolic critical field curve. The critical field curves of Nb_3Sn and V_3Si are close to parabolic curves. The critical field values of Nb_3Al and V_3Ga are above and those for NbN are below the corresponding parabolic curves represented by straight lines in Fig. 8b.

In Table I the values for the following critical fields and related quantities are compiled for the five compounds Nb_3Al, Nb_3Sn, V_3Si, V_3Ga, and NbN:

    1. The critical temperature T_c.

    2. The experimental values of the upper critical field $H_{c2(4.2)}$ at 4.2 °K.

    3. The values of the upper critical field $H_{c2(0)}$ extrapolated to zero temperature from the experimental values for higher temperatures.

    4. The slope $-(dH_{c2(T)}/dT)_{T=T_c}$ of the critical field curves at $T = T_c$.

    5. The linear extrapolated upper critical field $H_{01} = T_c(dH_{c2(T)}/dT)_{T=T_c}$ for zero temperature.

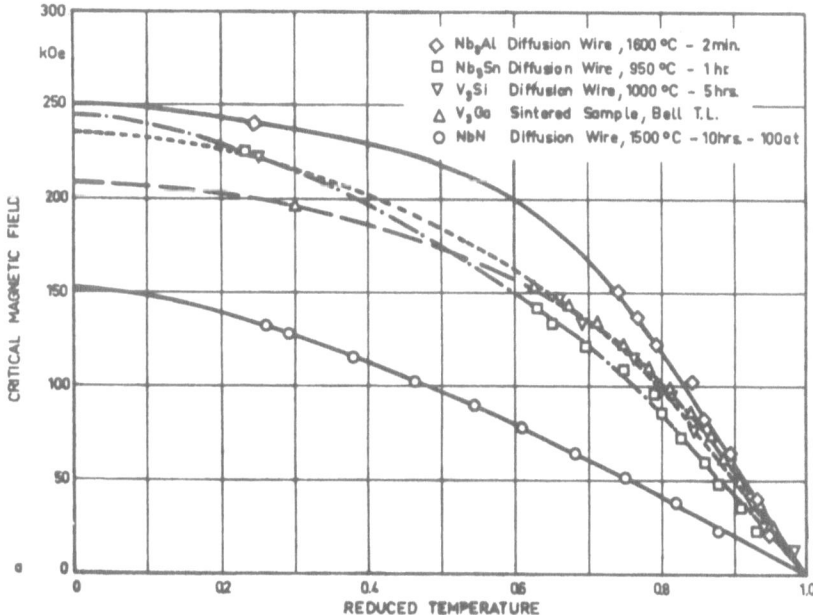

Fig. 8a. Critical field curves of different high-field type-II superconductors as a function of reduced temperature.

TABLE I

Comparison of Experimental and Theoretical Values of Critical Fields

	Nb$_3$Al	Nb$_3$Sn	V$_3$Si	V$_3$Ga	NbN
1. Critical temperature T_c (°K)	17.2	18.1	16.9	14.1	16.1
2. Upper critical field $H_{c2(4.2)}$ at 4.2 °K (kOe)	240	225	220	196	132
3. Upper critical field $H_{c2(0)}$, extr. to 0 °K (kOe)	250	245	235	208	153
4. Slope of critical field curve at $T = T_c$: $-(dH_{c2(T)}/dT)_{T=T_c}$ (kOe/°K)	39	22.5	29	39	13.7
5. Linear extrapolated zero upper critical field: $H_{01} = T_c(dH_{c2(T)}/dT)_{T=T_c}$ (kOe)	670	406	490	550	220
6. Parabolic extrapolated zero upper critical field H_{02} (kOe)	335	203	245	275	110
7. Zero upper critical field $H_{c2(0)}^*$ (kOe)	427	280	338	380	152
8. Clogston zero upper critical field $H_{p(0)}$ (kOe)	320	337	314	262	300
9. Maki parameter α	2.08	1.18	1.52	2.06	0.72
10. Maki zero upper critical field $H_{c2(0)}^{**}$ (kOe)	200	181	186	166	124

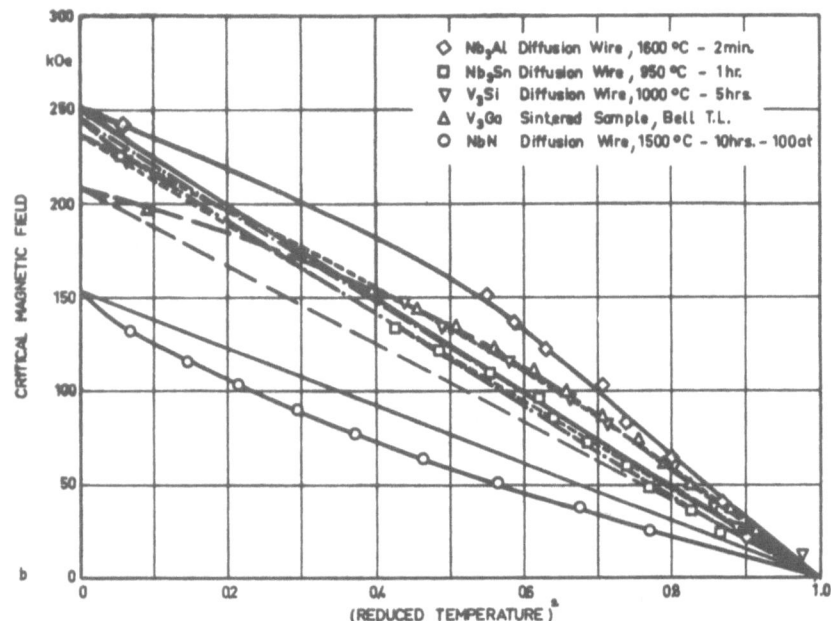

Fig. 8b. Critical field curves as a function of reduced temperature squared.

6. The parabolic extrapolated upper critical field H_{02} for zero temperature.

7. The upper critical field $H^*_{c2(0)1} = 0.69H_{01}$ at zero temperature according to the GLAG theory ([1]).

8. The upper critical field $H_{p(0)} = 18.6T_c$ at zero temperature according to Clogston ([13]).

9. The Maki parameter α ([14]) for influence of spin paramagnetism.

10. The values of the upper critical field $H^{**}_{c2(0)} = H^*_{c2(0)}/(1 + \alpha^2)^{1/2}$ according to Maki ([14]).

For NbN the GLAG value $H^*_{c2(0)}$ of the zero upper critical field is in good agreement with the experimental extrapolated value $H_{c2(0)}$, indicating that there is no suppression of the field by the influence of spin paramagnetism. For the β–W type compounds there is such an influence, and the extrapolated experimental values $H_{c2(0)}$ are between the values $H^*_{c2(0)}$ as an upper limit and the values $H^{**}_{c2(0)}$ as the lower limit.

ACKNOWLEDGMENTS

I would like to thank Professor Dr. B. Lax for the generous and excellent opportunity wich he gave Dr. H. Wizgall and K. Hechler to perform these measurements in his laboratory, and Dr. D. B. Montgomery and L. Rubin for kind advice and continuous assistence during the whole experimental procedure. The financial support by the Deutsche Forschungsgemeinschaft, the Stiftung Volkswagenwerk, and the Fraunhofer-Gesellschaft is also gratefully acknowledged.

REFERENCES

1. P. G. de Gennes, *Superconductivity of Metals and Alloys*, W. A. Benjamin, New York, 1966.
2. B. B. Goodman, *Rept. Progr. Phys.* **XXIX**: 445 (1966).
3. R. R. Hake, *Appl. Phys. Letters* **10**: 189 (1967); *Phys. Rev.* **158**: 356 (1967).
4. D. B. Montgomery and H. Wizgall, *Phys. Letters* **22**: 48 (1966).
5. J. Babiskin and P. G. Siebenmann, *Rep. NRL Progr.* April 1962, p. 1.
6. K. Hechler, E. Saur and H. Wizgall, *Z. Physik* **205**: 400 (1967).
7. E. Maxwell, B. B. Schwarz and H. Wizgall, *Phys. Letters* **25A**: 139 (1964); also, in: *Proceedings of the Xth International Conference on Low Temperature Physics*, Moscow, 1966.
8. J. E. Kunzler, J. H. Wernick, F. J. Morin, F. S. Hsu, D. Dorsi, and J. P. Maita, in: *High Magnetic Fields*, John Wiley and Sons, New York, 1962, p. 609.

9. E. Saur and J. Wurm, Naturwiss. **49**: 127 (1962); also, in: *High Magnetic Fields*, John Wiley and Sons, New York, 1962, p. 589.

10. W. DeSorbo, *Cryogenics* **4**: 218 (1964).

11. D. Koch, G. Otto, and E. Saur, *Phys. Letters* **4**: 292 (1963).

12. D. Koch, H. Speidel, G. Otto, and E. Saur, *Z. Physik* **180**: 476 (1964); J. Babiskin, P. G. Siebenmann, G. Otto, and E. Saur, *Z. Physik* **180**: 483 (1964).

13. A. M. Clogston, *Phys. Rev. Letters* **9**: 266 (1962).

14. K. Maki, *Physics* **1**: 127 (1964).

Chapter 25

MEGAOERSTED FIELDS AND THEIR RELATION TO THE PHYSICS OF HIGH ENERGY DENSITY

H. Knoepfel

Laboratori Gas Ionizzati
(Euratom–CNEN)
Frascati, Italy

INTRODUCTION

The interaction of very high magnetic fields with metallic conductors is a very interesting but complicated physical process which has only received certain attention in recent years. Here we are interested in this problem mainly because of its energetical aspects, i.e., for its relation to a relatively new research field which we define as the "Physics of High Energy Density." By "high energy density" we mean densities in excess of some 10 kJ/cm³, i.e., larger than the chemical binding energies of most solids.

While such physical conditions can be reached by other experimental means, mostly developed in the last 20 years (Table I), with megaoersted fields generated by explosive driven flux compression (Fig. 1) ([1,7,8]) the highest densities have been reached [in excess of 1 MJ/cm³, corresponding to fields of the order of 20 MOe ([9])].

Transient megaoersted fields can contribute in three different ways to increasing the energy of condensed matter: (1) by ohmic heating; (2) by accelerating a conducting particle to a high kinetic energy; and (3) by compressing a solid and thereby increasing its internal energy.

The last possibility has been discussed elsewhere ([10]) and we only point out here that it is particularly of interest in the very high pressure range above 10 Mbar where the explosive-driven flying plate method is not applicable ([11]).

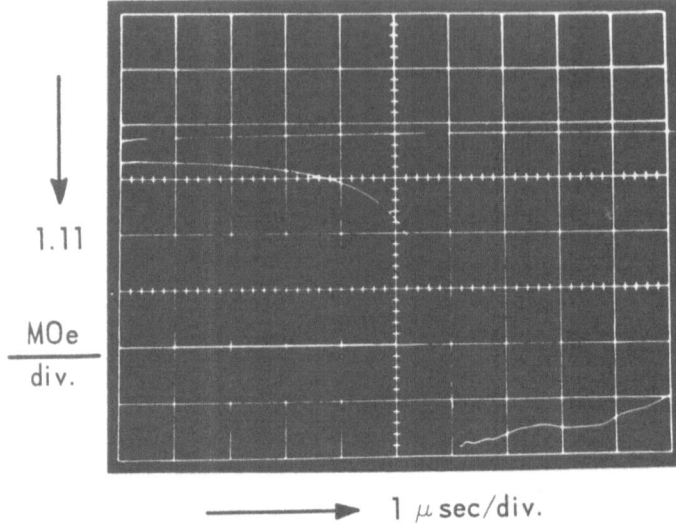

1.11

$\dfrac{\text{MOe}}{\text{div.}}$

1 μ sec/div.

Fig. 1. The magnetic field as recorded during the last phases of a typical magnetic flux compression experiment. In this case a final magnetic field of 6.2 MOe has been generated in a volume of about 0.5-cm diameter and 1.5-cm length.

THE DIFFUSION OF MAGNETIC FIELDS INTO A METALLIC CONDUCTOR WITH CONSTANT CONDUCTIVITY

For simplicity, we will limit our analysis, to the diffusion of a transient magnetic field into a conducting half space as shown in Fig. 2. Maxwell's first two equations in Gaussian units then reduce to

$$\partial H_z/\partial x = -(4\pi/c)j_y \tag{1}$$

$$\partial E_y/\partial x = -(1/c)\,\partial H_z/\partial t \tag{2}$$

since one can neglect displacement currents and at high magnetic fields the magnetic permeability can be taken $\mu = 1$. With the help of Ohm's law

$$j_y = \sigma E_y \tag{3}$$

and assuming $\sigma = \sigma_0 = $ const, we deduce the diffusion equation

$$\frac{\partial^2 H_z}{\partial x^2} - \frac{1}{\varkappa_0}\,\frac{\partial H_z}{\partial t} = 0 \tag{4}$$

TABLE I

Terrestrial Energy Sources for the Generation of High Energy Density

Primary source	Final form	Energy density (MJ/cm³)	Energy per atom involved (eV/element)	Total involved energy (MJ)	Ref.
Chemical explosives	—	8.10^{-3}	1 (H, N, O)	100	
	Metallic jets	1	1700 (Cu)	10^{-3}	[12]
	Metallic plates	0.8	60 (Fe)	3	[11]
	1 MOe	4.10^{-3}	0.3 (Cu)	5	[1]
	5 MOe	0.1	8 (Cu)	1	[1]
	25 MOe	2.5	200 (Cu)	1	[9]
Capacitor banks	—	10^{-8}	—	5	—
	Exploding wires	0.05	4 (Cu)	10^{-3}	[13]
	Plasma focus	0.01	1000 (D)	10^{-4}	[14]
Optical storage	—	10^{-6}	—	10^{-3}	[15]
	Focused laser beam	0.5	100 (D, T)	10^{-4}	[15]
Nuclear explosive (D, T)	—	10^{4}	9.10^{6} (D, T)	∞	—

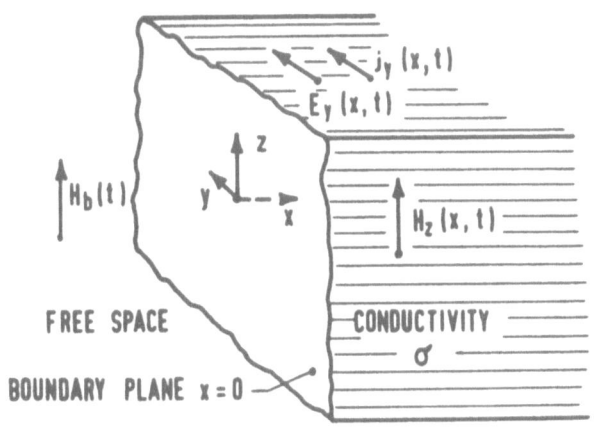

Fig. 2. The electrically conducting half-space to which a boundary field $H_b(t)$ is applied.

where the magnetic diffusivity is given by

$$\varkappa_0 = c^2/4\pi\sigma_0 \tag{5}$$

Equation (4) is well known in physics since it also describes, e.g., heat conduction and ordinary diffusion.

The solution of Eq. (4) dependent on the family of transient boundary fields

$$H_b = H_z(0, t; n) = H_0(t/t_0)^{n/2} \quad \text{for} \quad 0 \le t < \infty \tag{6}$$

with $n = 0, 1, 2, \ldots$, is known to be ([16])

$$H_z(x, t; n) = H_b 2^n \Gamma\left(\frac{n}{2} + 1\right) \{i^n \text{ erfc } \xi\}, \qquad \xi = \frac{x}{2\sqrt{\varkappa_0 t}} \tag{7}$$

where Γ is the gamma function and the expression in brackets is related to the error function, erf $\xi = 1 - \text{erfc } \xi$, through the repeated integration

$$\{i^n \text{ erfc } \xi\} = \int_\xi^\infty \{i^{n-1} \text{ erfc } \lambda\} \, d\lambda$$

From the general solution (7) different quantities which are characteristic for magnetic field diffusion can be calculated. It is very useful to define a magnetic flux skin depth

$$s_\varphi = \frac{1}{H_b} \int_0^\infty H_z \, dx \tag{8}$$

which in this case is

$$s_{\varphi.n} = \frac{\Gamma(\tfrac{1}{2}n + 1)}{\Gamma(\tfrac{1}{2}n + \tfrac{3}{2})} \sqrt{\varkappa_0 t} \tag{9}$$

The heating effects are calculated from Joule's law

$$\partial e/\partial t = j_y^2/\sigma \tag{10}$$

where for the internal energy e we take simply

$$e = c_v \theta \tag{11}$$

Assuming the specific heat c_v to be constant, one finds for the surface temperature

$$c_v \theta(0, t; n) = \frac{H_b^2}{8\pi} \frac{2}{n} \left[\frac{\Gamma(\tfrac{1}{2}n + 1)}{\Gamma(\tfrac{1}{2}n + \tfrac{1}{2})}\right]^2 \tag{12}$$

The case $n = 0$, corresponding to a step-function boundary condition, has to be treated separately, since the integration of Eq. (10) results in a logaritmic divergence. Taking heat conduction into account (which in all other cases plays no role), Kidder [25] obtained for the case of a copper conductor

$$c_v \theta(0, t \to \infty) \approx 3H_b^2/8\pi \qquad (13)$$

The total energy flow into the conductor per unity of surface is given by the Poynting vector

$$R_x = (c/4\pi)E_y H_z \qquad (14)$$

whose time integration gives for our solution (7) the simple result

$$W_T = \int_0^t P_x dt = \frac{H_b^2}{8\pi} s_{\varphi,n} \frac{n+1}{n+\frac{1}{2}} \qquad (15)$$

where the skin depth $s_{\varphi,n}$ was defined in Eq. (9).

In Fig. 3 we have plotted the functions defined in Eqs. (9), (12), and (15) versus the pulse form parameter n; we may conclude that (particularly for $n > 1$) the results are relatively insensitive to the form of the applied boundary field. As far as the temperature is concerned, we see that the surface of a copper conductor starts melting when the boundary field attains a value of between 0.5 and 0.8 MOe, depending on the pulse form applied.

THE INTERACTION OF MOe FIELDS WITH A METALLIC CONDUCTOR

As a consequence of Joule heating the temperature dependence of the conductivity cannot be neglected when considering high magnetic fields. It has been shown [17] that the simple expression

$$\sigma = \sigma_0/(1 + \beta\theta) \qquad (16)$$

with a conveniently chosen temperature coefficient, suffices to describe diffusion processes up to at least 5 MOe (see Fig. 7). The conductivity law at higher magnetic fields is uncertain. Some authors even suggest the possibility that at very high current densities one might attain a sort of superconducting state [18]; in any case, at high enough temperatures, when a metallic plasma is formed, the conductivity will gradually approach the classical $\theta^{3/2}$ law [6].

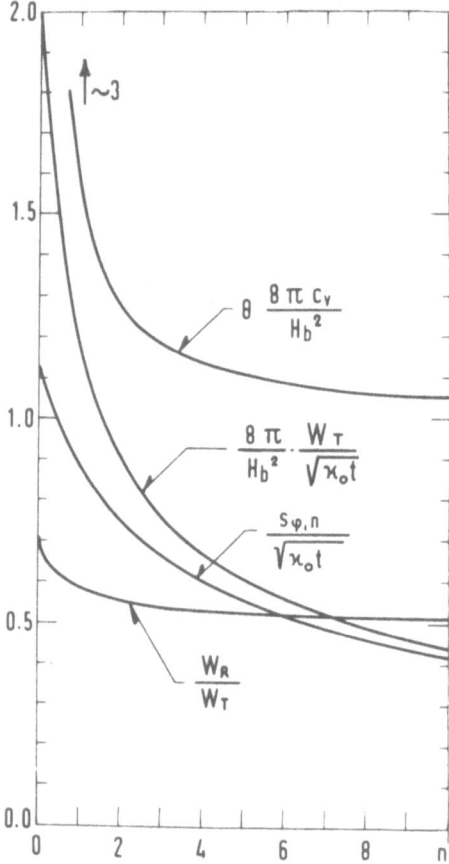

Fig. 3. Various dimensionless quantities corresponding to the boundary field $H_b = H_0 (t/t_0)^{n/2}$. From top to bottom: the surface temperature, the total absorbed energy, the flux skin depth, and the ratio between Joule heat and total absorbed energy.

The magnetic diffusion problem for an incompressible conductor is now described by the nonlinear system of Eqs. (1)–(3), (10), and (16), which in general can only be solved numerically. If we assume that the electromagnetic disturbance penetrates into the conductor at a constant velocity u and therefore is a function of $(ut - x)$ alone, one finds a physically acceptable solution ([4]). With this assumption we obtain from Eq. (2)

$$E_y = (u/c)H_z \tag{17}$$

and since Eq. (10) can be written with the help of Eqs. (1) and (3) in the form

$$c_v \frac{\partial \theta}{\partial t} = \frac{1}{4\pi} H_z \frac{\partial H_z}{\partial t}$$

we find by integration (and assuming as always $c_v = $ const) the interesting result

$$c_v \theta(x, t) = \gamma H_z^2(x, t)/8\pi \tag{18}$$

where here $\gamma = 1$. This means that for this particular solution the Joule energy anywhere in the conductor is equal to the magnetic field density, independently of the parameters σ_0 and β [Eq. (16)].

The conductivity (16) can now be rewritten in the form

$$\sigma = \frac{\sigma_0}{1 + (H_z^2/h_c^2)} \tag{19}$$

where we define a critical magnetic field

$$h_c = (8\pi c_v/\beta)^{1/2} \tag{20}$$

characterizing a field region above which the nonlinear magnetic diffusion process becomes predominant.

For $H_z \gg h_c$ our steady wave solution takes the simple form [10]

$$H_z(x, t) \approx h_c \left[\frac{t}{t_0} - x \left(\frac{2}{\varkappa_0 t_0} \right)^{1/2} \right]^{1/2} \tag{21}$$

where t_0 is a time parameter. For the flux skin depth (8) we now find

$$s_\varphi = \frac{\sqrt{2}}{3} \left(\frac{\varkappa_0}{t_0} \right)^{1/2} t \tag{22}$$

and for the total absorbed energy (15) one calculates easily

$$W_T = \frac{H_z^2(0, t)}{8\pi} \left(\frac{\varkappa_0}{2 t_0} \right)^{1/2} t \tag{23}$$

Note from Eq. (21) that the boundary field $H_z(0, t)$ is equal to that given in Eq. (6) with $n = 1$; one can therefore compare the results (22) and (23) [depending on the conductivity law (16)] with the corresponding constant conductivity results, Eqs. (9) and (15).

The equipartition law (18) has been confirmed by numerical integration [2-4] for quite different boundary fields up to 15 MOe and more, with

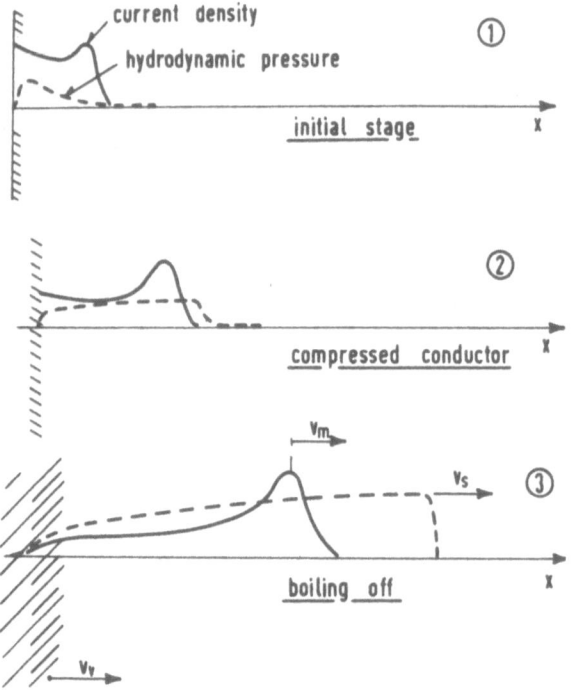

Fig. 4. The schematic interaction of a megaoersted field with
a metallic conductor at three successive times. (1) The boundary field may have reached 2 MOe: the diffusion is nonlinear,
with the typical current density peak in front; the surface
layers are melted, but the conductor is uncompressed. (2) The
surface, receded through compression, just starts boiling off
($H_b \approx 3$ MOe). (3) ($H_b \gtrsim 10$ MOe) the interaction process
has clearly developed into three waves: in front the pressure
pulse has built up a shock wave followed by a current "piston,"
while the surface is boiled off by a penetrating vaporization
wave.

γ varying between 0.5 and 1.5. With the measured magnetic fields in the
range from 6 to 25 MOe, the thermal energy density level of 1 MJ/cm³
has therefore been reached.

As was mentioned above and is sketched in Fig. 4, the interaction process
at those very high fields must take into account the compressibility and the
phase changes of the metal. In this magnetohydrodynamic problem one
must also formulate the boundary conditions on the free surface, where a
"vaporization wave" creeps into the metal. According to (^{19}) the velocity V_v
of this wave in a copper conductor and in the temperature range 5000–

10,000 °K (corresponding to about 3 to 4 MOe) increases from 0.7 to 3.10^4 cm/sec. At the upper end of this magnetic field range the shock speed V_s is typically 50×10^4 cm/sec, whereas the magnetic flux penetrates with a speed V_m of the order of 5×10^4 cm/sec ([10]).

THE EXPLODING METAL TUBE

Another arrangement in which very high energy densities can be generated is represented by a hollow cylindrical conductor to whose outer surface a megaoersted field is applied (Fig. 5).

Assuming that the wall thickness is small compared to the penetration depth (i.e., the current density is constant over d), one can easily establish the differential equation

$$H_e - H_i = \tau \, dH_i/dt \tag{24}$$

where

$$\tau = (2\pi/c^2)r_0 d\sigma = \tau_0/(1 + \beta\theta) \tag{25}$$

and the other quantities are defined in Fig. 5. Since

$$j_\vartheta d = (c/4\pi)(H_i - H_e) \tag{26}$$

the temperature Eq. (10) can easily be integrated, giving

$$1 + \beta\theta = \exp \frac{2}{\tau_0 H_c^2} \int_0^t (H_e - H_i)^2 \, dt \tag{27}$$

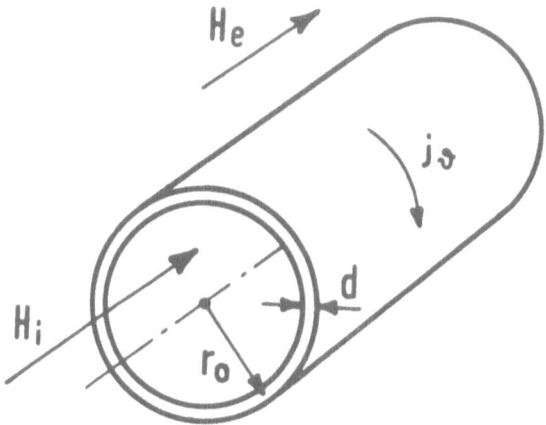

Fig. 5. The thin-walled cylindrical conductor.

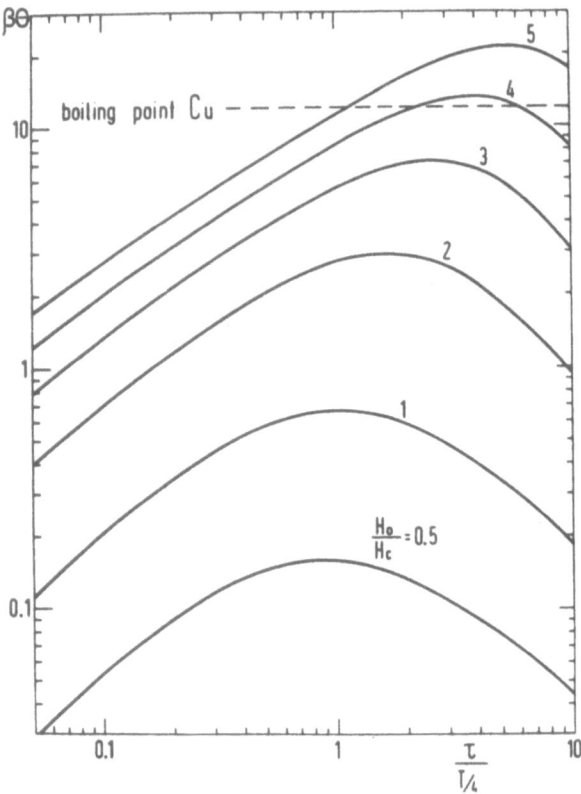

Fig. 6. The dimensionless temperature $\beta\theta$ plotted as a function of the dimensionless time constant for the case of an
external field $H_e = H_0 \sin(2\pi t/T)$.

where the critical field H_c is related to that given in Eq. (20) by

$$H_c = h_c[2(d/r_0)]^{1/2} \qquad (28)$$

As an example, we have solved numerically the system of Eqs. (24), (25),
and (27) for the external field

$$H_e = H_0 \sin(2\pi t/T) \qquad (29)$$

and plotted in Fig. 6 the dimensionless temperature $\beta\theta$ at the maximum
of the inner field as a function of the variable $\tau_0/\tfrac{1}{4}T$ and the parameter H_0/H_c.

In comparison with the problem treated in the last section, the temperature depends not only on the material constant h_c, but increases with the
geometrical parameter $(r_0/d)^{1/2}$. However, it is not possible to exceed the

boiling point appreciably, since the thin conductor then simply explodes. Because of the effect this experiment is intimately connected with the field of "exploding wires" (see Table I).

As an example of an application, we note that inversely it is possible to determine the temperature dependence of the conductivity if H_e, H_i, and dH_i/dt are measured simultaneously (Fig. 7).

THE ACCELERATION OF METALLIC CONDUCTORS THROUGH MEGAOERSTED FIELDS

The possibility of accelerating conducting particles with the dimensions of millimeters ("macroparticles") to velocities in excess of 1 cm/μsec has been studied by different authors [22,23,26]. We will consider this process here for its importance as an energy concentration effect, although little has been done up to now experimentally [24].

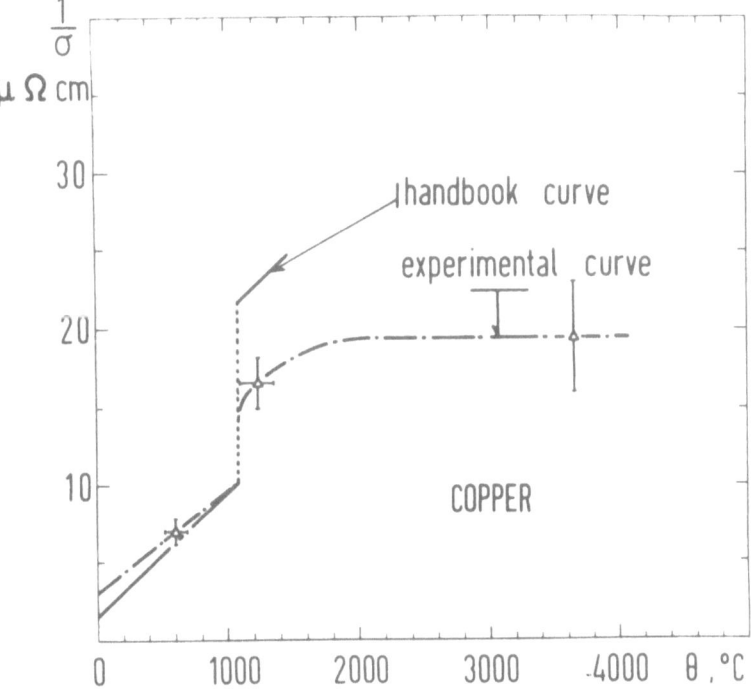

Fig. 7. The "transient" electrical resistivity of copper: the experimental curve has been determined in a conductor as shown in Fig. 5. The full line reproduces the usual resistivity law as determined in static conditions. From Knoepfel and Luppi [17].

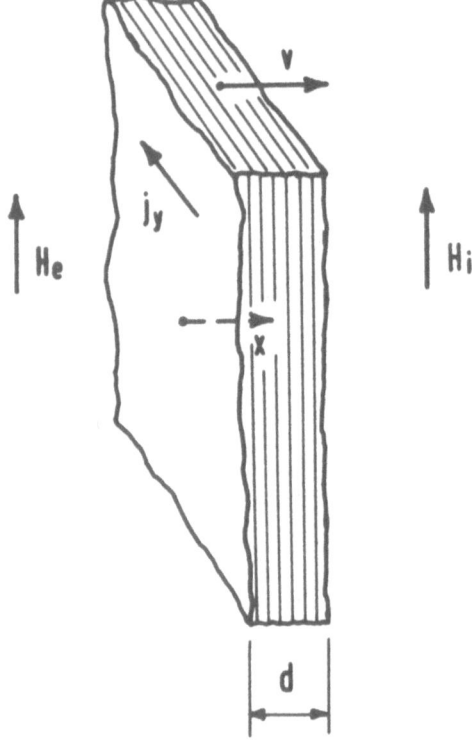

Fig. 8. The thin-sheet conductor.

The classical device for accelerating such a "macroparticle" is a gun. In this arrangement, however, the final velocity of the projectile is limited by the thermal speed of the combustion products to about 0.1 cm/μsec; the maximum velocity can be increased by an order of magnitude if hydrogen is used as a driving gas. No such limitation exists in this velocity range if magnetic fields are used as the driving medium. Since the acceleration time is limited, as we will see, by Joule heating effects, one is tempted to use megaoersted fields to do the work.

To start with, let us consider, as shown in Fig. 8, a conducting sheet which is accelerated by a magnetic field H_e (we suppose for simplicity $H_i \equiv 0$). To attain very high velocities, d will be reduced as much as possible, so that we can limit our analysis, as in the preceding section, to the study of a "thin" conductor [21].

By integrating the equation of motion

$$d\varrho\ddot{x} = H_e^2/8\pi \qquad (30)$$

and using Eq. (26), we find

$$v/d = (2\pi/\varrho)I \tag{31}$$

where I is the usual current integral (32) and ϱ is the specific gravity. From Joule's law, Eq. (10), assuming that the conductivity σ depends only on the internal energy e, we find through integration

$$I = \int_0^t (j/c)^2 \, dt = (1/c^2) \int_{e_0}^{e_t} \sigma \, de \tag{32}$$

This means that the current integral is independent of the current pulse and is related only to the electrical property of a particular conductor material. Taking the usual conductivity law (16), one finds, for example,

$$I = \frac{h_c^2}{8\pi} \frac{\sigma_0}{c^2} \ln(1 + \beta\theta) \tag{33}$$

The maximum velocity v_m/d which can be attained by a compact thin sheet is therefore obtained by introducing into Eq. (31) the value of I corresponding to the boiling point; this value has been deduced from exploding wire experiments ([20]), and the result is given in Table II. For a copper sheet of 0.1-cm thickness a kinetic energy density of 340 kJ/cm³ corresponds to this maximum velocity.

These considerations have a practical importance in experiments where a thin-walled metallic cylinder is imploded by an azimuthal field for the purpose of converting electrostatic energy stored in capacitor banks into kinetic energy ([20,27]).

The acceleration of discrete particles is, however, more important from the point of view of energy concentration. Consider, for simplicity, the system drawn in Fig. 9, containing a cylindrical, perfectly conducting projectile.

TABLE II

The Interaction of Megaoersted Fields with Metallic Conductors

Metal	h_c (kOe)	v_{max}/d (μsec^{-1})	$\varkappa_0$ (cm² sec^{-1})	β (°C)$^{-1}$
Cu	410	8.8	126	5.2×10^{-3}
Al	330	13.7	208	5.6×10^{-3}
Fe	290	2	680	$\sim 10 \times 10^{-3}$

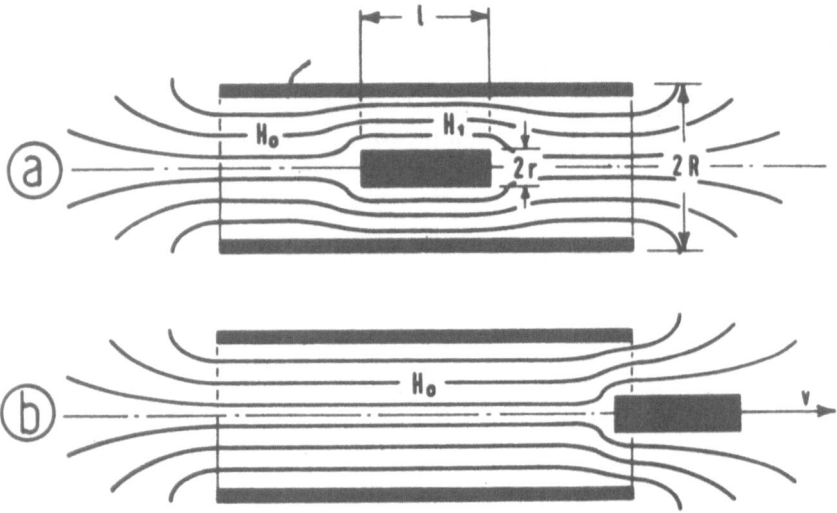

Fig. 9. The expulsion of a conducting projectile from a solenoidal field configuration.

If end effects are neglected, a simple energy balance before and after the bullet is extracted from the magnetic-flux-conserving solenoid gives

$$(H_1{}^2/8\pi)\pi(R^2 - r^2)l - (H_0{}^2/8\pi)\pi R^2 l = \tfrac{1}{2}\pi r^2 l \varrho v^2$$

and for the velocity we therefore find

$$v = \frac{H}{2(\pi\varrho)^{1/2}} \; \frac{1}{[1 - (r^2/R^2)]^{1/2}} \tag{34}$$

If $r/R \ll 1$, this velocity is independent from the particle size and is equal to the Alfvén velocity: at $H = 5$ MOe, for a copper conductor it amounts to 0.5 cm/μsec.

Higher velocities can be attained if the particle is resonantly pushed by a traveling magnetic wave, a situation described by Fig. 9b, if the field pattern follows the particle with the same velocity v. For the magnetic force $\mathbf{K} = (\mathbf{p}\nabla)\mathbf{H}$, where $\mathbf{p}$ is the dipole moment induced in the diamagnetic projectile by the outer magnetic field, we write

$$K = \delta\pi r^2 \, H^2/8\pi \tag{35}$$

where δ is a geometrical quality factor: if a spherical particle with radius r is placed at the border of a solenoid as in Fig. 9b, one has, for example,

$\delta \approx 4\,r/R$. From the energy balance for a spherical particle accelerated by a constant force K along a path L

$$\tfrac{1}{2}\,\tfrac{4}{3}\pi r^3 \varrho v^2 = L\delta r^2\, H^2/8$$

we find its final velocity to be

$$v = \frac{H}{2(\pi\varrho)^{1/2}}\left(\frac{3}{4}\,\delta\,\frac{L}{r}\right)^{1/2} \tag{36}$$

Larger velocities can be attained if instead of a sphere one chooses for the projectile a more convenient geometrical form, i.e., a larger dipole moment per unit of mass, as in the case of a torus or a solenoid ([22]).

From the foregoing sections we know that the magnetic field diffuses into the metallic projectile, eventually starting a shock and a vaporization wave. Let us consider the diffusion process first. When a constant magnetic field H_0 is applied to a conductor with conductivity law as in Eq. (16) the flux skin depth s_φ [Eq. (8)] can for large enough fields be expressed as ([25])

$$s_\varphi \simeq \alpha H_0 \sqrt{\varkappa_0 t}\,, \qquad \alpha = 1.8 \times 10^{-6}\quad [\text{Oe}]^{-1}, \qquad H_0 \gtrsim 1\quad \text{MOe} \tag{37}$$

with $\varkappa_0$ defined in Èq. (5). If we require that $s_\varphi < r$, with r being half of the projectile's thickness, we obtain a limitation on the acceleration time

$$t < r^2/\alpha^2 H_0^2 \varkappa_0$$

and similarly a limitation on the maximum length L and the velocity v. After some straightforward calculations one finds for a spherical conductor

$$v < \frac{3}{32\pi\alpha^2}\frac{r\delta}{\varkappa_0\varrho} \qquad (H_0 \gtrsim 1\quad \text{MOe}) \tag{38}$$

This result is remarkable, since it is independent of the field H_0. For a copper projectile, taking $\delta = 1$ and $r = 0.5$ cm we calculate $v < 4$ cm/μsec. On the other hand, for lower fields ($H_0 < h_c$), when the conductivity remains constant, we have from Eq. (9) with $n = 0$

$$s_\varphi \simeq \eta\sqrt{\varkappa_0 t}\,, \qquad \eta = 2/\sqrt{\pi} \tag{39}$$

and in this region the condition (38) changes to

$$v < \frac{3}{32\pi\eta^2}\frac{r\delta}{\varkappa_0\varrho}\,H_0^2 \qquad (H_0 \lesssim 300\quad \text{kOe}) \tag{40}$$

We conclude that as far as the diffusion effects are concerned, if super-conducting projectiles are excluded ([22]), it is convenient to use fields in excess of 1 MOe, but that in this range one is free to choose the most suitable value.

The boiling-off effect is more difficult to treat quantitatively, as we pointed out above. However, since the velocity of the vaporization wave, at least up to about 5 MOe, and in accordance with the results of ([19]), is lower than or comparable to that of the flux diffusion, no further limitation is added to our problem.

CONCLUSIONS

We have seen that the interaction of megaoersted fields with metallic conductors results in the liberation of thermal or kinetic energies which may attain a density of 1 MJ/cm^3. By the acceleration of metallic projectiles it is possible to go even further, but in practice it will be extremely difficult to surpass the velocity of about 5 cm/μsec for condensed matter.

Other more indirect processes not mentioned in here, exist which allow one to convert or even further concentrate the already large energy density of megaoersted fields. As an example we may mention the compression and heating of a hydrogen plasma through these fields ([5]) or the acceleration of extremely powerful electron beams in betatron-like magnetic flux compression devices.

REFERENCES

1. Proceedings of the *Conference on Megagauss Magnetic Field Generation by Explosives and Related Experiments*, EUR 2750.e (H. Knoepfel and F. Herlach, eds.), Euratom, Brussels, 1966.
2. J. P. Somon, *ibid.*, p. 235.
3. R. E. Kidder, *ibid.*, p. 37.
4. A. R. Bryant, *ibid.*, p. 183.
5. J. G. Linhart, *ibid.*, p. 387.
6. G. Lehner, *ibid.*, p. 55.
7. C. M. Fowler, W. B. Garn, and R. S. Caird, *J. Appl. Phys.* **31**: 588 (1960).
8. F. Herlach and H. Knoepfel, *Rev. Sci. Instr.* **36**: 1088 (1965).
9. A. D. Sakharov, *Soviet Phys.—Usp.* (*English Transl.*) **9**: 294 (1966).
10. H. Knoepfel, *Megagauss Fields as a Transmitting Medium for Very High Pressures*, presented at the High Dynamic Pressure Symposium, Paris, 1967.
11. L. V. Al'tshuler, *Soviet Phys.—Usp.* (*English Transl.*) **8**: 52 (1965).
12. F. H. Harlow and W. E. Pracht, *Phys. Fluids* **9**: 1951 (1966).
13. *Proceedings of the Exploding Wires Conference III* (W. G. Chace and H. K. Moore, eds.), Plenum Press, New York, 1964.

14. R. E. Marshak, *Phys. Fluids* **1**: 24 (1958).
15. A. K. Levine, *Lasers, Vol. I*, Marcel Dekker, New York, 1966.
16. H. S. Carslaw and J. C. Jaeger, *Conduction of Heat in Solids*, 2nd ed., Oxford University Press, London, 1959, p. 63.
17. H. Knoepfel and R. Luppi, in *Exploding Wires*, Vol. IV (W. G. Chace and H. K. Moore, eds.), Plenum Press, New York, 1968, p. 233.
18. R. H. Parmenter, *Phys. Rev.* **116**: 1390 (1959).
19. F. D. Bennett, *Phys. Fluids* **8**: 1425 (1965).
20. E. C. Cnare, *J. Appl. Phys.* **37**: 3812 (1966).
21. G. Lehner, J. G. Linhart, Ch. Maisonnier, CNEN report RTI/FI (64) 2 (1964).
22. Ch. Maisonnier, *Nuovo Cimento* **42**: 332 (1966).
23. J. G. Linhart, unpublished report 1967.
24. R. L. Chapman, in (¹), p. 107.
25. R. E. Kidder, UCRL 5467 (1959).
26. F. Winterberg, *J. Nucl. Energy, Part C* **8**: 541 (1966).
27. G. Schenk and J. G. Linhart, in (¹³), p. 223.